Handbook of
LUMINESCENT SEMICONDUCTOR MATERIALS

Handbook of

LUMINESCENT SEMICONDUCTOR MATERIALS

Edited by
Leah Bergman
Jeanne L. McHale

CRC Press
Taylor & Francis Group
Boca Raton London New York

CRC Press is an imprint of the
Taylor & Francis Group, an **informa** business

CRC Press
Taylor & Francis Group
6000 Broken Sound Parkway NW, Suite 300
Boca Raton, FL 33487-2742

© 2012 by Taylor & Francis Group, LLC
CRC Press is an imprint of Taylor & Francis Group, an Informa business

First issued in paperback 2019

No claim to original U.S. Government works

ISBN 13: 978-0-367-44593-5 (pbk)
ISBN 13: 978-1-4398-3467-1 (hbk)

Visit the Taylor & Francis Web site at
http://www.taylorandfrancis.com

and the CRC Press Web site at
http://www.crcpress.com

Contents

Preface

In broad terms, photoluminescence is the science of light. The term *luminescence* means light emission, and *photoluminescence* is luminescence that is excited by a photon source. Photoluminescence spectroscopy is a versatile technique enabling the study of light dynamics in matter, and it is an important approach for exploring the optical interactions in semiconductors and optical devices with the goal of gaining insight into material properties. This book is intended as a detailed examination of photoluminescence properties of semiconductors with applications to semiconductor-based devices.

Chapter 1 provides the reader with an overview of basic semiconductor theory. The chapter presents the formalisms of semiconductor aspects such as bandgap, doping, and p–n junctions; these concepts are the fundamentals that underlie light emission and photoluminescence. In addition, it gives an outline of the radiative transition mechanisms in semiconductors. The following six chapters focus on the optical properties of wide-bandgap semiconductors that include AlN, GaN, and ZnO. The bandgaps of this family of materials are in the range of ~ 3 eV–6.2 eV, which is well into the UV spectral range. In particular, Chapter 2 addresses the electronic band structure and radiative recombination of AlN, as well as doping issues and application to devices. The topic of GaN and GaN-based optical devices is presented in Chapter 3. That chapter describes the fundamentals of GaN-based blue light-emitting diodes and lasers. Chapter 4 provides a comprehensive overview of near-UV and visible photoluminescence of ZnO. Chapter 5 considers the applications of ZnO photoluminescence, including random lasing. The topic of optical alloys is presented in Chapter 6, where the issue of bandgap-engineered $Mg_xZn_{1-x}O$ is addressed. Chapter 7 covers luminescence studies of impurities and defects in GaN, AlN, and InN. In particular, that chapter focuses on donors, acceptors, intrinsic point defects, and structural defects of the III-Nitride group.

Chapters 8 and 9 present research on the topics of narrow-bandgap semiconductors and solid-state lighting, respectively. Chapter 8 gives a comprehensive description of the optical and electronic properties of narrow-bandgap semiconductors and their application to infrared (IR) detectors. Among the narrow-bandgap materials discussed are the HgCdTe ternary alloys, InAsSb, PbS, PbSe, and InGaAs. It covers the properties of various optical devices such as photodiodes and IR detectors, as well as their manyfold applications to defense technologies as well as IR astronomy. Solid-state lighting involves materials in the visible spectrum, and is the topic of Chapter 9. The chapter covers the material and optical characteristics of low- and high-brightness light-emitting diodes and solid-state lamps. The fundamentals of photometry, which is the science of luminosity, and colorimetry, which is the science of measurement of color, are discussed in detail.

The next six chapters (Chapters 10 through 15) focus on the optical properties of semiconductors in the nanoscale regime. Chapter 10 covers the fundamental aspects of quantum effects unique to nanoparticles. Chapter 11 discusses the consequences of quantum confinement in selenide and sulfide quantum dots and nanocrystals. Chapter 12 presents the formalism and experiments of radiative cascade in semiconductor quantum dots. Chapter 13 considers the photoluminescence of nanocrystalline TiO_2 and its relation to the carrier transport properties that are important in solar energy applications.

Chapter 14 continues the discussion of TiO_2 and other semiconductor nanoparticles, as revealed by the spectroscopy of individual nanoparticles. Finally, Chapter 15 reveals how the photoluminescence of semiconductor nanoparticles is proving useful in biological imaging applications.

This handbook demonstrates that photoluminescence is a powerful and practical analytical tool for the study of the optical properties of semiconductors. The knowledge gained through photoluminescence spectroscopy encompasses both the fundamentals of light interaction as well as valuable technological applications.

Leah Bergman
Jeanne L. McHale

Editors

Leah Bergman is an associate professor of physics at the University of Idaho, Moscow, Idaho. She received her PhD in materials science and engineering in 1995 from North Carolina State University, Raleigh, North Carolina. She is a recipient of a CAREER award from the National Science Foundation division of DMR and was a postdoctoral fellow for the National Research Council. Dr. Bergman's research is in the field of optical materials with a focus on wide-bandgap luminescent semiconductors.

Jeanne L. McHale is a professor of chemistry and materials science at Washington State University, Pullman, Washington. She received her PhD in physical chemistry in 1979 from the University of Utah, Salt Lake City, Utah. She is the author of *Molecular Spectroscopy* and a fellow in the American Association for the Advancement of Science. Dr. McHale's research focuses on spectroscopic studies of semiconductor nanoparticles and chromophore aggregates relevant to solar energy conversion.

Contributors

Leah Bergman
Department of Physics
University of Idaho
Moscow, Idaho

Lekhnath Bhusal
National Renewable Energy Laboratory
Golden, Colorado
and
Philips Lumileds Lighting Company
San Jose, California

Hui Cao
Department of Applied Physics
Yale University
New Haven, Connecticut

Robert P.H. Chang
Department of Materials Science and
    Engineering
Northwestern University
Evanston, Illinois

John Collins
Department of Physics
Wheaton College
Norton, Massachusetts

Steven P. DenBaars
Materials Department
and
Electrical and Computer Engineering
    Department
University of California
Santa Barbara, California

Baldassare Di Bartolo
Department of Physics
Boston College
Chestnut Hill, Massachusetts

Martin Feneberg
Institute for Experimental Physics
University of Magdeburg
Magdeburg, Germany

David Gershoni
Department of Physics
The Technion—Israel Institute of Technology
Haifa, Israel

Jesse Huso
Department of Physics
University of Idaho
Moscow, Idaho

Hongxing Jiang
Department of Electrical and Computer
    Engineering
Texas Tech University
Lubbock, Texas

Patanjali Kambhampati
Department of Chemistry
McGill University
Montreal, Quebec, Canada

Fritz J. Knorr
Department of Chemistry
Washington State University
Pullman, Washington

Oleg Kovtun
Department of Chemistry
Vanderbilt University
Nashville, Tennessee

Jingyu Lin
Department of Electrical and Computer
    Engineering
Texas Tech University
Lubbock, Texas

Tetsuro Majima
The Institute of Scientific and Industrial Research
    (SANKEN)
Osaka University
Osaka, Japan

Angelo Mascarenhas
National Renewable Energy Laboratory
Golden, Colorado

Jeanne L. McHale
Department of Chemistry
Washington State University
Pullman, Washington

Bo Monemar
Department of Physics, Chemistry and Biology
Linköping University
Linköping, Sweden

John L. Morrison
Department of Physics
University of Idaho
Moscow, Idaho

Andrea M. Munro
Department of Chemistry
Pacific Lutheran University
Tacoma, Washington

Shuji Nakamura
Materials Department
and
Electrical and Computer Engineering
    Department
University of California
Santa Barbara, California

M. Grant Norton
School of Mechanical and Materials Engineering
Washington State University
Pullman, Washington

Hiroaki Ohta
Materials Department
University of California
Santa Barbara, California

Plamen P. Paskov
Department of Physics, Chemistry and Biology
Linköping University
Linköping, Sweden

Eilon Poem
Department of Physics
The Technion—Israel Institute of Technology
Haifa, Israel

Antoni Rogalski
Institute of Applied Physics
Military University of Technology
Warsaw, Poland

Sandra J. Rosenthal
Departments of Chemistry, Physics and
    Astronomy, Chemical and Biomolecular
    Engineering, and Pharmacology
Vanderbilt University School of Medicine
and
Vanderbilt Institute of Nanoscale Science and
    Engineering
Vanderbilt University
Nashville, Tennessee
and
Oak Ridge National Laboratory
Oak Ridge, Tennessee

Ashok Sedhain
Department of Electrical and Computer
    Engineering
Texas Tech University
Lubbock, Texas

Takashi Tachikawa
The Institute of Scientific and Industrial Research
    (SANKEN)
Osaka University
Osaka, Japan

Klaus Thonke
Institute for Quantum Matter
University of Ulm
Ulm, Germany

1

Principles of Photoluminescence

Baldassare Di
Bartolo
Boston College

John Collins
Wheaton College

1.1 Introduction

Luminescence is the spontaneous emission of light from the excited electronic states of physical systems.

The emission is preceded by the process of excitation, which may be produced by a variety of agents. If it is achieved by the absorption of light it is called *photoluminescence*, if by the action of an electric field *electroluminescence*, if by a chemical reaction *chemiluminescence*, and so on.

Following the excitation, if the system is left alone without any additional influence from the exciting agent, it will emit spontaneously.

Even in absolute vacuum, an excited atom devoid of any external influence will emit a photon and return to its ground state. The spontaneity of the emission presents a conceptual problem. A tenet of physical science, expressed by the so-called fluctuation-dissipation theorem sets forth the fact that any dissipation of energy from a system is the effect of its interaction with some external entity that provides

1

the perturbation necessary for the onset of the process. Such an entity seems to be missing in the case of an isolated excited atom. If we hold the classical view of natural phenomena, we cannot explain the presence of spontaneous emission.

In the quantum world things are different. The harmonic radiative oscillators that populate the vacuum and that classically hold no energy when in their ground state have, each in this state, the energy $1/2\ h\nu$ and may produce a fluctuating electric field at the site of the atom, setting the perturbation necessary for the onset of spontaneous emission.

The first reported observation of luminescence light from glow worms and fireflies is the Chinese book *Shih-Ching* or *Book of Poems* (1200–1100 BC). Aristotle (384–322 BC) reported the observation of light from decaying fish. The first inquiry into luminescence dates ca. 1603 and was made by Vincenzo Cascariolo. Cascariolo's interests were other than scientific: he wanted to find the so-called philosopher's stone that would convert any metal into gold. He found some silvery white stones (barite) on Mount Paderno, near Bologna, which, when pulverized, heated with coal, and cooled, showed a purple-blue glow at night. This process amounted to reducing barium sulfate to give a weakly luminescent barium sulfide:

$$BaSO_2 + 2C \ \rightarrow \ BaS + 2CO_2$$

Cascariolo observed that the glow could be restored by exposing the powder to sunlight. News of this material, the *Bologna stone*, and some samples of it reached Galileo, who passed them to Giulio Cesare Lagalla. Lagalla called it *lapis solaris* and wrote about it in his book *De Phenomenis in Orbe Lunae* (1612).

It is rare when one can insert a literary citation in a scientific article. This happens to be the case here. In Goethe's *The Sorrows of Young Werther,* the protagonist is unable to see the woman he loves because of an engagement he cannot refuse and so sends a servant to her, "only so that I might have someone near me who had been in her presence...." This is then his reaction when the servant comes back [1]:

It is said that the *Bologna stone*, when placed in the sun, absorbs the sun's rays and is luminous for a while in the dark. I felt the same with the boy. The consciousness that her eyes had rested on his face, his cheeks, the buttons of his jacket and the collar of his overcoat, made all these sacred and precious to me. At that moment I would not have parted with him for a thousand taler. I felt so happy in his presence.

The second important investigation on luminescence is credited to Stokes and dates back to the year 1852. Stokes observed that the mineral fluorspar (or fluorite), when illuminated by blue light gave out yellow light. Fluorite is CaF_2, colorless in its purest form, but it absorbs and emits light when it contains such impurities as *Mn, Ce, Er,* etc. The term "fluorescence" was coined by Stokes and has continued to be used to indicate short-lived luminescence.

1.2 Photoluminescent Solid Systems

A *Stokes' law* has been formulated, according to which the wavelength of the emitted light is always longer than or equal to the wavelength of the absorbed light. The reason for this difference is the transformation of the exciting light, to a great or small extent, into a nonradiating vibrational energy of atoms or ions.

When the intense radiation from a laser is used or when sufficient thermal energy contributes to the excitation process [2], the wavelength of the emission may be shorter than the wavelength of the absorption (*anti-Stokes radiation*).

It is convenient to subdivide the luminescent system into two categories, *localized* and *delocalized*. For the first category, the absorption and emission processes are associated with quantum states of optically active centers that are spatially localized at particular sites in the solid. For the second category, these processes are associated with the quantum states of the entire solid.

The most important classes of localized luminescent centers are *Transition Metal Ions* and *Rare Earth Ions* that are generally intentionally doped into ionic insulating host materials. The luminescence properties of these systems depend on both the dopant ion and the host. Another class of localized centers is that of defects in solids. One such center is an electron trapped at a vacant lattice site. These defects often absorb in the optical region giving the crystal color and, for this reason, are called *color centers*.

The category of delocalized luminescent centers includes semiconductor systems to which we shall now dedicate our attention.

1.3 Classification of Crystalline Solids

Crystalline solids are arranged in a repetitive 3D structure called a *lattice*. The basic repetitive unit is the *unit cell*. Prototypes of crystalline solids are (i) copper-metals, (ii) diamond-insulators, and (iii) silicon semiconductors. We can classify the solid according to three basic properties:

1. Resistivity ρ at room temperature

$$\rho = \frac{E}{J} \qquad (\Omega\,\text{m})$$

where
E is the electric field
J is the current density

2. Temperature coefficient of resistivity

$$\alpha = \frac{1}{\rho}\frac{d\rho}{dt} \qquad (\text{K}^{-1})$$

3. Number density of charge carriers, n (m^{-3})

The resistivity of diamond is greater than 10^{24} times the resistivity of copper. Some typical parameters for metals and undoped semiconductors are reported in Table 1.1.

If we assemble N atoms, each level of an isolated atom splits into N levels in the solid. Individual energy levels of the solid form *bands*, adjacent bands being separated by *gaps*. A typical band is only a few eV wide. Since the number of levels in one band may be on the order of $\sim 10^{24}$, the energy levels within a band are very close.

1.3.1 Insulators

The electrons in the filled upper band have no place to go. The vacant levels of the band can be reached only by giving an electron enough energy to bridge the gap. For diamond the gap is 5.5 eV, and the

TABLE 1.1 Comparison of the Properties of Metals and Semiconductors

	Unit	Copper (Metal)	Silicon (Semiconductor)
n	m^{-3}	9×10^{28}	1×10^{16}
ρ	Ω m	2×10^{-8}	3×10^{3}
α	K^{-1}	4×10^{-3}	-70×10^{-3}

possibility that one electron occupies a quantum level at the bottom of the *conduction band* (see Equation 1.5) at room temperature is on the order of 10^{-46}, and as such is negligible.

1.3.2 Metals

The feature that defines a metal is that the highest occupied energy level falls near the middle of an energy band. Electrons have empty levels they can go to!

A classical free electron model can be used to deal with the physical properties of metals. This model predicts the functional form of Ohm's law and the connection between the electrical and thermal conductivity of metals, but does not give correct values for the electrical and thermal conductivities. This deficiency can be remedied by taking into account the wave nature of the electron.

1.3.3 Semiconductors

In this section, we shall treat semiconductors that do not contain any impurities, and that are generally called *intrinsic semiconductors*. We shall see later how the presence of impurities greatly affects the properties of semiconductors. The band structure of a semiconductor is similar to that of an insulator. The main difference is that a semiconductor has a much smaller energy gap E_g between the top of the highest filled band (*valence band*) E_v and the bottom of the lowest empty band (*conduction band*) E_c above it. For diamond $E_g = 5.5\,\mathrm{eV}$, whereas for *Si*, $E_g = 1.1\,\mathrm{eV}$.

The charge carriers in *Si* arise only because at thermal equilibrium, thermal agitation causes a certain (small) number of valence band electrons to jump over the gap into the conduction band. They leave an equal number of vacant energy states called *holes*. Both electrons in the conduction band and holes in the valence band serve as charge carriers and contribute to the conduction. The resistivity of a material is given by

$$\rho = \frac{m}{e^2 n \tau} \tag{1.1}$$

where

     m is the mass of the charge carrier
     n is the number of charge carriers/V
     τ is the mean time between collisions of charge carriers

Now, $\rho_{Cu} = 2 \times 10^{-8}\,\Omega\,\mathrm{m}$, $\rho_{Si} = 3 \times 10^3\,\Omega\,\mathrm{m}$, and $n_{Cu} = 9 \times 10^{28}\,\mathrm{m}^{-3}$, $n_{Si} = 1 \times 10^{16}\,\mathrm{m}^{-3}$, so that

$$\frac{\rho_{Si}}{\rho_{Cu}} \approx 10^{11} \quad \text{and} \quad \frac{n_{Cu}}{n_{Si}} \approx 10^{13}$$

The vast difference in the density of charge carriers is the main reason for the great difference in ρ.

We note than the temperature coefficient of resistivity is positive for *Cu* and negative for *Si*. The atom *Si* has the following electronic configuration:

$$Si : \underbrace{1s^2 2s^2 2p^6}_{core} 3s^2 3p^2$$

Each *Si* atom has a core containing 10 electrons and contributes its $3s^2 3p^2$ electrons to form a rigid two-electron covalent bond with its neighbors. The electrons that form the *Si*–*Si* bonds constitute the valence band of the *Si* sample. If an electron is torn from one of the four bonds so that it becomes free to wander through the lattice, we say that the electron has been raised from the valence to the conduction band.

1.4 Density of One-Electron States

Given a volume $V = L^3$, the number of one-particle states in the range $dp_x dp_y dp_z$ is

$$\frac{V}{8\pi^3} dk_x dk_y dk_z = \frac{V}{h^3} dp_x dp_y dp_z \tag{1.2}$$

The number of one-particle states in the range $(p, p + dp)$ is

$$g(p)dp = \frac{V}{h^3} \int_0^\pi \int_0^{2\pi} p^2 \sin\theta \, d\theta \, d\varphi \, dp = \frac{4\pi V}{h^3} p^2 dp \tag{1.3}$$

and if the particles are electrons, taking the spin into account

$$2g(E)dE = 2g(p)\frac{dp}{dE}dE$$

$$= 2\frac{4\pi V}{h^3} p^2 \frac{1}{2}\sqrt{\frac{2m}{E}} dE = \frac{4\pi V}{h^3} 2mE \frac{1}{2}\sqrt{\frac{2m}{E}} dE$$

$$= \frac{4\pi V}{h^3} (2m)^{3/2} E^{1/2} dE$$

$$= \frac{8\sqrt{2}\pi m^{3/2}}{h^3} VE^{1/2} dE \tag{1.4}$$

Given a system of Fermions at temperature *T*, the probability distribution that specifies the occupancy probability is

$$F(E) = \frac{1}{e^{(E-E_F)/kT} + 1} \tag{1.5}$$

In metals E_F, the *Fermi energy* is the energy of the most energetic quantum state occupied at $T = 0$. At $T \neq 0$, E_F is the energy of a quantum state that has the probability 0.5 of being occupied. The number of available states in $(E, E + dE)$ for a system of electrons is given by Equation 1.4. The Fermi energy at $T = 0$ is determined by

$$N = \int_0^{E_F} 2g(E)dE = \frac{8\sqrt{2}\pi m^{3/2}}{h^3} V \int_0^{E_F} E^{1/2} dE$$

$$= \frac{16\sqrt{2}\pi m^{3/2}}{3h^3} VE_F^{3/2} \tag{1.6}$$

Then

$$E_F = \frac{\hbar^2}{2m}\left(3\pi^2 \frac{N}{V}\right)^{2/3} = \frac{0.121h^2}{m} n^{2/3} \tag{1.7}$$

where $n = N/V$.

1.5 Intrinsic Semiconductors

We shall now present a model for an intrinsic semiconductor. In general, the number of electrons per unit volume in the conduction band is given by

$$n_c \int_{E_c}^{top} N(E)F(E)dE \tag{1.8}$$

where
  $N(E)$ is the density of states
  E_c is the energy at the bottom of the conduction band

We expect E_F to lie roughly halfway between E_v and E_c: the Fermi function $F(E)$ decreases strongly as E moves up in the conduction band. To evaluate the integral in Equation 1.8, it is sufficient to know $N(E)$ near the bottom of the conduction band and to integrate from $E = E_c$ to $E = \infty$. Near the bottom of the conduction band, according to Equation 1.4, the density of states is given by

$$N(E) = \frac{4\pi}{h^3}\left(2m_e^*\right)^{3/2}(E - E_c)^{1/2} \tag{1.9}$$

where m_e^* is the effective mass of the electron near E_c. Then

$$n_c = \frac{4\pi}{h^3}\left(2m_e^*\right)^{3/2}\int_{E_c}^{\infty}\frac{\left(E - E_c\right)^{1/2}}{e^{(E-E_F)/kT}+1}dE \; \rightarrow$$

$$\xrightarrow{(E_c - E_F)\gg kT} \frac{4\pi}{h^3}\left(2m_e^*\right)^{3/2}\int_{E_c}^{\infty}\frac{\left(E - E_c\right)^{1/2}}{e^{(E-E_F)/kT}}dE \tag{1.10}$$

The integral may then be reduced to one of type

$$\int_0^{\infty} x^{1/2}e^{-x}dx = \frac{\pi^{1/2}}{2} \tag{1.11}$$

and we obtain the number of electrons per unit volume in the conduction band:

$$n_c = 2\left(\frac{2\pi m_e^* kT}{h^3}\right)^{3/2}e^{(E_F - E_c)/kT} \tag{1.12}$$

Let us now consider the number of holes per unit volume in the valence band:

$$n_h \int_{bottom}^{E_v} N(E)[1-F(E)]dE \tag{1.13}$$

where E_v is the energy at the top of the valence band. $1-F(E)$ decreases rapidly as we go down below the top of the valence band (i.e., holes reside near the top of the valence band). Therefore, in order to evaluate n_h, we are interested in $N(E)$ near E_v

$$N(E) = \frac{4\pi}{h^3}\left(2m_h^\star\right)^{3/2}(E_v - E)^{1/2} \tag{1.14}$$

where $m_h^\star$ is the effective mass of a hole near the top of the valence band. For $E_F - E_v \gg kT$

$$1 - F(E) = 1 - \frac{1}{e^{(E-E_F)/kT}+1} \approx e^{(E-E_F)/kT} \tag{1.15}$$

Substituting (1.14) and (1.15) into (1.13), we obtain

$$
\begin{aligned}
n_h &= \int_{bottom}^{E_v} N(E)\left[1-F(E)\right]dE \\
&= \frac{4\pi}{h^3}\left(2m_h^\star\right)^{3/2}\int_{-\infty}^{E_v}(E_v-E)^{1/2}e^{(E-E_F)/kT}dE \\
&= 2\left(\frac{2\pi m_h^\star kT}{h^2}\right)^{3/2}e^{(E_v-E_F)/kT}
\end{aligned}
\tag{1.16}
$$

We now use the fact that

$$n_c = n_h \tag{1.17}$$

and equate the two expressions for n_c and n_h given by Equations 1.12 and 1.16, respectively. We find

$$E_F = \frac{E_c + E_v}{2} + \frac{3}{4}kT\ln\frac{m_h^\star}{m_e^\star} \tag{1.18}$$

If $m_e^\star = m_h^\star$, E_F lies exactly halfway between E_c and E_v. Replacing the expression (1.18) in Equation 1.16, we find

$$n_c = n_h = 2\left(\frac{2\pi kT}{h^2}\right)^{3/2}\left(m_e^\star m_h^\star\right)^{3/4}e^{-\frac{E_g}{2kT}} \tag{1.19}$$

At room temperature,

$$2\left(\frac{2\pi kT}{h^2}\right)^{3/2} m^{3/2} \approx 10^{19} \text{ cm}^{-3},$$

where m is the mass of the electron.

1.6 Doped Semiconductors

1.6.1 *n*-Type Semiconductors

Consider the phosphorus atom's electronic configuration:

$$P \; : \; 1s^2 2s^2 2p^6 3s^2 3p^3 \qquad (Z = 15)$$

If a P atom replaces an Si atom, it becomes a *donor*. The fifth (extra) electron is only loosely bound to the P ion core. It occupies a localized level with energy $E_d \ll E_g$ below the conduction band. By adding donor atoms, it is possible to greatly increase the number of electrons in the conduction band. Electrons in the conduction band are *majority carriers*. Holes in the valence band are *minority carriers*.

Example

In a sample of pure *Si*, the number of conduction electrons is $\approx 10^{16}$ m^{-3}. If we want to increase this number by a factor 10^6, we should dope the system with P atoms creating an *n*-type semiconductor. At room temperature, the thermal agitation is so effective that practically every P atom donates its extra electron to the conduction band. The number of P atoms that we want to introduce in the system is given by

$$10^6 n_0 = n_0 + n_P,$$

where
  n_0 is the number density of conduction electrons of pure *Si* ($\sim 10^{16}$ m^{-3})
  n_P is the number density of P atoms

Then

$$n_P = 10^6 n_0 - n_0 \approx 10^6 n_0 \approx 10^6 \times 10^{16} = 10^{22} \text{ m}^{-3}$$

The number density of *Si* atoms in a pure *Si* lattice is

$$n_{Si} = \frac{N_a \rho}{A} = 5 \times 10^{28} \text{ m}^{-3}$$

where
  N_a is the Avogadro number
  ρ is the density of $Si = 2330$ kg/m^3
  A is the molar mass $= 28.1$ g/mol $= 0.028$ kg/mol

The fraction of *P* atoms we seek is approximately

$$\frac{n_P}{n_{Si}} = \frac{10^{22}}{5 \times 10^{28}} = \frac{1}{5 \times 10^6} \qquad (1.20)$$

Therefore, if we replace only one *Si* atom in five million with a phosphorous atom, the number of electrons in the conduction band will be increased by a factor of 10^6.

1.6.2 *p*-Type Semiconductors

Consider the electronic configuration of an aluminum atom

$$Al \,:\, 1s^2 2s^2 2p^6 3s^2 3p \qquad (Z = 13)$$

If an *Al* atom replaces an *Si* atom, it becomes an *acceptor*. The *Al* atom can bond covalently with only three *Si* atoms; there is now a missing electron (a hole) in one *Al–Si* bond. With a little energy, an electron can be torn from a neighboring *Si–Si* bond to fill this hole, thereby creating a hole in that bond. Similarly, an electron from some other bond can be moved to fill the second hole. In this way, the hole can migrate through the lattice. It has to be understood that this simple picture should not be taken as indicative of a hopping process since a hole represents a state of the whole system. Holes in the valence band are now *majority carriers*. Electrons in the conduction band are *minority carriers*. We compare the properties of an *n*-type semiconductor and of a *p*-type semiconductor in Table 1.2.

1.7 Models for Doped Semiconductors

Most semiconductors owe their conductivity to impurities, i.e., either to foreign atoms put in the lattice or to a stoichiometric excess of one of its constituents. Energy level schemes for an *n*-type semiconductor and a *p*-type of semiconductor are shown schematically in Figure 1.1.

1.7.1 *n*-Type Semiconductors

At $T = 0$, all the donor levels are filled. At low temperatures, only a few donors are ionized: the Fermi level is halfway between donor levels and the bottom of the conduction band. If we assume that E_F is

TABLE 1.2 Comparison of the Properties of an *n*-Type and a *p*-Type Semiconductor

Property	Type of Semiconductor	
	n	*p*
Matrix material	Silicon	Silicon
Matrix nuclear charge	14 e	14 e
Matrix energy gap	1.2 eV	1.2 eV
Dopant	Phosphorus	Aluminum
Type of dopant	Donor	Acceptor
Majority carriers	Electrons	Holes
Minority carriers	Holes	Electrons
Dopant energy gap	0.045 eV	0.067 eV
Dopant valence	5	3
Dopant nuclear charge	+15 e	+13 e

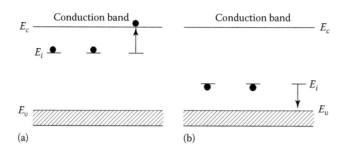

FIGURE 1.1   Energy level scheme for (a) an *n*-type semiconductor and (b) a *p*-type semiconductor. E_i is the energy of the donor level (a) or the acceptor level (b).

below the bottom of the conduction band by more than a few *kT*, then we can use, in this case, formula (1.12), that we rewrite as

$$n_c = 2\left(\frac{2\pi m_e^* kT}{h^2}\right)e^{(E_F - E_c)/kT} \tag{1.21}$$

This density is equal to the density of the ionized donors.

If E_F lies more than a few *kT* above the donor level at E_i, the density of the empty donors is equal to

$$n_d\left[1 - F(E_i)\right] \approx n_d e^{(E_i - E_F)/kT} \tag{1.22}$$

Equating (1.21) and (1.22), we obtain

$$E_F = \frac{1}{2}(E_i + E_c) + \frac{kT}{2}ln\left[\frac{n_d}{2}\left(\frac{2\pi m_e^* kT}{h^2}\right)^{-3/2}\right] \tag{1.23}$$

At *T* = 0, E_F lies halfway between the donor level and the bottom of the conduction band. As *T* increases, E_F drops (see Figure 1.2). Using the expression for E_F from Equation 1.23 in n_c given by Equation 1.21, we find

$$n_c = (2n_d)^{1/2}\left(\frac{2\pi m_e^* kT}{h^2}\right)^{3/4} e^{-\frac{E_c - E_d}{2kT}} \tag{1.24}$$

1.7.2 *p*-Type Semiconductors

The case of *p*-type semiconductors can be treated in a similar way as the *n*-type semiconductors. n_h has an expression similar to that for n_c. The Fermi level lies halfway between the acceptor level and the top of the valence band at *T* = 0. As *T* increases, E_F rises.

Figure 1.3 represents schematically the behavior of the Fermi level for an *n*-type and for a *p*-type semiconductor. The figure illustrates the fact that as the temperature increases, the Fermi level for an *n*-type semiconductor does not drop indefinitely as indicated by Equation 1.23. As the temperature increases, the intrinsic excitations of the semiconductor become more important and the Fermi level tends to set in the middle of the gap. Similar effects take place for the *p*-type semiconductor. For additional considerations, the reader is referred to the book by Dekker (see Bibliography).

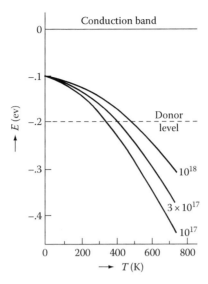

FIGURE 1.2 The variation of the position of the Fermi level with temperature with a donor level 0.2 eV below the bottom of the conduction band for three different values of n_d. (With kind permission from Springer Science+Business Media: *Handbook of Applied Solid State Spectroscopy*, 2006, Vij, D.R.)

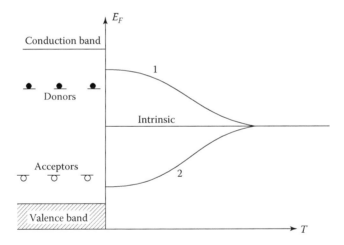

FIGURE 1.3 The variation of the position of the Fermi level with temperature. Curve 1 relates to insulators with donors and curve 2 relates to insulators with acceptors.

1.8 Direct Gap and Indirect Gap Semiconductors

The energy of the band gap of a semiconductor determines the spectral region in which the electronic transitions, both in absorption and emissions, take place. For visible or near-infrared transitions, we need materials with gaps of ~1–1.7 eV. A list of such materials is provided in Table 1.3.

Direct gap transitions take place when the maximum energy of the valence band and the minimum energy of the conduction band both occur in correspondence to a value of the linear momentum equal to zero or at the same $\vec{k} \neq 0$. Such semiconductors are called *direct gap semiconductors*.

In other materials, the maximum of the valence band and the minimum of the conduction band occur at different values of $\vec{k}$. Such materials are called *indirect gap semiconductors*.

TABLE 1.3   List of Typical
Semiconductors

Material	Type	Band Gap (eV)
Si	Indirect	1.16
InP	Direct	1.42
GaAs	Direct	1.52
GaP	Indirect	2.3
AlP	Indirect	2.5
SiC	Indirect	3.0

It is interesting to consider the case of the semiconductor *GaAs*. By changing the chemical composition of this material according to the formula $GaAs_{1-x}P_x$, it is possible to change the band gap from 1.52 eV with $x = 0$ to 2.3 eV with $x = 1$. In addition, for $x > 0.4$, the material changes its character from a direct gap to an indirect gap semiconductor. Mixtures of *InP* and *AlP* can also yield gaps from 1.42 to 2.5 eV.

1.9 Excitation in Insulators and Large Band Gap Semiconductors

If a beam of light, with photons exceeding in energy the energy gap, goes through an insulator or a semiconductor, it raises an electron from the valence band into the conduction band for each photon absorbed, leaving behind a hole. The electron and the hole may move away from each other contributing to the *photoconductivity* of the material. On the other hand, they may combine producing an *exciton*, a hydrogen-like or a positron-electron pair-like structure. Excitons are free to move through the material. Since the electron and the hole have opposite charge, excitons are neutral, and as such, are difficult to detect. When an electron and a hole recombine, the exciton disappears and its energy may be converted into light or it may be transferred to an electron in a close-by atom, removing the electron from this atom and producing a new exciton.

Excitons are generally more important in insulators and in semiconductors with large gaps, even if some excitonic effects in small gap materials have been observed. Excitons do not obey the Fermi–Dirac statistics and, therefore, it is not possible to obtain a filled band of excitons. Excitons may also be created in doped semiconductors. In these, however, the free charges provided by the impurities tend to screen the attraction between electrons and holes and excitonic levels are difficult to detect.

Two models are generally used to deal with excitons in solids. There are more than two different ways of looking at the same problem, but, rather, they reflect two extreme physical situations:

1. A model in which the electron, after its excitation, continues to be bound to its parent atom.
2. A model where the electron loses the memory of its parent atom and binds together with a hole.

The first case corresponds to the so-called *Frenkel exciton* and the second case to the *Wannier exciton*. Experimentally, the Frenkel exciton is in principle recognizable because the optical transitions responsible for the production of the exciton occur in the same spectral region of the atomic transitions. Experimentally, the transitions responsible for the production of a Wannier exciton fit a hydrogen-like type of behavior.

1.10 Radiative Transitions in Pure Semiconductors

1.10.1 Absorption

The absorption optical spectra of pure semiconductors generally present the following features (see Figure 1.4):

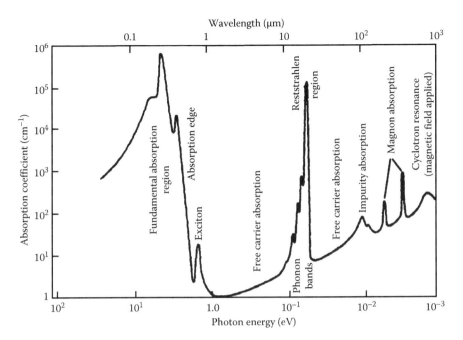

FIGURE 1.4 Absorption spectrum of a hypothetical semiconductor. (With kind permission from Springer Science+Business Media: *Handbook of Applied Solid State Spectroscopy*, 2006, Vij, D.R.)

1. A region of strong absorption is present in the ultraviolet with a possible extension to the visible and infrared due to electronic transitions from the valence to the conduction band. These *interband transitions* produce mobile electrons and holes that contribute to photoconductivity. The value of the absorption coefficient is typically 10^5–10^6 cm^{-1}. On the high energy side, the absorption band (~20 eV) decreases in value smoothly in a range of several eV. On the low energy side, the absorption decreases abruptly and may decrease by several orders over a range of a few tenths of an eV. In semiconductors, this region of the absorption spectrum is referred to as the *absorption edge*.

2. The low energy limit of the absorption edge corresponds to the photon energy necessary to move an electron across the minimum energy gap E_g. The exciton structure appears in the absorption edge region. It is more evident in insulators such as ionic crystals than in semiconductors.

3. At longer wavelengths, the absorption rises again due to *free-carrier absorption*, i.e., electronic transitions within the conduction or valence bands. This absorption extends to the infrared and microwave regions of the spectrum.

4. A set of peaks appear at energies 0.02–0.05 eV ($\lambda = 50$–$20\,\mu$m), due to the interaction between the photons and the vibrational modes of the lattice. In ionic crystals, the absorption coefficient may reach 10^5 cm^{-1}; in homopolar crystals the absorption coefficient is generally much lower, such as 10–10^2 cm^{-1}.

5. Impurities, if present in the semiconductor, may be responsible for absorption in the region of, say, 10^{-2} eV or so. This absorption is observable for kT lower than the ionization energy.

6. If the semiconductor contains paramagnetic impurities, the absorption spectrum will present absorption lines in the presence of a magnetic field that splits the Zeeman levels.

7. An absorption peak in the long wavelength region may be present in the presence of a magnetic field due to the *cyclotron resonance* of the mobile carriers.

We want to make some additional considerations regarding the features (1) and (2) of the absorption spectrum. Interband transitions can take place subject to the two conditions of energy and conservation of the wave vector:

$$\begin{cases} E_f - E_i = \hbar\omega(\vec{k}') \\ \vec{k}_f - \vec{k}_i = \vec{k}' \end{cases} \qquad (1.25)$$

where

    the subscripts *f* and *i* refer to the final and initial one electron states
    $\vec{k}'$ is the wave vector of the absorbed photon of energy $\hbar\omega(\vec{k}')$

Since the wavelength of the radiation is much longer than the lattice constant, $\vec{k}'$ is much smaller than the size of the reciprocal lattice and $\vec{k}'$ can then be neglected in the second Equation 1.25. This means that in an $(E, \vec{k})$ diagram, we should rely on *vertical* transitions.

The interband absorption is restricted by the conditions of Equation 1.25 and shows a structure that depends on the density of the final states. The peaks can be presumably associated with values of $\vec{k}$ about which the empty and the filled bands run parallel:

$$\vec{\nabla}_k E_v(\vec{k}) = \vec{\nabla}_k E_c(\vec{k}) \qquad (1.26)$$

In such a case, there is a large density of initial and final states available for the transitions in a small range of energies.

In Section 1.8, we have made a distinction between direct gap and indirect gap semiconductors. For the former, the maximum of the valence band energy and the minimum of the conduction band energy occur frequently at $\vec{k} = 0$ (but not always, e.g., not for *Ta*-doped halides of lead salts), whereas for the latter, they occur at different values of $\vec{k}$. In indirect gap semiconductors, the absorption transitions at the band edge are phonon-assisted and have probability smaller than that of the direct gap absorption transitions. The absorption edge of the indirect gap absorption may show features related to the available phonon energies.

We now turn our attention to the excitonic structure of the absorption band. We have two models at our disposal—the Frenkel model and the Wannier model. In the Frenkel model, an excited electron describes an orbit of atomic size around an atom with a vacant valence state; this model is more appropriate for ionic insulators. The Wannier model represents an exciton as an electron and a hole bound by the Coulomb attraction, but separated by several lattice sites. This model is more appropriate for semiconductors.

An example of the Frenkel exciton is given by the crystal MnF_2 [5] in which the excited state may be considered to consist of an electron and a hole residing in the same ion. The excitation can travel throughout the system via the energy transfer mechanism. A good example of the Wannier exciton is given by Cu_2O, which presents absorption lines up to $n = 11$ [6].

1.10.2 Emission

Following the absorption process, an excited electron can decay radiatively by emitting a photon (possibly accompanied by a phonon) or nonradiatively by transforming its excitation energy entirely into heat (phonons). The following reasons make the emission data relevant:

1. Emission is not simply the reversal of absorption. In fact, the two phenomena are thermodynamically irreversible and, therefore, emission spectroscopy furnishes data not available in absorption.

2. Emission is easier to measure than absorption, since its intensity depends on the intensity of excitation.
3. The applications of emission from solids, such as those of fluorescent lights and television, far outnumber the application of absorption.

In Section 1.12, we discuss the photon emission processes in solids.

1.11 Optical Behaviors of Doped Semiconductors

Two types of impurities are particularly important when considering the optical behavior of semiconductors.

Donors: As we have seen in Section 1.6, when a material made of group IV atoms, like *Si* or *Ge*, is doped with a small amount of group V atoms, like *As*, the extra electrons of these atoms continue to reside in the parent atoms, loosely bound to them. The binding energy, called E_D, is typically around 0.01 eV. It is in fact 0.014 eV for *As*, 0.0098 eV for *Sb*, and 0.0128 eV for *P*. E_D is also called the *ionization energy of the donor atom*. The electrons which, because of thermal excitation, a donor puts in the conduction band cannot produce luminescence, because this process needs, besides an excited electron, a hole where the electron can go, and the valence band, being filled with electrons, has no holes.

Acceptors: If a material made of group IV atoms, such as *Si* or *Ge*, is doped with group III atoms, such as *Ga* or *Al,* a hole for each of these atoms forms and remains loosely bound to the parent atom. The amount of energy necessary to move an electron from the top of the valence band to one of these holes is labeled E_A and is typically around 0.03 eV. E_A may also be called the *ionization energy of the acceptor*.

Both types of impurities can be doped into the same crystal, deliberately, or they may be due to the fact that it is practically impossible to fabricate semiconductor crystals of perfect purity.

1.12 Radiative Transitions across the Band Gap

We shall now examine, following Elliott and Gibson [7], the radiative processes that can take place across the band gap of a semiconductor (see Figure 1.5).

Processes A and B

An electron excited to a level in the conduction band will thermalize quickly with the lattice and reside in a region $\sim kT$ wide at the bottom of the conduction band. Thermalization is generally achieved by phonon emission, but also, less frequently, by phonon-assisted radiative transitions. If such photons have energy exceeding E_g, they can be reabsorbed and promote another electron to the conduction band.

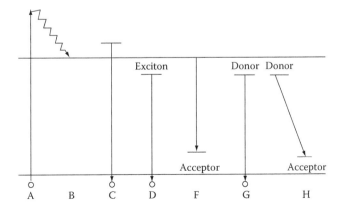

FIGURE 1.5 Transitions producing emission of photons in solids.

Process C

The recombination of electrons and holes with photon emission, the reverse process to absorption, is possible but not very likely because of competing processes. It may be present only in high purity single crystals. The widths of the related emission bands are expected to be ~kT because the thermalized electrons and holes reside at the band edges in this range of energy.

Process D

The radiative decay of the exciton can be observed at a low temperature in very pure crystals. There are two types of decay:

1. The decay of the free exciton
2. The decay of an exciton bound to an impurity

Transitions of the first type are observed at low temperatures. Since the exciton levels are well defined, a sharply structured emission can be expected.

As for the transitions of the second type, they may be observed in materials of high purity into which impurities are purposely doped. An exciton may bind itself to one such impurity; the energy of the *bound exciton* is lower than the $n=1$ energy by the binding energy of the exciton to the impurity. It may be noted that emission from bound excitons in indirect gap materials can take place without the assistance of phonons because the localization of a bound exciton negates the requirement of wave vector conservation. Bound electron emission is observed at a low temperature and is generally much sharper than free electron emission.

Processes F and G

The transitions related to these processes are between the band edges and donors and acceptors and are commonly observed in solids. In particular, we may have a conduction band to a neutral acceptor (F) and a neutral donor to valence band (G) transitions. They may be phonon assisted.

Process H

In the transitions related to these processes, an electron leaves a neutral donor and moves to a neutral acceptor. After such a transition, both donor and acceptor are ionized and have a binding energy equal to

$$E_b = -\frac{e^2}{4\pi\varepsilon_o kr} \tag{1.27}$$

where r is the donor–acceptor distance. The energy of the transition is then

$$\hbar\omega(r) = E_g - E_A - E_D + \frac{e^2}{4\pi\varepsilon_o kr} \tag{1.28}$$

An example of such a transition is given by *GaP* containing sulfur donors and silicon acceptors, both set in phosphorous sites.

1.13 Nonradiative Processes

In the great majority of cases, a recombination of electrons and holes takes place by the emission of phonons. Since the probability of such processes decreases with the number of phonons emitted, these processes are favored by the presence of intermediate levels between the valence and the conduction bands produced by impurities or defects.

An additional mechanism, known as the *Auger process*, could be responsible for the nonradiative recombination of electrons and holes. In an Auger process, an electron undergoes an interband transition and gives the corresponding energy to another conduction band electron, which is then brought to a higher level in the same band. The latter electron decays then to the bottom of the band with the phonon emission facilitated by the near continuum of states. In most cases, however, the mechanism of nonradiative decay has not been identified with certainty.

1.14 *p–n* Junctions

1.14.1 Basic Properties

A *p–n* junction consists of a semiconductor crystal doped in one region with donors and in an adjacent region with acceptors. Assuming, for simplicity's sake, that the junction has been formed mechanically by pushing toward each other a bar of *n*-type semiconductor and a bar of *p*-type semiconductor, a *junction plane* divides the two regions (see Figure 1.6).

Let us now examine the motion of the electrons (majority carriers of the *n*-type bar) and of holes (majority carriers of the *p*-type bar). Electrons on the *n*-side of the junction plane tend to diffuse (from right to left in the figure) across this plane and go to the *p*-side where there are only very few electrons. On the other hand, holes on the *p*-side tend to diffuse (from left to right in the figure) and go to the *n*-side where there are only very few holes.

The *n*-side region is full with positively charged donor ions. If this region is isolated, the positive charge of each donor ion is compensated by the negative charge of an electron in the conduction band. But, when an *n*-side electron diffuses towards the *p*-side, a donor ion, having lost its compensating electron, remains positively charged, thus introducing a fixed positive charge near the junction plane.

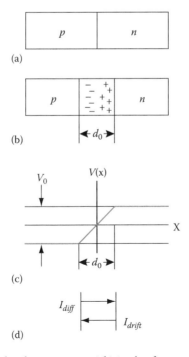

FIGURE 1.6 (a) An *n*-type material and a *p*-type material joined to form a *p–n* junction, (b) space charge associated with uncompensated donor ions at the right of the junction plane and acceptor ions at the left of the plane, (c) contact potential difference associated with the space charge, (d) diffusion current I_{diff} made up by majority carriers, both electrons and holes, compensated in an isolated *p–n* junction by a current I_{drift} made up by minority carriers.

An electron arriving to the *p*-side quickly combines with an acceptor ion and introduces a fixed negative charge near the junction plane on the *p*-side. Holes also diffuse, moving from the *p*-side to the *n*-side, and have the same effect as the electrons.

Both electrons and holes, with their motion, contribute to a *diffusion current* I_{diff}, that is conventionally directed from the *p*-side to the *n*-side. An effect of the motion of electrons and holes across the junction plane is the formation of two space charge regions, one negative and one positive. These two regions together form a *depletion zone* of width d_o in Figure 1.6, so called because it is relatively free of mobile charge carriers. The space charge has associated with it a *contact potential difference*, V_o, across the depletion zone, which limits the further diffusion of electrons and holes.

Let us now examine the motion of the minority carriers: electrons on the *p*-side and holes in the *n*-side. The potential V_o set by the space charges represents a barrier for the majority carriers, but favors the diffusion of minority carriers across the junction plane. Together, both types of minority carriers produce with their motion a drift current I_{drift} across the junction plane in the sense contrary to that of I_{diff}. An isolated *p–n* junction in equilibrium presents a contact potential difference V_o between its two ends. The average diffusion current I_{diff} that moves from the *p*-side to the *n*-side is balanced by the average drift current I_{drift} that moves in the opposite direction.

Note the following:

1. The net current due to holes, both majority and minority carriers, is zero.
2. The net current due to electrons, both majority and minority carriers, is zero.
3. The net current due to both holes and electrons, both majority and minority carriers included, is zero.

1.14.2 Junction Rectifier

When a potential difference is applied across a *p–n* junction, with such a polarity that the higher potential is on the *p*-side and the lower potential on the *n*-side, an arrangement called *forward-bias connection* (Figure 1.7a), a current flows through the junction. The reason for this phenomenon is that the *p*-side becomes more positive than it was before and the *n*-side more negative, with the result that the potential barrier V_o decreases, making it easier for the majority carriers to move through the junction plane and increasing considerably the diffusion current I_{diff}. The minority carriers sense no barrier and are not affected, and the current I_{drift} does not change.

Another effect that accompanies the setting of a forward bias connection is the narrowing of the depletion zone, due to the fact that the lowering of the potential barrier must be associated with a smaller space charge. The space charge is due to ions fixed in their lattice sites, and a reduction of their number produces a reduction of the width of the depletion zone. If the polarity is reversed in a *backward-bias* connection (Figure 1.7b) with the lower potential on the *p*-side and the higher potential on the *n*-side of the *p–n* junction, the applied voltage increases the contact potential difference and, consequently, I_{diff} decreases while I_{drift} remains unchanged. The result is a very small back current I_B.

1.14.3 Radiative Processes in *p–n* Junctions and Applications

In a simple semiconductor, one electron–hole pair may combine with the effect of releasing an energy E_g corresponding to the band gap. This energy in silicon, germanium, and other simple semiconductors is transformed into thermal energy, i.e., the vibrational energy of the lattice. In certain semiconductors, such as *GaAs*, the energy of a recombined electron–hole pair can be released as a photon of energy E_g. However, due to the limited number of electron–hole possible recombinations at room temperature, pure semiconductors are not apt to be good emitters.

Doped semiconductors also do not provide an adequate number of electron–hole pairs, with the *n*-type not having enough holes and the *p*-type not enough electrons.

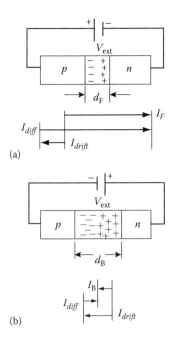

(a)

(b)

FIGURE 1.7 (a) Forward-bias connection of a p–n junction, showing the narrowed depletion zone and the large forward current I_F; (b) backward-bias connection of a p–n junction, showing the widened depletion zone and the small back-current.

A semiconductor system with a large number of electrons in the conduction band and a large number of holes in the valence band can be provided by a heavily doped p–n junction. In such systems, a current can be used in a forward-bias connection to inject electrons in the n-type part of the junction and holes in the p-part. With large dopings and intense currents, the depletion zone becomes very narrow, perhaps a few microns wide, and a great number of electrons are in the n-type material and a large number of holes in the p-type material. The radiative recombination of electrons and holes produces a light emission called *electroluminescence*, or more aptly, *injection electroluminescence*.

The materials used for *light emitting diodes* (LEDs) comprise such alloys as $GaAs_{1-x}P_x$, in which the band gap can be varied by changing the concentration x of the P atoms. For $x \doteq 0.4$, the material is a direct-gap semiconductor and emits red light. Almost pure GaP produces green light, but since it is an indirect-gap semiconductor, it has a low transition probability.

The passage of current through a properly arranged p–n junction can generate light. The reverse process is also possible, where a beam of light impinging on a suitable p–n junction can generate a current. This principle is at the basis of the *photo-diode*.

A remote TV control consists of an LED that sends a coded sequence of infrared light pulses. These pulses are detected by a photo-diode that produces the electrical signals that perform such various tasks as change of volume or channel.

In a forward biased p–n junction, a situation may be created in which there are more electrons in the conduction band of the n-type material than holes in the valence band of the p-type material. Such a situation of *population inversion* is essential for the production of *laser action*. Of course, in addition to this condition, the appropriate geometry for the p–n junction is necessary in order to allow the light to be reflected back and forth and produce the chain reaction of stimulated emission. In this way, a p–n junction can act as a p–n *junction laser* with a coherent and monochromatic light emission.

Acknowledgments

We want to acknowledge the kind permission granted by Springer Science and Business Media to use in this article part of the Chapter entitled "Luminescence spectroscopy" that we contributed to the *Handbook of Applied Solid State Spectroscopy*, D. R. Vij, Editor, published by Springer in 2006.

We want to thank Professor Claus Klingshirn for very helpful discussions and clarifications.

References

1. Goethe J. W. 1984, *The Sorrows of Young Werther*, Trans Mayer E. and Brogan L. (New York: The Modern Library).
2. Alam A. S. M. and Di Bartolo B. 1967, *Phys. Rev. Lett.* 19, 1030.
3. Dekker A. J. 1957, *Solid State Physics* (Englewood Cliffs, NJ: Prentice Hall), p. 312.
4. Elliott R. J. and Gibson A. F. 1974, *An Introduction to Solid State Physics and Its Applications* (London, U.K.: MacMillan), p. 208.
5. Flaherty J. M. and Di Bartolo B. 1973, *Phys. Rev. B* 8, 5232.
6. Baumesteir P. W. 1961, *Phys. Rev.* 121, 359.
7. Elliott R. J. and Gibson A. F. 1974, *An Introduction to Solid State Physics and Its Applications* (London and Basingstoke: MacMillan), p. 229.
8. Vij D.R. 2006, *Handbook of Applied Solid State Spectroscopy* (Berlin, Germany: Springer).

Bibliography

Dekker A. J. 1957, *Solid State Physics* (Englewood Cliffs, NJ: Prentice Hall).
Di Bartolo B. 1968, *Optical Interactions in Solids* (New York: John Wiley & Sons).
Elliott R. J. and Gibbon A. F. 1974, *An Introduction to Solid State Physics and Its Applications* (London, U.K.: Macmillan).
Henderson B. and Imbusch G. F. 1989, *Optical Spectroscopy of Inorganic Solids* (Oxford, U.K.: Clarendon Press).
Kaplianskii A. A. and Macfarlane R. M. 1987, *Spectroscopy of Solids Containing Rare Earth Ions* (Amsterdam, the Netherlands: North Holland).
Klingshirn C. F. 1997, *Semiconductor Optics* (Berlin, Germany: Springer).
Liu G. K. and Jacquier B. Eds. 2005, *Spectroscopic Properties of Rare Earths in Optical Materials* (Berlin: Springer).
Lumb M. D., Ed. 1978, *Luminescence Spectroscopy* (New York: Academic).
Rebane K. 1970, *Impurity Spectra of Solids* (New York: Plenum).
Yu P. Y. and Cardona M. 2010, *Fundamentals of Semiconductors Physics and Materials Properties* (Berlin, Germany: Springer).

2

AlN: Properties and Applications

Ashok Sedhain
Texas Tech University

Jingyu Lin
Texas Tech University

Hongxing Jiang
Texas Tech University

2.1 Introduction

Solid-state compact ultraviolet (UV) and deep UV (DUV) light sources are useful in various sectors including medical research/health care, water/air purification, equipment/personal decontamination, high-resolution photolithography, and white light generation by phosphor excitation [1,2]. AlN and Al-rich AlGaN alloys are highly suitable for these applications and are not easily replaced by any other semiconductors for various reasons. With a direct bandgap of ~6.1 eV, any emission wavelength between 362 and 207 nm can be obtained by simply tuning the alloy composition of AlGaN alloys. Furthermore, its wide bandgap and radiation tolerance permit AlN-based photodetectors to intrinsically suppress the visible background, so that the detectors can operate at room temperature without cooling. AlN has very low electron affinity, so that the electrons in its conduction band can be emitted easily into the vacuum, which makes AlN a promising material in field emission (FE) applications. Properties like DUV transparency, high thermal conductivity, and low electrical conductivity make AlN an ideal substrate material for the growth of high-quality DUV semiconductor device structures. The large band offset of AlN/GaN/InN heterostructures makes AlN unique for optoelectronic device implementation with superb performance. Therefore, AlN and Al-rich AlGaN alloys are technologically important materials for various applications, and they have been the subject of intense research effort in the last decade.

Several theoretical calculations on band structure of wurtzite (WZ) AlN near the Γ-point and many experiments confirmed that the near band edge emission from AlN is predominantly polarized in *c*-axis direction, so that emitted light propagates parallel to the surface in *c*-plane-oriented devices and epilayers. Therefore, efficient extraction of light is critical in the light-emitting devices based on these materials. Nanostructure incorporation, such as photonic crystals, has been suggested to improve the extraction efficiency of the emitters based on AlN and Al-rich AlGaN alloys. Polarization-resolved photoluminescence (PL) measurements revealed how the preferred polarization of emitted light changes continuously with Al-content in AlGaN alloys and also the possibility of future TM mode semiconductor lasers.

Rapid progress has been made in the last 10 years on AlN epilayer growth using metal-organic chemical vapor deposition (MOCVD) [3,4] and molecular beam epitaxy (MBE) [5–10]. Insertion of a low-temperature AlN buffer layer has made it possible to grow crack-free layers of AlN with low dislocation densities on foreign substrates such as sapphire and SiC. AlN epilayers with superior material quality have been achieved from either technique, as confirmed by narrow (~50 arcsec) full width at half maximum (FWHM) of (0002) x-ray diffraction (XRD) rocking curves, atomically flat surface with root mean square (rms) roughness of a few Å and strong excitonic transition peaks with negligible impurity emissions in their luminescence spectra. Relatively good quality AlN epilayers grown on sapphire and SiC typically contain threading dislocation densities (TDD) on the order of $10^7 – 10^9$ cm^{-2} arising primarily from the lattice and thermal expansion coefficient mismatches with the substrate. Lateral overgrowth on patterned substrates has been suggested for the growth of hetero-epitaxial layers on foreign substrates in order to further decrease TDD. Growth of AlN and Al-rich AlGaN-based device structures or epilayers on AlN bulk single crystals has the potential to significantly improve the material quality and device performance. More recently, growth and processing mechanisms that enable the attainment of sizeable AlN wafers with XRD rocking curve linewidths below 30 arcsec and TDD below 10^4 cm^{-2} have been reported, which represents a significant step forward to address the foreign substrates–related issues in III-nitrides.

Due to the wide band gap of AlN (6.1 eV at 10 K), undoped samples are highly resistive in nature, and there are only a few reports on electrical characteristics of doped AlN. N-type AlN with reasonable conductivity can be achieved by Si doping. Recently, room temperature *n*-type conductivity of up to 1 (Ω cm)$^{-1}$ and electron mobility up to 426 cm^2 V^{-1} s^{-1} have been achieved by Si doping [6,7,11–13]. For *p*-type doping of AlN, only a few experimental results have been reported, suggesting p-type conduction by Mg doping achievable only at elevated temperatures. This is attributed to the very high activation energy of Mg in AlN (~0.5–0.6 eV) and also partly, to a lesser degree, to low formation energies of acceptor compensating centers such as nitrogen vacancies (V_N) [14–16]. Postgrowth thermal annealing for acceptor activation, which was beneficial to achieve *p*-type conduction in GaN, could not help to derive room temperature p-type conduction in AlN. At this point, finding novel p-type doping strategies and/or identifying an alternate shallower *p*-type dopant is one of the most challenging issues in III-nitride R&D and hence for the further development of nitride DUV photonic devices. Several ongoing research efforts have been devoted in this direction [17–19]. Rapid progress has also been made on AlN nanostructures. Highly stable FE with current density more than 20 mA/cm^2 and a turn on electric field as low as ~1.8 V/μm has been observed in AlN nanoneedle arrays, which is bright enough for commercial applications in FE display [20].

More recently, AlN-based photodetectors with a sharp cutoff wavelength at 207 nm and extremely low dark current [21], p–i–n homojunction light-emitting diode (LED) with emission peak at 210 nm [15], and optically pumped room temperature stimulated emission at 214 nm [22] have been reported.

Excitons in AlN seem to possess the largest binding energy and shortest PL decay lifetime among all known semiconductors, indicating the inherently high radiative efficiency of AlN and its potential for optoelectronic applications. The electrical characterization of AlN is still limited due to highly resistive nature of the material and poor doping efficiencies. Therefore, optical studies have been critically important in evaluating the quality of AlN and Al-rich AlGaN alloys. The DUV time-resolved PL system in author's laboratory and various other optical systems around the world such as PL, cathodoluminescence (CL), absorption, reflectance, and transmission have been employed to explore the important physical parameters of these materials, which are essential in the development of DUV optoelectronic devices. Some of the existing optical probing systems are listed in Table 2.1 [23–34]. Excitonic emission energies in undoped, Si-doped n-type, and Mg-, Zn-, or Be-doped AlN, their decay kinetics, electron and hole-effective masses, exciton binding energy, variation of excitonic emission energy with strain in the epilayer, and optical polarization have all been investigated.

Understanding the native point defects in AlN and Al-rich AlGaN is the key to achieve good quality material and improve the performance of the device. Aluminum vacancy (V_{Al}) and its complex, such as V_{Al} and O_N complex ($V_{Al} - O_N$), of various charge states in undoped and Si-doped AlN and nitrogen vacancies (V_N) in Mg, Zn, or Be-doped AlN have been found to be the dominant native point defects, which offer energy levels deep in the bandgap and show parasitic behavior to device efficiency and lifetime, acting as carrier traps. Oxygen and carbon are the most abundant background impurities to affect the optical properties and are typically incorporated during the growth in relatively less pure materials.

Recent progress on optical studies in AlN is summarized in this chapter. Fundamental band structure of AlN and its effect on the polarization of emitted light and on the light extraction efficiency of nitride UV and DUV emitters have been discussed in Section 2.2. Various recombination mechanisms in AlN are summarized in Section 2.3 based on reported experimental results. These mechanisms include free and bound excitonic transitions, band-to-impurity transitions, donor–acceptor pair (DAP) transitions, and carrier–phonon interactions in AlN. Optical transitions of intentionally doped or unintentionally incorporated impurities and transitions of native point defects in AlGaN alloys such as V_{Al} ($V_{Al} - O_N$) in undoped, and Si-doped n-type AlN and V_N in Mg and Zn-doped AlN are the topics for Section 2.4. This section also discusses the effects of threading dislocations (TDs) on optical transitions and some viable approaches currently pursued in order to reduce the TDs in this material and hence to improve the overall material quality. It includes how we can improve the optical quality of these materials using LT buffer, high-quality AlN templates, lateral epitaxial overgrowth, and application of AlN bulk substrate to grow stress-free homoepitaxial layers. Modifications of the optical properties and manifestations of completely new optical features by nanostructuring the AlN are the topics in Section 2.5. AlN QW

TABLE 2.1 Common Optical Characterization Systems Used to Study III-Nitride Semiconductors

Method	Excitation Source	Detection System	References
Raman	532 nm solid state laser	JY-HR 800 laser raman spectrometer	[23]
Raman	Ar$^+$ laser at 514.53 nm	T6400 Jobin Yvon spectrometer and CCD	[24]
PL	300 nm Xe lamp (450 W)	Fluoro tau-3 fluorescence spectrometer (Jobin Yvon company)	[25]
		Hitachi f-4500 spectrofluorometer	
PL	ArFexcimer at 193 nm (pulsed)	UV optimized LN$_2$-cooled CCD	[22,25–29]
TR PL	Freq. quad. mode-locked Al$_2$O$_3$:Ti laser	Hamamarsu PMT/streak camera	[30]
		Streak camera	[31]
CL	Electron gun at high vacuum	UV sensitive GaAs photomultiplier	[32,33]
Abs. & OR	150 W deuterium lamp	Solar-blind photomultiplier	

Raman, Raman spectroscopy; PL, photoluminescence; TR PL, time-resolved PL; OR, optical reflectivity; CL, cathodoluminescence; Abs., absorption.

structures (mainly as a barrier layer), optical properties of nanowires, and the enhanced light extraction using the photonic crystal structure are summarized. Finally, various applications of AlN are discussed in Section 2.6.

2.2 Band Structure, Transition Probability, and Polarization Properties

2.2.1 AlN Band Structure and Selection Rules

III-nitrides with WZ crystal structure are anisotropic along and perpendicular to their *c*-axis, which induces crystal-field splitting of the valence band (Γ_{15} symmetry) into the twofold degenerate Γ_6 state and nondegenerate Γ_1 state [35,36]. Spin-orbit interaction further splits Γ_6 state into Γ_9 and Γ_7 states as illustrated in Figure 2.1. Earlier reports on band structure calculations for AlN suggest that crystal-field splitting parameter (Δ_{CF}) is negative in WZ AlN, GaN, and InN (Γ_{1v} above Γ_{6v}) [37]. However, more recent reports from both calculations [38–40] and experiments [40–42] confirmed that Δ_{CF} is positive for GaN and InN, while it is negative for AlN. Negative Δ_{CF} in AlN has huge consequences on electrical and optical properties and therefore on device performance. The fundamental parameters that mainly determine the magnitude and nature of crystal-field splitting are the deviations of the ratio of two lattice constants (*c/a*) and cell internal structure parameter (*u*) from their values in an ideal WZ structure [*c/a* = (8/3)$^{1/2}$ and *u* = 0.375] as described by the following equation [43]:

$$\Delta_{CF} = \Delta_{CF}^0 + \alpha(u - 0.375) + \beta\left(\frac{c}{a} - 1.633\right),\qquad(2.1)$$

where Δ_{CF}^0 represents the value of Δ_{CF} for an ideal WZ structure and $\alpha = -17\,eV$ and $\beta = 2\,eV$ for GaN and assumed to be very similar for AlN. AlN, being more ionic, has a much smaller *c/a* ratio (1.601 vs. 1.626 for GaN) and a much larger *u* parameter (0.3819 vs. 0.3768 for GaN). These structural parameters in AlN result in a large negative value of Δ_{CF} (−219 meV) instead of a positive value in GaN (+38 meV) and in all other III–V and II–VI binary semiconductors [35]. Values of Δ_{CF} reported so far for AlN based on various calculations [35,43,44] and experimental results [26,45,46] scattered in a range from −206 to −237 meV (summarized in Table 2.2), possibly due to different amounts of strain involved in AlN bulk or epilayers grown on different substrates with varying thicknesses. However, it is widely accepted now that AlN has the different ordering of the valence bands compared to that in GaN owing to its negative Δ_{CF}. The valence bands, given in an increasing order of their transition energies, are Γ_{7vbm} (A), Γ_{9v} (B), and Γ_{7v} (C) for AlN as shown in Figure 2.2, whereas in GaN, the order is Γ_{9vbm}, Γ_{7v}, and Γ_{7v} [40].

The immediate consequence of difference in the ordering of the three valence bands (known as A-, B-, and C- in the order of increasing emission energy) is the significant difference in optical properties of AlN compared to that of GaN. The fundamental optical transitions near the Γ point and the transport

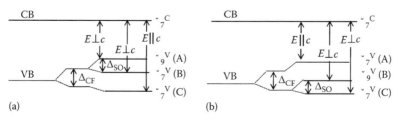

FIGURE 2.1 Schematic diagram (not to the scale) showing the crystal-field (Δ_{CF}) and spin-orbit (Δ_{SO}) splittings of the valence bands of WZ III-nitride for (a) positive (GaN) and (b) negative (AlN) Δ_{CF}. (After Taniyasu, Y. et al., *Appl. Phys. Lett.*, 90, 261911, 2007.)

TABLE 2.2 Theoretically Calculated and Experimentally Determined Valence Band Splitting Parameters for Wz AlN Bulk and Epilayers Measured by Various Methods

Δ_{SO} (meV)	Δ_{CF} (meV)	ΔE_{AB} (meV)	ΔE_{BC} (meV)	Notes
	−219	213	13	Calc.
19	−217	211	13	Calc.
19	−224	218	19	Calc.
	−237	$E_A = 5.981$ (5.985) eV		OR, CL, and PL
		$E_B, E_C = 6.22$ eV not resolved		AlN:Si/SiC
	−230	$E_A = 6.024$ eV		OR
		$E_B, E_C = 6.25$ eV not resolved		m-Face AlN bulk
36	−225	234	25	OR, bulk AlN
20	−206	199	13	PL, AlN/Al$_2$O$_3$

Source: After Sedhain, A. et al., *Appl. Phys. Lett.*, 92, 04114, 2008.
Calc., calculation; OR, optical reflectivity; CL, cathodoluminescence.
E_A, E_B, E_C, emission energies corresponding to the A, B, and C valence bands.

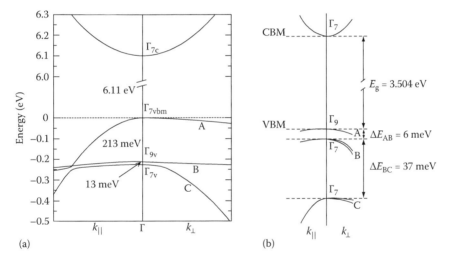

FIGURE 2.2 Calculated band structures of (a) AlN and (b) GaN together with the experimentally measured energy bandgap and exciton binding energies. (After Li, J. et al., *Appl. Phys. Lett.*, 83, 5163, 2003; Chen, G.D. et al., *Appl. Phys. Lett.*, 68, 2784, 1996.)

properties of the free holes in AlN are mainly determined by the top Γ_{7vbm} band instead of the top Γ_{9vbm} band in GaN due to the large energy separation between the top most (A) valence band and the second and third (B and C) valence bands in AlN. The majority of photogenerated holes populate the topmost valence band, which dominate the fundamental optical and electrical properties observed in experiments. The second valence band may influence the PL only when it is close to the topmost band as in the case of GaN. The square of the dipole transition matrix elements $I = |\langle\psi_v|\mathbf{p}|\psi_c\rangle|^2$ between the conduction band and all of three valence bands at the Γ-point of WZ AlN for light polarized parallel ($\parallel$) and perpendicular ($\perp$) to the c-axis were calculated and are listed in Table 2.3 [35,40]. For an arbitrary light polarization, the square of the dipole transition matrix element is expressed as: $I(\theta) = \cos^2\theta\, I(\mathbf{E}\parallel\mathbf{c}) + \sin^2\theta\, I(\mathbf{E}\perp\mathbf{c})$, where θ is the angle between $\mathbf{E}$ (the electric field component of the light) and the crystallographic c-axis. Our results showed that the recombination between electrons in the conduction band and the

TABLE 2.3 Calculated Square of the Dipole
Transition Matrix Elements, I (in Arbitrary Unit,
a.u.) of Wz AlN and GaN for Light Polarized
Parallel (‖) and Perpendicular ($\perp$) to the c-Axis

	AlN		GaN	
Transition	$I(E\|c)$	$I(E\perp c)$	$I(E\|c)$	$I(E\perp c)$
E_A	0.4580	0.0004	0	1
E_B	0	0.2315	0.053	0.974
E_C	0.0007	0.2310	1.947	0.026

Sources: After Li, J. et al., *Appl. Phys. Lett.*, 83, 5163,
2003; Chen, G.D. et al., *Appl. Phys. Lett.*, 68, 2784,
1996.

holes in the topmost valence band (Γ_{7vbm} or A) is almost prohibited for **E**$\perp$**c**, whereas the recombination between the electrons and holes in the Γ_{9v} (or B) and Γ_{7v} (or C) valence bands is almost forbidden for **E**‖**c**. This is in sharp contrast to the case of GaN in which the theoretical and experimental results have shown that the recombination of electrons and the holes in the topmost valence band Γ_{9vbm} (or A) is almost prohibited for **E**‖**c** [40,46].

2.2.2 Experimental Confirmation of Predicted Band Structure

Several experimental results [43,47–50] devoted to optical polarization properties of AlGaN epilayers, AlGaN/AlN multiple quantum wells (MQWs), and blue and UV LEDs confirmed the unique feature of the valence band structure of AlN due to large negative Δ_{CF}. These results are listed in Table 2.2. Figure 2.3 shows the low temperature (10 K) PL spectra of $Al_xGa_{1-x}N$ epilayers covering the entire alloy composition range measured in the author's laboratory [50]. The inset shows the PL measurement geometry to collect either **E**$\perp$**c** or **E**‖**c** polarized component by selecting the proper orientation of a polarizer, which was placed in front of the monochromator. PL signal polarized in **E**$\perp$**c** (**E**‖**c**) direction is shown with dotted (solid) lines. We observed that the emission intensity (I_{PL}) of **E**$\perp$**c** component decreases

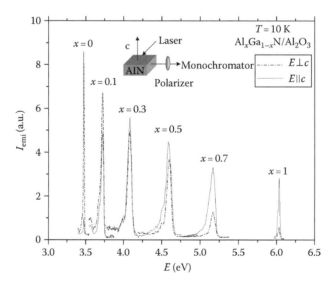

FIGURE 2.3 Low temperature (10 K) PL spectra of $Al_xGa_{1-x}N$ alloys of varying x from $x = 0$ to 1. The inset shows the experimental geometry, where the electrical field of PL emission (**E**) can be selected either parallel (**E**‖**c**) or perpendicular (**E**$\perp$**c**) to the c-axis. (After Nam, K.B. et al., *Appl. Phys. Lett.*, 84, 5264, 2004.)

with increasing x, while I_{PL} of the **E∥c** component remains almost independent of x except for GaN. The PL emission evolves from **E⊥c** being dominant in GaN to **E∥c** being dominant in AlN. The critical Al composition in AlGaN alloys for the abrupt switch of dominant light polarization from **E⊥c** to **E∥c** was determined to be $x \approx 0.25$ by measuring the degree of polarization, $p = (I_\perp - I_\parallel)/(I_\perp + I_\parallel)$, from the experimental values of $I_\perp$ and $I_\parallel$.

Polarization-resolved PL spectra of AlN epilayers have been measured [51]. An A-exciton emission peak at 6.05 eV predominantly polarized in **E∥c** and an additional peak at 6.26 eV polarized in **E⊥c** direction were observed, as shown in Figure 2.4. In the high resolution spectrum, the higher energy peak is resolved into two peaks at 6.249 and 6.262 eV, which were assigned to the B and C valence band–related free exciton (FX) recombinations, FX$_B$ and FX$_C$. **E⊥c** polarized component still showed relatively weak A valence band–related emission line at 6.05 eV in addition to the expected higher energy peaks at 6.249 and 6.262 eV. This is primarily due to the fact that the exact selection rule only applies to the free-hole transition at $\Gamma = 0$, without considering the excitonic effect. The transition is not totally forbidden under the influence of the excitonic effect, but appears with weaker emission intensity [52]. The absorption spectra for AlN were calculated for two different light polarization configurations, **E⊥c** and **E∥c**, and compared with that of GaN [35]. These results are consistent with the polarization-resolved PL measurements [51] in the sense that both reveal the fundamental energy gap of AlN is ~6.1 eV with polarization orientation of **E∥c**. For AlN epilayer grown on *c*-plane sapphire, measured absorption spectra are always in the **E⊥c** configuration. Therefore, excitation of holes from B and C valence bands is more likely than that from the topmost A valence band resulting in an apparent energy gap of $E_g + \Delta E_{AB} \approx 6.3$ eV, about 0.2 eV larger than the fundamental gap.

Taniyasu et al. [27] reported FX emission from a *c*-plane AlN epilayer grown on SiC characterized by angle-resolved (θ_R) PL measurement, where θ_R is an angle between *c*-axis (surface normal) and detected PL signal. It was observed that the PL emission intensity from c-AlN (0001) surface ($\theta_R = 0$) is weak and increases remarkably as the emission direction inclines from surface normal toward in-plane direction ($\theta_R = 90$) as shown in Figure 2.5 (solid circles). For GaN (0001) surface, emission has a maximum at $\theta_R = 0$ and decreases with θ_R (solid squares). These results agree quite well with polarization-resolved

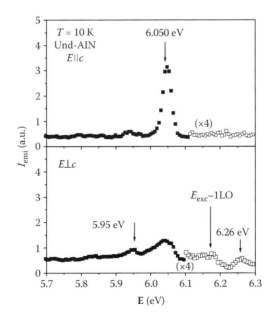

FIGURE 2.4 Low temperature (10 K) PL spectra of AlN epilayer measured under different polarization configurations with (a) **E∥c** and (b) **E⊥c**. (After Sedhain, A. et al., *Appl. Phys. Lett.*, 92, 041114, 2008.)

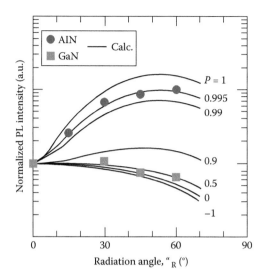

FIGURE 2.5 Normalized PL intensity of free-exciton emission from c-AlN surface (solid circles) and c-GaN (solid squares) as a function of radiation angle, θ_R. Solid lines represent the calculated intensities for different polarization ratios, P. (With permission from Taniyasu, Y., Kasu, M., and Makimoto, T., *Appl. Phys. Lett.*, 90, 261911, 2007. Copyright 2007, American Institute of Physics.)

PL measurements and are expected from AlN unique band structure. Stimulated edge-emission from pulsed lateral overgrown (PLOG) AlN epilayer with emission strongly polarized in **E‖c** (TM mode) direction, as shown in Figure 2.6, also agrees well with earlier reports [22].

The energy bandgap (mainly the splitting of valence band edge) varies systematically across the entire $Al_xGa_{1-x}N$ alloy composition, and the crossover takes place between Γ_9 and Γ_7 valence bands around $x = 0.25$ [53]. The positive value of Δ_{CF} in GaN decreases with x and becomes zero at $x = 0.25$ and negative for $x > 0.25$. It flips the topmost valence band having Γ_9 symmetry for $x < 0.25$ into Γ_7 for $x > 0.25$. Since the conduction band always has Γ_7 symmetry in the whole alloy composition range, the lowest bandgap (A-exciton) radiative emission being predominantly **E⊥c** polarized in the GaN side flips into **E‖c**

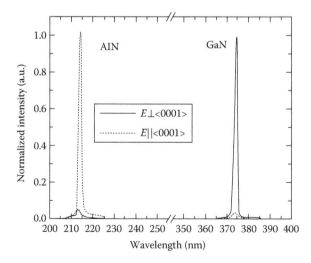

FIGURE 2.6 Stimulated edge emission spectra of AlN and GaN for two different configurations of **E⊥c** and **E‖c**. (After Shatalov, M. et al., *Jpn. J. Appl. Phys.*, 45, L1286, 2006. With permission from Japan Society of Applied Physics.)

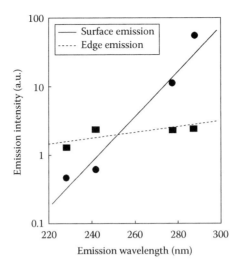

FIGURE 2.7 Surface (through *c*-plane) and edge (through *m*-plane) emission intensities of *c*-plane $Al_xGa_{1-x}N$ alloys ($x = 0 - 0.76$) MQWs as functions of emission wavelength. (With permission from Kawanishi, H., Senuma, M., Yamamoto, M., Niikura, E., and Nukui, T., *Appl. Phys. Lett.*, 89, 081121, 2006. Copyright 2006, American Institute of Physics.)

polarized in the AlN side. Polarization-resolved EL spectra of 458 nm blue and 333 nm UV LEDs with fixed sample orientation and driving current [47] also demonstrated the preferential emission properties. Light emission with $E \perp c$ component is predominant in blue LED (GaN like) in contrast to the $E \| c$ component being predominant in UV LED (AlN like), which further confirms the unique polarization properties of AlGaN alloys. More recent studies on optical polarization properties of AlGaN alloys and QWs showed that the critical Al composition for the polarization switch, in fact, can be finely tuned all the way upto $x = 0.82$ by engineering the valence band structure through the compressive strain and quantum confinement in the active layer [28].

PL spectra of two polarization components from surface (through the *c*-plane) and edge (through the *m*-plane) are shown in Figure 2.7 for four different *c*-plane $Al_xGa_{1-x}N$ ($x = 0 - 0.76$) MQWs [49]. The solid and dashed lines indicate surface and edge emissions, respectively. At 287 nm, surface emission intensity was 30–40 times higher than that of edge emission, which decreased continuously with emission wavelengths. Only extremely weak surface emissions, even smaller than edge emission, were detected at 228 and 240 nm. This further verifies the polarization anisotropy with increasing Al-content in AlGaN alloys.

2.2.3 Consequences on Performance of DUV Emitters

Al-rich AlGaN alloys are promising for compact DUV emitters; however, their light output efficiency for the wavelengths shorter than ~300 nm decreases with an increase in Al composition in the conventional *c*-plane AlGaN-based devices. The optical transition probability is high for polarization $E \| c$ due to negative Δ_{CF}, making it difficult to emit from the surface. Review article by Khan et al. [54,55] summarized the maximum quantum efficiencies of III-nitride DUV LEDs at various wavelengths achieved by research groups worldwide up to 2007. The highest external quantum efficiencies (EQEs) were 0.14% and 1.5% at emission wavelengths of 255 and 280 nm, respectively. Very recently, Amano et al. [56] have reported AlGaN-based DUV LEDs with an improved EQE of 3% at emission wavelengths in the range of 255–280 nm using HT-AlN thick template and flip-chip bonding. The efficiency drops quite alarmingly (~10^{-6}%) as the emission wavelength approaches 210 nm in pure AlN-based *p–n* junction emitters [15]. Reported results from Refs. [15,55,56] are plotted together in Figure 2.8, which illustrates

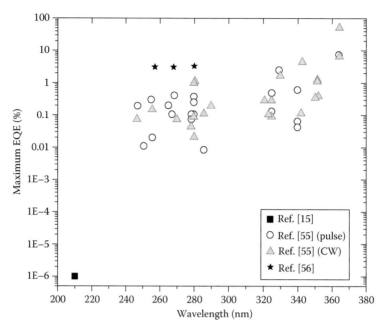

FIGURE 2.8 Maximum EQEs of III-nitride-based UV emitters reported by different research groups worldwide. (After Taniyasu, Y. et al., *Nature*, 441, 325, 2006; Khan, A. et al., *Nat. Photon.*, 2, 77, 2008; Pernot, C. et al., *Appl. Phys. Exp.*, 3, 061004, 2010.)

the current scenario of the performance of III-nitride-based DUV emitters. Due to internal reflection in conventional LEDs, light emits from top and bottom surfaces when it is within a cone of $\theta_c \approx 20°$, where θ_c is the critical angle. Only photons polarized perpendicular to the *c*-axis ($\mathbf{E} \perp \mathbf{c}$) can be extracted from the light escaping cone. For DUV LEDs using Al-rich $Al_xGa_{1-x}N$ as active layers, the $\mathbf{Z}$-like character of the topmost valence band makes the emission polarized in the *c*-axis direction. The emitted photons propagate in the *c*-plane and cannot be extracted easily from the active layer. This unfavorable optical polarization is one of the factors limiting the efficiency of Al-rich AlGaN-based emitters in addition to the other factors, such as increased defect formation, p-type doping issues, and foreign substrate-induced TDs.

As the extraction of emitted light becomes a concern, fabrication of nanostructures, μ-LEDs, and photonic crystals (PC) are suggested as the approaches to extract the light of transverse propagation [57–59] and to achieve high-power nitride emitters in UV and DUV wavelengths. With 2D AlN PC formation, a 20-fold enhancement in the band edge emission intensity at 208 nm over unpatterned AlN epilayer has been observed in the author's laboratory by DUV PL measurement [60].

Since AlN and Al-rich AlGaN-based emitters emit most of the light perpendicularly to the *c*-axis direction, for surface-emitting devices, one approach to improve light extraction efficiency is to perform the growth on a nonpolar *a*-plane or *m*-plane substrate. Their surface orientations are perpendicular to the *c*-axis direction. Although AlGaN QWs on nonpolar or semipolar orientations [49,61] are promising to enhance the surface emission, this approach faces difficulty with respect to crystal growth. Growing nitrides on the *c*-plane is much easier compared to other crystal planes for any substrates including sapphire, SiC, etc. Thus, it is desirable to explore other possibilities to enhance the surface emission from AlGaN active layers.

More recently, engineering the valence band structure of active layer through the variation of compressive strain and quantum confinement [28,61,62] has been suggested. Using thin well width (~1 nm), AlGaN QWs were found to maintain $\mathbf{E} \perp \mathbf{c}$ configuration being the dominant emission with Al content as high as 83%. Banal et al. [28] have demonstrated that the cutoff region for sharp efficiency drop of UV

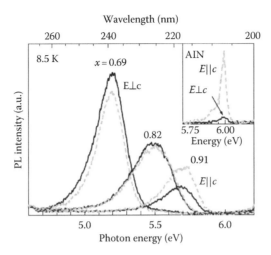

FIGURE 2.9 Polarization-resolved PL spectra of *c*-plane oriented Al$_x$Ga$_{1-x}$N (1.5 nm)/AlN (13.5 nm) MQW with different Al compositions: 0.91, 0.82, and 0.69. The inset shows two polarization components of PL spectra of AlN template used for QW growth. (With permission from Banal, R.G., Funato, M., and Kawakami, Y., *Phys. Rev. B*, 79, 121308, 2009. Copyright 2009 by the American Physical Society.)

LEDs can be brought to a shorter wavelength by pushing the critical Al composition of polarization switch to a higher value than that previously reported for AlGaN epilayers [50] and QWs [47–49]. Figure 2.9 shows the polarization-resolved PL spectra of *c*-plane Al$_x$Ga$_{1-x}$N/AlN QWs (well width = 1.5 nm) for different Al mole fraction (*x*) in well region. As *x* decreases from 0.91 to 0.69, predominant polarization switches from **E||c** to **E⊥c** at *x* ~ 0.82 suggesting a strong surface emission even with such a high Al content with a corresponding emission wavelength at 227 nm. The inset of Figure 2.9 shows the PL spectrum of AlN epilayer being used as template in those structures. By designing the c-AlGaN QW structure (well width = 1.3 nm) to include strain and/or quantum confinement effects and modify the topmost valence band structure, surface emission at 227 nm has been remarkably enhanced with an output power of 0.15 mW at 30 mA and EQE of 0.2%. Such a modification switches the unfavorable *c*-axis polarized emission into in-plane polarized emission, which is favorable for the extraction from *c*-plane-based DUV LEDs [63].

These new results indicate that the valence band structure of AlGaN alloys can be finely tuned through the variation of compressive strain and the quantum confinement, which is schematically shown in Figure 2.10 for (a) unstrained Al$_x$Ga$_{1-x}$N, (b) strained Al$_x$Ga$_{1-x}$N on unstrained AlN epilayer, and (c) Al$_x$Ga$_{1-x}$N/AlN QWs. In unstrained Al$_x$Ga$_{1-x}$N, the critical Al composition was estimated to be *x* = 0.044, which is increased to *x* = 0.60 for strained Al$_x$Ga$_{1-x}$N due to the effect of strain on the valence bands. The previously determined critical Al composition of *x* = 0.25 [50] is probably due to the influence of strain. The in-plane compressive strain in Al$_x$Ga$_{1-x}$N pushes the (*X* + *iY*)-related bands (Γ_9 and lower Γ_7) upward, but tensile strain along the growth direction pushes the (*Z*)-related band (upper Γ_7) downward, as shown in Figure 2.10b. Consequently, the energy separation between the topmost Γ_7 and Γ_9 bands decreased, and, thus, a larger Al composition was needed for the polarization switch. Figure 2.10c also shows the effect of quantum confinement on the valence band structure. Because the hole-effective mass of the topmost Γ_7 band in AlN is much lighter than that of Γ_9 band (0.26m_0 vs 3.57m_0) [42], the energy of Γ_7 is lowered by quantum confinement and eventually causes the crossover between Γ_7 and Γ_9 bands. It indicates that the quantum confinement has a similar effect on polarization switch as that of in-plane compressive strain. Consequently, Γ_9^{QW} could be the topmost valence band even with Al compositions much higher than 0.60 as shown in Figure 2.10b.

Figure 2.11b shows the well-width dependence of η, the index of in-plane polarization, or surface emission for unstrained Al$_{0.8}$Ga$_{0.2}$N/AlN QWs with various crystal orientations calculated by using 6 × 6

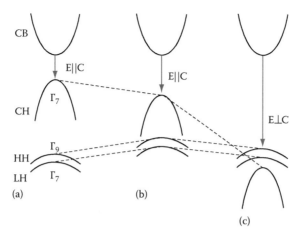

FIGURE 2.10   Schematic diagrams of the band structures for (a) unstrained $Al_xGa_{1-x}N$, (b) strained $Al_xGa_{1-x}N$ on AlN, and (c) $Al_xGa_{1-x}N/AlN$ QWs near the Γ-point indicating that in-plane compressive strain and the quantum confinement effect have the similar effect on shifting the critical Al composition for polarization switch. (With permission from Banal, R.G., Funato, M., and Kawakami, Y., *Phys. Rev. B*, 79, 121308, 2009. Copyright 2009 by the American Physical Society.)

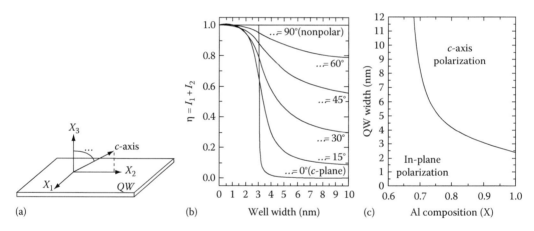

FIGURE 2.11   (a) Schematic diagram of coordinate system; (b) QW-width dependence of index of in-plane optical polarization or surface emission, η, for unstrained $Al_{0.82}Ga_{0.18}N/AlN$ QWs with various substrate orientations; and (c) the "phase diagram" of the emission polarization in AlGaN QWs coherently grown on *c*-plane AlN substrate. (After Yamaguchi, A.A.: *Phys. Stat. Sol. (c)*, 2008. 5. 2364. Copyright Wiley-VCH Verlag GmbH & Co. KGaA. With permission.)

k·p Hamiltonian [62]. The substrate orientation was described by θ, the angle between the *c*-plane and the substrate plane, as shown in Figure 2.11a. For QWs grown on nonpolar substrates ($\theta = 90°$), light is always polarized in the plane (*c*-axis direction), and the value of η is unity, independent of the well width. For QWs on *c*-plane substrates ($\theta = 0°$), light is polarized in the *c*-axis direction in wider wells (>4 nm) indicated by $\eta = 0$. However, the crossing of the topmost Γ_7 and Γ_9 valence bands occurs at the critical well width (~3 nm), and η becomes unity as the polarization changes abruptly for narrower QWs due to the quantum confinement effect on the valence band structure. The polarization switch is also observed in QWs on semipolar substrates, where critical QW thickness for polarization switch and extent of surface emission strongly depends on θ. Figure 2.11c shows the calculated phase diagram of emission polarization in AlGaN QWs coherently grown on *c*-plane AlN substrates [62]. The critical QW width becomes larger, and effect of confinement becomes less important as Al composition decreases. The polarization switch finally takes place by strain effects for Al composition of ~0.65. Experimental investigation on well-width dependence of the polarization-resolved PL spectra on $Al_{0.82}Ga_{0.18}N$/AlN MQWs clearly demonstrates that wider QWs (>8 nm) show **E∥c** polarized emission predominantly just like epilayers, whereas the well width (L_w) in the range from 3 to 8 nm promotes **E⊥c** polarized emission with critical Al composition ~0.80, which is independent of L_w. However, for the wells thinner than 3 nm, quantum confinement comes into play, and critical Al composition shifts toward an even higher value x ~ 0.82 [28].

2.3 Recombination Processes in AlN

Optical transitions observed in AlN include recombination of free and bound excitons involving different valence bands, band-to-impurity transitions, and DAP transitions.

2.3.1 Free and Bound Excitons

Photogenerated electrons and holes form pairs or FX in typically undoped high purity and high crystalline quality AlN due to the Coulomb interaction, and their recombination results in narrow emission lines. In FX recombination, the energy of the emitted photon is

$$h\nu = E_g - E_x,\qquad(2.2)$$

where E_x denotes the binding energy of the FX. In the materials with WZ structure, such as AlN and GaN, three types of excitons are distinguished depending on whether the hole participating in the exciton formation belongs to A, B, or C valence band.

In the presence of impurities, doped or unintentionally incorporated during the growth, excitons can be bound to these impurities and form bound excitons. In PL spectra, recombination of a bound exciton is characterized by an emission line at a lower energy than that of the FX, and the energy difference measures the binding energy of impurity bound exciton. In an impurity (donor or acceptor) bound exciton recombination, the energy of the emitted photon is

$$h\nu = E_g - E_x - E_{bx},\qquad(2.3)$$

where E_{bx} is the binding energy of impurity bound excitons. In p-type materials ($N_A > N_D$), FX can be bound to neutral acceptors (A^0), and the corresponding PL emission line is called the acceptor bound exciton (A^0X) or I_1 transition. In *n*-type materials ($N_D > N_A$), however, FX is bounded to a neutral donors (D^0), and the corresponding PL emission lines are called D^0X (or I_2) transition. Some unintentional donor impurities are usually incorporated during the growth of undoped AlN. Therefore, PL spectra of undoped AlN typically show two peaks for FX and D^0X or I_2 transitions just as the case in the undoped GaN.

There have been many experimental studies on the near band edge optical transitions of AlN bulk single crystal and epilayers in recent years by various groups using different probing methods such as PL, time-resolved PL, CL, and optical reflectivity (OR) spectroscopy [3,26,27,29,30,32,35,45,64–74]. Some of the reported results from literature are summarized in Table 2.4. The energy bandgap of AlN was first measured by optical absorption spectra and suggested to be 6.2 eV at 300 K [33] and 6.28 eV at 5 K in the late 1970s [75]. Progress on material synthesis improved the sample quality, and their optical spectra appeared with dominant excitonic features. In general, the near band edge emission spectra of

TABLE 2.4   Observed Near Band Edge (NBE) Emission Properties of AlN by Different Methods

Peak Position	FWHM (meV)	Exciton BE (meV)	Assignment	Sample/Probe-Method	References
6.023	1.0	63	FX_A	AlN/AlN bulk	[7]
6.008	1.5		I_2	(Misaligned by 10^0 from c-axis) CL	
5.984	9.4		I_1		
6.030		63	FX_A	AlN single crystal (CL/Trans./OR)	[3]
6.010			I_2		
6.025	13.1	74	FX_A	AlN homoepilayers	[69]
6.008	11.6		I_2	(a- and c-plane) PL	
6.063 (6.030)			FX_A	AlN/c-Al$_2$O$_3$ (AlN/r-Al$_2$O$_3$)	
6.045 (6.013)			I_2		
6.085		48	$FX_{A,n=2}$	AlN/Al$_2$O$_3$ (AlN/SiC)	[3]
6.048 (6.052)			FX_A	PL	
6.035 (6.039)			I_2		
(6.025)			XX		
6.012			P_2 band		
6.031	12.5	80	FX_A	AlN/Al$_2$O$_3$	[30]
6.015	15.5		I_2	PL	
6.050			FX_A	AlN/Al$_2$O$_3$	[51]
6.249			FX_B	PL	
6.262			FX_C		
6.177			$FX_{A,n=2}$	AlN/Al$_2$O$_3$	[31]
6.138			FX_A	CL	
6.124	2.9		I_2		
6.27			$FX_{B,C}$		

PL, photoluminescence; CL, cathodoluminescence; trans., transmission; OR, optical reflectivity.

high quality samples involve FX and their phonon replicas at high temperatures and the bound excitons at low temperatures. Under strong light excitation, a dense electron–hole (e–h) plasma forms, and a new broad band appears in the near band edge region due to the radiative recombination of the e–h plasma (P band), which is usually located on the longer wavelength side of the FX lines due to bandgap renormalization [31]. The values of the assigned peak positions of FX, I_2, and I_1 transition lines vary due to the different probe techniques used and amount of stresses present in different samples. The amount of stress varies depending on the substrate used and amount of impurities incorporated. Figure 2.12 shows the low temperature (6–7.5 K) CL, OR, and transmission spectra measured on the same piece of AlN bulk single crystal in the near band edge spectral region [72]. It is interesting to note that three different optical measurements consistently demonstrate the free A-excitonic (FX_A) feature at 6.030 eV, which also agrees with the PL results obtained from strain-free c-, a-, and m-plane AlN homoepilayers [67]. With FX_A binding energy of 50–70 meV reported earlier [32], fundamental bandgap of strain-free AlN is determined to be ~6.09 ± 0.01 eV at 10 K.

Figure 2.13a shows the temperature-dependent band edge PL spectra of an undoped AlN epilayer (~1-μm thick) grown on a c-plane sapphire substrate. Low-temperature (10 K) PL spectra exhibit dominant I_2 emission line at 6.015 eV, which is about 16 meV below FX_A peak at 6.031 eV giving I_2 binding energy of 16 meV in AlN. As temperature increases, the relative intensity of the I_2 transition peak at 6.015 eV decreases, while that of the FX transition at 6.031 eV increases, which resembles the behavior of I_2 and FX transitions observed in GaN [40,76]. This is expected since donor bound excitons dissociate at higher temperatures into FX and neutral donors D^0, ($D^0X \rightarrow FX + D^0$). Figure 2.13b shows the Arrhenius plot of the PL intensity of the FX and I_2 transition lines. The solid lines in Figure 2.13b are the

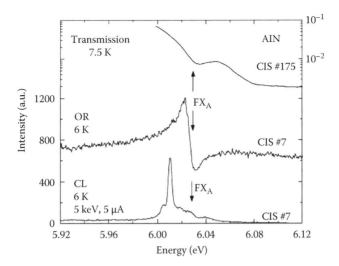

FIGURE 2.12 Low-temperature cathodoluminescence (CL), optical reflectivity (OR), and transmission spectra of AlN bulk crystals in the near band edge energy range. FX$_A$ indicates the position of the free A-exciton. (After *J. Cryst. Growth*, 310, Silveira, E., Freitas, J.A., Schujman, S.B., and Schowalter, L.J., 4007, Copyright 2008, with permission from Elsevier.)

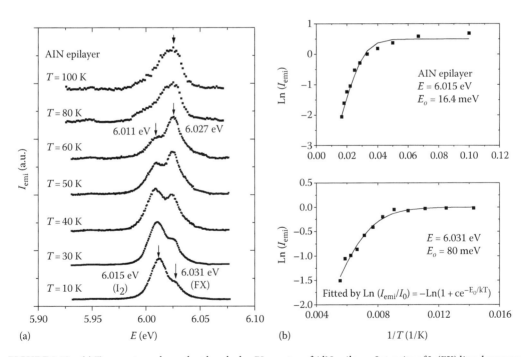

FIGURE 2.13 (a) Temperature-dependent band edge PL spectra of AlN epilayer. Intensity of I$_2$ (FX) line decreases (increases) with increasing T due to dissociation of D^{0}X into FX and D^0. (b) Arrhenius plot of the PL intensity of the FX and I$_2$ transition lines in AlN, which gives their binding energies as 16.4 and 80 meV, respectively. (After Nam, K.B. et al., *Appl. Phys. Lett.*, 82, 1694, 2003; Li, J.K. et al., *Appl. Phys. Lett.*, 83, 5163, 2003.)

least squares fit of the measured data to the equation below, which describes the thermal dissociation (activation) of free and donor-bound excitons:

$$I_{emi}(T) = I_0[1 + Ce^{(-E_0/KT)}]^{-1}, \tag{2.4}$$

where $I_{emi}(T)$ and I_0 are, respectively, the PL intensities at a finite temperature T and 0 K, while E_0 is the activation energy. Binding energies of $E_0 = 80$ and 16.4 meV have been obtained for FX and D^0X, respectively, from the fitting, which also agree with the values determined from the temperature dependence of the FX decay lifetime [30] and the difference of the I_2 and FX spectral peak positions. The energy gap at 10 K is thus determined to be 6.033 + 0.080 eV = 6.11 (±0.01) eV for our epilayer grown on sapphire substrate, which is about 20 meV higher than that of strain-free AlN bulk due to the compressive strain. The FX binding energy of 80 meV obtained from the Arrhenius plot of PL intensity may be overestimated compared to the results published later from different groups.

The temporal responses of the I_2 and FX transitions were measured at their respective spectral peak positions at 10 K as displayed in Figure 2.14. The decay lifetimes were found to be around 80 ps for I_2 and 50 ps for FX transition. A much shorter FX decay lifetime of 11.3 ps at 10 K has recently been reported for c-AlN epilayer [74]. The fast PL decay has been attributed to the purity and better crystalline quality of the materials, as seen from the narrower PL line-width. The effective radiative lifetime of the near band edge transition in most common III–V and II–VI semiconductors plotted as a function of energy gap (E_g) shows that AlN has the fastest radiative recombination rate, which indicates its inherent nature of vary fast radiative recombination [74]. The recombination lifetime of the I_1 transition measured on AlN:Mg at 6.02 eV at 10 K was about 130 ps [77].

The radiative recombination lifetime of FXs, τ_{rad}, can be calculated from the experimentally measured decay lifetimes (τ_{eff}) and quantum efficiency (η) [78]

$$\tau_{rad} = \frac{\tau_{eff}}{\eta}, \tag{2.5}$$

where $\eta = I_{emi}(T)/I_{emi}(0)$ and $I_{emi}(T)$ and $I_{emi}(0)$ are the PL emission intensities at temperature T and 0 K, respectively. Assuming $I_{emi}(0) \approx I_{emi}(10\,K)$, the temperature dependence of τ_{rad} for AlN shows that τ_{rad}

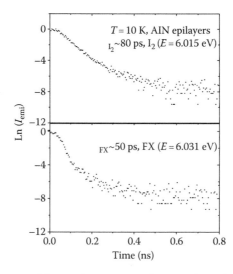

FIGURE 2.14 Temporal responses of the I_2 and FX transition lines in AlN epilayer measured at 10. (After Nam, K.K.B. et al., *Appl. Phys. Lett.*, 82, 1694, 2003.)

increases with T according to $T^{3/2}$ in the temperature range of $100 < T < 200\,\text{K}$ for AlN, a well-known feature of FXs or free carriers in semiconductors [79]. The temperature range for the relation $\tau_{\text{rad}} \propto T^{3/2}$ to hold in AlN is higher than that in GaN [78], which is attributed to a larger binding energy of the bound exciton in AlN. The observed $T^{3/2}$ dependence of τ_{rad} gives a FX binding energy of $E_x = 80\,\text{meV}$, which agrees with the thermal activation energy of the FX obtained from temperature-dependent FX emission intensity as shown in Figure 2.13b. However, this value lies in the higher side compared to values reported by other groups as seen in Table 2.4. FX in AlN is thus very robust and survives at room temperature due to its much larger binding energy compared to ~25 meV in GaN. The enhanced binding energies of free and bound excitons in AlN are attributed to the larger effective masses of electrons and holes in AlN.

More recent results from AlN epilayers grown on different substrates (Al_2O_3, SiC, Si, and AlN bulk) clearly demonstrate that FX_A peak position may differ by several tens of meV depending on the amount of strain in the epilayers [65]. The in-plane stress on each of the epilayers can be calculated from the relation:

$$\sigma_{\parallel} = \frac{a - a_0}{a_0}\left(C_{11} + C_{12} - 2\frac{C_{13}}{C_{33}}\right), \tag{2.6}$$

where

C_{ij} is the elastic constant of AlN ($C_{11} = 410\,\text{GPa}$, $C_{12} = 140\,\text{GPa}$, $C_{13} = 100\,\text{GPa}$, and $C_{33} = 390\,\text{GPa}$)
a_0 is the in-plane lattice constant of strain-free bulk AlN (3.112 Å)

AlN homoepilayer is perfectly lattice matched to the AlN substrate, and the FX emission peak position in AlN homoepilayers lines up with that of unstrained AlN. The shift of FX emission peak position for each layer relative to that of homoepilayer as a function of the in-plane stress is plotted in Figure 2.15, and a linear relationship is evident. The experimental value of the linear coefficient for stress-induced FX peak position (or bandgap) shift in AlN epilayers is 45 meV/GPa. It is about 88% higher than that in GaN (24 meV/GPa) [77], which is expected due to a smaller lattice constant as well as higher mechanical strength of AlN than that in GaN. The strain-free bandgap of AlN is ~6.09 eV at 10 K, which increases

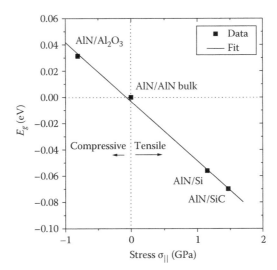

FIGURE 2.15 Stress-induced shift of the FX emission peak position in AlN epilayers as a function of the in-plane stress. The solid line is the least-squares linear fit of the experimental data. The deduced linear coefficient of stress-induced bandgap shift in AlN epilayers is 45 meV/GPa. (After Pantha, B.N. et al., *Appl. Phys. Lett.*, 91, 121117, 2007.)

in AlN hetero-epilayers grown on sapphire substrates due to a compressive strain, whereas it decreases in hetero-epilayers grown on SiC and Si substrates due to a tensile strain.

2.3.2 Band-to-Impurity Transitions

Band-to-impurity transitions, also called free-to-bound transitions, could be shallow or deep transition. In typical semiconductors, a deep transition may be either from the transition of a free electron to an acceptor or from a donor level to the valance band. The shallow transition of free electrons (holes) to donors (acceptors) is typically in the infrared (IR) or far IR region, which is difficult to probe optically in many cases. However, due to its wide energy gap of ~6.1 eV, shallow transitions in AlN involving several defect levels deep inside the forbidden gap have been observed, such as the transition from Al vacancy or complex to the valence band. In direct bandgap materials, such a transition emits a photon with an energy:

$$h\nu = E_g - E_{D(A)}, \tag{2.7}$$

where $E_{D(A)}$ represents the donor (acceptor) energy level. Band-to-impurity transitions have a recombination lifetime on the order of 1 ns [80].

Some of the impurities and crystal defects, such as Mg, V_N, or V_{Al} in AlN, have large ionization energies and form relatively deep levels in the energy gap. Low temperature (10 K) PL emission spectra of m-AlN bulk single crystal shows a 2.76 eV peak in blue spectral region, which is attributed to a band-to-impurity transition from Al vacancy ($V_{Al}^{3-/2-}$) to the valence band [81]. Band-to-impurity transitions can also be observed in PL spectra of Mg-doped AlN measured at 10 K as shown in Figure 2.16a. In addition to the I_1 transition at 6.02 eV, band-to-impurity transition at 5.54 eV involving the conduction band and Mg (Mg^0) acceptor has been observed [14]. A similar transition has been observed in Zn-doped AlN epilayer at 5.40 eV [17]. The temporal response of the 5.54 eV line in Mg-doped AlN is shown in Figure 2.16b, which has a double exponential decay kinetics with a faster component of decay time constant of about 300 ps suggesting a band-to-impurity type of transition.

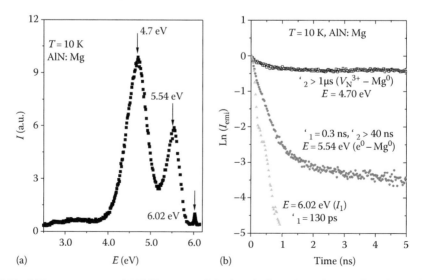

FIGURE 2.16 (a) Low temperature (10 K) PL spectra of Mg-doped AlN epilayer showing the I_1 (6.02 eV), band-to-impurity (5.54 eV), and DAP (4.7 eV) type transitions and (b) their respective PL decay lifetimes. (After Nam, K.B. et al., *Appl. Phys. Lett.*, 83, 878, 2003; Nepal, N. et al., *Appl. Phys. Lett.*, 89, 192111, 2006.)

2.3.3 Donor–Acceptor Pair Transitions

When the wave functions of an electron bound to a donor and a hole bound to an acceptor overlap, they recombine with a donor–acceptor pair (DAP) type transition. The energy of a photon resulting from such a radiative transition is given by

$$h\nu = E_g - E_D - E_A + \frac{e^2}{\varepsilon r},$$

(2.8)

where
E_D (E_A) is the donor (acceptor) energy level
ε is the dielectric constant
r is the distance between donor and acceptor

With increasing excitation intensity, the number of neutral donors and acceptors increases, which reduces the average distance between DAPs and hence the wavelength of the emitted photons.

In AlN, DAP-type transitions are mainly observed in (a) highly resistive Mg- and Zn-doped samples involving the shallow and deep donor levels due to the presence of V_N^{1+} or V_N^{3+} and neutral Mg or Zn acceptor and (b) undoped or Si-doped samples involving the shallow donor and deep acceptors [82,83]. Deep acceptor includes isolated Al vacancy $(V_{Al})^{3-/2-}$ and the vacancy impurity complex of different charge states $[(V_{Al} - O_N)^{2-/1-}$ and $(V_{Al} - 2O_N)^{1-/0}]$ or complex with Si_{Al} instead of O_N. Three different groups of DAP transitions involving shallow donors such as O or Si, and these deep acceptors have been identified in AlGaN alloys. These acceptors act as free electron traps and hence are responsible for high resistivity of these alloys. PL decay lifetime of these transitions is on the order of 1 μs.

2.3.4 Exciton–Phonon Interaction

The vibrational spectra of the phonon waves in solids occupy a wide range of frequencies from acoustic to the terahertz region. With four atoms in the unit cell of WZ AlN, group theory predicts eight sets of phonon normal modes: $2A_1$, $2B_1$, $2E_1$, and $2E_2$ at near zero wave vectors ($\vec{k} \approx 0$). Among these, one from each of the A_1 and E_1 modes and two of the E_2 modes are Raman active, the remaining modes of A_1 and E_1 are acoustic phonons, and the B_1 modes are silent [84,85]. Raman measurements of high-quality strain-free AlN in the backscattering geometry were reported by Tischler and Freitas [86] with Raman shift frequencies for E_2^1, E_2^2, $E_1(TO)$, $E_1(LO)$, $A_1(TO)$, and $A_1(LO)$ modes found to be 246.1, 655.1, 667.2, 909.6, 608.5, and 888.9 cm^{-1}, respectively [86]. The dynamics of phonons and their interactions with free carriers can affect the performance of III-nitride-based high-speed optoelectronic devices [87,88].

Due to the ionic nature of AlN, where the two constituent atoms (Al and N) are oppositely charged, the electric field of a photon generates zone-center longitudinal-optic (LO) modes of vibration in AlN resulting in strong exciton-LO phonon Frohlich interaction. It is manifested by the appearance of the phonon replicas accompanying the excitonic emission lines in their near band edge PL spectra. The emitted photon energy $h\nu$ of the nth order phonon replica is given by

$$h\nu = E_0 - nE_p,$$

(2.9)

where
$n = 0, 1, 2, \ldots$ indicates the number of phonons involved
E_p is the phonon energy
E_0 is the energy of the main emission peak

Such exciton–phonon interaction strongly influences the optical and transport properties of III-nitrides. Relative intensity of the phonon replica ($n = 1, 2, \ldots$) to the main emission peak (zero phonon line, $n = 0$) gives a measure of exciton–phonon coupling strength of the material, which is expressed as Huang–Rhys factor (S) [89–91]:

$$I_n = I_0 \left(\frac{S_n}{n!} \right) \tag{2.10}$$

Phonon replicas in PL spectra of AlN and GaN have been reported by several groups [92,93]. There have been several studies on carrier–phonon interaction in GaN [92,93], InGaN/GaN, and GaN/AlGaN QWs [94].

Phonon replicas of the A-exciton transition (FX_A) line have been identified using polarization-resolved PL measurement, and the exciton–phonon coupling strength (S) has been determined in AlN epilayers [95]. Low temperature (10 K) PL spectra of an AlN epilayer grown on c-plane sapphire are shown in Figure 2.17a for both $\vec{E} \| \vec{c}$ and $E \perp \vec{c}$ configurations [95]. The PL signal collected in the $\vec{E} \| \vec{c}$ configuration exhibits the FX_A line at 6.06 eV and its 1LO and 2LO phonon replicas at 5.95 and 5.84 eV, respectively [96,97]. The observed phonon lines have an energy separation corresponding to the A_1 (LO) phonon in AlN (110 meV), and no other phonon replica line was observed. The measurement set-up was analogous to the Raman measurement geometry $X(ZZ)Y$ in Porto's notation, which allows A_1 (LO) phonon to propagate along Y direction. Despite the much lower intensity of the $\vec{E} \perp \vec{c}$ polarized PL component of FX_A emission due to unique band structure of AlN, the relative intensities of phonon-assisted emission lines were much higher and LO phonon replicas up to $n = 3$ are well resolved. It results in a much higher S-parameter of 0.78 in AlN in $\vec{E} \perp \vec{c}$ configuration compared to 0.11 in $\vec{E} \| \vec{c}$ configuration as shown in Figure 2.17b. The higher value of S-parameter in $\vec{E} \perp \vec{c}$ configuration has been attributed to the anisotropic ratio of the effective mass of hole to that of electron (much larger in $\vec{E} \perp \vec{c}$ configuration).

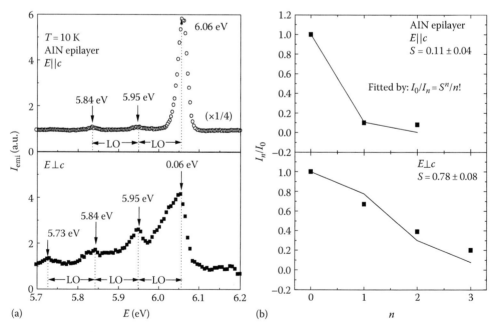

FIGURE 2.17 (a) Polarization-resolved PL spectra of an undoped AlN epilayer measured at 10 K. (b) Normalized PL intensities of nth order phonon replicas relative to the zero phonon line measured in an AlN epilayer in two different polarization configurations. (After Sedhain, A. et al., *Appl. Phys. Lett.*, 95, 061106, 2009.)

2.4 Optical Transitions of Defects and Impurities in AlN

2.4.1 Native Point Defects, Unintentional Impurities, and Defect–Impurity Complexes

Native point defects such as vacancies and vacancy-impurity complexes provide bound states, which are more localized and offer energy levels deep in the bandgap acting as carrier traps or recombination centers. Defect-induced transitions in AlGaN alloys are commonly observed as broadband emissions in PL spectra (linewidth upto 1 eV). They are responsible for the reduced efficiency and lifetime of AlGaN-based light emitters. This fact becomes critical in laser devices, where the parasitic components in the emission spectrum are highly undesirable. Under Al-rich growth conditions, cation vacancy (V_{III}) and its complex with substitutional donor impurities (O_N or Si_{III}) are the most easily formed defects for n-type or undoped AlN or Al-rich AlGaN alloys. The formation energies of nitrogen vacancy (V_N) and substitutional oxygen and silicon (O_N and Si_{III}) are relatively low for p-type materials [98–100]. In order to achieve the device quality p- and n-type layers of AlN and Al-rich AlGaN, in-depth understanding of the growth mechanism is essential in order to suppress these defects.

Low temperature (10 K) optical studies of Co, Mn, and Cr ion-implanted AlN epilayers revealed that the singly charged state of V_N (V_N^{1+}) acts as a donor with an activation energy E_0 of 260 meV, which agrees quite well with calculated energy level of 300 meV [97,101,102]. Figure 2.18a shows the room temperature (300 K) PL spectra of (a) an undoped AlN epilayer, (b) a Mg-doped AlN epilayer with higher resistivity, and (c) a Mg-doped AlN epilayer with lower resistivity [103]. Strong emission at 4.7 eV is dominant in highly resistive Mg-doped samples as shown in the second panel of Figure 2.18a, which is due to a DAP

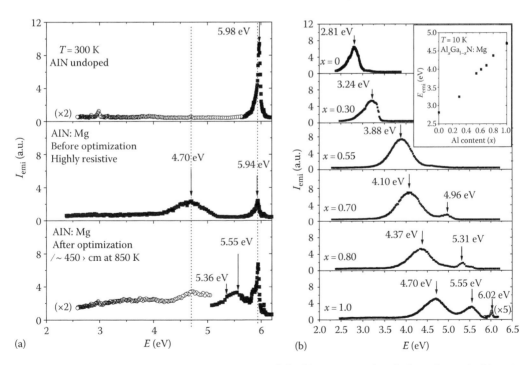

FIGURE 2.18 (a) Room temperature (300 K) PL spectra of a highly resistive undoped AlN epilayer, a highly resistive Mg-doped AlN epilayer, and an Mg-doped AlN epilayer with improved conductivity. (b) Low temperature (10 K) PL spectra of Mg-doped $Al_xGa_{1-x}N$ alloys for Al-content between 0 and 1. The inset shows the PL peak position of the transition of electron bound to V_N^{3+} and Mg^0 transition in the whole composition range. (After Nakarmi, M.L. et al., 89, 152120, 2006; Nakarmi, M.L. et al., *Appl. Phys. Lett.*, 94, 091903, 2009.)

transition from electrons bound to V_N^{3+} to Mg^0. It revealed a donor level of V_N^{3+} in AlN to be ~0.86 eV, which is very close to calculated value of 0.9 eV [101,104]. Impurity emission at 4.7 eV was significantly suppressed, and the I_1 transition at 5.94 eV was enhanced by optimizing the growth conditions to reduce the number of acceptor compensating defects (V_N^{3+}) as shown in the bottom panel of Figure 2.18a [103,105]. This sample has a p-type resistivity of ~450 Ω cm at 800 K. This is correlated with the PL spectrum, which shows reduced intensity of V_N^{3+}-related emission line. Two additional Mg-related lines at 5.36 and 5.55 eV have been assigned to DAP transitions involving V_N^{1+} and Mg^0 and band-to-impurity transition from conduction band to Mg^0, respectively. Low temperature (10 K) PL spectra of Mg-doped $Al_xGa_{1-x}N$ alloys of varying x (0, 0.3, 0.55, 0.7, 0.8, and 1.0) are shown in Figure 2.18b. In all samples, the observation of a group of impurity transitions with peak positions at 2.81 eV in GaN [106] to 4.7 eV in AlN [105], similar to the shifting of band edge transitions at 3.42 eV in GaN to 5.96 eV in AlN, clearly demonstrates that the triply charged nitrogen vacancies (V_N^{3+}) are the favorable native defects in p-type AlN and AlGaN alloys in the entire alloy composition range.

In undoped and Si-doped AlGaN alloys, calculations have indicated that the cation vacancy and its complex have low formation energies, which further decrease with an increase in Al-content [99,100,107,108]. In addition to the deepening of donor level with increasing Al-content, an increased number of defects is one of the major problems to achieve highly conductive n-type AlN and Al-rich AlGaN alloys. A systematic experimental study has demonstrated the existence of three groups of DAP transitions in the PL spectra of $Al_xGa_{1-x}N$ alloys [83,109]. These are the transitions from shallow donors (Si_{III} or O_N) to three different groups of deep acceptors [$(V_{III})^{3-/2-}$, $(V_{III}-O_N)^{2-/1-}$, and $(V_{III}-2O_N)^{1-/0}$]. Quite a broad nature of the spectral line and the PL, decay lifetime > 1 μs are the typical characteristics of these transitions. The concentration of isolated V_{Al} or V_{Al}-complexes depends on many factors including growth temperature, pressure, and V/III ratio.

The room temperature (300 K) PL spectra of a set of undoped $Al_xGa_{1-x}N$ epilayers ($0 \leq x \leq 1$) with relatively low impurity concentrations (<10^{18} cm^{-3}) is shown in Figure 2.19a [109]. It shows that the yellow line (YL) in GaN is a special case of a group of DAP transitions from shallow donor to vacancy

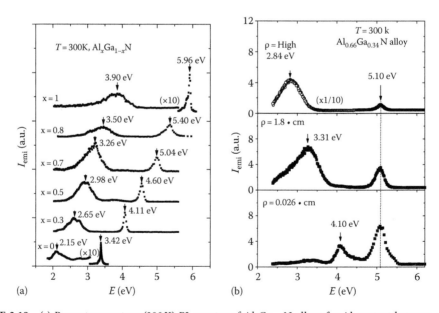

FIGURE 2.19 (a) Room temperature (300 K) PL spectra of $Al_xGa_{1-x}N$ alloys for Al-content between $x=0$ to 1 showing the band edge and deep impurity transitions. (b) Room temperature (300 K) PL spectra of an $Al_{0.66}Ga_{0.34}N$ epilayer with different resistivities. (After Nam, B. et al., *Appl. Phys. Lett.*, 86, 222108, 2005; Nepal, N. et al., *Appl. Phys. Lett.*, 89, 092107, 2006.)

complex $(V_{III} - O_N)^{2-/1-}$ in undoped and Si-doped AlGaN alloys, which blueshifts from 2.15 eV (YL) in GaN to 3.90 eV in AlN. In $Al_xGa_{1-x}N$ alloys grown under different conditions with relatively higher impurity concentrations and high Al content ($x > 0.58$), a set of deep impurity transitions exhibits a blueshift from 2.56 eV for $Al_{0.58}Ga_{0.42}N$ to 3.40 eV for AlN. It indicates that the presence of high impurity concentration ($>10^{18}$ cm^{-3}) favors the violet line (VL) at 3.40 eV over the 3.90 eV line in AlN. The relative intensity between the band edge and deep impurity transitions depends strongly on the growth condition, similar to the YL in GaN.

The third group of impurity transitions with emission energies even higher than that in the above two cases was observed with their spectral peak position blueshifted from 2.86 eV in GaN to 4.71 eV in AlN. The deep acceptor involved in these transitions is singly ionized/neutral cation vacancy complex, $(V_{III} - 2O_N)^{1-/0}$ or $(V_{III} - 2Si_{III})^{1-/0}$ [83]. The blue band (2.8 eV) has been observed in Si-doped and undoped GaN by PL, CL, and photoconductivity measurements [106,109–111]. Assuming the ionization energies of the shallow donors (E_D^0) to increase linearly from 25 to 86 meV with x from 0 to 1 and neglecting the Coulomb interaction between the ionized donors and acceptors, the deep acceptor level of $(V_{III})^{3-}$, $(V_{III} - O_N)^{2-/1-}$, and $(V_{III} - 2O_N)^{1-/0}$ can be deduced from corresponding PL peaks of DAP transitions by using $E_A = E_g(x) - h\nu_{emi} - E_D$. In AlN, these values were 1.25, 1.95, and 2.45 eV, with energy separations of 0.5 and 0.7 eV between the levels of $(V_{III})^{3-/2-}$ and $(V_{III} - O_N)^{2-/1-}$ and $(V_{III} - O_N)^{2-/1-}$ and $(V_{III} - 2O_N)^{1-/0}$, respectively, in perfect agreement with calculations [98,108].

Since $(V_{III} - 2O_N)^{1-/0}$ captures only one electron, its presence will not be as detrimental to the conductivity as the presence of $(V_{III})^{3-}$ and $(V_{III}-O_N)^{2-}$ defects, which are triple and double acceptors, respectively. As shown in Figure 2.19b, the resistivity of insulating $Al_{0.66}Ga_{0.34}N$ alloy has been significantly reduced to 0.026 Ω cm by optimizing the growth conditions to suppress the formation of double and triple electron traps [83]. However, an impurity transition involving $(V_{III} \text{complex})^{1-/0}$ at 4.10 eV is still evident. Therefore, it is believed that the conductivities of Al-rich AlGaN alloys and AlN will continue to improve by further suppressing the densities of these intrinsic defects.

2.4.2 AlN Homoepilayers

One of the major challenges in the development of UV and DUV devices based on AlN and Al-rich AlGaN alloys is the presence of high dislocation density. Due to the unavailability of the sizeable native bulk substrate, the vast majority of the III-nitride films and device structures, so far, have been grown on foreign substrates such as sapphire and SiC. Lattice and thermal expansion coefficient mismatch with the substrate results in the formation of large number of TDs. Dislocations, acting as non-radiative recombination centers, reduce the lifetime and emission efficiency of light-emitting devices, such as LEDs and LDs, as shown in Figure 2.20 [112,113]. In electronic devices, TDs act as a scattering center and reduce the carrier mobility [12]. Dislocations increase the leakage current, generate excessive heat, and lead to premature failure of the devices.

Significant research effort has been devoted in the recent years to reduce the dislocations in III-nitrides, especially in Al-rich AlGaN alloys, to improve the overall efficiency and the operational lifetime of DUV devices. Insertion of thin AlN [114,115] or GaN [116] buffer layer grown at low temperature (~500°C) and AlGaN/GaN and AlGaN/AlN short period superlattices (SP SLs) as dislocation filter [117,118] has been utilized to reduce the TDs. More recently, epitaxial lateral overgrowth (ELO) has been adopted to grow AlGaN epilayers with almost dislocation free area in the laterally grown wings region. The vertically grown region, however, contains large number of dislocations propagated from the underneath starting material as shown in Figure 2.21. TD density as low as 10^6–10^7cm^{-2} in thick AlN epilayer by pulsed lateral epitaxial overgrowth (PLOG) process over patterned sapphire substrate has been demonstrated [22,119]. Significantly improved luminescence properties of these high-quality AlN layers were further confirmed with the demonstration of room temperature stimulated emission at 214 nm using these layers.

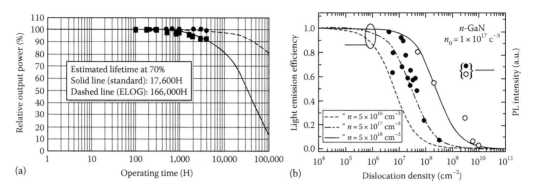

FIGURE 2.20 (a) Lifetime-test results of 365 nm UV LED grown on GaN/sapphire template with TDD of 1×10^8 cm^{-2} and on ELO GaN template with TDD of 1×10^7 cm^{-2}. (After Mukai, T., Morita, D., Yamamoto, M., Akaishi, K., Matoba, K., Yasutomo, K., Kasai, Y., Sano, M., and Nagahama, S.: *Phys. Stat. Sol. (c).* 2006. 3. 2211. Copyright Wiley-VCH Verlag GmbH & Co. KGaA. With permission.) (b) Light emission efficiency (lines, left axis) and the PL intensity in (circles, right axis) GaN as functions of dislocation density. (With permission from Karpov, S.Y. and Makarov, Y.N., *Appl. Phys. Lett.*, 81, 4721, 2002. Copyright 2002, American Institute of Physics.)

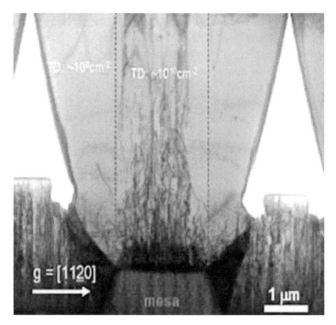

FIGURE 2.21 Bright field cross-section TEM images of a partially coalesced PLOG-AlN film. (With permission from Chen, Z., Fareed, R.S.Q., Gaevski, M., Adivarahan, V., Yang, J.W., Khan, A., Mei, J., and Ponce, F.A., *Appl. Phys. Lett.*, 89, 081905, 2006. Copyright 2006, American Institute of Physics.)

Recent advances on AlN bulk growth are quite promising toward the development of Al-rich AlGaN-based DUV photonic devices. The availability of bulk AlN wafers with the TDD less than 10^4 cm^{-2} and XRD linewidths of both the symmetric and asymmetric plane reflections <30 arcsec has been reported [120,121]. AlN homoepitaxy has the potential to produce excellent photonic devices as there will be no lattice and thermal expansion mismatch between the substrate and subsequent layers. The high thermal conductivity of AlN (270 vs. 35 W/mK in sapphire) is also advantageous to manage the heat generated during the operation of the device. Band edge emissions from AlN homoepilayers

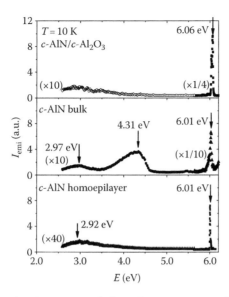

FIGURE 2.22 Low temperature (10 K) PL spectra of AlN epilayers grown on *c*-plane sapphire, *c*-plane AlN bulk, and AlN homoepilayer. FX peak positions from AlN bulk and homoepilayers overlap and homoepilayers are stress free. (After Sedhain, A. et al., *Appl. Phys. Lett.*, 93, 041905, 2008.)

have been studied using DUV PL [67]. The optical absorption coefficient in AlN for the above bandgap excitation is $>10^5$ cm^{-1} [42], which provides a typical optical absorption depth of $<0.1\,\mu$m to ensure that the measured PL signal was purely from the homoepilayer rather than from the underneath bulk substrate. Figure 2.22 compares the low temperature (10 K) PL spectra of AlN homoepilayers with that of bulk substrate and the best AlN heteroepilayer grown on sapphire substrate. All three spectra show that the intensities of the deep level native defects and impurity-related transitions are less than 1% of the corresponding band edge emissions, and the linewidth of the band edge emission is quite narrow. The dominant I_2 transition in homoepilayer spectra at 10 K indicates the presence of a high unintentional donor concentration, which is most likely due to the diffusion of oxygen from the AlN bulk substrate to homoepilayers. Impurity peak observed at 4.31 eV from AlN bulk substrates also indicates high concentration of O.

The band edge emissions of all *a*-, *c*-, and *m*-plane homoepilayers have an energy position at about 5.961 eV at 300 K, which also lines up with the FX peak position of AlN bulk substrate. X-ray diffraction (XRD) measurements also revealed that the spectral peak positions of θ/2θ scans of both the symmetric (002) and asymmetric (102) reflection peaks of a *c*-plane AlN homoepilayer line up exactly with that of AlN bulk substrate giving an in-plane lattice constant of $a = 3.112$ Å. Thus, it indicates that the AlN homoepilayer is perfectly lattice matched to the AlN bulk substrate and is almost strain free [45].

Native AlN bulk substrate for III-nitride DUV photonic devices is required to be transparent from the visible to DUV region. However, all the sizeable available bulk AlN substrates exhibit a yellow or dark amber color due to the broad absorption band starting at around 2 eV with a maximum at around 2.8 eV. We have shown recently through PL measurement that this absorption band is due to the excitation of electrons from the valence band to the V_{Al}^{2-} charge state of $(V_{Al})^{3-/2-}$ vacancy [81]. Therefore, optimizing the growth parameters to suppress the formation of native defects like $(V_{Al})^{3-/2-}$ will be important toward achieving high-quality native substrate for III-nitride-based UV and DUV optoelectronic devices.

Nitride UV laser diodes grown on low dislocation density AlN bulk substrates have been demonstrated [122]. Figure 2.23a shows the room temperature emission spectra obtained from a series of optically pumped laser devices using InAlGaN and AlGaN MQW active regions with their peak emission

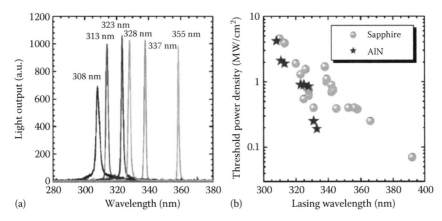

FIGURE 2.23 (a) Room temperature emission spectra obtained from a series of optically pumped laser devices with different InAlGaN and AlGaN MQW active regions grown on bulk AlN substrates. (b) Threshold power densities for optically pumped lasers grown on sapphire (red dots) and AlN substrates (blue stars). (With permission from Kneissl, M., Yang, Z., Teepe, M., Knollenberg, C., Schmidt, O., Kiesel, P., Johnson, N.M., Schujman, S., and Schowalter, L.J., *J. Appl. Phys.*, 101, 123103, 2007. Copyright 2007, American Institute of Physics.)

wavelengths from 308 to 355 nm as indicated. As seen from Figure 2.23b, the threshold power densities were consistently lower for lasers grown on AlN substrates as compared to lasers grown on sapphire, in particular for lasers emitting in the mid- to deep-UV wavelength range.

2.4.3 N-Type AlN Epilayers by Si Doping

Achieving highly conductive n-type $Al_xGa_{1-x}N$ alloys with increasing Al content becomes difficult due to deepening of the donor energy level and an enhanced formation of acceptor like native defects. Si is a universal n-type dopant in all of the nitride materials. Previously, the effect of persistent photoconductivity has been observed in Si-doped AlN, suggesting that Si may undergo a DX-like metastability [123]. However, more recent experimental results suggest that Si is an effective mass type impurity, in agreement with calculations. Progress has been made over the last decade in terms of the n-type conductivity control in AlN and Al-rich $Al_xGa_{1-x}N$ by Si-doping both by MOCVD and MBE growth [6,7,11–13,123–127].

Efficient doping has been achieved through In-Si co-doping [128]. Highly conductive n-type $Al_{0.7}Ga_{0.3}N$ epilayers (resistivity, $\rho = 0.0075\,\Omega$ cm) was demonstrated with high Si doping ($N_{Si} = 3.5 \times 10^{19}$ cm^{-3}) and using high-quality AlN template [128]. Taniyasu et al. [12] have recently reported n-type conductivity in AlN by Si doping with a much lower doping concentration ($N_{Si} \sim 3 \times 10^{17}$ cm^{-3}) and attributed their success to reduced TDD in their samples. Room temperature electron mobility (μ) of Si-doped n-type AlN as a function of N_{Si} is shown in Figure 2.24 for a set of samples with different TDD from 10^6 to 10^{11} cm^{-2}. It indicates that μ is limited by the neutral impurity scattering at higher N_{Si} and by TDD at lower N_{Si}. N_{Si} required to achieve μ_{max} depends on TDD; the lower N_{Si} yields μ_{max} in samples with lower TDD. These authors have reported 300 (220) K electron mobility of 426 (730) cm^2/Vs in Si-doped AlN homoepilayers, which are the highest values ever reported for AlN [12]. The maximum electron mobilities of 242 and 285 cm^2/Vs measured at 300 and 240 K were reported for heteroepitaxial n-type Si-doped AlN grown on SiC. Ive et al. [6] have also achieved n-type AlN with free electron concentration $\sim 7.4 \times 10^{17}$ cm^{-3} and conductivity of ~ 1 $(\Omega$ cm$)^{-1}$ for $N_{Si} = 1 \times 10^{20}$ cm^{-3}.

N_{Si}-dependent optical studies of Si-doped AlN epilayers by time-resolved PL and CL have been reported for the near band edge and defect-related emissions [7,13,129]. Room temperature (300 K) PL spectra of Si-doped AlN epilayers of different N_{Si} from 5×10^{17} to 1×10^{19} cm^{-3} exhibited a relatively weak I_2 emission at ~ 5.93 eV and strong impurity emission around 3.5 eV as shown in Figure 2.25a, due to

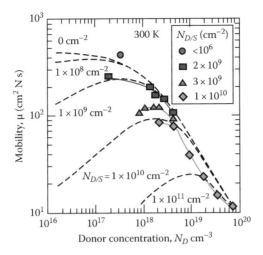

FIGURE 2.24 Experimental and calculated mobilities of AlN:Si as functions of N_{Si} for different dislocation densities, $N_{D/S}$. (With permission from Taniyasu, Y., Kasu, M., and Makimoto, T., *Appl. Phys. Lett.*, 89, 182112, 2006. Copyright 2006, American Institute of Physics.)

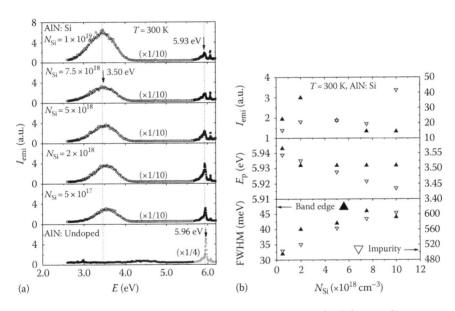

FIGURE 2.25 (a) Room temperature (300 K) PL spectra of AlN:Si epilayers for different Si doping concentration (N_{Si}) in a wide spectral range from 2 to 6.2 eV. (b) PL properties (intensity, peak position, and line width) of I_2 line (solid triangles) and impurity band (open triangles) as a function of N_{Si}. (After Pantha, B.N. et al., *Appl. Phys. Lett.*, 96, 131906, 2010.)

isolated Al vacancies, $V_{Al}^{3-/2-}$ [130]. N_{Si}-dependent PL properties of AlN:Si are summarized in Figure 2.25b. The top panel shows that the I_2 (3.5 eV impurity line) emission intensity decreases (increases) with increasing N_{Si} due to an increased V_{Al} density ($V_{Al}^{3-/2-}$) and reduced material quality. This is consistent with AFM results, which show surface roughness increased from 0.7 nm for $N_{Si} = 0$ to 6 nm for $N_{Si} = 1.5 \times 10^{18}$ cm^{-3} [129]. Reduced material quality with an increase of N_{Si} also agrees with increased screw dislocation density (N_{screw}) with N_{Si}, which was almost twice (>8×10^8 cm^{-2}) in heavily doped layers

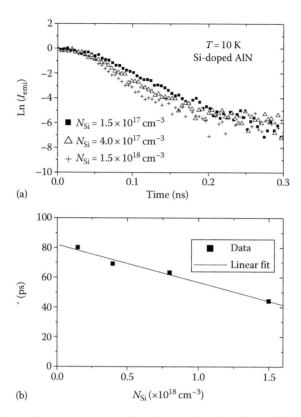

FIGURE 2.26 (a) PL temporal responses measured at the spectral peak position of the I_2 line in AlN:Si for three different Si-doping concentrations (N_{Si}). (b) PL decay lifetime as a function of N_{Si}. (After Nam, K.B. et al., *Appl. Phys. Lett.*, 83, 2787, 2003.)

compared to that in undoped layers (4.2×10^8 cm^{-2}) as measured by XRD. Consequently, the FWHM of both the I_2 and impurity transition lines increased continuously with increasing N_{Si} and hence with N_{screw} at a rate of $\sim 3.3 \pm 0.7$ meV/10^8 cm^{-2} for I_2 and $\sim 26.5 \pm 4$ meV/10^8 cm^{-2} for the 3.5 eV impurity transitions. This result also revealed that PL characteristics are affected less by edge-type dislocations (N_{edge}) but very sensitive to N_{screw} in AlN epilayers.

The PL temporal responses of AlN epilayers with different N_{Si} have been measured at their respective spectral peak positions, and the results are shown in Figure 2.26a. Figure 2.26b shows that the PL decay lifetime (τ) decreases linearly with increasing N_{Si}, where the solid line is a linear fit of the experimental data. The fitted value of $\tau = 82$ ps at $N_{Si} = 0$ also agrees with decay lifetime (80 ps) of the I_2 transition in undoped AlN as shown in Figure 2.14a. It was believed that the linear decrease of PL decay lifetime with N_{Si} is associated with an increased non-radiative recombination rate with increasing N_{Si}, which corroborates the band edge PL intensity decreasing with an increase in N_{Si} as shown in Figure 2.25.

2.4.4 P-Type Doping of AlN with Mg, Zn, and Be

Conductive *p*-type AlN and Al-rich AlGaN alloys are critical to develop solid-state UV and DUV light sources (LEDs and LDs). Mg is the most commonly used *p*-type dopant in all of the III-nitride semiconductors. However, p-type conductivity has been achieved only in GaN and AlGaN (InGaN) alloys in the low Al (In) side. The resistivity of Mg-doped AlGaN layers is found to increase with Al content and becomes extremely high in Mg-doped AlN (1 Ω cm in GaN to 10^7 Ω cm in AlN at 300 K) due to increased Mg energy level (0.16 meV in GaN versus 0.51–0.63 eV in AlN) along with lower formation

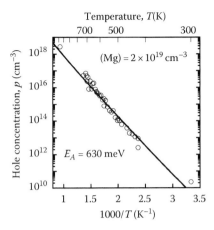

FIGURE 2.27 Temperature-dependent free hole concentration measured in Mg-doped AlN. The solid line shows the least-squares fit by equation $\exp\{-E_A/k_B T\}$, which gives a Mg energy level of 630 meV. (By permission from Macmillan Publishers Ltd. *Nature*, Taniyasu, Y., Kasu, M., and Makimoto, T., 441, 325, Copyright 2006.)

energy of donorlike native defects (V_N^{3+} and V_N^{1+}) that act as hole compensating centers [14,131–133]. Several groups have reported MOCVD growth and electrical properties of Mg-doped $Al_xGa_{1-x}N$ alloys with low Al contents [131,134–136]. Earlier, we have achieved p-type conduction at room temperature (RT) in $Al_xGa_{1-x}N$ epilayers for x up to 0.27 with Mg doping [132]. P-type conductivities in Mg-doped $Al_{0.70}Ga_{0.30}N$ alloy and pure AlN were measured at elevated temperatures (>700 K) [105,133]. Taniyasu et al. [15] have achieved p-type conductivity at elevated temperatures in Mg-doped AlN with heavy Mg doping of ~2×10^{19} cm^{-3} and postgrowth thermal annealing at 800°C. A Mg acceptor energy level of 0.63 eV has been reported by electrical characterization as shown in Figure 2.27, which is close to a value of 0.51 eV deduced from PL measurement [14].

The feasibility of using Be and Zn as alternate p-type dopants in AlN has been studied [17,18], however, all as-grown and postgrowth annealed layers were highly resistive at RT. PL spectra of Be, Mg, and Zn-doped and undoped AlN epilayers measured at 10 K are shown in Figure 2.28 [14,17,18]. The PL spectrum of Mg-doped AlN comprises an I_1 transition at 6.02 eV, which is observed at 6.03 (6.01) eV in Be (Zn)-doped AlN. Therefore, the binding energies of the I_1 transition in Be, Mg, and Zn-doped AlN are roughly 30, 40, and 50 meV, respectively, as measured from the energy difference between the FX peak position in undoped AlN and the corresponding I_1 peak positions. According to Haynes' rule, the BE of the exciton-neutral impurity complex is about 10% of the impurity binding energy, neglecting the central cell correction [137], which indicates that Be is shallower in AlN than Mg while Zn is deeper than Mg.

In addition to the I_1 transition, Mg (Zn)-doped AlN exhibits additional emission lines at 4.70 and 5.54 (4.50 and 5.40) eV, which are absent in undoped AlN. The measured recombination lifetime of the 4.70 and 5.54 eV emission lines in Mg-doped AlN were >1 μs and 300 ps, respectively, as shown in Figure 2.16b. The spectral peak positions, relatively slower PL decay lifetimes, and thermal activation energy measurements suggest that the 4.70 eV line is a DAP type transition involving V_N^{3+} and Mg^0. However, a much faster PL decay lifetime of 5.54 eV line in Mg-doped AlN suggests that it is a band to impurity type transition involving the conduction band and Mg^0. The PL spectrum of Zn-doped AlN is very similar to that of Mg-doped AlN, which also shows both the DAP and band-to-impurity transitions but at lower energies (4.50 and 5.40 eV vs. 4.70 and 5.54 eV). These results support that the energy level of Zn is deeper than that of Mg in AlN, which is also consistent with the larger I_1 binding energy in Zn-doped AlN compared to that in Mg-doped AlN [17].

The acceptor levels of Be, Mg, and Zn, energy levels of V_N^{3+} and V_N^{1+}, and the corresponding transitions in AlN are shown in Figure 2.29, where the downward arrows represent the experimentally

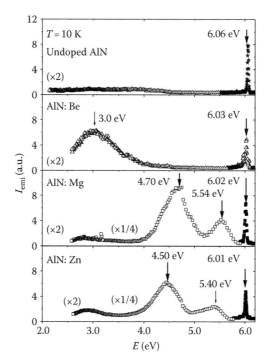

FIGURE 2.28 Low temperature (10 K) PL spectra of undoped and Mg, Zn, and Be-doped AlN epilayers in a wide spectral range from 2 to 6.2 eV. (After Nam, K.B. et al., *Appl. Phys. Lett.*, 83, 878, 2003; Nepal, N. et al., *Appl. Phys. Lett.*, 89, 192111, 2006; Sedhain, A. et al., *Appl. Phys. Lett.*, 93, 141104, 2008.)

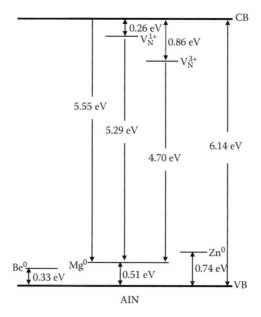

FIGURE 2.29 Schematic diagram showing the donor levels of nitrogen vacancies (VN^{3+} and VN^{1+}) and acceptor levels of Be^0, Mg^0, and Zn^0 in AlN.

observed transition lines. As a consequence of the large value of $E_A = 0.5-0.6$ eV, only a very small fraction $(e^{-E_a/kT-Ea/kT} = e^{-0.51/0.025} - e^{-0.6/0.025} = 10^{-9}-10^{-11})$ of the Mg dopants can contribute free holes at room temperature in Mg-doped AlN. For instance, for a Mg-dopant concentration (N_A) of 10^{20} cm^{-3} and a hole mobility (μ_h) of 10 cm^2/Vs, the resistivity of Mg-doped AlN is estimated to be as high as $\rho \approx 3-300$ MΩ cm with a free hole concentration of $p \approx 10^{11} - 10^9$ cm^{-3}.

2.5 Optical Properties of AlN in Low-Dimensional Quantum Structures

2.5.1 AlN/AlGaN Quantum Wells

Photonic devices employing quantum well (QW) and/or superlattice (SL) structures exhibit enhanced performance in several ways, such as increased quantum efficiency, reduced threshold current density, and reduced sensitivity to the temperature. Low temperature (10 K) PL spectra of *c*-plane AlN/Al$_{0.65}$Ga$_{0.35}$N QWs of various well width ($L_w = 1$, 1.5, 2, 2.5, and 3 nm) is shown in Figure 2.30 (solid squares). PL emission energy, assigned to the localized exciton recombination, was redshifted for QWs with $L_w = 2.5$ and 3 nm and blueshifted for QWs with $L_w = 1$, 1.5, and 2 nm relative to the band edge emission at 4.969 eV in Al$_{0.65}$Ga$_{0.35}$N epilayers [138]. The blueshift of the peak position in thinner wells is expected from the quantum confinement of photoexcited carriers, whereas the redshift in wider wells is due to the induced polarization (F) field in the wells, which is in the range of 3.4–4.0 MV/cm depending on L_w. The effective transition energy in QWs is given by

$$E = E_g + E_{con} - eFL_w, \tag{2.11}$$

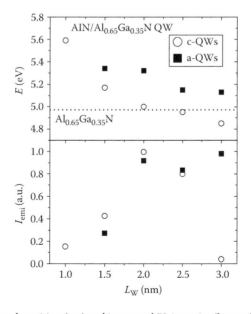

FIGURE 2.30 PL emission peak position (top) and integrated PL intensity (bottom) of AlN/Al$_{0.65}$Ga$_{0.35}$N QWs as functions of well width in polar *c*-plane (open circle) and nonpolar *a*-plane (solid squares) orientations. (After Al Tahtamouni, T.M. et al., *Appl. Phys. Lett.*, 89, 131922, 2006; Al Tahtamouni, T.M. et al., *Appl. Phys. Lett.*, 90, 221105, 2007.)

where

E_g is the bandgap energy of the well material
E_{con} is the confinement energy
eFL_w accounts for the effect of polarization field

However, in nonpolar a-plane AlN/Al$_{0.65}$Ga$_{0.35}$N QWs grown on r-plane sapphire (open circles), PL peak positions are always above that of the band edge transition of Al$_{0.65}$Ga$_{0.35}$N epilayer indicating a weak polarization field [139]. As shown in the bottom panel of Figure 2.30, c-plane QWs showed the highest QE when L_w is between 2 and 2.5 nm. As L_w increases, the induced polarization field increases the spatial separation of electron and hole wave functions, which reduces the radiative recombination rate and hence the emission intensity [140]. On the other hand, for the thinner wells with $L_w < 2$ nm, emission intensity decreases with decreasing L_w due to enhanced carrier leakage into the barrier region [140].

Excitation intensity (I_{exc})–dependent PL of highly excited Al$_{0.79}$Ga$_{0.21}$N/AlN MQWs demonstrated a clear transition of emission mechanism from FXs to electron–hole plasma (EHP) (Mott transition) as I_{exc} is increased [141,142]. The PL intensity is observed to be proportional to I_{exc}^2 under low excitation and becomes linearly dependent on I_{exc} under stronger excitation [143]. Balakrishnan et al. [144] have demonstrated a semi-polar (11$\bar{2}$2) Al$_{0.38}$Ga$_{0.62}$N/Al$_{0.29}$Ga$_{0.71}$N MQW DUV LED operating at 307 nm with on-wafer light output power of ~20 μW at 20 mA. It hardly showed any shift in peak emission wavelength ($\Delta\lambda \sim 0.5$ nm) for more than a 16-fold increase in excitation power, whereas a red shift of ~1.9 nm was observed in comparable c-plane LED. Horita et al. [145] have also demonstrated the feasibility of producing efficient DUV LEDs on nonpolar 4H-SiC (1$\bar{1}$00). They observed high-intensity band edge emission from CL measurement of AlN/Al$_{0.93}$Ga$_{0.07}$N (15 nm/4 nm) MQW with peak emission at 229 nm (5.41 eV), which did not show any blue shift with variation of excitation e-beam current by one order of magnitude.

2.5.2 AlN Nanowires

Various 1D AlN nanostructures such as nanocones [146], nanotips [147,148], nanorods [20], and nanoneedles [149–151] have been synthesized [20,149–162]. Byeun et al. [20] were able to control the diameter of well-aligned AlN nanorods grown on a Si substrate by varying the growth temperature and the N$_2$ carrier gas flow and maintained the smooth surface via continuous NH$_3$ flow during the cooling process. Homogeneous size distribution of AlN nanorods with a length of 900 nm and tip diameter less than 20 nm were prepared as shown in SEM image in Figure 2.31a. The magnified TEM image of a single nanorod is shown in Figure 2.31b. Figure 2.31c shows the XRD pattern of (002) reflection peak with

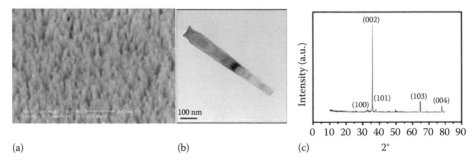

(a)   (b)   (c)   2°

FIGURE 2.31   (a) FESEM image of AlN nanorod array. (b) TEM image of a single AlN nanorod with a length of 700 nm and smooth surface. (c) XRD pattern of AlN nanorod array grown on a Si (100) substrate. (After Byeun, Y.K., Telle, R., Jung, S.H., Choi, S.C., and Hwang, H.I.: *Chem. Vap. Dep.*, 2010. 16. 72. Copyright Wiley-VCH Verlag GmbH & Co. KGaA. With permission.)

TABLE 2.5 FE, Raman, and Impurity Band CL Properties of AlN Nanorods; Sample-A: As-Grown, Sample-B (Sample-C): Treated at 875°C for 1 H with a Constant Ammonia Flow (in Vacuum)

	E_{th} (V/μm)	Work Function (eV)	FWHM of (meV)		CL Intensity (Relative)
			A_1 (TO)	E_2 (High)	
Sample-B	21.2	3.4	15.3	12.3	0.545
Sample-A	25.2	3.7	16.1	14.2	0.636
Sample-C	—	3.9	27.9	23.9	1

Source: After Yablonovitch, E., *Phys. Rev. Lett.*, 58, 2059, 1987.

a strong intensity. It confirms the WZ structure of these nanorods and also indicates that the crystal growth is preferentially along the *c*-axis.

Zhao et al. reported large area highly ordered single-crystalline AlN NW arrays synthesized by physical vapor deposition (PVD) on sapphire with an average diameter of the stem (tip) less than 100 (10) nm and average length ~3.5 μm [20]. They observed a blue-shift of the frequency of the E_{2h} mode by 1 cm^{-1} compared to that in bulk AlN as seen from Raman spectroscopy, which was due to a compressive strain of ~0.295 GPa in their NW arrays. Transmission spectra of as-synthesized AlN NW showed a minimum at 6.47 eV, corresponding to its absorption edge. It is blue-shifted relative to the absorption edge of the bulk AlN (5.95–6.2 eV) at 300 K due to quantum confinement. However, the strain in the MBE grown AlN nanowires on SiO_2/Si (100) template was relaxed as seen from the near band edge PL emission energy, which is similar to that from AlN bulk but red-shifted relative to that grown on AlN/Al_2O_3 templates. This conclusion is consistent with the results from Raman spectroscopy and HRTEM images [163].

Emission lines between 2 and 3.1 eV have been reported for AlN nanostructures under the below bandgap excitation with energies in the range of 3.7–4.2 eV [164–167]. Impurity bands at 2.1, 3.4 and near band edge at 6.2 eV were observed in quasi-aligned single crystalline AlN nanotips by CL, PL, and TL measurements [166]. Zhao et al. [168] also reported a wide CL band ~3 eV in a set of three AlN nanorod samples. The nanorod with the strongest defect emission peak has the largest FWHM of both A_1(TO) and E_2(high) modes in the Raman spectra, which are correlated with the FE properties as summarized in Table 2.5. The increased defect densities, which act as electron traps, lead to the deterioration of the electron FE characteristics of nanorods. PL spectra (300 K) of AlN nanobelts grown on Si and excited with sub-band gap excitation from a 325 nm He–Cd laser were reported by Tang et al. [160–162,169], showing that impurities are incorporated, and structural defects are introduced more easily in 1D AlN nanostructures due to the large specific area. Ji et al. [24] have reported strong band edge emission at 6.03 eV from Fe-doped AlN nanorods grown on Si substrate by catalyst-free vapor phase method.

2.5.3 AlN Photonic Crystals

Though still low, the internal quantum efficiency of solid-state UV emitters has been improved in the last decade. Improving light extraction efficiency as well as the EQE is also critical, especially for Al-rich AlGaN-based light sources. The band-to-band transition (topmost valence band) in Al rich AlGaN alloys is partially prohibited for $\mathbf{E} \perp c$ polarization. The allowed $\mathbf{E} \| c$ polarized emission propagates parallel to the surface, so that most of the emitted light is laterally guided and cannot be extracted out of the active layer [50]. As discussed in Section 2.2, this is partly responsible for the observed rapidly deteriorating EQE with decreasing emission wavelength, particularly in DUV region.

With the realization that light extraction is a critical issue in nitride DUV emitters, various schemes have been suggested, such as surface roughening [23,170–172], μ-LED arrays [57], PCs [58–60,173–182],

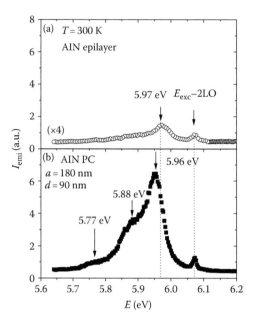

FIGURE 2.32 Room temperature (300 K) PL spectra of (a) AlN epilayer and (b) AlN PCs. The PL emission intensity (I_{emi}) of AlN PCs is 20 times higher than that of unpatterned AlN epilayer. (After Nepal, N. et al., *Appl. Phys. Lett.*, 88, 133113, 2006.)

and integration with microlens arrays [183]. Properties of nitride PCs in the visible and UV regions have been studied [58,59,177–182], and a 20-fold enhancement in the PL emission intensity of an InGaN/GaN MQW emitting at 475 nm was observed [58]. More recently, AlN PCs with triangular lattice of circular air holes with varying diameter (d) from 75 to 300 nm and periodicity or lattice constant (a) from 150 to 600 nm were fabricated. Room temperature PL spectra of AlN epilayers with and without PC formation are shown in Figure 2.32 [60]. The dominant band edge peak emission energy decreased from 5.970 eV in the AlN epilayer ($a = \infty$) to 5.952 in AlN with PC ($a = 150$ nm) as compressive stress is released due to etched air holes (PC). This effect is more pronounced for smaller holes. PL intensity increased with a decrease in the PC lattice constant, and a maximum of a 20-fold enhancement was observed upon PC formation as shown in Figure 2.32. Enhancement of light extraction with PC formation is due to forbidden propagation of light in the lateral direction. The lattice dimension of the PCs is designed in such a way that emission wavelength would lie within the photonic bandgap region. Picosecond time-resolved EL studies have shown that the incorporation of PCs into III-nitride LEDs also improves their modulation speed due to enhanced surface recombination velocity [179].

2.6 Applications of AlN

2.6.1 AlN Epilayers as Templates

Direct growth of $Al_xGa_{1-x}N$ epilayer on Al_2O_3 or SiC leads to cracking and pit formation due to lattice and thermal expansion coefficient mismatch. Two-step growth using a low temperature (LT) AlN or GaN buffer layer [114,116] and lateral overgrowth significantly improve material quality of AlGaN epilayers. Instead of a LT buffer layer, the use of high quality AlN deposited directly on the sapphire substrate at high temperature as templates has been suggested by Sakai et al. [184] and Shibata et al. [185,186]. Asai et al. [187] have reported that very high dislocation density (over 10^{11} cm^{-2}) is generated at the initial growth stage of $Al_xGa_{1-x}N$ ($x < 0.5$) on AlN/sapphire template. However, most of the dislocations continuously bend and terminate by forming loops within 1 μm layer due to large compressive stress.

AlN epi-templates act as dislocation filters [184–189] for the subsequent growth of epilayers and device structures. GaN layers grown on AlN/sapphire templates demonstrated much better quality compared to those grown on sapphire with a LT buffer layer, although the AlN template contained a large number of TDs. It was previously shown that GaN epilayers grown on AlN/sapphire templates comprise a lower TDD (5×10^7 cm^{-2}) as probed by TEM [184]. Compared to the *n*-GaN grown on sapphire using a LT GaN buffer, CL measurements revealed a 50%–60% reduction in dark spot density in *n*-GaN grown on AlN/sapphire template resulting from lower TDs. Smooth surface morphology, no pit termination, and clear step formation in GaN grown on AlN/sapphire templates were observed by AFM measurement.

A higher output optical power as well as a better thermal stability was demonstrated in InGaN/GaN MQW blue LEDs using AlN/sapphire template relative to LEDs grown on sapphire using a LT GaN buffer layer [185]. The output power at 200 mA decreased by 7.3% for LEDs on the AlN/sapphire template upon increasing temperature from 25°C to 95°C, while those on sapphire decreased by 23.9%. The maximum EQE was quite low at that time and decreased from 0.23% to 0.22% and from 0.15% to 0.10% for LEDs grown on the AlN/sapphire template and on sapphire, respectively. The EL spectral peak position at 20 mA shifted to lower energy by 17.2 meV for the LED grown on AlN/sapphire template upon increasing temperature from 25°C to 95°C, while that for the LED grown on sapphire shifted by 32.7 meV.

Enhanced band edge emission was observed from $Al_{0.67}Ga_{0.33}N$ alloys grown on high-quality AlN/sapphire template due to reduced parasitic recombination [190]. High-quality AlN template plays a crucial role in suppressing V_{III} and its related complexes. The overall quality, including crystalline quality, surface morphology, PL intensity, and the conductivity of the *n*-AlGaN epilayer grown on AlN/sapphire templates exhibited a remarkable improvement compared to *n*-AlGaN epilayers grown on sapphire with LT AlN buffer layer [190]. Using AlN/sapphire templates, improved efficiency of AlGaN DUV LED structures emitting at 305 and 290 nm has also been demonstrated [188]. More recently, record high EQE of DUV emitters (3% at 255–280 nm) has been reported, and the success was attributed partly to the improved quality of AlGaN DUV LED structures by using HT-AlN/Al_2O_3 templates with thick AlN layers [56]. It is now well recognized that AlN/sapphire template can improve the overall performance of DUV LEDs. Moreover, the efficiency improvement increases with an increase in AlN layer thickness.

2.6.2 AlN-Based Light Emitters and Photodetectors

A current injected homojunction LED based on pure AlN has been demonstrated with an EL emission peak at 210 nm, as shown in Figure 2.33a, which is the shortest wavelength ever reported among any kind of LED [15]. However, the emission efficiency of this LED is extremely low (10^{-6}%) [15]. Despite the extremely low efficiency, it is a clear indication of the feasibility to realize AlGaN-based emitters in the entire alloy composition range. Very low hole concentration in p-type AlN, undesired recombination in p-type layer due to leakage of injected electrons, high-density of TDs, and absorption by SiC substrate have been suggested to be responsible for the measured low efficiency. Large Mg acceptor ionization energy (0.5–0.6 eV) results room-temperature free hole concentrations that are too low, on the order of 10^9–10^{11} cm^{-3}. Only very small numbers of free holes are injected into the active region resulting extremely low efficiency. The output power of this PIN LED was 0.02 μW with 40 mA current [15].

Khan et al. [22] have reported optically pumped room temperature stimulated emission at 214 nm using high quality AlN grown on patterned sapphire by PLOG, which represents the shortest wavelength ever reported for stimulated emission as shown in Figure 2.33b. Further optimizing the growth conditions to improve the material quality by reducing the dislocation density, improving the p-type doping efficiency to reduce p-layer resistivity, and novel designs of device structures to enhance the light extraction efficiency are necessary to utilize the full potential of these materials.

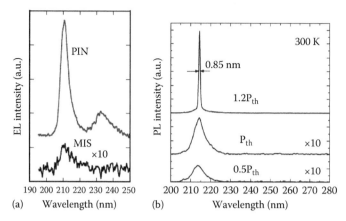

FIGURE 2.33 (a) EL spectra of AlN PIN homojunction LED and MIS LED under a d.c. bias with peak emission at 210 nm. (By permission from Macmillan Publishers Ltd. *Nature*, Taniyasu, Y., Kasu, M., and Makimoto, T., 441, 325, Copyright 2006.) (b) Edge emission PL spectra from PLOG AlN at different excitation showing the stimulated emission at 214 nm under pulsed optical pumping. (From Shatalov, M. et al., *Jpn. J. Appl. Phys.*, 45, L1286, 2006. With permission from Japan Society of Applied Physics.)

Traditional UV detectors use photomultiplier tubes (PMTs) or UV-enhanced Si detectors. Since these detectors are also sensitive in the visible region of the spectrum, they impose the additional requirement of costly UV filters. PMTs also require very high operating voltages (>1 kV). Special power supply and cooling hardware make these detectors bulky and inappropriate for many applications. Photodetectors based on various wide bandgap semiconductors have already demonstrated their superiority for many applications. Among them, SiC [191], GaN [192], and II–VI compound-based detectors [193] show a cutoff wavelength longer than 300 nm, whereas AlGaN [194] and diamond-based [195] devices present a significantly shorter cutoff wavelength at 229 and 225 nm, respectively. The robustness of AlGaN films due to strong ionic bonds inherently provides radiation hardness, which makes AlGaN alloy–based detectors the most suitable for space applications. Photodetectors based on pure AlN would overcome many of the limitations of Si detectors. The 6.1 eV bandgap permits the visible background to be intrinsically suppressed, and detectors can be operated at room temperature. It removes the requirements of optical filters and cooling hardware.

Recently, high-quality AlN metal–semiconductor–metal (MSM), Schottky, and avalanche photodiodes have been demonstrated in the author's laboratory [21,196,197]. The I–V characteristics of a pure AlN-based MSM photodetector exhibited a very low dark current (~100 fA at 200 V) with a breakdown field of >2 MV/cm, features which are attributed to the exceptional physical properties of AlN, including large energy bandgap, dielectric constant, and mechanical strength. The spectral response of an MSM detector demonstrated a peak responsivity at 200 nm, a sharp cut off at 207 nm, and more than four orders of magnitude of DUV to UV–vis rejection ratio. Previous work has shown that the AlGaN photodetectors significantly outperform GaN photodetectors in the vacuum UV (VUV) and extreme UV (EUV) spectral region due to the larger energy bandgap of AlGaN than GaN [194,195,198]. These results indicate that AlN-based photodetectors can also be used for the detection in VUV/EUV wavelengths. Dark current density as a function of bias voltage is shown in Figure 2.34 for MSM photodetectors based on diamond, c-BN, and AlN [195]. The lowest dark current measured in the AlN-based device at the highest bias voltage also confirms the superiority of AlN-based detectors.

2.6.3 AlN for Surface Acoustic Wave Devices

In a surface acoustic wave (SAW) device, the acoustic wave is electrically excited on a piezoelectric crystal plate by using a metallic interdigital transducer (IDT), and the influence on acoustic wave

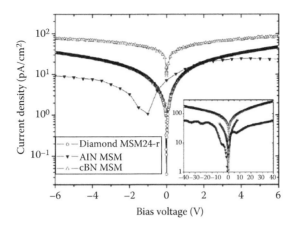

FIGURE 2.34 Dark current density versus voltage characteristics of the diamond MSM24-r, c-BN, and AlN MSM photodetectors at room temperature between −6 and +6 V. The inset shows a large view of the dark current density between −40 and +40 V. (a: From *Diamond Relat. Mater.*, 18, BenMoussa, A., Soltani, A., Schühle, U., Haenen, K., Chong, Y.M., Zhang, W.J., Dahal, R., Lin, J.Y., Jiang, H.X., Barkad, H.A., BenMoussa, B., Bolsee, D., Hermans, C., Kroth, U., Laubis, C., Mortet, V., De Jaeger, J.C., Giordanengo, B., Richter, M., Scholze, F., and Hochedez, J.F., 8600, Copyright 2009, with permission from Elsevier; (b): *Appl. Phys. Lett.*, 92, BenMoussa, A., Hochedez, J.F., Dahal, R., Li, J., Lin, J.Y., Jiang, H.X., Soltani, A., De Jaeger, J.-C., Kroth, U., and Richter, M., 022108, Copyright 2008, with permission from Elsevier.)

phase propagation speed (v) is measured. In addition to their conventional use in communication systems as filters, resonators, or delay lines, passive and remotely requestable (wireless) SAW devices are used as physical and chemical sensors with the advantage of high temperatures/radiation operation. Most of the existing SAW devices utilize bulk piezoelectric crystals such as quartz [199] and lithium niobate ($LiNbO_3$), which cannot be used at high temperatures [200,201]. Rapid growth in wireless communication technology requires low-loss thermally stable SAW resonators operating in the GHz range. Langasite ($La_3Ga_5SiO_{14}$) was perceived as the common material for high-temperature SAW applications due to its superior stability up to its melting point of 1473°C [202], however, relatively high acoustic propagation losses prohibits its use to operate above 1 GHz at high temperatures. The quest of reducing the antenna size of the wireless devices and increase the sensor sensitivity led to the research interest on new piezoelectric thin films such as ZnO [203,204], GaN [205,206], and AlN [200–211]. SAW characteristics of various materials and device are summarized in Table 2.6.

TABLE 2.6 SAW Characteristics of Various Materials

Material	V_{SAW} (m/s)	Electromechanical Coupling Coefficient (K^2) (%)	TCF (PPM/°C)	References
Quartz	3158	0.11	−32	[199]
$LiNbO_3/Al_2O_3$ (lithium niobate)	3918	5.3	−76.32	[217]
$La_3Ga_5SiO_{14}$ (langasite or LGS)	2742	0.32		[202]
ZnO/LGS	2740.8	1.3		[204]
GaN/Al_2O_3	5243		−49.2	[150]
AlN/Al_2O_3	5700	0.65	−40.9	[213]
AlN/diamond	10400	0.5–2.7	−13.4	[218]

V_{SAW} = surface acoustic wave velocity and TCF = temperature coefficient of frequency.

The operating frequency of SAW device (f_0) depends on SAW velocity (V_{SAW}) and wavelength (λ):

$$f_0 = \frac{V_{SAW}}{\lambda}, \tag{2.12}$$

where λ is determined by line and space widths (L/S) of the IDT geometry, which is limited by lithographic capability. AlN thin film has emerged as a promising material with excellent SAW velocity, piezoelectricity coupling, stability against harsh environment, and superior stop-band rejection ratio.

Over the last 15 years, AlN films grown on various substrates such as diamond, LiNbO$_3$, Si, Al$_2$O$_3$, and GaN/Al$_2$O$_3$ have been investigated as a potential substrate for the SAW devices operating at higher temperatures of up to 950°C [200,201,207–211]. SAW devices fabricated on AlN films of different thicknesses sputtered on GaN/sapphire template at 300°C for various IDTs wavelengths were characterized by measuring their frequency response, temperature coefficient of frequency (TCF), etc. [200]. The center frequency of the SAW device fabricated on AlN/GaN/sapphire with an IDT wavelength of 16 μm was 317.9 MHz (V_{SAW} = 5086 m/s) and was quite stable with a 1 kHz shift for change of relative humidity from 50% to 80%, which is much smaller relative to ~20 kHz shift in SAW filters on GaN/sapphire. These authors also reported that the acoustic wave attenuation factor increases with the conductivity of the film. AlN/GaN/sapphire-based SAW devices with higher resistivity are found to have smaller insertion loss and better stop band rejection ratio compared to such devices on GaN/sapphire with higher conductivity.

The correlation between the mechanical and acoustical properties of AlN film on Si has been investigated by measuring the effect of residual strain on SAW velocity (V_{SAW}) [210]. Fourier transform IR (FTIR) absorption spectroscopy has shown absorption bands at 678 and 620 cm^{-1} corresponding to the E_1(TO) and A_1(TO), respectively, due to vibrational modes of Al-N bonds. AlN/Si SAW filter were realized on low stress films, which exhibit fundamental and third harmonics at resonance frequencies of 212 and 629 MHz, respectively. XRD revealed that the AlN films were highly *c*-axis oriented perpendicular to the surface. Figure 2.35 shows the correlation between SAW velocity (dark curve) and residual stress (gray curve) for various AlN film thicknesses. Residual stress tends to be stabilized to around −3 GPa for AlN film with thickness >2 μm and consequently, V_{SAW} was increased significantly as desired. Wu et al. [207] have reported that AlN SAW devices sputtered on LiNbO$_3$ enhance the SAW velocity and improve the TCF, but reduce K^2. All of these recent reports on SAW characteristics of AlN point out the technological benefits of fabricating SAW device on AlN over that on traditional SAW materials.

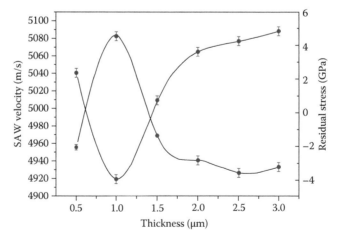

FIGURE 2.35 SAW velocity (dark) and residual stress (grey) of AlN films on Si as a function of AlN layer thickness. (After Assouar, M.B. et al., *Integr. Ferroelectr.*, 82, 45, 2006. With permission from Taylor & Francis.)

2.6.4 AlN Field Emission Devices

AlN is promising for electron field emission (FE) application due to its very low electron affinity, which means that free electrons can be emitted easily into the vacuum [212,213]. AlN field emitters have advantages of high mechanical, thermal, and chemical stabilities, excellent thermal conductivity, and low coefficient of thermal expansion [214].

One way of improving the FE properties of any material is to increase the free electron concentration by impurity doping, which has been achieved in AlN with Si doping. A triode-type FE display has been demonstrated [215,216]. Improvement in electron FE properties has also been demonstrated by Si doping in AlN nanoparticle thin films [217]. The turn-on electric field (E_{to}) was significantly reduced from 15 (undoped) to 7.8 V/μm and field enhancement factor (β) increased from 727 (undoped) to 957 when the nanocrystalline AlN films were doped with Si (5 at.%). The possible mechanism of the FE enhancement with Si doping could be due to the formation of a Si impurity band about 86 meV below the conduction band [215]. A similar conclusion concerning the lowest E_{to} has been reported for heavily Si doped ($N_{Si} = 1 \times 10^{21}$ cm^{-3}) AlN [218]. A decrease in E_{to} with an increase in N_{Si} is explained by the decreased energy barrier for the FE as a result of ridge formation. In Si-doped AlN, ridge structures form spontaneously with nanometer sizes and sharp tops.

Another way to improve the FE properties is to use a well-aligned array of 1D nanostructures, which possess sharp apexes and a large aspect ratio to efficiently generate an extremely high local electric field and thus a large current with a low applied voltage. In the past several years, various 1D AlN nanostructures have been synthesized, and their FE properties have been investigated [20,146–151]. Some of the representative results are summarized in Table 2.7. E_{to} varies from 1.8 to 12 V/μm for the AlN film, nanocrystalline films, and 1D nanostructures. Tang et al. [150] have reported values of E_{to} and the threshold electric field (E_{th}) of 1.8 and 4.6 V/μm, respectively, on a Si-doped AlN nanoneedle array on Si substrate with Co catalyst, which are the lowest values ever reported for AlN. As shown in Figure 2.36a, a maximum FE current density higher than 20 mA/cm^2 was observed, which is high enough for commercial FE display applications.

The inset of Figure 2.36a shows a Fowler–Nordheim (F–N) plot, which exhibits roughly linearly varying current density at high electric fields. The FE characteristics of the material can be described by the F–N theory:

$$J = A\left(\frac{\beta^2 E^2}{\phi}\right)\exp\left(\frac{-B\phi^{3/2}}{\beta E}\right), \tag{2.13}$$

where

$A = 1.54 \times 10^{-10}$ AV^{-2} eV

$B = 6.83 \times 10^3$ V eV$^{-3/2}$ μm^{-1}

$\phi = 3.7$ eV is the work function for AlN [219]

TABLE 2.7 FE Properties of AlN Films, Nanoparticles, and 1D Nanostructure Arrays

Sample	E_{to} (V/μm)	E_{th} (V/μm)	β	Fluctuation	References
Si:AlN nanoneedle	1.8	4.6	3271	<5%/5 h	[150]
AlN nanorod	2.25		784		[20]
AlN nanoneedle	3.1				[149,151]
AlN nanotip	4.7	10.6	1175–1889	0.74%/4 h	[147]
Patterned AlN nanocones	4.8	11.2	1561		[153]
AlN nanotip 300 (573) K	7.7(3.9)	7.9(4.1)	483(1884)		[148]
Nanocrystalline Si-AlN film	7.8		957		[217]
Si:AlN film	11	23		5.5%/1 h	[216]

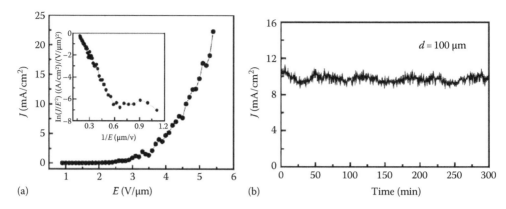

FIGURE 2.36 (a) FE current density as a function of applied electric field for Si-doped AlN nanoneedle array. The inset shows the corresponding F–N plot. (b) FE current stability of the AlN nanoneedles at an applied field of 4.6 V/μm and sample-anode distance of 100 μm. (With permission from Tang, Y.B., Cong, H.T., Wang, Z.M., and Cheng, H.M., *Appl. Phys. Lett.*, 89, 253112, 2006. Copyright 2006, American Institute of Physics.)

Here, β is the field enhancement factor as defined by

$$E_{local} = \beta E = \beta \frac{V}{d}, \tag{2.14}$$

where E_{local} is the local electric field at the tip of emitter. From the slope of the F–N plot, the field enhancement factor (β) was estimated to be 3271 for the Si-doped AlN nanoneedle. Figure 2.36b shows the FE stability test for more than 5 h, which exhibits a stable emission with current fluctuation lower than 5% at an applied field of 4.6 V/μm [150]. Highly stable FE of AlN nanoneedles has been attributed to high mechanical, chemical, and thermal stabilities of AlN.

2.7 Concluding Remarks

Application of AlN and Al-rich AlGaN alloys in DUV photonics has been the major driving force for the research interest in these materials. However, the emission efficiency of these devices is much lower than the InGaN-based visible and violet-emitting devices. In addition, the threshold current density in UV lasers increases with a decrease in emission wavelength. The major hurdles in achieving high power UV and DUV emitters are high dislocation density, very poor p-type conductivity, lack of suitable substrate, and poor light extraction due to the unique optical polarization property.

Considerable progress has been made on AlN bulk growth by sublimation-recondensation and physical vapor transport (PVT) resulting in AlN boules with diameters from 10 to 50 mm, XRD (002) rocking curve FWHM < 30 arcsec, and dislocation density < 10^4 cm^{-2}. Improved growth technology and possible mass production are expected to reduce the cost of these bulk substrates. Realization of a fourfold increase in the output power of 300 nm AlGaN-based LED on bulk AlN substrate relative to that of a comparable LED on sapphire has already been demonstrated [220]. UV laser diodes based on AlGaInN heterostructures grown on near-c plane AlN bulk substrates emitting at 308–373 nm have also been achieved [122].

Improving p-type conductivity in AlN and AlGaN alloys is the most challenging task for the nitride community to further extend nitride photonic devices into DUV and EUV wavelengths. Future research must be intensified to search for alternate acceptor elements with low enough ionization energy in Al$_x$Ga$_{1-x}$N with high x or for alternative p-layer approaches to exploit the full potential of these materials for the applications in DUV emitters and detectors. Extraction of emitted light for DUV-emitting devices has become another critical issue due to their unique optical polarization properties originated from the fundamental band structure of AlN. Different methods to enhance light extraction in DUV emitters including photonic crystals should be further explored. Overall, in the last decade, rapid progress has been made in terms of AlN material growth, fundamental understanding, and device implementation.

Acknowledgments

We would like to acknowledge our former and current group members Dr. J. Li, Dr. K. B. Nam, Dr. K. H. Kim, Dr. T. N. Order, Dr. Z. Y. Fan, Dr. J. Shakya, Dr. M. L. Nakarmi, Dr. N. Nepal, Dr. T. A. Tahtamouni, Dr. B. Pantha, and Dr. R. Dahal for their contributions to this work. H. X. Jiang and J. Y. Lin are grateful to the AT&T Foundation for their support of Edward Whitacre and Linda Whitacre Endowed Chair positions. This work was supported by grants from DARPA, DOE, ARO, and NSF.

References

1. J. R. Lakowicz, *Principles of Fluorescence Spectroscopy*, 2nd edn. (Kluwer Academic Publishers, New York, 1999).
2. Bergh, G. Craford, A. Duggal, and R. Haitz, *Phys. Today*, December 2001, p. 42.
3. J. Li, K. B. Nam, M. L. Nakarmi, J. Y. Lin, and H. X. Jiang, *Appl. Phys. Lett.* **81**, 3365 (2002).
4. Y. Taniyasu, M. Kasu, and T. Makimoto, *J. Cryst. Growth* **298**, 310 (2007).
5. R. Boger, M. Fiederle, L. Kirste, M. Maier, and J. Wagner, *J. Phys. D* **39**, 4616 (2006).
6. T. Ive, O. Brandt, H. Kostial, K. J. Friedland, L. Däweritz, and K. H. Ploog, *Appl. Phys. Lett.* **86**, 024106 (2005).
7. E. Monroy, J. Zenneck, G. Cherkashinin, O. Ambacher, M. Hermann, M. Stutzmann, and M. Eickhoff, *Appl. Phys. Lett.* **88**, 071906 (2006).
8. S. T. B. Goennenwein, R. Zeisel, O. Ambacher, M. S. Brandt, M. Stutzmann, and S. Baldovino, *Appl. Phys. Lett.* **79**, 2396 (2001).
9. N. Onojima, J. Suda, and H. Matsunami, *Jpn. J. Appl. Phys.* **42**, L445 (2003).
10. A. Y. Polyakov, N. B. Smirnov, A. V. Govorkov, R. M. Frazier, J. Y. Liefer, G. T. Thaler, C. R. Abernathy, S. J. Pearton, and J. M. Zavada, *Appl. Phys. Lett.* **85**, 4067 (2004).
11. S. B. Thapa, J. Hertkorn, F. Scholz, G. M. Prinz, R. A. R. Leute, M. Feneberg, K. Thonke, R. Sauer, O. Klein, J. Biskupek, and U. Kaiser, *J. Csyst. Growth* **310**, 4939 (2008).
12. Y. Taniyasu, M. Kasu, and T. Makimoto, *Appl. Phys. Lett.* **89**, 182112 (2006).
13. M. L. Nakarmi, K. H. Kim, K. Zhu, J. Y. Lin, and H. X. Jiang, *Appl. Phys. Lett.* **85**, 3769 (2004).
14. K. B. Nam, M. L. Nakarmi, J. Li, J. Y. Lin, and H. X. Jiang, *Appl. Phys. Lett.* **83**, 878 (2003).
15. Y. Taniyasu, M. Kasu, and T. Makimoto, *Nature (London)* **441**, 325 (2006).
16. C. Stampfl and C. G. Van de Walle, *Appl. Phys. Lett.* **72**, 459 (1998).
17. N. Nepal, M. L. Nakarmi, H. U. Jang, J. Y. Lin, and H. X. Jiang, *Appl. Phys. Lett.* **89**, 192111 (2006).
18. A. Sedhain, T. M. Al Tahtamouni, J. Li, J. Y. Lin, and H. X. Jiang, *Appl. Phys. Lett.* **93**, 141104 (2008).
19. A. Janotti, E. Snow, and C. G. Van de Walle, *Appl. Phys. Lett.* **95**, 172109 (2009).
20. Y. K. Byeun, R. Telle, S. H. Jung, S. C. Choi, and H. I. Hwang, *Chem. Vap. Dep.* **16**, 72 (2010).
21. J. Li, Z. Y. Fan, R. Dahal, M. L Nakarmi, J. Y. Lin, and H. X. Jiang, *Appl. Phys. Lett.* **89**, 213510 (2006); A. BenMoussa, A. Soltani, U. Schühle, K. Haenen, Y.M. Chong, W.J. Zhang, R. Dahal et al., *Diamond Relat. Mater.* **18**, 860 (2009).
22. M. Shatalov, M. Gaevski, V. Adivarahan, and A. Khan, *Jpn. J. Appl. Phys.* **45**, L1286 (2006).
23. W. B. Joyce, R. J. Bachrach, R. W. Dixon, and D. A. Sealer, *J. Appl. Phys.* **45**, 2229 (1974).
24. X. H. Ji, S. P. Lau, S. F. Yu, H. Y. Yang, T. S. Herng, A. Sedhain, J. Y. Lin, H. X. Jiang, K. S. Teng, and J. S. Chen, *Appl. Phys. Lett.* **90**, 193118 (2007).
25. M. Boroditsky and E. Yablonovitch, *Proc. SPIE* **3002**, 119 (1997).

26. G. I. M. Prinz, A. Ladenburger, M. Schirra, M. Feneberg, K. Thonke, R. Sauer, Y. Taniyasu, M. Kasu, and T. Makimoto, *J. Appl. Phys.* **101**, 023511 (2007).
27. Y. Taniyasu, M. Kasu, and T. Makimoto, *Appl. Phys. Lett.* **90**, 261911 (2007).
28. R. G. Banal, M. Funato, and Y. Kawakami, *Phys. Rev. B* **79**, 121308 (2009).
29. E. Kuokstis, J. Zhang, Q. Fareed, J. W. Yang, G. Simin, M. A. Khan, R. Gaska, M. Shur, C. Rojo, and L. Schowalter, *Appl. Phys. Lett.* **81**, 2755 (2002).
30. K. B. Nam, J. Li, M. L. Nakarmi, J. Li, J. Y. Lin, and H. X. Jiang, *Appl. Phys. Lett.* **82**, 1694 (2003).
31. W. F. Brinkman and T. M. Rice, *Phys. Rev. B* **7**, 1508 (1973).
32. E. Silveria, J. A. Freitas, Jr., O. J. Glembocki, G. A. Slack, and L. J. Schowalter, *Phys. Rev. B* **71**, 041201 (2005).
33. W. M. Yim, E. J. Stofko, P. J. Zanzucchi, J. I. Pankove, M. Ettenbergm, and S. L. Gilbert, *J. Appl. Phys.* **44**, 292 (1973).
34. R. A. R. Leute, M. Feneberg, R. Sauer, K. Thonke, S. B. Thapa, F. Scholz, Y. Taniyasu, and M. Kasu, *Appl. Phys. Lett.* **95**, 031903 (2009).
35. J. Li, K. B. Nam, M. L. Nakarmi, J. Li, J. Y. Lin, H. X. Jiang, P. Carrier, and S. H. Wei, *Appl. Phys. Lett.* **83**, 5163 (2003).
36. A. R. Chourasia, D. R. Chopra, and T. K. Hatwar, *Surf. Sci. Spectra* **1**, 75 (1992).
37. W. R. L. Lambrecht and B. Segal, in *Properties of Group III Nitrides*, eds. J. H. Edgar (Inspec, London, 1994), pp. 135–156.
38. M. Suzuki and T. Uenoyama, *Jpn. J. Appl. Phys.* **34**, 3442 (1995).
39. M. Palummo, C. M. Bertoni, L. Reining, and F. Finocchi, *Physica B* **185**, 404 (1993).
40. G. D. Chen, M. Smith, J. Y. Lin, H. X. Jiang, S.-H. Wei, M. A. Khan, and C. J. Sun, *Appl. Phys. Lett.* **68**, 2784 (1996).
41. W. Shan, T. J. Schmidt, X. H. Yang, S. J. Hwang, and J. J. Song, *Appl. Phys. Lett.* **66**, 985 (1995).
42. I. Vurgaftman and J. R. Meyer, *J. Appl. Phys.* **94**, 3675 (2003).
43. S. H. Wei and A. Zunger, *Appl. Phys. Lett.* **69**, 2719 (1996).
44. P. Carrier and S.-H. Wei, *J. Appl. Phys.* **97**, 033707 (2005).
45. L. Chen, B. J. Skromme, R. F. Dalmau, R. Schlesser, Z. Sitar, C. Chen, W. Sun, J. Yang, M. A. Khan, M. L. Nakarmi, J. Y. Lin, and H. X. Jiang, *Appl. Phys. Lett.* **85**, 4334 (2004).
46. L. Eckey, *Inst. Phys. Conf. Ser.* **142**, 943 (1996).
47. J. Shakya, K. Knabe, K. H. Kim, J. Li, J. Y. Lin, and H. X. Jiang, *Appl. Phys. Lett.* **86**, 091107 (2005).
48. H. Kawanishi, M. Senuma, and T. Nukui, *Appl. Phys. Lett.* **89**, 041126 (2006).
49. H. Kawanishi, M. Senuma, M. Yamamoto, E. Niikura, and T. Nukui, *Appl. Phys. Lett.* **89**, 081121 (2006).
50. K. B. Nam, J. Li, M. L. Nakarmi, J. Y. Lin, and H. X. Jiang, *Appl. Phys. Lett.* **84**, 5264 (2004).
51. A. Sedhain, J. Y. Lin, and H. X. Jiang, *Appl. Phys. Lett.* **92**, 041114 (2008).
52. Y. Zhang, A. Mascarenhas, P. Ernst, F. A. J. M. Driessen, D. J. Friedman, K. A. Bertness, and J. M. Olson, *J. Appl. Phys.* **81**, 6365 (1997).
53. B. Monemar, P. P. Paskov, J. P. Bergman, A. A. Toropov, and T. V. Shubina, *Phys. Stat. Sol. (b)* **244**, 1759 (2007).
54. M. A. Khan, M. Shatalov, H. P. Maruska, H. M. Wang, and E. Kuokstis, *Jpn. J. Appl. Phys.* **44**, 7191 (2005).
55. A. Khan, K. Balakrishnan, and T. Katona, *Nat. Photonics* **2**, 77 (2008).
56. C. Pernot, M. Kim, S. Fukahori, T. Inazu, T. Fujita, Y. Nagasawa, A. Hirano et al., *Appl. Phys. Exp.* **3**, 061004 (2010).
57. S. X. Jin, J. Li, J. Y. Lin, and H. X. Jiang, *Appl. Phys. Lett.* **77**, 3236 (2000).
58. T. N. Oder, J. Shakya, J. Y. Lin, and H. X. Jiang, *Appl. Phys. Lett.* **83**, 1231 (2003).
59. T. N. Oder, K. H. Kim, J. Y. Lin, and H. X. Jiang, *Appl. Phys. Lett.* **84**, 466 (2004).
60. N. Nepal, J. Shakya, M. L. Nakarmi, J. Y. Lin, and H. X. Jiang, *Appl. Phys. Lett.* **88**, 133113 (2006).
61. A. A. Yamaguchi, *Appl. Phys. Lett.* **96**, 151911 (2010).
62. A. A. Yamaguchi, *Phys. Stat. Sol. (c)* **5**, 2364 (2008).

63. H. Hirayama, N. Noguchi, T. Yatabe, and N. Kamata, *Appl. Phys. Exp.* **1**, 051101 (2008).

64. K. B. Nam, J. Li, J. Y. Lin, and H. X. Jiang, *Appl. Phys. Lett.* **85**, 3489 (2004).

65. B. N. Pantha, N. Nepal, T. M. Al Tahtamouni, M. L. Nakarmi, J. Li, J. Y. Lin, and H. X. Jiang, *Appl. Phys. Lett.* **91**, 121117 (2007).

66. Y. Yamada, K. Choi, S. Shin, H. Murotani, T. Taguchi, N. Okada, and H. Amano, *Appl. Phys. Lett.* **92**, 131912 (2008).

67. A. Sedhain, N. Nepal, M. L. Nakarmi, T. M. Al tahtamouni, J. Y. Lin, H. X. Jiang, Z. Gu, and J. H. Edgar, *Appl. Phys. Lett.* **93**, 041905 (2008).

68. T. Onuma, S. F. Chichibu, T. Sota, K. Asai, S. Sumiya, T. Shibata, and M. Tanaka, *Appl. Phys. Lett.* **81**, 652 (2002).

69. J. A. Freitas, Jr., G. C. B. Braga, E. Silveira, J. G. Tischler, and M. Fatemi, *Appl. Phys. Lett.* **83**, 2584 (2003).

70. E. Silveira, J. A. Freitas, Jr., M. Kneissl, D. W. Treat, N. M. Johnson, G. A. Slack, and L. J. Schowalter, *Appl. Phys. Lett.* **84**, 3501 (2004).

71. T. Koyama, M. Sugawara, T. Hoshi, A. Uedono, J. F. Kaeding, R. Sharma, S. Nakamura, and S. F. Chichibu, *Appl. Phys. Lett.* **90**, 241914 (2007).

72. E. Silveira, J. A. Freitas, S. B. Schujman, and L. J. Schowalter, *J. Cryst. Growth* **310**, 4007 (2008).

73. T. Onuma, T. Shibata, K. Kosaka, K. Asai, S. Sumiya, M. Tanaka, T. Sota, A. Uedono, and S. F. Chichibu, *J. Appl. Phys.* **105**, 023529 (2009).

74. T. Onuma, K. Hazu, A. Uedono, T. Sota, and S. F. Chichibu, *Appl. Phys. Lett.* **96**, 061906 (2010).

75. P. B. Perry and R. F. Rutz, *Appl. Phys. Lett.* **33**, 319 (1978).

76. H. X. Jiang and J. Y. Lin, in *III-Nitride Semiconductors—Optical Properties*, eds. by M. O. Manasreh and H. X. Jiang (Taylor & Francis, New York, 2002), Chapter 2.

77. N. Nepal, M. L. Nakarmi, K. B. Nam, J. Y. Lin, and H. X. Jiang, *Appl. Phys. Lett.* **85**, 2271 (2004).

78. J. S. Im, A. Moritz, F. Steuber, V. Harle, F. Scholz, and A. Hangleiter, *Appl. Phys. Lett.* **70**, 631 (1997).

79. G. Lasher and F. Stern, *Phys. Rev.* **133**, 554 (1964).

80. M. Smith, G. D. Chen, J. Y. Lin, H. X. Jiang, A. Salvador, W. Kim, O. Aktas, A. Botchkarev, and H. Morkoc, *Appl. Phys. Lett.* **68**, 1883 (1996).

81. A. Sedhain, L. Du, J. H. Edgar, J. Y. Lin, and H. X. Jiang, *Appl. Phys. Lett.* **95**, 262104 (2009).

82. H. C. Yang, T. Y. Lin, M. Y. Huang, and Y. F. Chen, *J. Appl. Phys.* **86**, 6124 (1999).

83. N. Nepal, M. L. Nakarmi, J. Y. Lin, and H. X. Jiang, *Appl. Phys. Lett.* **89**, 092107 (2006).

84. A. T. Collins, E. C. Lightowlers, and P. J. Dean, *Phys. Rev.* **158**, 833 (1967).

85. H. Siegle, G. Kaczmarczyk, L. Filippidis, A. P. Litvinchuk, A. Hoffman, and C. Thomsen, *Phys. Rev. B* **55**, 7000 (1997).

86. J. G. Tischler and J. A. Freitas, *APL* **85**, 1943 (2004).

87. M. A. Stroscio, *J. Appl. Phys.* **80**, 6864 (1996).

88. M. V. Kisin, V. B. Gorfinkel, M. A. Stroscio, G. Belenky, and S. Luryi, *J. Appl. Phys.* **82**, 2031 (1997).

89. B. D. Bartolo and R. Powell, in *Phonons and Resonances in Solids* (Wiley, New York, 1976), Chapter 10.

90. K. W. Böer, in *Survey of Semiconductor Physics* (Van Nostrand Reinhold, New York, 1990), Chapter 20.

91. H. Zhao and H. Kalt, *Phys. Rev. B* **68**, 125309 (2003).

92. G. Popovici, G. Y. Xu, A. Botchkarev, W. Kim, H. Tang, A. Salvador, H. Morkoç, R. Strange, and J. O. White, *J. Appl. Phys.* **82**, 4020 (1997).

93. D. C. Raynolds, D. C. Look, B. Jogai, and R. J. Molnar, *Solid State Commun.* **108**, 49 (1998).

94. M. Smith, J. Y. Lin, H. X. Jiang, A. Khan, Q. Chen, A. Salvador, A. Botchkarev, W. Kim, and H. Morkoc, *Appl. Phys. Lett.* **70**, 2882 (1997).

95. A. Sedhain, J. Li, J. Y. Lin, and H. X. Jiang, *Appl. Phys. Lett.* **95**, 061106 (2009).

96. L. Bergman, M. Dutta, C. Balkas, R. F. Davis, J. A. Christman, D. Alexson, and R. J. Nemanich, *J. Appl. Phys.* **85**, 3535 (1999).

97. N. Nepal, K. B. Nam, M. L. Nakarmi, J. Y. Lin, H. X. Jiang, J. M. Zavada, and R. G. Wilson, *Appl. Phys. Lett.* **84**, 1090 (2004).

98. T. Mattila and R. M. Nieminen, *Phys. Rev. B* **55**, 9571 (1997).
99. C. Stampfl and C. G. Van de Walle, *Phys. Rev. B* **65**, 155212 (2002).
100. C. G. Van de Walle and J. Neugebauer, *J. Appl. Phys.* **95**, 3851 (2004).
101. R. M. Frazier, J. Stapleton, G. T. Thaler, C. R. Abernathy, S. J. Pearton, R. Rairigh, J. Kelly et al., *J. Appl. Phys.* **94**, 1592 (2003).
102. D. W. Jenkins and J. D. Dow, *Phys. Rev. B* **39**, 3317 (1989).
103. M. L. Nakarmi, N. Nepal, C. Ugolini, T. M. Altahtamouni, J. Y. Lin, and H. X. Jiang, *Appl. Phys. Lett.* **89**, 152120 (2006).
104. T. L. Tansley and R. J. Egan, *Phys. Rev. B* **45**, 10942 (1992).
105. M. L. Nakarmi, N. Nepal, J. Y. Lin, and H. X. Jiang, *Appl. Phys. Lett.* **94**, 091903 (2009).
106. Y. H. Kwon, S. K. Shee, G. H. Gainer, G. H. Park, S. J. Hwang, and J. J. Song, *Appl. Phys. Lett.* **76**, 840 (2000).
107. J. Neugebauer and C. G. Van de Walle, *Appl. Phys. Lett.* **69**, 503 (1996).
108. I. Gorczyca, N. E. Christensen, and A. Svane, *Phys. Rev. B* **66**, 075210 (2002).
109. K. B. Nam, M. L. Nakarmi, J. Y. Lin, and H. X. Jiang, *Appl. Phys. Lett.* **86**, 222108 (2005).
110. V. Fomin, A. E. Nikolaev, I. P. Nikitina, A. S. Zubrilov, M. G. Mynbaeva, N. I. Kuznetsov, A. P. Kovarsky, B. Ja. Ber, and D. V. Tsvetkov, *Phys. Stat. Sol. (a)* **188**, 433 (2001).
111. I. Shai, C. E. M. de Oliviera, Y. Shapira, and J. Salzman, *Phys. Rev. B* **64**, 205313 (2001).
112. T. Mukai, D. Morita, M. Yamamoto, K. Akaishi, K. Matoba, K. Yasutomo, Y. Kasai, M. Sano, and S. Nagahama, *Phys. Stat. Sol. (c)* **3**, 2211 (2006).
113. S. Y. Karpov and Y. N. Makarov, *Appl. Phys. Lett.* **81**, 4721 (2002).
114. H. Amano, N. Sawaki, I. Akasaki, and Y. Toyoda, *Appl. Phys. Lett.* **48**, 353 (1985).
115. I. Akasaki and H. Amano, *Jpn. J. Appl. Phys.* **45**, 9001 (2006).
116. S. Nakamura, *Jpn. J. Appl. Phys.* **30**, L1705 (1991).
117. H. M. Wang, J. P. Zhang, C. Q. Chen, Q. Fareed, J. W. Yang, and M. A. Khan, *Appl. Phys. Lett.* **81**, 604 (2002).
118. J. R. Gong, C. W. Huang, S. F. Tseng, T. Y. Lin, K. M. Lin, W. T. Liao, Y. L. Tsai, B. H. Shi, and C. L. Wang, *J. Cryst. Growth* **260**, 73 (2004).
119. Z. Chen, R. S. Q. Fareed, M. Gaevski, V. Adivarahan, J. W. Yang, A. Khan, J. Mei, and F. A. Ponce, *Appl. Phys. Lett.* **89**, 081905 (2006).
120. S. G. Mueller, R. T. Bondokov, K. E. Morgan, G. A. Slack, S. B. Schujman, J. Grandusky, J. A. Smart, and L. J. Schowalter, *Phys. Stat. Sol. (a)* **206**, 1153 (2009).
121. R. T. Bondokov, S. G. Mueller, K. E. Morgan, G. A. Slack, S. Schujman, M. C. Wood, J. A. Smart, and L. J. Schowalter, *J. Cryst. Growth* **310**, 4020 (2008).
122. M. Kneissl, Z. Yang, M. Teepe, C. Knollenberg, O. Schmidt, P. Kiesel, N. M. Johnson, S. Schujman, and L. J. Schowalter, *J. Appl. Phys.* **101**, 123103 (2007).
123. R. Zeisel, M. W. Bayerl, S. T. B. Goennenwein, R. Dimitrov, O. Ambacher, M. S. Brandt, and M. Stutzmann, *Phys. Rev. B* **61**, R16283 (2000).
124. C. Skierbiszeski, T. Suski, M. Leszczynski, M. Shin, M. Skowronski, M. D. Bremser, and R. F. Davis, *Appl. Phys. Lett.* **74**, 3833 (1999).
125. V. Adivarahan, G. Simin, G. Tamulaitis, R. Srinivasan, J. Yang, and M. Asif Khan, *Appl. Phys. Lett.* **79**, 1903 (2001).
126. K. B. Nam, J. Li, M. L. Nakarmi, J. Y. Lin, and H. X. Jiang, *Appl. Phys. Lett.* **81**, 1038 (2002).
127. P. Cantu, S. Keller, U. K. Mishra, and S. P. DenBaars, *Appl. Phys. Lett.* **82**, 3683 (2003).
128. K. Zhu, M. L. Nakarmi, K. H. Kim, J. Y. Lin, and H. X. Jiang, *Appl. Phys. Lett.* **85**, 4669 (2004).
129. K. B. Nam, M. L. Nakarmi, J. Li, J. Y. Lin, and H. X. Jiang, *Appl. Phys. Lett.* **83**, 2787 (2003).
130. B. N. Pantha, A. Sedhain, J. Li, J. Y. Lin, and H. X. Jiang, *Appl. Phys. Lett.* **96**, 131906 (2010).
131. T. Tanaka, A. Watanabe, H. Amano, Y. Kobayashi, I. Akasaki, S. Yamazaki, and M. Koike, *Appl. Phys. Lett.* **65**, 593 (1994).
132. J. Li, T. N. Order, M. L. Nakarmi, J. Y. Lin, and H. X. Jiang, *Appl. Phys. Lett.* **80**, 1210 (2002).
133. M. L. Nakarmi, K. H. Kim, M. Khizar, Z. Y. Fan, J. Y. Lin, and H. X. Jiang, *Appl. Phys. Lett.* **86**, 092108 (2005).

134. I. Akasaki and H. Amano, *Mater. Res. Soc. Symp. Proc.* **242**, 383 (1991).
135. M. Suzuki, J. Nishio, M. Onomura, and C. Hongo, *J. Cryst. Growth* **189/190**, 511 (1998).
136. L. Sugiura, M. Suzuki, J. Nishio, K. Itaya, Y. Kokubun, and M. Ishikawa, *Jpn. J. Appl. Phys.* Part 1 **7**, 3878 (1998).
137. R. A. Mair, J. Li, S. K. Duan, J. Y. Lin, and H. X. Jiang, *Appl. Phys. Lett.* **74**, 513 (1999).
138. T. M. Al Tahtamouni, N. Nepal, J. Y. Lin, H. X. Jiang, and W. W. Chow, *Appl. Phys. Lett.* **89**, 131922 (2006).
139. T. M. Al Tahtamouni, A. Sedhain, J. Y. Lin, and H. X. Jiang, *Appl. Phys. Lett.* **90**, 221105 (2007).
140. K. C. Zeng, J. Li, J. Y. Lin, and H. X. Jiang, *Appl. Phys. Lett.* **76**, 3040 (2000); E. Shin, J. Li, J. Y. Lin, and H. X. Jiang, *Appl. Phys. Lett.* **77**, 1170 (2000).
141. T. Oto, R. G. Banal, M. Funato, and Y. Kawakami, *Phys. Stat. Sol. (c)* **7**, 1909 (2010).
142. F. Binet, J. Y. Duboz, J. Off, and F. Scholz, *Phys. Rev. B* **60**, 4715 (1999).
143. K. Kazlauskas, G. Tamulaitis, A. Zukauskas, T. Suski, P. Perlin, M. Leszczynski, P. Prystawko, and I. Grzegory, *Phys. Rev. B* **69**, 245316 (2004).
144. K. Balakrishnan, V. Adivarahan, Q. Fareed, M. Lachab, B. Zhang, and A. Khan, *J. Appl. Phys.* **49**, 040206 (2010).
145. M. Horita, T. Kimoto, and J. Suda, *Appl. Phys. Exp.* **3**, 051001 (2010).
146. N. Liu, Q. Wu, C. He, H. Tao, X. Wang, W. Lei, and Z. Hu, *Appl. Mater. Interface* **1**, 1927 (2009).
147. Y. Tang, H. Cong, Z. Chen, and H. Cheng, *Appl. Phys. Lett.* **86**, 233104 (2005).
148. X. H. Ji, Q. Y. Zhang, S. P. Lau, H. X. Jiang, and J. Y. Lin, *Appl. Phys. Lett.* **94**, 173106 (2009).
149. X. Song, Z. Guo, J. Zheng, X. Li, and Y. Pu, *Nanotechnology* **19**, 115609 (2008).
150. Y. B. Tang, H. T. Cong, Z. M. Wang, and H. M. Cheng, *Appl. Phys. Lett.* **89**, 253112 (2006).
151. Q. Zhao, J. Xu, X. Y. Xu, Z. Wang, and D. P. Yu, *Appl. Phys. Lett.* **85**, 5331 (2004).
152. J. A. Haber, P. C. Gibbons, and W. E. Buhro, *J. Am. Chem. Soc.* **119**, (1997).
153. J. A. Haber, P. C. Gibbons, and W. E. Buhro, *Chem. Mater.* **10**, 4062 (1998).
154. C. Liu, Z. Hu, Q. Wu, X. Z. Wang, Y. Chen, H. Sang, J. M. Zhu, S. Z. Deng, and N. S. Xu, *J. Am. Chem. Soc.* **127**, 1318 (2005).
155. Y. B. Tang, H. T. Cong, Z. G. Zhao, and H. M. Cheng, *Appl. Phys. Lett.* **86**, 153104 (2005).
156. Q. Wu, Z. Hu, X. Wang, Y. Hu, Y. Tian, and Y. Chen, *Diamond Relat. Mater.* **13**, 38 (2004).
157. Z. W. Pan, Z. R. Dai, and Z. L. Wang, *Science* **291**, 1947 (2001).
158. Y. J. Liang and M. X. Che, in *Handbook for Thermodynamic Data of Inorganic Compounds* (Press of Northeast University, China, 1993).
159. C. Liang, Y. Shimizu, T. Asaki, H. Umehara, and N. Koshizaki, *J. Phys. Chem. B* **108**, 9728 (2004).
160. J. Zhang and L. Zhang, *Chem. Phys. Lett.* **363**, 293 (2002).
161. J. S. Jeong, J. Y. Lee, C. J. Lee, S. J. An, and G. C. Yi, *Chem. Phys. Lett.* **384**, 246 (2004).
162. H. W. Seo, S. Y. Bae, J. Park, H. Yang, M. Kang, S. Kim, J. C. Park, and S. Y. Lee, *Appl. Phys. Lett.* **82**, 3752 (2003).
163. O. Landre, V. Fellmann, P. Jaffrennou, C. Bougerol, H. Renevier, A. Cros, and B. Daudin, *Appl. Phys. Lett.* **96**, 061912 (2010).
164. H. T. Chen, X. L. Wu, X. Xiong, W. C. Zhang, L. L. Xu, J. Zhu, and P. K. Chu, *J. Phys. D* **41**, 025101 (2008).
165. M. Lei, H. Yang, Y. F. Guo, B. Song, P. G. Li, and W. H. Tang, *Mater. Sci. Eng. B* **143**, 85 (2007).
166. S. C. Shi, C. F. Chen, S. Chattopadhyay, K. H. Chen, B. W. Ke, L. C. Chen, L. Trinkler, and B. Berzina, *Appl. Phys. Lett.* **89**, 163127 (2006).
167. Q. Zhao, H. Zhang, X. Xu, Z. Wang, J. Xu, D. Yu, G. Li, and F. Su, *Appl. Phys. Lett.* **86**, 193101 (2005).
168. Q. Zhao, S. Feng, Y. Zhu, X. Xu, X. Zhang, X. Song, J. Xu, L. Chen, and D. Yu, *Nanotechnology* **17**, S351 (2006).
169. Y. Tang, H. Cong, F. Li, and H. M. Cheng, *Diamond Relat. Mater.* **16**, 537 (2007).
170. I. Schnitzer, E. Yablonovitch, C. Caneau, T. J. Gmitter, and A. Scherer, *Appl. Phys. Lett.* **63**, 2174 (1993).

171. R. Windisch, C. Rooman, S. Meinlschmidt, P. Kiesel, D. Zipperer, G. H. Döhler, B. Dutta, M. Kuijk, G. Borghs, and P. Heremans, *Appl. Phys. Lett.* **79**, 2315 (2001).
172. T. Fujii, Y. Gao, R. Sharma, E. L. Hu, S. P. DenBaars, and S. Nakamura, *Appl. Phys. Lett.* **84**, 855 (2004).
173. E. Yablonovitch, *Phys. Rev. Lett.* **58**, 2059 (1987).
174. M. Boroditsky, R. Vrijen, T. F. Krauss, R. Coccioli, R. Bhat, and E. Yablonovitch, *J. Lightwave Technol.* **17**, 2096 (1999).
175. H. Y. Ryu, Y. H. Lee, R. L. Sellin, and D. Bimberg, *Appl. Phys. Lett.* **79**, 3573 (2001).
176. A. A. Erchak, D. J. Ripin, S. Fan, P. Rakich, J. D. Joannopoulos, E. P. Ippen, G. S. Petrich, and L. A. Kolodziejski, *Appl. Phys. Lett.* **78**, 563 (2001).
177. J. J. Wierer, M. R. Krames, J. E. Epler, N. F. Gardner, M. G. Craford, J. R. Wendt, J. A. Simmons, and M. M. Sigalas, *Appl. Phys. Lett.* **84**, 3885 (2004).
178. J. Shakya, K. H. Kim, J. Y. Lin, and H. X. Jiang, *Appl. Phys. Lett.* **85**, 142 (2004).
179. J. Shakya, J. Y. Lin, and H. X. Jiang, *Appl. Phys. Lett.* **85**, 2104 (2004).
180. L. Chen and A. V. Nurmikko, *Appl. Phys. Lett.* **85**, 3663 (2004).
181. A. David, C. Meier, R. Sharma, F. S. Diana, S. P. DenBaars, E. Hu, S. Nakamura, C. Weisbuch, and H. Benisty, *Appl. Phys. Lett.* **87**, 101107 (2005).
182. D. H. Kim, C. O. Cho, Y. G. Roh, H. Jeon, Y. S. Park, J. Cho, J. S. Im, C. Sone, Y. Park, W. J. Choi, and Q. H. Park, *Appl. Phys. Lett.* **87**, 203508 (2005).
183. M. Khizar, Z. Y. Fan, K. H. Kim, J. Y. Lin, and H. X. Jiang, *Appl. Phys. Lett.* **86**, 173504 (2005).
184. S. Arulkumaran, M. Sakai, T. Egawa, H. Ishikawa, T. Jimbo, T. Shibata, K. Asai, S. Sumiya, Y. Kuraoka, M. Tanaka, and O. Oda, *Appl. Phys Lett.* **81**, 1131 (2002).
185. B. Zhang, T. Egawa, H. Ishikawa, Y. Liu, and T. Jimbo, *J. Appl. Phys.* **95**, 3170 (2004).
186. N. Kuwano, T. Tsuruda, Y. Kida, H. Miyake, K. Hiramatsu, and T. Shibata, *Phys. Stat. Sol. (c)* **0**, 2444 (2003).
187. T. Asai, K. Nagata, T. Mori, K. Nagamatsu, M. Iwaya, S. Kamiyama, H. Aman, and I. Akasaki, *J. Cryst. Growth* **311**, 2850 (2009).
188. K. H. Kim, Z. Y. Fan, M. Khizar, M. L. Nakarmi, J. Y. Lin, and H. X. Jiang, *Appl. Phys. Lett.* **85**, 4777 (2004).
189. D. Morita, A. Fujioka, T. Mukai, and M. Fukui, *Jpn. J. Appl. Phys.* **46**, 2895 (2007).
190. M. L. Nakarmi, N. Nepal, J. Y. Lin, and H. X. Jiang, *Appl. Phys. Lett.* **86**, 261902 (2005).
191. J. A. Edmond, H. S. Kong, A. Survorov, and C. H. Carter, *Physica B* **185**, 453 (1993).
192. E. Monroy, F. Caller, E. Muñoz, F. Omnès, B. Beaumont, and P. Gibart, *J. Electron. Mater.* **28**, 240 (1999).
193. I. K. Sou, M. C. V. Wu, T. Sun, K. S. Wong, and G. K. L. Wong, *Appl. Phys. Lett.* **78**, 1811 (2001).
194. S. Butun, T. Tut, B. Butun, M. Gokkavas, H. Yu, and E. Ozbay, *Appl. Phys. Lett.* **88**, 123503 (2006).
195. (a) A. BenMoussa, A. Soltani, U. Schühle, K. Haenen, Y.M. Chong, W.J. Zhang, R. Dahal et al., *Diamond Relat. Mater.* **18**, 8600 (2009); (b) A. BenMoussa, J. F. Hochedez, R. Dahal, J. Li, J. Y. Lin, H. X. Jiang, A. Soltani, J.-C. De Jaeger, U. Kroth, and M. Richter, *Appl. Phys. Lett.* **92**, 022108 (2008).
196. R. Dahal, T. M. Al Tahtamouni, Z. Y. Fan, J. Y. Lin, and H. X. Jiang, *Appl. Phys. Lett.* **90**, 263505 (2007).
197. R. Dahal, T. M. Al Tahtamouni, J. Y. Lin, and H. X. Jiang, *Appl. Phys. Lett.* **91**, 243503 (2007).
198. A. Motogaito, K. Hiramatsu, Y. Shibata, H. Watanabe, H. Miyake, Y. Fukui, Y. Ohuchi, K. Tadatomo, and Y. Hamamura, *Mater. Res. Soc. Symp. Proc.* **798**, Y6.6.1 (2004).
199. W. Wallofer and P. W. Krempl, *IEEE Ultras. Symp.* **01**, 411 (1994).
200. H. L. Kao, W. C. Chen, W. C. Chien, H. F. Lin, T. C. Chen, C. Y. Lin, Y. T. Lin, J. I. Chyi, and C. H. Hsu, *Jpn. J. Appl. Phys. Part 1* **47**, 124 (2008).
201. T. Aubert, O. Elmazria, B. Assouar, L. Bouvot, and M. Oudich, *Appl. Phys. Lett.* **96**, 203503 (2010).
202. I. Shrena, D. Eisele, E. Mayer, L. M. Reindl, J. Bardong, and M. Schmitt, *IEEE Int. Conf. Signals Circuits Systems* 5414173 (2009).
203. H. Ieki and M. Kadota, *IEEE Ultrason. Symp. Proc.* **1**, 281 (1999).

204. S. Wu, G. J. Yan, M. S. Lee, R. Ro, and K. I. Chen, *IEEE Trans. Ultrason. Ferroelectr. Freq. Control* **54**, 2456 (2007).
205. S. Strite and H. Morkoc, *J. Vac. Sci. Technol. B* **10**, 1297 (1992).
206. D. Ciplys, R. Rimeika, M. S. Shur, R. Gaska, A. Sereika, J. Yang, and A. M. Khan, *Electron. Lett.* **38**, 134 (2002).
207. S. Wu, Y. C. Chen, and Y. S. Chang, *Jpn. J. Appl. Phys.* **41**, 4605 (2002).
208. O. Elmazria, V. Mortet, M. El Hakiki, M. Nesladek, P. Alnot, and U. H. Pomcare, *IEEE Trans. Ultrason. Ferroelectr. Freq. Control* **50**, 710 (2003).
209. K. Y. Uehara, C.M. Shibata, T. Kim, S. K. Kameda, S. Nakase, and K. H. Tsubouchi, *IEEE Ultrason. Symp. Proc.* **1**, 203 (2004).
210. M. B. Assouar, O. Elmazria, M. El Hakiki, and P. Alnot, *Integrat. Ferroelectron.* **82**, 45 (2006).
211. C. Caliendo, *Appl. Phys. Lett.* **92**, 033505 (2008).
212. M. C. Benjamin, C. Wang, R. F. Davis, and R. J. Nemanich, *Appl. Phys. Lett.* **64**, 3288 (1994).
213. S. P. Grabowski, M. Schneider, H. Nienhaus, W. Mönch, R. Dimitrov, O. Ambacher, and M. Stutzmann, *Appl. Phys. Lett.* **78**, 2503 (2001).
214. Y. B. Tang, H. T. Cong, and H. M. Cheng, *Nano-Brief Reports Rev.* **2**, 6, 307 (2007).
215. M. Kasu and N. Kobayashi, *Appl. Phys. Lett.* **78**, 1835 (2001).
216. Y. Taniyasu, M. Kasu, and T. Makimoto, *Appl. Phys. Lett.* **84**, 2115 (2004).
217. R. Thapa, B. Saha, and K. K. Chattopadhyay, *Appl. Surf. Sci.* **255**, 4536 (2009).
218. M. Kasu and N. Kobayashi, *Appl. Phys. Lett.* **79**, 3642 (2001).
219. Y. B. Li, Y. Bando, and D. Golberg, *Appl. Phys. Lett.* **84**, 3603 (2004).
220. Z. Ren, Q. Sun, S. Y. Kwon, J. Han, K. Davitt, Y. K. Song, A. V. Nurmikko, H. K. Cho, W. Liu, J. A. Smart, and L. J. Schowalter, *Appl. Phys. Lett.* **91**, 051116 (2007).

3

GaN-Based Optical Devices

Hiroaki Ohta
University of California

Steven P. DenBaars
University of California

Shuji Nakamura
University of California

InGaN-based blue light-emitting diodes (LEDs) play an essential role in high-efficiency, solid-state white light sources. To improve the efficiency of LEDs, photoluminescence (PL) measurements have been used extensively to understand internal material properties and to reveal the unique nature of nitride alloys, in particular, InGaN. Excellent review books and papers on basic nitride materials covering the exciting progress of the last two decades are now available (see, for example, Refs. [1,2]). In this chapter, we introduce key phenomena that govern optical device performance through PL measurement data together with typical examples of LEDs and laser diodes (LDs). These PL data are not necessarily typical, but rather are ones that may give insight about nitride alloy systems. Recent findings regarding nonpolar and semipolar GaN technology used to suppress the quantum confined stark effect (QCSE) are also covered.

3.1 GaN, AlGaN, InGaN

The wurtzite group-III nitrides consisting of GaN, AlN, InN, and their alloys have direct bandgaps that range from 0.7 eV (InN) through 3.4 eV (GaN) to 6.2 eV (AlN), as shown in Figure 3.1 (e.g., Refs. [3,4]). This enables transitions from the deep ultraviolet region (less than 300 nm) in Al-containing quantum wells (QWs) to the visible color regions of violet/blue/green/amber/red and even infrared by In-containing QWs. The detailed material properties of AlN and AlGaN alloys and device applications in UV LEDs and LDs as well as various sensors have been summarized by Kahn et al.[5] InN growth technology has rapidly progressed and high-quality single crystalline InN can be grown easily. Discussion of state-of-the-art growth, characterization, and physical/chemical properties of InN can be found in an excellent review paper by Bhuiyan et al.[6] In this chapter, we mainly describe Ga-rich InGaN alloys, which are applicable to QWs for visible color emissions.

Besides wide coverage of emission wavelengths, a distinctive feature of other compound semiconductors such as GaAs and AlInGaP is the huge lattice mismatch among heterostructures consisting of different compositional alloys. In Figure 3.1, the points at lattice constants of 3.11, 3.19, and 3.54 Å

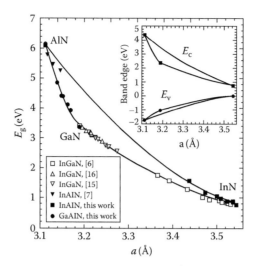

FIGURE 3.1   Bandgaps of group III-nitride alloys as a function of the in-plane lattice constant from the paper by Wu et al.[3] The references in the figure inset can be found in the original paper. (From *Solid State Commun.*, 127, Wu, T.J., Walukiewicz, W., Yu, K.M., Ager, III, J.W., Li, S.X., Haller, E.E., Lu, H., and Schaff, W.J., 411, Copyright 2003, with permission from Elsevier.)

represent AlN, GaN, and InN, respectively. The strain caused by this huge lattice mismatch affects optical properties and electronic properties such as bandgaps.[7,8] Even simple GaN films have different PL emission peaks because they are usually strained on foreign substrates.

Figure 3.2 shows a typical PL spectrum of hexagonal-GaN/sapphire measured at 10 K by Chichibu et al.,[9] where the free exciton peak [FE(A) and FE(B)], their E_2 phonon replica, and the peak corresponding to the recombination of excitons bound to a neutral donor (I_2) are identified. A PL spectrum of tensile-strained GaN films grown on 6H-SiC$^{(0001)Si}$ and Si(111) are also shown, where exciton-resonance energies and the PL peak energies depended on the substrate.[10] Another example can be found in the experiments by Aumer et al.,[11] where the lattice constant of a 3 nm $In_{0.08}Ga_{0.92}N$ QW was controlled by the relaxed $In_xGa_{1-x}N$ template layer underneath. It was observed that PL peak emission energies depend on the strain caused by lower layers. Generally, InGaN and AlGaN QWs are grown on a thick GaN template, causing them to suffer from strong compressive or tensile strains, respectively.

3.2 Material Growth

To grow crystalline group-III nitrides, a standard approach accepted in the industry called metal organic chemical vapor deposition (MOCVD) is employed. For academic studies, we can use the molecular beam epitaxial method, although high-quality InGaN growth for optical devices is difficult to achieve. The detailed description about MOCVD growths, obtained crystalline properties, and device performances are comprehensively reviewed in the book authored by Nakamura et al.[1]

The optimized MOCVD growth temperature for obtaining high-quality GaN films is around 1000°C ~ 1100°C. Trimethylgallium or triethylgallium are used as a Ga source while NH$_3$ is used as a N source. N$_2$ and/or H$_2$ gases are employed as carrier gases. A high N/Ga source ratio, typically more than 3000, is required to obtain high crystalline quality. To grow GaN films on foreign substrates such as *c*- or *a*-sapphire, a GaN buffer layer grown with a low temperature is used.[12,13] It was not only a key breakthrough in the early 1990s but is currently a standard for the fabrication of bright blue LEDs.

When GaN is grown on foreign substrates, threading dislocations (TDs) of the order of more than 10^8 cm^{-2} appear due to the huge lattice mismatch between the substrate and GaN layer. The presence of TDs affects material properties such as strain and the nonradiative recombination lifetime.

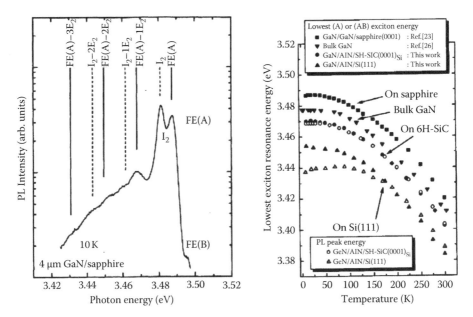

FIGURE 3.2 PL spectrum of hexagonal GaN/sapphire measured at 10 K[9] (left) and the lowest A or AB exciton resonance energies as well as the PL peak energies of tensile-strained GaN as a function of temperature[10] (right) by Chichibu et al. The references shown in the figure can be found in the original paper. (With permission from Chichibu, S., Azuhata, T., Sota, T., Amano, H., and Akasaki, I., *Appl. Phys. Lett.*, 70, 2085, 1997. Copyright 1997, American Institute of Physics.)

To characterize the radiative recombination rate, low-temperature (LT-) PL and time-resolved PL (TRPL) measurements can be applied (examples are shown in following sections). To reduce TD density, advanced growth techniques can be used, such as lateral epitaxial overgrowth (LEO),[14] thick GaN growth by hydride vapor phase epitaxy (HVPE),[15] and the ammonothermal method.[16] PL spectra of LEO-grown GaN and bulk GaN grown by the ammonothermal method can be seen in Refs. [17,18]. Optical properties such as the PL spectra are influenced by the densities of points defect and TDs, which eventually determines device performance.[19]

For InGaN and AlGaN growths, trimethylindium and trimethylaluminum are usually added to control composition. InGaN growths for QWs require low temperatures, typically around 800°C–900°C, to incorporate In atoms into alloys. AlGaN growth usually requires the same or higher temperatures than those for GaN growths. This wide range of growth temperatures from 800°C to more than 1000°C causes difficulties in MOCVD growth and the design of equipment. For donor (Si) and acceptor (Mg) doping, Si_2H_6 gas and a biscyclopentadienylmagnesium (Cp_2Mg) MO source, respectively, are usually included. To activate the Mg acceptor in GaN, H atoms should be expelled by thermal annealing after growth in an N_2-dominant ambient.[20] The PL data of Mg acceptor-doped GaN films are shown in Figure 3.3,[22] where two distinct emissions are observed: a free to bound recombination process (Mg^0, e) and blue emission around 2.8 eV caused by a nitrogen vacancy complex.[21-23] PL measurement played an important role in investigating the complicated nature of Mg-doped p-GaN.

3.3 InGaN-Based Light-Emitting Diodes

Here, a typical InGaN LED structure is introduced. The basic structure of visible wavelength InGaN-based LEDs consists of, at least, an n-type GaN layer, InGaN-based single or multi-QWs (SQW or MQWs), an AlGaN-based electron-blocking layer to prevent electron overflow, and a p-type GaN layer. The number of QWs is usually 5–10. To obtain wavelengths from violet to green, In composition in QWs

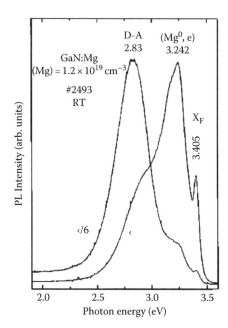

FIGURE 3.3 PL spectra of a GaN:Mg sample recorded with two different excitation densities and normalized to the same peak intensity. (With permission from Kaufmann, U., Kunzer, M., Maier, M., Obloh, H., Ramakrishnan, A., Santic, B., and Schlotter, P., *Appl. Phys. Lett.*, 72, 1326, 1998. Copyright 1998, American Institute of Physics.)

can be changed by altering growth temperature, though it is difficult to optimize growth conditions to obtain high internal quantum efficiency (IQE) for each wavelength region. In Figure 3.4, an LED structure on sapphire and the electroluminescence (EL) of blue/green/yellow LEDs are shown.[24–27] The widths of EL mission spectra become wider with increasing wavelength (In composition), as shown in Figure 3.4. This is mainly due to fluctuations in the In composition of InGaN alloys, as shown later.

The efficiency of LEDs is described as external quantum efficiency (EQE). EQE can be decomposed into the product of two components: IQE and the extraction efficiency of photons. IQE is determined by the nonradiative recombination lifetime (τ_{NR}) and radiative recombination lifetime (τ_R). IQE = $1/\tau_R/(1/\tau_R + 1/\tau_{NR})$ in QW when carrier injection efficiency is 1. Nonradiative recombination reduces the IQE of a near-band-edge (NBE) emission through the decrease of τ_{NR}. At less than 10 K, nonradiative recombination is prevented due to the lack of phonons ($\tau_{NR} \rightarrow \infty$). Hence, IQE at room temperature can be estimated by the ratio of PL intensity at room temperature [$I_{PL}(300\,K)$] to that at low temperature [e.g., $I_{PL}(8\,K)$]: IQE ~ $I_{PL}(300\,K)/I_{PL}(8\,K)$. A temperature dependent PL measurement of three MQWs consisting of $In_{0.12}Ga_{0.88}N$ (390–400 nm region) by Akasaka et al. is shown as an example in Figure 3.5.[28] In this experiment, the estimated IQE was 0.71 for the InGaN underlayer (UL) case. Recent technologies enable us to achieve IQE of more than 90%. Therefore, much effort has been devoted to improve extraction efficiency using advanced processing technologies rather than to material studies in blue regions because IQE is high enough already.[29] In green regions, IQE is still low mainly due to QCSE and other challenges such as nonpolar and semipolar planes, which are inevitable.

3.4 Piezo and Spontaneous Polarization

The optical properties in polar, for example, *c*-plane, InGaN QWs are governed by QCSE, which is caused by the polarization itself. Due to asymmetry in the wurtzite structure, the group III nitrides have strong internal spontaneous polarization.[30] Additionally, the strain due to the huge lattice mismatch causes additional piezoelectric polarization and interface charges.[31] In a standard blue LED structure

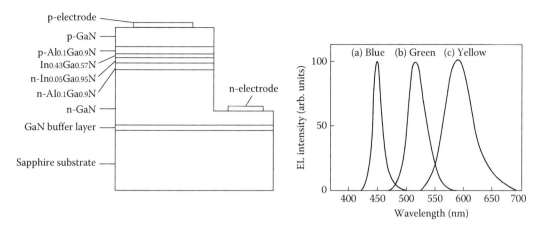

FIGURE 3.4 Typical InGaN-based LED structure (left) and ELs from single QW LEDs (right). Note that the recent standard for the active region design is not SQW but MQWs. Note that the active region area is typically 300 μm × 300 μm for small LEDs operated at a driven current of 20 mA while 1 mm × 1 mm is used for high power LEDs. (After Nakamura, S., Senoh, M., Iwasa, N., and Nagahama, S., *Jpn. J. Appl. Phys.*, 34, L797, 1995. With permission from Japan Society of Applied Physics.)

grown along the *c*-axis, the piezo and spontaneous polarization are along *c*-axis. The spontaneous polarization difference is not huge, but such alloys are coherently strained on GaN films and the large internal electric field caused by piezo polarization is in the *c*-plane of InGaN QWs. Schematic representation of polarization and energy band profiles are shown in Figure 3.6.[32] Electron and hole wave functions are separated in QWs, so radiative recombination life times become longer. Thus, the photon generation rate (IQE) is lower than the ideal case without QCSE. The internal electric field also causes a distorted band structure, which results in longer wavelength emission.

Chichibu et al. identified a unique phenomena in InGaN QWs called exciton localization though a series of precise, standard, LT-, and TRPL measurements.[33–37] Figure 3.7 shows an example of a PL measurement of amber SQW LED under a bias that was used to investigate the presence of QCSE.[37] As

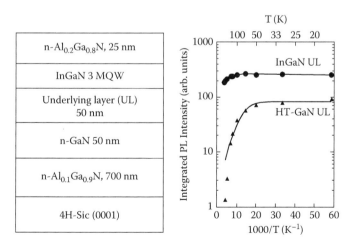

FIGURE 3.5 Arrhenius plots of integrated PL intensity for InGaN MQWs prepared using an InGaN UL (closed circles) and an high temperature (HT-) GaN UL (closed triangles); 378.3 nm light from a Ti:sapphire laser was used to excite QWs selectively. (With permission from Akasaka, T., Gotoh, H., Saito, T., and Makimoto, T., *Appl. Phys. Lett.*, 85, 3089, 2004. Copyright 2004, American Institute of Physics.)

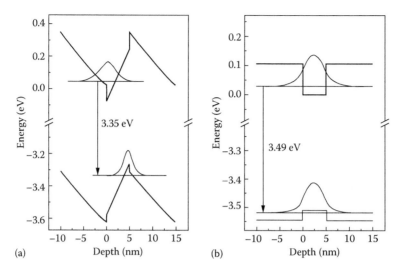

FIGURE 3.6 Calculated band profiles in (5 nm GaN)/(10 nm $Al_{0.1}Ga_{0.9}N$) QWs. These profiles were obtained by self-consistent effective mass Schrödinger–Poisson calculations. The transition energies given take into account both strain and Coulomb interaction. (a) The large electrostatic fields in the [0001] orientation result in a QCSE and poor electron–hole overlap. (b) The [1–1 00] orientation is free of electrostatic fields and true flat-band conditions are established. (With permission from Macmillan Publishers Ltd., *Nature*, Waltereit, P., Brandt, O., Trampert, A., Grahn, H.T., Menniger, J., Ramsteiner, M., Reiche, M., and Ploog, K.H., 406, 865, Copyright 2000.)

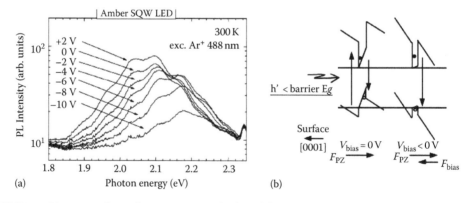

FIGURE 3.7 PL spectra of an amber SQW LED under bias (left) and schematic band profiles (right). (With permission from Chichibu, S.F., Azuhata, T., Sota, T., Mukai, T., and Nakamura, S., *J. Appl. Phys.*, 88, 5153, 2000. Copyright 2000, American Institute of Physics.)

the external bias increased, the emission peak shifted toward shorter wavelengths. The mechanisms to explain this phenomenon are shown in Figure 3.7 on the right.

3.5 InGaN-Based Laser Diodes

Another application of InGaN alloy is in a colored LD. The main commercial application at present uses a 405 nm light source for blue-ray DVD optical storage systems. Until the first success of InGaN-based violet LDs, the main route to short wavelength LDs has been Zn-Se-based II–VI materials.

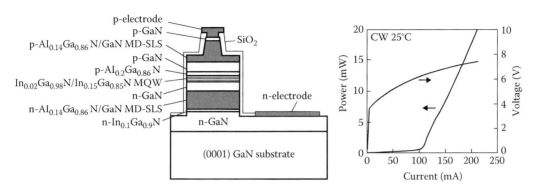

FIGURE 3.8 Structure of InGaN MQW LDs grown on a GaN substrate (left) and their performance (right). The lasing wavelength was around 393 nm. Note that the recent typical threshold for lasing is 1–2 kA cm^{-2}. (With permission from Nakamura, S., Senoh, M., Nagahama, S., Iwasa, N., Yamada, T., Matsushita, T., Kiyoku, H., Sugimoto, Y., Kozaki, T., Umemoto, H., Sano, M., and Chocho, K., *Appl. Phys. Lett.*, 72, 2014, 1998. Copyright 1998, American Institute of Physics.)

Nowadays, the regions from UV to green can be covered by III-nitride semiconductors. The first pulsed current injection operation at room temperature was achieved by Nakamura et al.,[38] where an InGaN MQW-based active region sandwiched by GaN/AlGaN waveguide structures was grown on a sapphire substrate. After about 2 years, the continuous wave (CW) operation of violet LDs was demonstrated using a similar structure grown on both sapphire and GaN substrates.[39,40]

The first device structure on a bulk GaN substrate is shown in Figure 3.8.[40] A typical active region structure is two or three MQWs sandwiched by GaN waveguiding layers. For the cladding layer, around 7% (averaged) AlGaN consisting of an AlGaN/GaN superlattice is used. For lateral optical confinement, the ridge type is commonly accepted. Its width and cavity length are around 2 μm and 300–1000 μm (depending on the output powers), respectively. The use of GaN substrates grown by the HVPE method, enables us to reduce TDs to the order of 10^5 cm^{-2}, which guarantees less than 1 TD on average in the active region area (~10^{-5} cm^2 = 2 μm × 500 μm). Hence, commercial LDs on GaN substrates do not suffer from TDs. This also contributes to sufficiently low thermal resistivity, which enables CW and high power operations. Furthermore, it also guarantees the formation of perfectly cleaved facets, due to the absence of a tough sapphire substrate. At present, a device stack structure grown along the *c*-axis with *m*-plane cleaved facets on a GaN bulk substrate is commonly used to mass-produce violet LDs. So far, the largest wavelength based on a *c*-plane InGaN-based LD was 515 nm.[41] On the other hand, the shortest wavelength from an LD based on a *c*-plane nitride was 342 nm. This UV LD consisting of AlGaN materials for all of its layers.[42] The blue and green LDs based on nonpolar and semipolar GaN will be shown in the following section.

3.6 PL Measurement of InGaN Alloys

In this section, important PL measurement data regarding InGaN alloys is introduced. Nitride systems are governed by several important effects, such as the strain due to lattice mismatch, QCSE, In-composition inhomogeneity, and extremely high defect densities compared to other compound semiconductors. These are taken into account in the analysis when we discuss material properties.

TRPL measurements can be used to identify the decay time (effective recombination lifetime), which is determined by $(1/\tau_R + 1/\tau_{NR})^{-1}$. An example of TRPL with micron spatial resolution is shown in Figure 3.9.[43] Here various types of GaN with different defect densities were characterized. Longer carrier lifetime can be seen in higher-quality GaN crystals, that is, in epitaxial lateral overgrowth (ELO) GaN

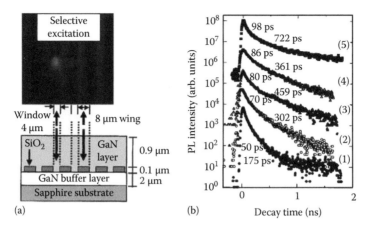

(a)　　　　　　　　　　　(b)　Decay time (ns)

FIGURE 3.9 (a) Image of ELO-GaN observed by an optical microscope and the cross figure of ELO-GaN. The window region is selectively excited by a focused laser beam. (b) Decay spectra of the PL intensity of E_{XA} band at RT taken by macroscopic photoexcitation (100 μm in diameter) from (1) 4 μm-GaN, (3) ELO-GaN, and (5) bulk GaN, and by microscopic photoexcitation (2 × 3 μm) from ELO-GaN (2) in the window region [dislocation densities (DD) = 10^8 cm^{-2}] and (4) in the wing region (DD = 10^6 cm^{-2}). The photoexcitation energy density was 3.4 μJ cm^{-2} for all measurements. (After *J. Lumin.*, 87, Izumi, T., Narukawa, Y., Okamoto, K., Kawakami, Y., Fujita, S., and Nakamura, S., 1196, Copyright 2000, with permission from Elsevier.)

or bulk GaN. This is because of the smaller number of TDs. This result indicates that the presence of TDs contributes to a high radiative recombination rate.

A TR measurement method (TR differential transmission spectroscopy) was also used to investigate carrier relaxation processes in InGaN QWs.[44] It was revealed that carriers are captured from barriers to the QWs in 1 ps, while carrier recombination rates increased with increased In composition. It is interesting to understand carrier relaxation and the transition mechanism in InGaN-based alloys.

The nature of recombination dynamics in (Al,In,Ga)N alloys was revealed by the comprehensive work of Chichibu et al.[45] In Figure 3.10, the PL spectra of various In-containing (Al,In,Ga)N alloys are shown. All alloys exhibit a NBE emission peak, though the value of the full width at half maximum (FWHM) increases with increasing molar fractions of InN or AlN. This is a unique feature in nitride systems. Judging from the large FWHMs for In-containing alloys, In-containing films have pronounced macroscopic compositional inhomogeneity. However, they also show that the (Al,In,Ga)N alloys suffer from large intrinsic alloy broadening, because the effective Bohr radius of excitons is as small as 3.4 nm in GaN and excitons are sensitive to local bandgap variations caused by naturally and randomly formed atomic-level inhomogeneity such as in the In-N-In-N chain. This nature also correlated to the defect-insensitive emission probability mechanism in In-containing (Al,In,Ga)N alloys, which is why InGaN blue-LEDs and green-LEDs emit brilliant light, even though the TD density generated due to lattice mismatch is six orders of magnitude higher than that in conventional LEDs.[46] The complicated emission mechanisms in such inhomogeneous alloys systems as InGaN in green regions is still an attractive academic topic.

Regarding In-compositional fluctuation, the anomalous phenomenon called *S*-shaped (decrease–increase–decrease) temperature dependence of the peak emission energy peaks (E_p) can be observed in temperature dependent PL measurements. The PL data obtained from 3 nm $In_{0.18}Ga_{0.92}N$ MQWs are shown in Figure 3.11.[46] E_p redshifts in the 10–70 K temperature range, blueshifts for 70–150 K, and then redshifts again for 150–300 K, with increasing temperature. The integrated PL intensities as a function of $1/T$ are also shown in the same figure. They indicate that radiative recombination is not dominant even at 100–200 K. Therefore, blueshifts for 70–150 K and the thermal quenching of InGaN-related PL can be explained by thermionic emission of photocarriers out of local potential minima into higher

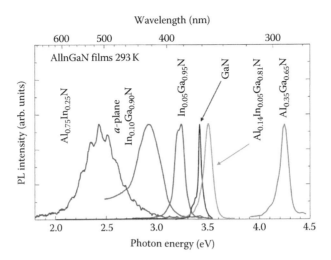

FIGURE 3.10 Normalized PL spectra of 3D films of In-containing (Al,In,Ga)N alloys at room temperature. (With permission from Chichibu, S.F., Azuhata, T., Sota, T., Mukai, T., and Nakamura, S., *J. Appl. Phys.*, 88, 5153, 2000. Copyright 2000, American Institute of Physics.)

energy states within the wells. In other words, this behavior is evidence of In compositional inhomogeneity and carrier localization in the InGaN/GaN QWs. Similar behavior in InGaN systems can also be found in a paper by Sasaki et al.[47] They discussed the effect of well widths and the correlation between S-behavior and radiation recombination time.

An anomalous temperature dependence of the PL spectrum was also found in a 7 nm $Ga_{0.72}In_{0.28}N_{0.028}As_{0.972}$/GaAs single QW, as shown in Figure 3.12.[48] At around 100 K, the blueshift of the emission peak and the broadening of the spectrum width can be observed simultaneously. This

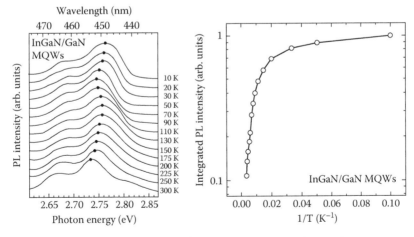

FIGURE 3.11 Typical InGaN-related PL spectra for $In_{0.18}Ga_{0.82}$N/GaN MQWs in the temperature range from 10 to 300 K (left). The main emission peak shows an S-shaped shift with increasing temperature (solid circles). All spectra are normalized and shifted in the vertical direction for clarity. A 235-nm He-Cd laser was used. Normalized and integrated PL intensity as a function of $1/T$ for the InGaN-related emission in the InGaN/GaN MQWs (right). Activation energy of ~35 MeV is obtained from the Arrhenius plot. (With permission from Cho, Y.H., Gainer, G.H., Fischer, A.J., Song, J.J., Keller, S., Mishra, U.K., and DenBaars, S.P., *Appl. Phys. Lett.* 73, 1370, 1998, Copyright 1998, American Institute of Physics.)

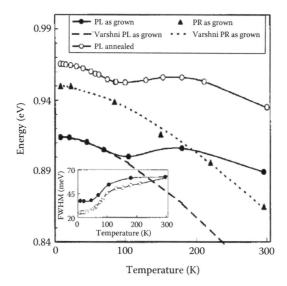

FIGURE 3.12 PL temperature dependence of the emission peak of the GaInNAs/GaAs SQW, before (closed circles) and after (open circles) annealing. Solid lines are a guide for the eyes. Closed triangles represent the optical transition energy between the confined electron level and the confined hole level of the GaInNAs/GaAs SQW before annealing determined by photoreflectance (PR) measurements; the dotted line is a Varshni fit of these points. The dashed line is a Varshni fit of the closed circles in the 0–100 K region. The evolution of the FWHM of the PL peak with temperature, before (closed circles) and after (open circles) annealing is shown in the inset. (With permission from Grenouillet, L., Bru-Chevallier, C., Guillot, G., Gilet, P., Duvaut, P., Vannuffel, C., Million, A., and Chenevas-Paule, A., *Appl. Phys. Lett.*, 76, 2241, 2000. Copyright 2000, American Institute of Physics.)

indicates a strong localization of carriers at low temperatures that could be induced by the presence of nitrogen. LT-PL can be a great route for characterizing alloy inhomogeneity.

Spatially resolved PL by confocal microscopy is a powerful approach to revealing in-plane spatial distribution of compositional variation in InGaN alloys. In Figure 3.13, PL mapping data of green-yellow LEDs by Okamoto et al. is shown.[49] The μm-scale macroscopic wavelength distribution can be seen due to in-plane inhomogeneity of In. Emission intensity dependence on wavelength was also observed. This indicates complicated carrier dynamics and recombination processes in the inhomogeneous potential distribution of QWs. One possible explanation can be found in the same paper.[49] A spatially resolved PL technique is essential to understanding the nature of InGaN alloys.

The final topic in this subsection is thermal damage of InGaN alloys. The study of thermal annealing effects on the size and composition of indium-aggregated clusters in $In_{0.4}Ga_{0.6}N$ thin films is shown in Figure 3.14.[50] As shown, the PL spectrum was dramatically broadened and blueshifted after thermal annealing. Under CL observation, the microscopic change of the emission distribution was confirmed. This indicates the structure of InGaN alloy can be changed by thermal stress. In typical LED and LD growth by MOCVD, the growth temperature of the active region (InGaN) is around 800°C, while that of the p-type layer (typically Mg doped p-GaN) is preferably more than 900°C. Hence, InGaN QWs usually suffer from heat stress, and their structure can be altered after device growth is complete. Also, for Mg activation, we should apply a relatively high temperature (>700°C) annealing process after the crystalline growth. This also helps modify InGaN microstructures. If we use a low growth temperature for the p-layer or low activation temperature, higher resistivity in the p-type layer results, which is a severe issue in LDs with smaller areas ($\sim 10^{-5}$ cm²) than LEDs. This effect can be insignificant in blue regions because IQE has reached close to 100% in present technologies. However, it can be a problem in the green region, where In composition is more than 20%. Generally, IQE becomes smaller after heat stress, so it is a challenge for MOCVD growers to suppress this heat damage in the InGaN-based active

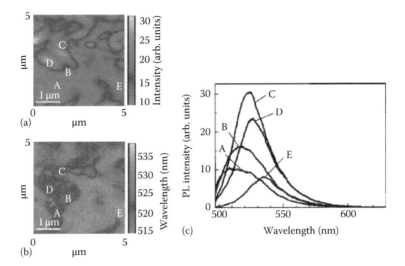

FIGURE 3.13 Peak intensity mapping (a) and peak wavelength mapping (b) of the PL over a $5 \times 5 \, \mu m^2$ area using a confocal microscope for an InGaN-based LED with yellow-green emission at room temperature. Spatially resolved PL spectra corresponding to the position A-E in the left figures (c). (With permission from Okamoto, K., Kaneta, A., Kawakami, Y., Fujita, S., Choi, J., Terazima, M., and Mukai, T., *J. Appl. Phys.*, 98, 064503, 2005. Copyright 2005, American Institute of Physics.)

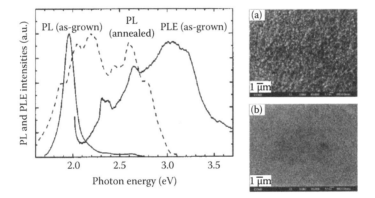

FIGURE 3.14 PL and PL excitation spectra at 10 K before and after thermal annealing of $In_{0.4}Ga_{0.6}N$ (left). The typical CL images before (right, a) and after (right, b) thermal annealing with excitation by 3 kV are also shown. (With permission from Feng, S.W., Tang, T.Y., Lu, Y.C., Liu, S.J., Lin, E.C., Yang, C.C., Ma, K.J., Shen, C.H., Chen, L.C., Kim, K.H., Lin, J.Y., and Jiang, H.X., *J. Appl. Phys.* 95, 5388, 2004. Copyright 2004, American Institute of Physics.)

region, particularly in green regions. This is one of the reasons why green LDs are more difficult to realize than blue ones.

3.7 Nonpolar and Semipolar GaN

As discussed in Section 3.4, QCSE is one factor that reduces IQE. This effect becomes more dominant with increasing In composition in InGaN QWs, that is, in the green region, where In composition is greater than 20%. Actually, IQE in the green region is far from 100%, while it reached close to 100% in blue and red regions using InGaN and AlInGaP, respectively. In commercial RGB (red-green-blue) LED modules, two green LEDs are sometimes used per one blue and red LED to compensate for their insufficient emission power. Therefore, we need to develop new materials for green emission to overcome what we call the green gap issue (see, for example, a recent review paper by Speck and Chichibu[51]). The most simple and straightforward approach is to use nonpolar and semipolar planes. Various planes are shown in Figure 3.15. As shown in the dependence of the piezoelectric polarization on the semipolar plane orientation, semipolar planes also have very little polarization.[52] In addition to LEDs, creating a direct semiconductor LD in the green region is also a challenge, for the same reason mentioned above.[53] Using not only c-plane but nonpolar m-plane and semipolar $(20\bar{2}1)$ planes, some groups created green LDs in 2009. The details will be outlined in this section.

To realize such materials, initially, nonpolar and semipolar GaN heteroepitaxial growth methods were tried using foreign substrates. In 2000, Waltereit et al. reported the first nonpolar $(10\bar{1}0)$ m-plane films on (100)-oriented γ-LiAlO$_2$ substrates.[32] In this study, they showed the absence of internal electric fields in GaN QWs with AlGaN barriers. Subsequently, various heteroepitaxies were achieved: a-plane GaN on r-plane sapphire, m-plane GaN on m-plane SiC, $(10\bar{1}1)$ and $(11\bar{2}2)$ semipolar planes on m-plane sapphire (see Refs. [54,55]). However, regardless of the theoretical prediction, the epitaxial growth of certain orientations on foreign substrates was extremely difficult. The nitride films grown on such substrates had an extremely high density of TDs and stacking faults (SFs).[54]

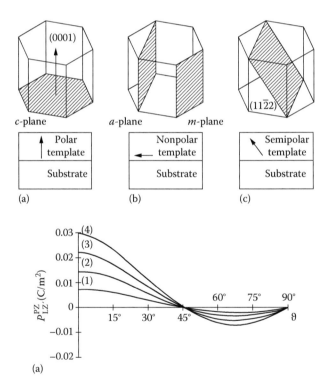

FIGURE 3.15 Crystallography and c-axis orientation for the growth of III-nitride layers (left): (a) polar growth with the c-axis normal to the layer surface; the growth plane is the (0001) c-plane; (b) nonpolar growth with c-axis parallel to the layer surface; possible growth planes include the $(11\bar{2}0)$ a-plane and the $(1\bar{1}00)$ m-plane; (c) semipolar growth with the c-axis inclined with respect to the layer surface. This example shows the $(11\bar{2}2)$ growth plane. Dependence of the piezoelectric polarization on the semipolar plane orientation for In$_x$Ga$_{1-x}$N layers on GaN (right). Composition $x = 0.05$ (1), 0.10 (2), 0.15 (3), and 0.20 (4). (With permission from Romanov, A.E., Baker, T.J., Nakamura, S., and Speck, J.S., *Appl. Phys. Lett.*, 100, 023522, 2006. Copyright 2006, American Institute of Physics.)

In 2006, the first mW class semipolar (11$\bar{2}$2) and nonpolar *m*-plane LEDs on high-quality GaN substrates were reported.[56,57] Because these substrates were sliced from thick *c*-plane GaN bulk crystals grown by HVPE,[15] almost no TDs and SFs exist in the substrates or device epilayers. Dislocation-free nonpolar nitride layers with smooth surfaces were obtained under growth conditions involving high (typically greater than 3000) N/Ga source ratios, which is quite similar to those in *c*-plane technology. After these achievements, any kind of nonpolar or semipolar planes can become a candidate to be used epitaxial growths. As examples, bright 8.9 mW *m*-plane and 20.6 mW semipolar (10$\bar{1}\bar{1}$) plane LEDs can be found in Refs. [58,59]. In Figure 3.16, an amber (563 nm) LED based on (11$\bar{2}$2) semipolar GaN by Sato et al. is shown.[60] This device was grown on a low extended defect density semipolar bulk substrate. The output power and EQE at drive currents of 20 and 200 mA under pulsed operation were 5.9 mW and 13.4% and 29.2 mW and 6.4%, respectively. Together with bright emission, the excellent temperature dependence of the InGaN LED output power being significantly smaller than that of AlInGaP LEDs was confirmed. Nowadays, owing to high-quality bulk substrates, nonpolar and semipolar GaN-based LEDs are comparable to *c*-plane GaN LEDs in the laboratories.

Nonpolar and semipolar LEDs also have the unique characteristics of polarized emission from QWs due to the in-plane anisotropic QWs nature (see, for examples, Refs. [61–64]). The reader also can see some theoretical papers by Ghosh et al.[65] and Yamaguchi[66] to understand the complicated valence band behavior in these materials, such as optical transition as a function of crystalline angle and biaxial strain. One example of PL data obtained from a nonpolar *m*-plane full LD structure is shown in Figure 3.17. As shown there, the PL spectrum depends on polarizer angles. We can identify two emissions (E$\perp$c and E||c) with different emission energies, which correspond to two different optical transitions originating from two different (top and second) valence bands. Detailed discussion should be found in Refs. [65,66]. As shown experimentally[64] and theoretically,[65,66] the polarization ratio linearly increases with increasing In composition (strain in QWs) for the nonpolar *m*-plane case. Its behavior is complex as a function of In composition for the semipolar cases. For (11$\bar{2}$2) semipolar LEDs, switching of the dominant polarization direction around In% = 30% was reported by Ueda et al.[67] Anisotropic strain (In mole fraction), optical confinement effects (well widths), and the crystalline angle (type of nonpolar and semipolar plane) contribute to valence band structures (the degree of polarization).[66,67] This is an interesting research paradigm.

The main goal at present in exploring nonpolar and semipolar nitrides is to create higher power direct semiconductor LDs in pure green regions (520–535 nm). After the first breakthrough with pulsed

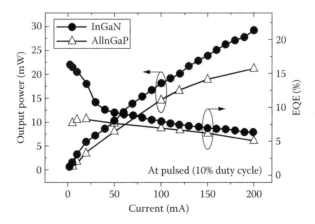

FIGURE 3.16 Dependence of the output power and EQE on the pulsed drive current for the (11$\bar{2}$2) InGaN (peak wavelength: 562.7 nm) and AlInGaP (peak wavelength: 589 nm) LEDs in the amber region. (With permission from Sato, H., Chung, R.B., Hirasawa, H., Fellows, N., Masui, H., Wu, F., Saito, M., Fujito, K., Speck, J.S., DenBaars, S.P., and Nakamura, S., *Appl. Phys. Lett.*, 92, 221110, 2008. Copyright 2008, American Institute of Physics.)

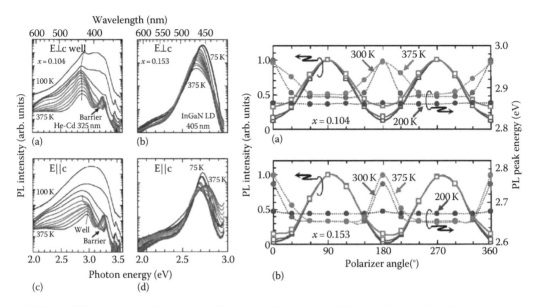

FIGURE 3.17 (See color insert.)   Polarized PL spectra of $In_{0.10}Ga_{0.90}N$ QWs for (a) $E \perp c$ and (b) $E\|c$, and those of $In_{0.15}Ga_{0.85}N$ QWs for (c) $E \perp c$ and (d) $E\|c$ as functions of T (left). Normalized PL intensity and PL peak energy as functions of the polarizer angle at various temperatures for InN molar fractions x of (a) 0.104 and (b) 0.153 (right). (With permission from Kubota, M., Okamoto, K., Tanaka, T., and Ohta, H., *Appl. Phys. Lett.*, 92, 011920, 2008. Copyright 2008, American Institute of Physics.)

and CW operation of nonpolar m-plane violet LDs,[68,69] some groups have reached green regions with nonpolar/semipolar planes. For the m-plane case,[53] the longest lasing wavelength saturated around 500 nm,[64,70–75] due to growth and material issues.[76] After extensive studies to apply ($11\overline{2}2$) planes to green LDs,[77–80] ($20\overline{2}1$) planes have helped realize lasing operations in green regions.[81,82] These breakthroughs are mostly due to the presence of high-quality bulk GaN substrates with arbitrary crystalline orientation.[15] An interesting feature of nonpolar and semipolar LDs is an anisotropic gain due to the asymmetric nature of the crystalline structure and biaxial strain.[83] Experimentally, the difference in threshold currents of c- and a-axes stripe m-plane LDs was reported in Ref. [68]. Direct gain measurement (amplified spontaneous emission spectra) of m-plane InGaN LDs was also reported by Onuma et al.[84] For LD device application, understanding the anisotropic gain is quite important.

In Figure 3.18, the spectra of nonpolar m-plane LDs before lasing are shown.[71] The m-plane blue LDs showed an extremely small emission wavelength shift from spontaneous emission to stimulated emission compared with that of c-plane blue LDs. This clearly demonstrates the absence of QCSE, which is an advantage of nonpolar gallium nitride for the fabrication of actual LD devices. As shown in Figure 3.19, small or no QCSE was observed through PL measurement with externally applied biases.[85] It was observed that the peak energy was nearly independent of the bias voltage for the case of m-plane nonpolar $In_{0.15}Ga_{0.85}N$ MQWs because of the absence of QCSE.

As described in this chapter, research paradigms of InGaN-based optical devices are now expanding due to various possibilities with crystalline planes. For material research, the nature of InGaN alloys governed by QCSE, alloy fluctuation, and strain due to the lattice mismatch are exciting topics, and PL measurement can be essential to exploring them. Together with commercialized UV/blue/green LEDs and violet/blue LDs based on c-plane GaN technologies, green LDs and longer wavelength LEDs based on nonpolar and semipolar GaN could become the new application of III-nitride materials. In particular, ~100% IQE in the green region by InGaN alloys is an essential challenge to overcome for high-power green LDs.

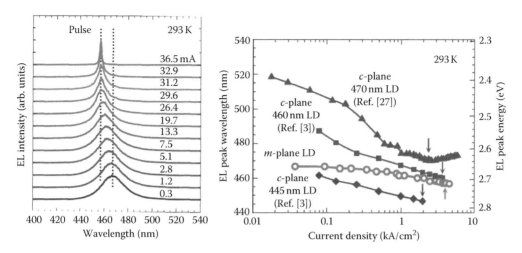

FIGURE 3.18 (See color insert.) EL spectra (left) and EL peak wavelength of *m*-plane LD (open circles) as a function of current density (right). Data for *c*-plane LDs with lasing wavelengths of 445 nm (closed diamonds), 460 nm (closed squares), and 470 nm (closed triangles) are also shown for comparison. The vertical arrows indicate the threshold current density for each LD. The references in the right figure can be found in the original paper.) (From Kubota, M., Okamoto, K., Tanaka, T., and Ohta, H. *Appl. Phys. Express*, 1, 011102, 2008. With permission from Japan Society of Applied Physics.)

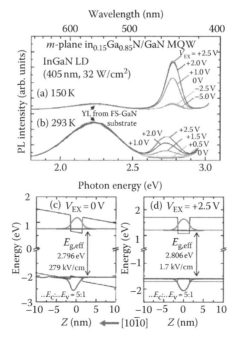

FIGURE 3.19 PL spectra of the *m*-plane $In_{0.15}Ga_{0.85}N$/GaN MQW blue LED under external bias (V_{ex}) at (a) 150 K and (b) 293 K. PL from the wells was selectively excited with the 405.0 nm line of a cw InGaN LD. Calculated band diagrams (c) under thermal equilibrium and (d) $V_{ex} = +2.5$ V. (With permission from Onuma, T., Amaike, H., Kubota, M., Okamoto, K., Ohta, H., Ichihara, J., Takasu, H., and Chichibu, S.F., *Appl. Phys. Lett.*, 91, 181903, 2007. Copyright 2007, American Institute of Physics.)

References

1. S. Nakamura, S. Pearton, and G. Fasol: *The Blue Laser Diode: The Complete Story*, 2nd edn. (Springer, Berlin, Germany, 2000).
2. S. Strite and H. Morkoç, *J. Vac. Sci. Technol. B* **10**, 1237 (1992).
3. T. J. Wu, W. Walukiewicz, K. M. Yu, J. W. Ager, III, S. X. Li, E. E. Haller, H. Lu, and W. J. Schaff, *Solid State Commun.* **127**, 411 (2003).
4. P. Rinke, M. Winkelnkemper, A. Qteish, D. Bimberg, J. Neugebauer, and M. Scheffler, *Phys. Rev. B* **77**, 075202 (2008).
5. M. A. Khan, M. Shatalov, H. P. Matuska, H. M. Wang, and E. Kuokstis, *Jpn. J. Appl. Phys.* **44**, 7191 (2005).
6. A. G. Bhuiyan, A. Hashimoto, and A. Yamamoto, *J. Appl. Phys.* **94**, 2779 (2003).
7. Q. Yan, P. Rinke, M. Scheffler, and C. G. Van de Walle, *Appl. Phys. Lett.* **95**, 121111 (2009).
8. S. Chichibu, A. Shikanai, T. Axuhata, T. Sota, A. Kuramata, K. Horino, and S. Nakamaura, *Appl. Phys. Lett.* **68**, 3766 (1996).
9. S. Chichibu, H. Okumura, S. Nakamura, G. Feuillet, T. Azuhata, T. Sota, and S. Yoshida, *Jpn. J. Appl. Phys.* **36**, 1976 (1997).
10. S. Chichibu, T. Azuhata, T. Sota, H. Amano, and I. Akasaki, *Appl. Phys. Lett.* **70**, 2085 (1997).
11. M. E. Aumer, S. F. LeBoeuf, and S. M. Bedair, M. Smith, J. Y. Lin, and H. X. Jiang, *Appl. Phys. Lett.* **77**, 821 (2000).
12. S. Nakamura, *Jpn. J. Appl. Phys.* **30**, L1705 (1991).
13. S. Nakamura, M. Senoh, and T. Mukai, *Jpn. J. Appl. Phys.* **30**, L1708 (1991).
14. S. Nakamura, M. Senoh, S. Nagahama, N. Iwasa, T. Yamada, T. Matsushita, H. Kiyoku, Y. Sugimoto, T. Kozaki, H. Umemoto, M. Sano, and K. Chocho, *Jpn. J. Appl. Phys.* **36**, L1568 (1997).
15. K. Fujito, S. Kubo, and I. Fujimura, *MRS Bull.* **34**, 313 (2009).
16. T. Hashimoto, F. Wu, J. S. Speck, and S. Nakamura, *Nat. Mater.* **6**, 568 (2007).
17. S. F. Chichibu, H. Marchand, M. S. Minsky, S. Keller, P. T. Fini, J. P. Ibbetson, S. B. Fleischer et al., *Appl. Phys. Lett.* **74**, 1460 (1999).
18. S. F. Chichibu, T. Onuma, T. Hashimoto, K. Fujito, J. S. Speck, and S. Nakamura, *Appl. Phys. Lett.* **91**, 251911 (2007).
19. M. A. Reshchikov and H. Morkoç, *J. Appl. Phys.* **97**, 061301 (2005).
20. S. Nakamura, T. Mukai, M. Senoh, and N. Iwasa, *Jpn. J. Appl. Phys.* **31**, L139 (1992).
21. W. Götz, N. M. Johnson, J. Walker, D. P. Bour, and R. A. Street, *Appl. Phys. Lett.* **68**, 667 (1996).
22. U. Kaufmann, M. Kunzer, M. Maier, H. Obloh, A. Ramakrishnan, B. Santic, and P. Schlotter, *Appl. Phys. Lett.* **72**, 1326 (1998).
23. F. Shahedipour and B. W. Wessels, *Appl. Phys. Lett.* **76**, 3011 (2000).
24. S. Nakamura, M. Senoh, N. Iwasa, and S. Nagahama, *Appl. Phys. Lett.* **67**, 1868 (1995).
25. S. Nakamura, M. Senoh, N. Iwasa, and S. Nagahama, *Jpn. J. Appl. Phys.* **34**, L797 (1995).
26. S. Nakamura, M. Senoh, N. Iwasa, and S. Nagahama, *Jpn. J. Appl. Phys.* **34**, L1332 (1995).
27. T. Mukai, M. Yamada, and S. Nakamura, *Jpn. J. Appl. Phys.* **38**, 3976 (1999).
28. T. Akasaka, H. Gotoh, T. Saito, and T. Makimoto, *Appl. Phys. Lett.* **85**, 3089 (2004).
29. T. Fujii, Y. Gao, R. Sharma, E.L. Hu, S. P. DenBaars, and S. Nakamura, *Appl. Phys. Lett.* **84**, 855 (2004).
30. F. Bernardini and V. Fiorentini, *Phys. Rev. B* **56**, R10024 (1997).
31. G. Martin, A. Botchkarev, A. Rockett, and H. Morkoç, *Appl. Phys. Lett.* **68**, 2541 (1996).
32. P. Waltereit, O. Brandt, A. Trampert, H. T. Grahn, J. Menniger, M. Ramsteiner, M. Reiche, and K. H. Ploog, *Nature* **406**, 865 (2000).
33. S. Chichibu, T. Azuhata, T. Sota, and S. Nakamura, *Appl. Phys. Lett.* **69**, 4188 (1996).
34. S. Chichibu, T. Azuhata, T. Sota, and S. Nakamura, *Appl. Phys. Lett.* **70**, 2822 (1997).
35. S. F. Chichibu, A. C. Abare, M. S. Minsky, S. Keller, S. B. Fleischer, J. E. Bowers, E. Hu, U. K. Mishra, L. A. Coldren, and S. P. DenBaars, *Appl. Phys. Lett.* **73**, 2006 (1998).

36. S. Chichibu, T. Sota, K. Wada, and S. Nakmaura, *J. Vac. Sci. Technol. B* **16**, 2204 (1998).
37. S. F. Chichibu, T. Azuhata, T. Sota, T, Mukai, and S. Nakamura, *J. Appl. Phys.* **88**, 5153 (2000).
38. S. Nakamura, M. Senoh, S. Nagahama, N. Iwasa, T. Yamada, T. Matsushita, H. Kiyoku, and Y. Sugimoto, *Jpn. J. Appl. Phys.* **35**, L74 (1995).
39. S. Nakamura, M. Senoh, S. Nagahama, N. Iwasa, T. Yamada, T. Matsushita, Y. Sugimoto, and H. Kiyoku, *Appl. Phys. Lett.* **70**, 1417 (1997).
40. S. Nakamura, M. Senoh, S. Nagahama, N. Iwasa, T. Yamada, T. Matsushita, H. Kiyoku et al., *Appl. Phys. Lett.* **72**, 2014 (1998).
41. T. Miyoshi, S. Masui, T. Okada, T. Yamamoto, T. Kozaki, S. Nagahama, and T. Mukai, *Appl. Phys. Express* **2**, 062201 (2009).
42. H. Yoshida, Y. Yamashita, M. Kuwabara, and H. Kan, *Nat. Photon.* **2**, 561 (2008).
43. T. Izumi, Y. Narukawa, K. Okamoto, Y. Kawakami, S. Fujita, S. Nakamura, *J Lumin.* **87**, 1196 (2000).
44. Ü. Özgür, H. Everitt, S. Keller, and S. P. DenBaars, *Appl. Phys. Lett.* **82**, 1416 (2003).
45. S. F. Chichibu, A. Uedono, T. Onuma, B. A. Haskell, A. Chakraborty, T. Koyama, P. T. Fini et al., *Nat. Mater.* **5**, 810 (2006).
46. Y. H. Cho, G. H. Gainer, A. J. Fischer, J. J. Song, S. Keller, U. K. Mishra, and S. P. DenBaars, *Appl. Phys. Lett.* **73**, 1370 (1998).
47. A. Sasaki, K. Nishizuka, T. Wang, S. Sakai, A. Kaneta, Y. Kawakami, and S. Fujita, *Solid State Commun.* **129**, 31 (2004).
48. L. Grenouillet, C. Bru-Chevallier, G. Guillot, P. Gilet, P. Duvaut, C. Vannuffel, A. Million, and A. Chenevas-Paule, *Appl. Phys. Lett.* **76**, 2241(2000).
49. K. Okamoto, A. Kaneta, Y. Kawakami, S. Fujita, J. Choi, M. Terazima, and T. Mukai, *J. Appl. Phys.* **98**, 064503 (2005).
50. S. W. Feng, T. Y. Tang, Y. C. Lu, S. J. Liu, E. C. Lin, C. C. Yang, K. J. Ma et al., *J. Appl. Phys.* **95**, 5388 (2004).
51. J. S. Speck and S. F. Chichibu, *MRS Bull.* **34**, 304 (2009).
52. A. E. Romanov, T. J. Baker, S. Nakamura, and J. S. Speck, *Appl. Phys. Lett.* **100**, 023522 (2006).
53. H. Ohta and K. Okamoto, *MRS Bull.* **34**, 324 (2009).
54. M. D. Craven, S. H. Lim, F. Wu, J. S. Speck, and S. P. DenBaars, *Appl. Phys. Lett.* **81**, 469 (2002).
55. T. Baker, B. A. Haskell, F. Wu, J. S. Speck, and S. Nakamura, *Jpn. J. Appl. Phys.* **45**, L154 (2006).
56. M. Funato, M. Ueda, Y. Kawakami, Y. Narukawa, T. Kosugi, M. Takahashi, and T. Mukai, *Jpn. J. Appl. Phys.* **45**, L659 (2006).
57. K. Okamoto, H. Ohta, D. Nakagawa, M. Sonobe, J. Ichihara, and H. Takasu, *Jpn. J. Appl. Phys.* **45**, L1197 (2006).
58. K. Iso, H. Yamada, H. Hirasawa, N. Fellows, M. Saito, K. Fujito, S. P. DenBaars, J. S. Speck, and S. Nakamura, *Jpn. J. Appl. Phys.* **46**, L960 (2007).
59. A. Tyagi, H. Zhong, N. N. Fellows, M. Iza, J. S. Speck, S. P. DenBaars, and S. Nakamura, *Jpn. J. Appl. Phys.* **46**, L129 (2007).
60. H. Sato, R. B. Chung, H. Hirasawa, N. Fellows, H. Masui, F. Wu, M. Saito, K. Fujito, J. S. Speck, S. P. DenBaars, and S. Nakamura, *Appl. Phys. Lett.* **92**, 221110 (2008).
61. H. Masui, A. Chakraborty, B. A. Haskell, U. K. Mishra, J. S. Speck, S. Nakamura, and S. P. DdenBaars, *Jap. J. Appl. Phys. Lett.* **44**, L1329 (2005).
62. H. Tsujimura, S. Nakagawa, K. Okamoto, and H. Ohta, *Jpn. J. Appl. Phys.* **46**, L1010 (2007).
63. S. Nakagawa, H. Tsujimura, K. Okamoto, M. Kubota, and H. Ohta, *Appl. Phys. Lett.* **91**, 171110 (2007).
64. M. Kubota, K. Okamoto, T. Tanaka, and H. Ohta, *Appl. Phys. Lett.* **92**, 011920 (2008).
65. S. Ghosh, P. Waltereit, O. Brandt, H. T. Grahn, and K. H. Ploog, *Phys. Rev. B* **65**, 075202 (2002).
66. A. A. Yamaguchi, *Jpn. J. Appl. Phys.* **46**, L789 (2007).
67. M. Ueda, M. Funato, K. Kojima, Y. Kawakami, Y. Narukawa, and T. Mukai, *Phys. Rev. B* **78**, 233303 (2008).

68. M. C. Schmidt, K. C. Kim, R. M. Farrell, D. F. Feezell, D. A. Cohen, M. Saito, K. Fujito, J. S. Speck, S. P. DenBaars, and S. Nakamura, *Jpn. J. Appl. Phys.* **46**, L190 (2007).
69. K. Okamoto, H. Ohta, S. F. Chichibu, J. Ichihara, and H. Takasu, *Jpn. J. Appl. Phys.* **46**, L187 (2007).
70. K. Okamoto, T. Tanaka, M. Kubota, and H. Ohta, *Jpn. J. Appl. Phys.* **46**, L820 (2007).
71. M. Kubota, K. Okamoto, T. Tanaka, and H. Ohta, *Appl. Phys. Express* **1**, 011102 (2008).
72. Tsuda, M. Ohta, P. O. Vaccaro, S. Ito, S. Hirukawa, Y. Kawaguchi, Y. Fujishiro, Y. Takahira, Y. Ueta, T. Takakura, and T. Yuasa, *Appl. Phys. Express* **1**, 011104 (2008).
73. K. Okamoto, T. Tanaka, and M. Kubota, *Appl. Phys. Express* **1**, 072201 (2008).
74. K. M. Kelchner, Y. D. Lin, M. T. Hardy, C. Y. Huang, P. S. Hsu, R. M. Farrell, D. A. Haeger et al., *Appl. Phys. Express* **2**, 071003 (2009).
75. Y. D. Lin, M. T. Hardy, P. S. Hsu, K. M. Kelchner, C. Y. Huang, D. A. Haeger, R. M. Farrell et al., *Appl. Phys. Express* **2**, 082102 (2009).
76. A. M. Fischer, Z. Wu, K. Sun, Q. Wei, Y. Huang, R. Senda, D. Iida, M. Iwaya1, H. Amano, and F. A. Ponce, *Appl. Phys. Express* **2**, 041002 (2009).
77. A. Tyagi, Y. D. Lin, D. A. Cohen, M. Saito, K. Fujito, J. S. Speck, S. P. DenBaars, and S. Nakamura, *Appl. Phys. Express* **1** 091103, (2008).
78. H. Asamizu, M. Saito, K. Fujito, J. S. Speck, S. P. DenBaars, and S. Nakamura, *Appl. Phys. Express* **1**, 091102 (2008).
79. H. Asamizu, M. Saito, K. Fujito, J. S. Speck, S. P. DenBaars, and S. Nakamura, *Appl. Phys. Express* **2**, 021002 (2009).
80. D. S. Sizov, R. Bhat, J. Napierala, C. Gallinat, K. Song, and C. Zah, *Appl. Phys. Express* **2**, 071001 (2009).
81. Y. Enya, Y. Yoshizumi, T. Kyono, K. Akita, M. Ueno, M. Adachi, T. Sumitomo, S. Tokuyama, T. Ikegami, K. Katayama, and T. Nakamura, *Appl. Phys. Express* **2** 082101 (2009).
82. A. Tyagi1, R. M. Farrell1, K. M. Kelchner, C. Y. Huang, P. S. Hsu, D. A. Haeger, M. T. Hardy et al., *Appl. Phys. Express* **3**, 011002 (2010).
83. S. H. Park and D. Ahn, *Appl. Phys. Lett.* **90**, 013505 (2007).
84. T. Onuma, K. Okatomo, H. Ohta, and S. F. Chichibu, *Appl. Phys. Lett.* **93**, 091112–1 (2008).
85. T. Onuma, H. Amaike, M. Kubota, K. Okamoto, H. Ohta, J. Ichihara, H. Takasu, and S. F. Chichibu, *Appl. Phys. Lett.* **91**, 181903 (2007).

4

Photoluminescence of ZnO: Basics and Applications

Klaus Thonke
University of Ulm

Martin Feneberg
University of Magdeburg

ZnO is presently a very "fashionable" material in semiconductor research for several reasons: It could provide an alternative material for light-emitting diodes (LEDs) in the green to UV spectral range. Heterostructures are possible with MgZnO for higher bandgap and CdZnO for lower bandgap. It is very easy to grow nanostructures from ZnO with a broad variety of morphologies. It is highly sensitive to chemicals, which makes it useful for sensors. It is a candidate for high-T_C ferromagnetic semiconductors.

Generally, the photoluminescence from ZnO is quite intense and shows a very rich spectrum of sharp excitonic lines, making it an interesting material for optical spectroscopy. It is very easily grown under a multitude of differing conditions and can be produced in good quality even with rather simple methods. Even from early research several decades ago, single crystals with superior quality were available. There are meanwhile good quality bulk substrates commercially available, providing a basis for (opto-) electronic applications. The main problem for broader usability is the still unsolved p-type doping issue.

4.1 Introduction to ZnO: Material Parameters of Bulk Crystals and Layers

4.1.1 History of Research

Zinc oxide (ZnO) is a II–VI compound semiconductor for which, under normal conditions, the hexagonal wurtzite crystal structure is the stable one. It rarely can be found in nature* as zincite, where its

---

* For example, in the Franklin and Sterling Hill Mines in New Jersey.

color depends on the impurity content, with manganese as a typical impurity. In the early development of radios (circa 1900–1940), it served in so-called crystal detectors for rectifying purposes. A thin copper wire spring was pressed onto the surface of a small natural or synthetic zincite crystal, forming an unstable Schottky-type metal/semiconductor diode contact, which frequently had to be reconstituted by trial and error. Electroluminescence was discovered in investigations of zinc sulfide and zinc oxide. Many tried to develop green and yellow phosphors from ZnO. By the 1930s, Wagner and Schottky (1930) proposed that imperfections control the semiconductor behavior of ZnO. Another early electronic application of ZnO was its use in "varistors," where ZnO grains in a matrix of other metal oxides serve to suppress overvoltage spikes. With the availability of synthetic ZnO bulk crystals, intense studies of electrical properties started (see, e.g., Heiland 1954). During the late-1950s to early-1970s, a first period of intense research on the optical properties of ZnO crystals—mainly in the form of high quality needles grown by CVD processes—began. Andress and Mollwo (1959) reported the first data on the luminescence of ZnO in the band edge region—after earlier reports on green and yellow visible emissions. Thomas (1960) studied the bandgap region in reflection and absorption. Reynolds et al. (1965) reported the first Zeeman data on bound excitons. The interest in ZnO declined in the later 1970s due to the failure of obtaining stable p-type doping—very similar to that of gallium–nitride at that time. The next intense phase of optical studies on ZnO started again in the mid-1990s when different kinds of nanostructures were grown and investigated. Meanwhile, thousands of papers appeared on the luminescence, absorption, reflection, and other optical properties of ZnO.

4.1.2 Crystal Structure

The normal crystal structure of the binary II–VI compound zinc oxide is the hexagonal 2H wurtzite structure or, rarely, under very special growth conditions on substrates which enforce cubic growth, also the metastable cubic 3C zinc blende structure. Under high pressure, ZnO in rocksalt structure can also be stabilized (Desgreniers 1998). In both structures, wurtzite and zinc blende, each anion is surrounded by four next neighbor cations in a tetrahedral arrangement, and vice versa, as a consequence of covalent sp^3 bonding. The energy required to form the sp^3 hybrid bonds from the initial s and p orbitals is overcompensated by the gain in energy when bonding is established. The character of the bonds of course is not purely covalent, but has a strong ionic contribution. The relative difference in electronegativity of Zn and O is 2.9 units on the Pauling scale (compare to 0.9 units for the polar HCl bond).

The symmetry point group is 6mm or C_{6v}, and the space group $P6_3mc$ or C_{6v}. The wurtzite structure can alternatively be considered as an arrangement of two hexagonal close-packed sublattices of only oxygen or only zinc atoms, which are shifted relative to each other by u = 3/8c. As shown in Figure 4.1,

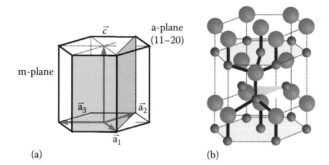

(a)   (b)

FIGURE 4.1 (a) Hexagonal conventional unit cell with the extended base vectors $\vec{a}_1$ to $\vec{a}_3$ in the hexagonal base plane and the vector $\vec{c}$ in the main symmetry direction. Most prominent planes are the basal hexagonal c-plane, the "side wall" m-plane, and the a-planes, which are perpendicular to the m-planes. (b) Ball and stick model of the crystal structure with Zn atoms (bigger) and oxygen atoms (smaller). Thick black lines mark some of the sp^3 bonds in tetrahedral arrangement.

the hexagonal (conventional) unit cell has the lattice parameters a = 0.32498 nm (for the hexagonal edge length) and c = 0.52066 nm for the height (Desgreniers 1998). The ratio of c/a is 1.6021 and thus slightly deviates from to the ideal ratio $\sqrt{8/3} = 1.633$. Since there is no center of inversion symmetry, the slight deviation from the ideal c/a factor results in spontaneous polarization (pyroelectric field) and polarization induced under strain (piezoelectric response) of the crystal. The polarity of the crystal end faces (Zn or O terminated) influences chemical properties, etching rates, growth speed, defect formation, etc. Conventionally, the c lattice vector [0001] points from the O terminated plane to the Zn plane.

4.1.3 Band Structure

ZnO is a semiconductor with a direct bandgap of 3.36 eV at room temperature and has therefore a very high potential for optical applications in the near UV range. Together with the ternary compounds (Zn,Mg)O with larger and (Cd,Zn)O with smaller bandgap heterostructures and quantum wells (QWs) can be formed—within a limited admixture range of Cd and Mg to ZnO, since for higher content of these metals the rocksalt crystal structure becomes more stable. Due to the axial symmetry of the wurtzite crystal, the valence band (VB) at k=0 is split by crystal field and spin-orbit interaction into three separate Kramers (i.e., spin degenerate) doublets labeled A, B, and C from low to high hole energy. There is a long-lasting dispute about the correct order of these VBs (see discussion in Klingshirn 2007b, Appendix A). Park et al. (1966) were the first who concluded a conventional order of the VBs in the sequence $\Gamma_9/\Gamma_7/\Gamma_7$ as in other II–VI semiconductors, whereas Segall (1967) suggested a reversed order for the topmost VBs. This reversed order was confirmed by Blattner et al. (1982) from their high-field Zeeman measurements, whereas Reynolds et al. (1999) argue that the order is the "classical" one. Lambrecht et al. (2002) find from their theoretical calculations a reversed order $\Gamma_7/\Gamma_9/\Gamma_7$, which is due to the strong repulsive interaction of the zinc 3d orbitals with the oxygen 2p orbitals. From interactions observed in the Zeeman splitting of donor-bound excitons, Thonke et al. (2003) concluded Γ_7 symmetry of the hole involved. Ding et al. (2007) performed angular dependent magneto-PL studies on the free A exciton and concluded a reversed VB order. Wagner et al. (2009) recently reported a fourfold Zeeman splitting for donor-bound excitons in Voigt configuration, which only can be explained for holes in the topmost VB having Γ_7 character, but not for holes with Γ_9 character. Therefore, the correct VB order is $\Gamma_7/\Gamma_9/\Gamma_7$ (see Figure 4.2).

The effective mass of the electron is nearly isotropic and was determined by Baer (1967) by Faraday rotation to be $0.29m_0$ in the low frequency limit (i.e., the polaron mass) and $0.24m_0$ in the high frequency

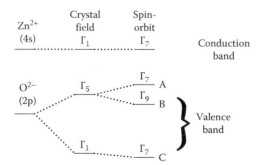

FIGURE 4.2 (a) Schematic band structure of ZnO. The conduction band originates mainly from the 4 s states of Zn^{2+}, whereas the valence band is derived from the O^{2-} 2p states, which are partially mixing with the Zn 3d states. (b) The band structure of ZnO modeled with a $6 \times 6 \mathrm{k} \cdot \mathrm{p}$ band calculation, using the parameters of Lambrecht et al. (2002). The energies quoted were determined by Schildknecht on bulk ZnO from Eagle-Picher. (From Schildknecht, A., *Optische Spektroskopie von Halbleitern mit großer Bandlücke: ZnO und AlGaN* (in German), Diploma thesis, University of Ulm, Germany, 2003b.)

(band mass) limit. Similar values were determined by Imanaka et al. (2001) in a cyclotron resonance experiment: $0.29m_0$ and $0.23m_0$, respectively. Kim et al. (2008) confirmed in their measurement using the method of four coefficients the polaron mass of $0.29m_0$ and quote for the bare band mass at the conduction band minimum $m^*_\perp = 0.247m_0$.

The hole-effective masses for ZnO can be considered mostly isotropic around $k = 0$ and are similar for all three VBs. For $m_{A\perp}$ values ranging from 0.45 to $0.66\,m_0$ can be found and $\sim 0.59\,m_0$ for $m_{A\parallel}$. For the B VB, perpendicular effective masses $m_{B\perp}$ range from 0.59 to $0.64\,m_0$ and the parallel mass $m_{B\parallel}$ is $0.59\,m_0$. The higher VB "C" is seemingly slightly more anisotropic with $m_{h\parallel} = 0.31\,m_0$ and $m_{h\perp} \approx 0.53$–$0.55\,m_0$ (Dinges et al. 1970; Hümmer 1973; Oshikiri et al. 2002; Syrbu et al. 2004). A small k-linear term for $k\perp c$ can occur for bands of symmetry Γ_7 resulting in a further minor splitting for $k \neq 0$. For higher values of $k_\perp$, significant non-parabolicities and mixing of the bands is expected (see, e.g., Figure 4.4 in Lambrecht et al. 2002).

The excitonic bandgap related to the topmost "A" VB of ZnO, E_{XA}, decreases from its value of $E_{XA,0} = 3.375\,eV$ at low temperature to $\sim 3.30\,eV$ at room temperature. Hauschild et al. (2006) traced it by PL and transmission measurements up to 800 K. Using the fit formula of Viña et al. (1984).

$$E(T) = E(0) - \frac{\alpha\Theta}{\exp(\Theta/T) - 1}, \tag{4.1}$$

based on a Bose–Einstein model, Boemare et al. (2001) derive from their PL data up to room temperature a parameter set of $E_{XA,0} = 3.3767\,eV$, $\alpha = 0.25\,meV/K$, and $\Theta = 203\,K$ for the energy position of the free A exciton X_A. Similarly, Wang and Giles (2003) arrive at $E_{XA,0} = 3.380\,eV$, $\alpha = 0.38\,meV/K$, and $\Theta = 240 \pm 5\,K$. Reflectance data recorded from liquid He temperature to $\approx 450°C$ (see Figure 4.3) on bulk ZnO (from Eagle Picher) were fitted to the modified Bose–Einstein model developed by Pässler (2002), which takes electron coupling to both acoustic and optical phonons into account:

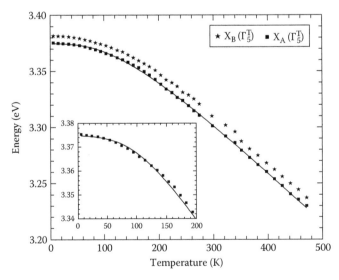

FIGURE 4.3 Temperature dependence of the bandgap of ZnO measured by reflectance measurements on a bulk ZnO crystal grown by the hydrothermal method. The solid line marks a fit using the Pässler formula (see text). (From Schildknecht, A., *Optische Spektroskopie von Halbleitern mit großer Bandlücke: ZnO und AlGaN* (in German), Diploma thesis, University of Ulm, Germany, 2003b.)

$$E(T) = E(0) - \alpha\Theta\left[\frac{1-3\Delta^2}{\exp(\Theta/T)-1} + \frac{3\Delta^2}{2}\left(\sqrt[6]{1 + \frac{\pi^2}{3(1+\Delta^2)}\left(\frac{2T}{\Theta}\right)^2 + \frac{3\Delta^2-1}{4}\left(\frac{2T}{\Theta}\right)^3 + \frac{8}{3}\left(\frac{2T}{\Theta}\right)^4 + \left(\frac{2T}{\Theta}\right)^6} - 1\right)\right]$$

$$(4.2)$$

Here, the parameter Θ means the average phonon temperature $\hbar\bar{\omega}/kT$, Δ is the phonon dispersion coefficient $\Delta \equiv \sqrt{\langle(\hbar\omega - \hbar\bar{\omega})^2\rangle}/\hbar\bar{\omega}$ ranging from 0 to 1, and α is the high temperature limit of the slope dE/dT. For the ZnO case, optimal parameters $E_{XA,0} = 3.3747$ eV, $\alpha = 0.47$ meV/K, $\Delta = 0.38$, and $\Theta = 372$ K were determined (Schildknecht 2003b). To obtain the single-particle bandgap energy E_g, an exciton binding energy of 60 meV (Reynolds et al. 1999) must be added.

Another set of parameters frequently needed to analyze optical spectra are the anisotropic dielectric constants for high and low frequencies. From ellipsometry measurements at room temperature in the range 1.5–5 eV (Yoshikawa et al. 1997), almost isotropic high-frequency limit values of $\varepsilon_{\infty,\perp} = 3.68$ and $\varepsilon_{\infty,\parallel} = 3.72$ were reported. The static dielectric constants (i.e., for frequencies below the optical phonon frequencies) are $\varepsilon_{s,\perp} = 7.4$ and $\varepsilon_{s,\parallel} = 8.49$. For a rather comprehensive collection of data for the refractive index, especially in the range of the bandgap, see Özgür et al. (2005), Tab. XIII therein.

4.1.4 n- and p-Type Doping

Whereas n-type doping of ZnO is easily achieved and there is widespread use in applications for transparent conductive oxide (TCO), p-type doping is still an issue that is not satisfactorily resolved. Nominally undoped ZnO crystals are all n-type, with doping levels ranging from a few 10^{17} to 10^{21} cm^{-3}, depending on the growth conditions. Intrinsic defects such as oxygen vacancies (V_O) or zinc interstitials (Zn_i) are commonly considered as the likely candidates for the uncontrolled donors—but recent density functional calculations hint to V_O merely being a deep donor with "negative U" character (i.e., that the 2+/+ transition level lies above the +/0 level) (Janotti and van de Walle 2005). Alternatively, interstitial hydrogen was suggested in addition to traces of group III-atoms as the source of the uncontrolled n-type conductivity (van de Walle 2000). For intentional n-type doping, group III atoms like aluminum, gallium, or indium replacing Zn yield shallow donors, which can be incorporated up to concentrations above 10^{20} cm^{-3}, i.e., above the Mott density yielding a degenerate electron gas. The binding energy of these donors is close to the effective mass value of ~55 meV. n-Type doping with the group VII elements chlorine or fluorine on oxygen site is much less prominent and less investigated. Another shallow donor is hydrogen with ~35 meV thermal activation energy (Hofmann et al. 2002), which is highly abundant, especially in crystals, which were grown by the hydrothermal method, by metal–organic vapor phase epitaxy (MOVPE) or other methods that involve a hydrogen-containing growth atmosphere.

In contrast, stable p-type doping is still not available, either reliably or reproducibly. The binding energy of acceptors found in experiments is generally quite high (typically around 150 meV), and the hole concentrations at room temperature obtained are limited to the low 10^{18} cm^{-3}. Together with the low Hall mobility of ~0.1–2 cm^2/Vs, this results in relatively high resistances of such p-doped layers (~2–5 Ωcm), which hampers the production of efficient LEDs or electrically driven laser diodes. Theoretical calculations predict shallower levels for group I elements on Zn sites (0.09 eV for Li_{Zn}, 0.17 eV for Na_{Zn}, and 0.32 eV for K_{Zn}) than for group V elements on O site (0.4 eV for N_O to 1.15 eV for As_O) (Park et al. 2002). Generally, nitrogen is considered to be the best candidate for p-type doping because of its small ionic radius. However, recent theoretical work predicts much deeper acceptor states of 1.3–1.6 eV for N_O (Lyons et al. 2009; Lany et al. 2010). The doping under equilibrium conditions is difficult because of the formation of the compensating intrinsic donors V_O and Zn_i. Experimental results reported after doping with the different group-V atoms N, P, As, and Sb do not differ too much: Effective ionization energies reported from Hall or PL experiments are always around 150 meV, and maximum acceptor concentrations and mobilites are also quite similar—which casts some doubt on the nature of the acceptor centers created. Indeed, electron emission channeling studies on radioactive As and β-channeling studies on implanted Sb show that most of the As and Sb atoms, respectively, reside on the Zn site instead of the desired O site (Wahl et al. 2005, 2009).

Quite frequently, the conductivity type of nominally p-doped samples is reported to be unstable and to fluctuate from p- to n-type and back after different annealing steps (see, e.g., Barnes et al. 2005; Allenic et al. 2007; Hwang et al. 2008; Xiao et al. 2008). High concentrations of stacking faults and other crystal defects were observed after attempts to obtain p-type by group V doping (Oh et al. 2008). A low-temperature PL band at 3.31 eV, frequently observed after group V doping, was shown to be linked to basal plane stacking faults (BSFs) and (presumably intrinsic) acceptors with an activation energy of 130 meV located there (Schirra et al. 2008). The combination of these observations raises the question whether it is really substitutional atoms which act as acceptors, and not acceptor states introduced by (complex) intrinsic or structural defects.

Several authors tried using co-doping to obtain stable acceptors, e.g., with indium and nitrogen (Chen et al. 2006), with gallium and nitrogen (Joseph et al. 1999), or with zirconium and nitrogen (Kim et al. 2008b) and claim to obtain slightly higher hole concentrations of up to 5×10^{18} cm^{-3}. Recent theoretical calculations question the feasibility of this approach (Lany and Zunger 2010). For a review of doping of ZnO, see McCluskey and Jokela (2009).

4.1.5 Other Doping: Transition Metals, Rare Earth

One of the most prominent transition metal (TM) impurities in ZnO seemingly is copper, which is a common trace impurity in Zn and leads to the well-known "green luminescence band" peaking at around 2.45 eV investigated in detail by Dingle in 1969. Copper acts as a deep acceptor ($E_A \approx 0.4$ eV) (Dietz et al. 1963), thus compensating shallow donors (see Section 4.3.3.1).

Kang et al. (2006) report on ZnO doped with silver and find varying n- or p-type conductivity. In PL, again a band at 3.31 eV dominates, which might be linked to stacking faults (Schirra et al. 2008).

Driven by the interest in diluted magnetic high-gap semiconductors with high-ferromagnetic critical temperature for spintronic applications in combination with transparency in the visible, ZnO—mostly powders—doped by the TM such as Ni (Wakano et al. 2001; Singh et al. 2006; Yu et al. 2008), Mn (Fukumura et al. 1999; Cheng and Chien 2003; Jin et al. 2003), Co (Yan et al. 2004), V (Saeki et al. 2001), Cr, Ti, Sc, and Fe (Ando et al. 2001; Venkatesan 2004) were investigated. Jin et al. (2001) grew thin TM-doped layers by molecular beam epitaxy (MBE) to determine the solubility limit for different TM species. Typically, crystal quality is deteriorated when high amounts of TM are added. For ZnO:Ni, Wakano et al. (2001) report ferromagnetism for low temperatures, which turned into superparamagnetism for T > 30 K. Li et al. (2010) reported ferromagnetism up to 600 K for ZnNiO powders. Yan et al. (2004) report room temperature ferromagnetism for their ZnO:(Al,Co,Mn) co-doped ZnO layers. Values of T_C above room temperature have also been reported for Sc-, Ti-, and V-doped n-type ZnO powders. For ZnCoO powders, antiferromagnetic behavior was also reported (Liu and Morkoc 2005). Experimentally, it is not easy to distinguish intrinsic ferromagnetic behavior of TM-doped ZnO from ferromagnetism introduced by small clusters of contaminants (Quesada et al. 2006). Since there is often a large discrepancy in the type of magnetism and Curie temperatures reported for the same TM species, the origin of the magnetism is frequently not unambiguously clear. For ZnO intentionally doped with Cu, intrinsic defects were found to be involved in the ferromagnetic behavior (Xu et al. 2008).

In addition to doping with 3d TM, theoretical and experimental attempts to use rare earth (4f) ions to obtain ferromagnetism were started, e.g., for Nd and Gd (Ungureanu et al. 2006; Liu et al. 2008; Shi et al. 2009) or Dy (Bandyopadhyay et al. 2010). Generally, there are a lot of results yielding arguments both pro and con for the ferromagnetism of TM dopants in ZnO, see, e.g., the discussion in Liu and Morkoc (2005), Fukumura et al. (2004), and Pan et al. (2008).

Early attempts to prepare electroluminescent ZnO powders doped with different rare earth species (Bhushan et al. 1979) showed only broad, slightly shifted emission bands, which were assigned to donor-acceptor transitions. Kossanyi et al. (1990) found in their ZnO sintered with the rare earth ions Tm^{3+}, Ho^{3+}, Nd^{3+}, and Er^{3+} neither internal $4f$ nor donor-acceptor transitions in room temperature PL. Aziz et al. (2009) report $4f$ internal transitions for terbium and erbium-doped sol-gel prepared nanostructured ZnO. Che et al. (2007) found red emissions at $\sim$600 nm in their sol-gel prepared ZnO:Eu, and Zeng et al. (2008) observed sharp PL lines at around 400–600 nm after Eu^{3+} doping their "nanosheet" samples. Eu^{3+}-specific electroluminescence (EL) was also reported for sol-gel prepared Eu-doped ZnO films by Xue et al. (2008). Ishizumi and Kanemitsu (2005) found for their Eu-doped ZnO nanorods internal Eu^{3+} transitions only for quasi-resonant excitation into higher Eu^{3+} states, but not for above bandgap excitation (Ishizumi and Kanemitsu 2005). Minami et al. (2000) used scandium and yttrium to obtain transparent, conductive ZnO films on glass. Doping of ZnO nanobelts with cerium gave rather unspecific broad green defect bands in room temperature PL (Cheng et al. 2004).

4.1.6 Growth Processes

Bulk crystals are available from different suppliers grown with various techniques. By hydrothermal growth, hexagonal 50 mm wafers with good crystallinity (18″ line width in the (002) XRD reflex rocking curve) and high resistivity ($N_D \approx N_A \approx 10^{16}$ cm^{-3}; the compensation is presumably due to Li) can be grown (Maeda et al. 2005). Crystals of similar size were obtained by vapor-phase transport, which showed slightly higher donor concentration (low to mid 10^{17} cm^{-3}), but much lower compensation and higher low-temperature mobility (Look et al. 1998). Pressurized melt growth yields crystals with $\sim$25 mm diameter, 49″ rocking curve (0002) line width, and shows slightly higher N_D with very low compensation (Reynolds et al. 2004; Nause and Nemeth 2005). The typical electrical data of commercial bulk crystals were compared by Look (2007). In the early days of ZnO research, very pure and also doped crystals of high quality were grown directly in the gas phase by oxidation of Zn vapor (Helbig 1972).

For the growth of epitaxial layers—mostly on sapphire substrates despite the large lattice mismatch of $\sim$18%—several techniques have also been established. Because of their low costs and simplicity, dc sputtering, rf magnetron sputtering, and reactive sputtering are very popular—especially for deposition on silicon or glass substrates (Sernelius et al. 1988; Ryu et al. 2002; Fan et al. 2009).

Metal-organic chemical vapor deposition (MOCVD) makes use of several alternative precursors and allows the controlled growth of heterostructures (Gruber et al. 2002; Zhang et al. 2005; Heinze et al. 2007). MBE allows more precise control of the layers deposited and uses typically Zn metal and O_2 as sources (Chen et al. 1997; Iwata et al. 2000; Sakurai et al. 2000; Jiao et al. 2007; Yang et al. 2010). Meanwhile, pulsed-laser deposition (a technique established earlier for the deposition of high-T_C superconductors) has been used very successfully for the growth of ZnO layers and nanostructures. The material composition in the sintered target is more or less preserved in the deposition process (Kordi Ardakani 1996; Lorenz et al. 2004). With a process called "liquid phase epitaxy" using a chemical reaction of $ZnCl_2$ and K_2CO_3 in a continuous process at 630°C, nearly perfect homoepitaxial films could be grown (Ehrentraut et al. 2006). Polycrystalline ZnO films with $\sim$1 μm grains can even be grown at very low temperature by electrochemical deposition on metallic films (Ashida et al. 2008). Sol-gel processes at higher temperature yielded nanocrystalline films with preferential c-orientation on (100) silicon (Zhang et al. 2005b).

4.2 ZnO Nanostructures

ZnO has the tendency to grow under different conditions in an extremely large variety of morphologies. Even relatively simple growth procedures like CVD, vapor–liquid–solid (VLS), vapor–solid (VS), and simple wet chemistry can yield quite interesting and fancy structures. Some examples of these are shown in Figure 4.4. For future applications, presumably, the controlled, ordered growth of upright pillars on conductive or reflecting substrates is of highest interest, since only then either optical or electrical access to single ZnO structures can be managed.

Figure 4.4a shows ZnO nanopillars grown at high temperature by the carbothermal VLS process on a-plane sapphire substrate, showing the characteristic alloy droplet of Zn and the Au catalyst frozen on top. Pillars grown without catalyst (Figure 4.4b), but using Au or Zn nanoparticles as condensation seeds, show flat upper ends (Reiser et al. 2011). By pre-structuring the substrate with the help of polystyrene microspheres, arrays of hexagonally ordered pillars can be grown (Figure 4.4c). For nanolasers, straight pillars with well-defined facets are needed. For the improvement of the reflectivity of the lower end, these pillars were grown on an fcc iridium layer on silicon (Figure 4.4d).

Figure 4.4e and f shows ring- and ribbon-shaped ZnO nanobelts (Pan et al. 2001, Kong 2004). Even hollow polyhedral cages are possible (Figure 4.4g). The growth can proceed in a branched manner forming "nanopropellers" (Gao and Wang 2003) or creating "hedgehog"-like arrangements (Figure 4.4h). Tetrapods seem to form in the gas phase in a CVD-process and are then deposited on the substrate (Figure 4.4i). Comblike structures have also been reported (Figure 4.4k).

4.3 Photoluminescence Properties

4.3.1 Free Excitons

Holes from all three VBs can form excitons together with electrons from the conduction band. The excitonic behavior of ZnO luminescence was observed for the first time in 1960 (Thomas 1960). Due to thermal population arguments, in low temperature photoluminescence, mainly excitons with a hole from the A VB participate. While the C VB is separated 38.6 meV from the A VB (see Figure 4.2), excitons with holes from it are not observable in standard luminescence experiments. The distance from A to B VB is only 4.6 meV; therefore, excitons containing holes from the B VB may contribute to emission spectra especially at elevated temperatures. The free A exciton has an exact binding energy of $E_{xb} = 59.7$ meV (Reynolds et al. 1999), which is more than twice $k_B T = 26$ meV at room temperature. Thus, one expects excitons in ZnO to be stable even above room temperature, and exciton-based devices discussed in the current literature are controversial (e.g., Klingshirn et al. 2007). Because the conduction band is of Γ_7 symmetry, transitions from the conduction band to the VBs of either Γ_7 (A,C) or Γ_9 (B) symmetry lead to excitonic ground state symmetries of

$$\Gamma_7 \otimes \Gamma_7 \rightarrow \Gamma_1 + \Gamma_2 + \Gamma_5$$
$$\Gamma_7 \otimes \Gamma_9 \rightarrow \Gamma_5 + \Gamma_6$$

(4.3)

The fine structure is observable in high quality samples only. For the free A exciton (the VB having Γ_7 symmetry), the Γ_2 state is optically inactive, and Γ_1 is allowed in π polarization only which means E‖c, while the doubly degenerated Γ_5 states are split into longitudinal and transverse states for k⊥c and E⊥c and lead to transitions with E⊥c (σ polarization).

The latter splitting is characteristic for exciton-polaritons, which was first recognized by Hopfield and Thomas (1965) and by Park et al. (1966). These exciton-polaritons are the fundamental quantized (e,h) quasi-particle pair excitations in the solid, where the polarizable excitons couple to the particles of the electromagnetic field (photons). In a simplified picture, the coupling is mediated by the dipole momentum induced by the exciton, which influences the electromagnetic field and vice versa. The dispersion of these exciton-polaritons within semiconductors is shown in Figure 4.5

Two exciton-polariton branches exist, which, remote from the resonance, exhibit either clear excitonic (quadratic dispersion) or clear photonic (linear dispersion) behavior. The position at the lower polariton branch (LPB) closely above the resonance is often called the "bottleneck."

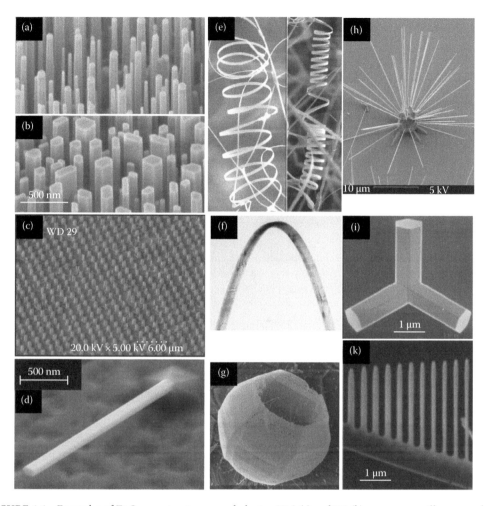

FIGURE 4.4 Examples of ZnO nanostructure morphologies: VLS (a) and VS (b) grown nanopillars on a-plane sapphire substrate (From Reiser, A., Ladenburger, A., Prinz, G., Schirra, M., Feneberg, M., Zayan, D.O., Röder, U., Leute, R.A.R., Sauer, R., and Thonke, K., Optimized growth conditions for well-defined zinc oxide nanopillars grown by the VLS process on different substrate materials, to be published 2011), (c) ordered arrays of ZnO pillars (From Madel, M., University of Ulm, private communication, 2010), (d) single nanolaser pillar on fcc Ir layer on Si substrate (With kind permission from Springer Science+Business Media: *Adv. Solid State Phys.*, ZnO nanostructures: Optical resonators and lasing, 48, 2009, 39, Thonke, K., Reiser, A., Schirra, M., Feneberg, M., Prinz, G.M., Röder, T., Sauer, R., Fallert, J., Stelzl, F., Kalt, H., Gsell, S., Schreck, M., and Stritzker, B., Copyright Springer), (e) rings (With permission from Kong, X.Y. and Wang, Z.L., Spontaneous polarization-induced nanohelixes, nanosprings, and nanorings of piezoelectric nanobelts, *Nano Lett.*, 3(12), 1625–1631. Copyright 2003 American Chemical Society), (f) nanobelts of ZnO (With permission from Kong, X.Y. and Wang, Z.L., Spontaneous polarization-induced nanohelixes, nanosprings, and nanorings of piezoelectric nanobelts, *Nano Lett.*, 3(12), 1625–1631. Copyright 2003 American Chemical Society), (g) hollow cage (With permission from Gao, P. and Wang, Z., Mesoporous polyhedral cages and shells formed by textured self-assembly of ZnO nanocrystals, *J. Am. Chem. Soc.*, 125, 11299. Copyright 2003 American Chemical Society), (h) nano-hedgehog (Madel 2010), (i) a large example of a tetrapod (From Yan, H., He, R., Pham, J. and Yang, P.: Morphogenesis of one-dimensional ZnO nano- and microcrystals. *Adv. Mater.* 2003. 15. 402–405. Copyright Wiley-VCH Verlag GmbH & Co. KGaA. With permission), and (k) a comb-like structure. (With permission from Yan, H., He, R., Johnson, J., Law, M., Saykally, R., and Yang, P., Dendritic nanowire ultraviolet laser array, *J. Am. Chem. Soc.*, 125, 4728–4729, 2003c. Copyright 2003 American Chemical Society.)

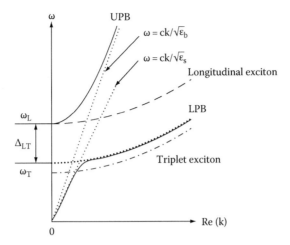

FIGURE 4.5 Dispersion relation of exciton-polaritons. LPB and UPB are the lower and upper polarition branches respectively. Δ_{LT} marks the longitudinal-transverse splitting of the polariton. Photon branches in the medium are marked by the dotted lines with different slope corresponding to different dielectric constants. The dashed line represents the longitudinal exciton, i.e., when electromagnetic field and wave vector are parallel to each other, the dash-dotted line marks the triplet exciton energy.

Exciton-polaritons located at this specific position are not easily scattered into the photon branch (where the energy finally can leave the crystal) due to a low density of final states and a small-scattering matrix element (Klingshirn 2007a). Due to this blockade in the recombination channel, a large amount of exciton-polaritons with long lifetime accumulates at the "bottleneck," leading to broadened free excitonic luminescence due to their dispersion. Additionally, luminescence from the upper polariton branch (UPB) is visible with an energy distance of $\Delta_{LT} \approx 2\,\text{meV}$ (Klingshirn 2007c) from the LPB (Figure 4.7). The absolute energy positions of both contributions are given in Table 4.1. The LT-splitting amounts to 10 meV for the B VB and 12 meV for the C VB (Klingshirn 2005) as deduced from reflectance spectra (Figure 4.3).

The broadening of the free exciton polaritons and the bandgap shrinkage due to increasing temperature results mainly from the electron–phonon interaction. It can be satisfactorily modeled by the expression developed by Lee et al. (1986) for the exciton interaction with acoustic and optical phonons:

$$\Gamma(T) = \Gamma_0 + \frac{\gamma}{\exp\left(\hbar\omega_A/k_BT\right)-1} + \frac{\gamma_{LO}}{\exp\left(\hbar\omega_{LO}/k_BT\right)-1}$$

$$\approx \Gamma_0 + \tilde{\gamma}T + \frac{\gamma_{LO}}{\exp\left(\hbar\omega_{LO}/k_BT\right)-1} \tag{4.4}$$

Here
   $\gamma(\gamma_{LO})$ is the coefficient of the exciton-acoustic (optic) phonon interaction
   $\hbar\omega_A$ ($\hbar\omega_{LO}$) the phonon energy

The initial line width Γ_0 is limited by the lifetime broadening and by the residual strain and electric field distribution within the radius of the exciton, which is sample dependent. A fit to the data points obtained from a bulk sample yields the parameters given in the caption of Figure 4.6 for the exciton polariton with a hole from the A (B) VB.

TABLE 4.1 Line Positions of Prominent Contributions to Luminescence Spectra in the Region of the Free and Bound Excitons of ZnO at 4.2 K

Label	E (eV)	E_{loc} (meV)	E_D (meV)	Identification
A_L	3.3785			Free A exciton-polariton (longitudinal)
A_T	3.3765			Free A exciton-polariton (transverse)
I_0	3.3730	3.5	54.5	(D^+,X)—aluminum
I_1	3.3724	4.1	56.7	(D^+,X)—gallium
I_2	3.3674	9.1	66.9	(D^+,X)—indium
I_3	3.3665	10.0	—	(D^+,X)—unknown donor
I_4	3.3628	13.7	50.2	(D^0,X)—hydrogen
I_5	3.3614	15.1	54.0	(D^0,X)—unknown donor
I_6	3.3612	15.3	54.5	(D^0,X)—aluminum
I_7	3.3607	15.8	55.9	(D^0,X)—unknown donor
I_8	3.3604	16.1	56.7	(D^0,X)—gallium
I_{8a}	3.3598	16.7	58.3	(D^0,X)—unknown donor
I_9	3.3566	19.9	66.9	(D^0,X)—indium
I_{10}	3.3530	23.5	76.7	(D^0,X)—unknown donor
I_{11}	3.3519	24.6	79.6	(D^0,X)—unknown donor

Note: The localization energy E_{loc} is the energy spacing from the transverse component of the free A exciton A_T. E_D is the donor single particle binding energy. The I_1–I_{11} notation is adopted from Reynolds et al. (1965). The assignment to specific donors is following the article of Meyer et al. (2004).

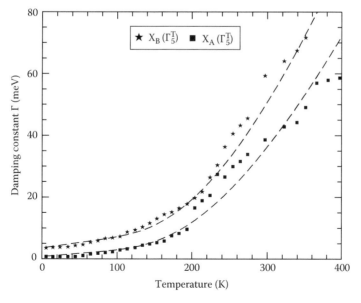

FIGURE 4.6 Temperature dependence of the damping constant Γ of the X_A and X_B polariton resonances in reflection. The dashed lines are a fit to the formula (Equation 4.4) with the parameters $\Gamma_0 = 0.75$ meV, $\tilde{\gamma} = 4.1 \cdot 10^{-5}$ eV K^{-1}, $\gamma L_O = 345$ meV for the A exciton, and $\Gamma_0 = 3.55$ meV, $\tilde{\gamma} = 4.5 \cdot 10^{-5}$ eV K^{-1}, $\gamma L_O = 558$ meV for the B exciton. (From Schildknecht, A., *Optische Spektroskopie von Halbleitern mit großer Bandlücke: ZnO und AlGaN* (in German), Diploma thesis, University of Ulm, Germany, 2003b.)

4.3.2 Bound Excitons

Free excitons as discussed above are able to bind to any local potential minimum present in the band structure. Such minima can be introduced by point or structural defects, interfaces, and impurities. Free (e,h) pairs can bind to neutral dopants (donors and acceptors) by exchange interaction of the impurity bound particle with the like partner in the (e,h) pair. Each of them gives rise to bound exciton emissions, as long as the decay of the exciton is still mainly radiative (Auger recombination is competing). Since for good crystal quality, the bound exciton lines can be extremely sharp (FWHMs below 100 μeV were observed in PL), an extremely large number of different emission lines can be observed, since every minor change of the crystal environment within the exciton radius by formation of complexes, etc., leads to different emission energies. Minor changes in strain shift the complete set of free and bound exciton lines on the energy scale, complicating definite assignments. Therefore, in ZnO, a very high number of different bound exciton lines is documented, making an exhaustive discussion within this work futile. Thus, we will limit ourselves to certain instructive examples and a compilation of frequently observed bound exciton energies (Table 4.1). For a comparison to other spectra, all energies should be aligned with that of the free X_A exciton at low temperature.

By time-resolved photoluminescence measurements, typical bound exciton lifetimes between 100 ps and 1 ns were determined for high quality material (Travnikov et al. 1990; Bertram et al. 2007; Klingshirn et al. 2007). These values are comparable to other wide bandgap semiconductors. The localization binding energy E_{loc} and the measured decay time τ are expected to be related with $\tau \sim E_{loc}^{3/2}$, a behavior that can be nicely experimentally reproduced in ZnO (Travnikov et al. 1990). But even in the very best crystals available, the nonradiative recombination processes are still the dominant mechanisms, as can be concluded from measurements of total quantum efficiencies (Klingshirn 2007a).

Under high optical pumping densities, ZnO is a textbook material for high excitation effects. Besides the biexciton (M-band), also exciton–exciton scattering processes (P-band), exciton-electron scattering, and the electron-hole plasma (EHP) are well-documented recombination channels (see also Figure 4.16). Their observation is facilitated by the high exciton binding energy of 60 meV and the availability of high-quality crystals. For a survey on high excitation effects in ZnO, see, e.g., Klingshirn (2007b).

4.3.2.1 Donors

In ZnO, shallow donors can be introduced by intentional or unintentional doping, e.g., with group III elements replacing a Zn atom, group VII elements on oxygen site, and even hydrogen that is a dominant impurity in hydrothermally grown crystals (see Section 4.1.4). Most of these give rise to extremely sharp donor-bound exciton (D^0,X) recombination observed as bright, sharp emission lines at low temperature (Figure 4.7). Full widths at half maximum below 100 μeV have been recorded frequently in very high quality crystals. The hole of the exciton can in principle originate from any of the three VBs, but again the A VB dominates in terms of population probability over the other bands. We therefore discuss only (D^0,X_A) recombination here. To be more accurate, the transversal A exciton is the most important one for conventional c-plane layers; we therefore use its energy $E(A_T) = 3.3765$ eV (Table 4.1) as a reference. Partially, the large number of (D^0,X) lines have been catalogued and referred to historically by I_0–I_{11} (see, e.g., Reynolds et al. 1965; Meyer et al. 2004), but due to their unmanageable number in ZnO and their extremely close spacing leading to overlapping spectral contributions, it cannot be guaranteed that from time to time the same bound exciton is named differently, while others remain unrecognized.

Some of the donor-bound excitons can be chemically identified. A generally accepted consensus has been established for hydrogen (I_4), aluminum (I_6), gallium (I_8), and indium (I_9). Their one-particle donor-ionization energies are known from independent experiments including observation of the two-electron satellite (TES) in PL (Figure 4.8). This is a side band of the donor-bound exciton, which is lower in energy by the amount of the energy separation of donor ground state 1 s from its first excited states

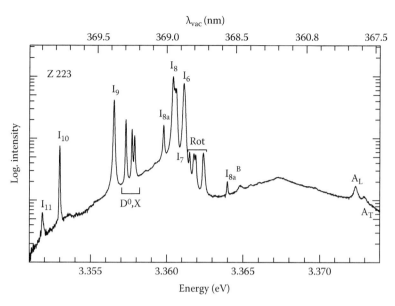

FIGURE 4.7 Low-temperature high-resolution PL spectrum of a high-quality ZnO crystal produced by the method of gas phase oxidation of Zn vapor (Helbig 1972). Besides the LT-split-free exciton X_A with hole from the A valence band, numerous very sharp donor bound exciton (D^0,X) lines I_6–I_{11} are observed. "Rot" marks excited rotator states of the hole in the (D^0,X) complex.

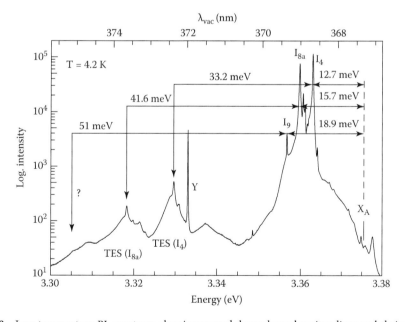

FIGURE 4.8 Low-temperature PL spectrum showing several donor-bound exciton lines and their related TES transitions. The arrows mark the energy differences relevant for Haynes' rule. X_A marks the position of the top-most valence band A related free exciton-polariton as determined from accompanying reflectance measurements. "Y" labels a sharp (D^0,X) transition introduced by an unknown deep donor at a structural defect (see also Section 5.3.3.2). (Reprinted from *Physica B*, 340–342, Schildknecht, A., Sauer, R., and Thonke, K., Donor-related defect states in ZnO substrate material, 205–209, Copyright 2003, with permission from Elsevier.)

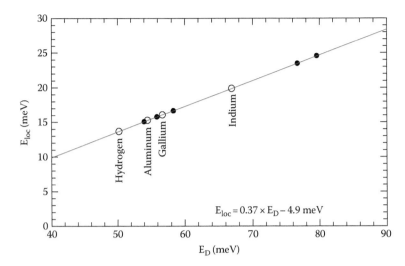

FIGURE 4.9 The relation between the donor bound exciton localization energy E_{loc} and the single particle ionization energy E_D of these donors. A fit to the data points yields the parameters for "Haynes rule" (see text).

2 s or 2 p. It occurs when after recombination of the bound exciton the donor is left behind in one of these excited states. Therefore, Hayne's rule can be verified for shallow donors in ZnO (Haynes 1960). It states a linear relationship between the donor single particle ionization energy E_D and the corresponding bound exciton localization energy E_{loc}. In ZnO, we find $E_{loc} = 0.37\,E_D − 4.9\,meV$ (Schildknecht et al. 2003) (Figure 4.9). The perfect match of Hayne's rule with these donors allows reliable assignments of further—up to now unidentified—donors: The lines I_5, I_7, I_{10}, and I_{11} are expected to be donor-bound exciton lines as well and should be related to donors with ionization energies of 54.0, 55.9, 76.7, and 79.6 meV, respectively.

The lines indexed $I_0 − I_3$ do not fit this Hayne's rule correlation line. Indeed, they recently were assigned to excitons bound to the same donors, but with these donors being in an ionized state D^+. More complicated features of donor-bound excitons have been reported, e.g., by Gutowski and co-workers (Gutowski et al. 1988), albeit there called (A^0,X). They probed by excitation spectroscopy excited states of donor-bound exciton recombinations. They occur between 1 and 3 meV above the donor-bound exciton ground state in agreement with Zeeman studies, in which excited states approximately 1.5 meV above the ground state were detected (Thonke et al. 2003). By excitation spectroscopy, donor-bound excitons with holes from the B VB can also be detected without increasing the temperature. They were found in good agreement with the energy spacing of A and B VBs approximately 4 meV higher in energy than the (D^0,X_A) (Gutowski et al. 1988), partly also visible in the spectrum shown in Figure 4.7.

4.3.2.2 Acceptors

No unambiguous proof of any acceptor-bound exciton (A^0,X) spectra has yet been presented in the available literature. However, many attempts to assign distinct excitonic features to acceptors have been made due to the desirable possibility of controlling p-type doping via monitoring (A^0,X) luminescence. So it is not very surprising that from time to time (A^0,X) luminescence has been claimed. Gutowski et al. (1988) assigned the same lines to be acceptor related, which were later shown to be (D^0,X) recombinations. Later, Look et al. (2002) presented an attempt to introduce nitrogen in ZnO to create p-type material and claimed (A^0,X) luminescence which was very weak and broad and was energetically located in the region of the 3.314 eV emission of the basal plane stacking fault (see Section 4.3.3.4)—which would

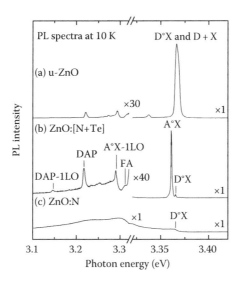

FIGURE 4.10 Low-temperature PL spectra of three differently doped samples. The sample in (a) was unintentionally doped, in (b) was co-doped with nitrogen and tellurium, while in (c) was only doped with nitrogen. The sharp and prominent line in (b) labeled (A^0,X) at 3.359 eV is a very likely candidate for an acceptor bound exciton. (Reproduced from Park, S., Minegushi, T., Oh, D., Lee, H., Taishi, T., Park, J., Jung, M., Chang, J., Im, I., Ha, J., Hong, S., Yonenaga, I., Chikyow, T., and Yao, T., High-quality p-type ZnO films grown by co-doping of N and Te on Zn-face ZnO substrates, *Appl. Phys. Exp.*, 3, 031103, 2010. With permission from Japanese Society of Applied Physics.)

be a possible alternative explanation for the spectra. However, very recently some groups succeeded in presenting p-type material exhibiting sharp and pronounced transitions in the region of the bound excitons, which they assign to (A^0,X) recombinations (see Figure 4.10) (Kim et al. 2003; Ryu et al. 2003; Park et al. 2010). These recombinations claimed (A^0,X) occur mostly in a very narrow energy range around 3.359 eV, no matter which dopant species—arsenic (Ryu et al. 2003), phosphorus (Kim et al. 2003), nitrogen (Park et al. 2010)—has been employed for acceptor doping.

So far, conclusive Magneto-PL data proving unambiguously the acceptor bound character of near-bandgap excitonic emissions are still lacking. PL signatures of acceptors being incorporated so far are donor–acceptor pair (DAP) bands, for which mostly up to now no definite chemical assignment of the acceptors species is available (see Section 4.3.3.3).

A different approach for the creation of acceptors instead of group V elements on the oxygen site makes use of group I elements on the zinc site. Especially lithium (Li), sodium (Na), and potassium (K) are possible candidates. However, they usually form only acceptor centers too deep in the forbidden gap to allow for any technical usage. Only in the case of Li and Na, an additional shallow acceptor binding energy of around 300 meV has been established giving rise to DAP transitions but no (A^0,X) recombinations (Meyer et al. 2007). Due to the high mobility of group I elements, reports of successful p-type doping should be taken with caution as long as no long-term stability is proven at the same time. The very near future will show if any of the investigated routes to reliable p-type ZnO will succeed in producing a technically useful acceptor.

4.3.3 Defect-Related PL

4.3.3.1 Green Band and Yellow Band

There are two prominent broad luminescence bands in ZnO: the aforementioned "green band" centered at ~2.45 eV (see Section 4.1.5) and the "yellow band" centered at ~2.2 eV. There seem to be at least two

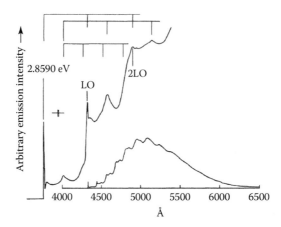

FIGURE 4.11 The characteristic green band due to copper in ZnO. It starts on the high energy side with a sharp doublet (not resolved here), which is followed by multiple phonon replicas of the LO phonon with 72 meV energy. The enlarged portion shows details of the no-phonon lines and the phonon replicas. (Reprinted with permission from Dingle, R., Luminescent transitions associated with divalent copper impurities and the green emission from semiconducting zinc oxide, *Phys. Rev. Lett.*, 23, 579–581, 1969. Copyright 1969 by the American Physical Society.)

different green bands: One band is introduced by copper and shows a sharp no-phonon line doublet with 0.1 meV spacing at 2.859 eV ascribed to the two stable Cu isotopes ^{63}Cu and ^{65}Cu. This band is accompanied by a multi-phonon sideband with pronounced modulation due to strong coupling to 72 meV LO phonons (Dingle 1969; Reynolds et al. 2001) (see Figure 4.11). By comparison of Zeeman measurements on the no-phonon doublet with EPR data, it was found that Cu^{2+} on zinc site must be the origin of the green band. The optical transition is a charge transfer between the split 3d states of Cu^{2+} and a shallow pseudo-acceptor state (i.e., Cu^+ + bound hole) ~0.4 eV above the VB edge (Dietz et al. 1963).

Depending on crystal growth, another broad green band might show up at the same energy position—but without the sharp doublet lines; and without phonon coupling modulation (see Leiter et al. 2001, and discussion in Kuhnert and Helbig 1981). This band (see Figure 4.12) was interpreted as a

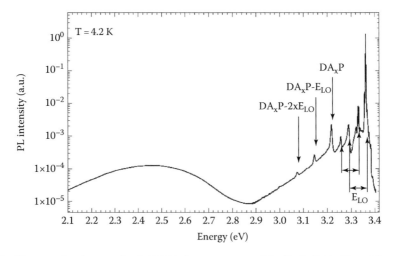

FIGURE 4.12 The green band centered at 2.45 eV in low-temperature PL of bulk ZnO. (From Meyer, B.K., Alves, H., Hofmann, D.M., Kriegseis, W., Forster, D., Bertram, F., Christen, J., Hoffmann, A., Straßburg, M., Dworzak, M., Haboeck, U., and Rodina, A.V.: Bound exciton and donor-acceptor pair recombinations in ZnO. *Phys. Status Solidi* (*b*). 2004. 241. 231–260. Copyright Wiley-VCH Verlag GmbH & Co. KGaA. With permission.)

DAP transition from shallow donors to Zn vacancies, which form double acceptors in ZnO (Leiter et al. 2001; Hofmann et al. 2007; Janotti and van de Walle 2007). It can be diminished by exposure of the sample to hydrogen—as expected for Zn vacancies, which can be passivated by hydrogen. Børseth et al. (2006) annealed nominally undoped, hydrothermally grown ZnO single crystal wafers in either O_2 or metallic Zn ambient and find either a deep band centered at 2.35 or 2.53 eV, which were ascribed to the presence of zinc vacancies or oxygen vacancies, respectively.

Block et al. (1982) ascribed the green band seen in their samples to donor–acceptor (DAP) transitions involving deep lithium acceptors and shallow donors. Similarly, Na was reported to introduce such a deep acceptor, which gives green DAP transitions (Zwingel and Gärtner 1974; Meyer et al. 2004). A yellow band at ~615 nm (2 eV) appearing as a low-energy tail broadening of the green band was attributed by Schirmer and Zwingel (1970) and Cox et al. (1978) to a DAP transition from a shallow donor into paramagnetic deep acceptor Li centers, similarly as assigned by Ohashi et al. (2005). Metastable behavior was reported for the yellow band by Özgür et al. (2005) in the sense that its intensity increases under above-bandgap irradiation at the expense of the green band. Özgür et al. (2005) report in their review article on a further broad red PL band at 1.75 eV, which they observed after annealing in air. For increasing sample temperatures, it is quenched, while the green band increases—presumably due to a competition for holes. Weber et al. (2010) suggest a complex of hydrogen with oxygen vacancies to be responsible for ZnO looking red in transmission.

4.3.3.2 Sharp Line Spectra from Point Defects and Extrinsic Atoms

For the TM dopants Ni, Co, and Cu, Schulz et al. showed sharp PL spectra originating from internal transitions between crystal field and spin-orbit split states (Schulz and Thiede 1987). For ZnO:Co, they find a series of lines around 0.45 eV [Co^{2+} $^4T_2(F) \rightarrow {}^4A_2(F)$] and 1.87 eV [$Co^{2+}$ $^4T_1(F)$, $^2T_1(F)$, $^2E(G) \rightarrow {}^4A_2(F)$]. For ZnO:Ni, a pair of emission lines was observed at ~0.74 eV [Ni^{3+} $^4T_2(F) \rightarrow {}^4A_2(F)$] and for codoped ZnO:Cu:Ni lines in the 0.85 eV range [Cu^{3+} $^3T_2(F) \rightarrow {}^3T_1(F)$]. In cathodoluminescence, Kimpel and Schulz observed a Cu^{2+} $^2E(D) \rightarrow {}^2T_2$ transition at 719 meV (Kimpel and Schulz 1991). Vanadium-doped ZnO was investigated by Heitz et al. (2002). They find an emission at 0.854 eV, which they assign with the help of Zeeman measurements to an internal $^3A_2 \rightarrow {}^3T_2$ transition at isolated substitutional V^{3+}. For isolated Fe^{3+} ions on Zn^{2+} site, Heitz et al. (1992) find a series of sharp lines in the range from 1.72 to 1.787 eV (see Figure 4.13). Magneto-PL data allow assigning these lines to the $^4T_1 \rightarrow {}^6A_1$ spin forbidden transition (Azamat et al. 2010).

4.3.3.3 Hydrogen

In 1954, Mollwo reported that ZnO shows a reversible change in conductivity when heated in hydrogen atmosphere (Mollwo 1954). Ohashi et al. (2002) found that hydrogenation passivates shallow impurities and increases the efficiency of the bandgap PL, depending on the impurity concentration. Ip et al. (2002) find the opposite effect that even after low implantation doses of hydrogen, the intensity of the bandgap PL is drastically reduced, presumably due to the recombination via competing nonradiative decay channels. Hydrogen can be observed in infrared absorption (McCluskey et al. 2002) and Raman spectra (Nickel and Fleischer 2003) monitoring local vibrational modes. Hofmann et al. (2002) found two shallow effective-mass like hydrogen donors in their electron paramagnetic resonance studies on bulk crystals (EPR). Lavrov et al. (2009) also identified two species of hydrogen donors: a bond centered version with ionization energy $E_D = 53$ meV, giving rise to a PL line at 3.3601 eV, and another species bound in an oxygen vacancy with $E_D = 47$ meV, leading to the "I_4" line at 3.3628 eV (see also Figure 4.8).

4.3.3.4 Donor–Acceptor Pair Bands

Besides the 2.45 eV band (see Section 4.3.3.1), several DAP or (D^0,A^0) bands have been reported. These are transitions between electrons bound at shallow donors and holes bound to—mostly not reliably

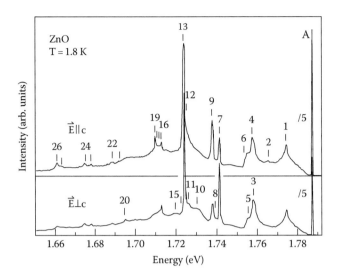

FIGURE 4.13   Polarized PL spectra of internal $^4T_1 \leftrightarrow {}^6A_1$ spin at Fe^{3+} in ZnO. (Reprinted with permission from Heitz, R., Hoffmann, A., and Broser, I., Fe^{3+} center in ZnO, *Phys. Rev. B*, 45, 8977–8988, 1992. Copyright 1992 by the American Physical Society.)

identified—acceptors. In nominally undoped ZnO bulk crystals grown by seeded chemical vapor transport, a DAP band at 3.22 eV and its associated band-acceptor transition (e, A^0) have been observed. From these, the acceptor binding energy E_A of ~195 meV was deduced (or 209 meV found in a similar work by wang et al. 2004). The acceptor involved might be nitrogen (Thonke et al. 2001). Zeuner et al. (2002) find after nitrogen implantation and annealing a DAP band at 3.235 eV accompanied by strong LO replicas. In a nitrogen-doped ZnO film grown by CVD at 600°C with ammonia as source, Meyer et al. (2004) find besides the 3.235 eV band also with half the intensity and weaker phonon coupling a second DAP band at 3.22 eV. On the same sample type, Monteiro et al. (2005) find two DAP bands at 3.22 and 3.238 eV, for which they quote acceptor energies of 250 and 232 meV, respectively, which they also tentatively assign to nitrogen. Xiong et al. (2005) observe an increase of their 3.232 eV band after nitrogen implantation; they determine an acceptor energy of 177 meV. Such differences in acceptor energies determined from DAP bands alone might come from different excitation densities, which shift the band maximum, and different assumptions concerning the density of the more abundant (i.e., donor) dopant species. Cui et al. (2009) observe for their layers grown by PLD in oxygen and nitrogen ambient a DAP band at 3.296 eV. Sun et al. (2007) find a DAP transition at 3.26 eV (at 80 K) in their MBE grown ZnO:N layers. For phosphorous-doped ZnO nanowires, two DAP bands at 3.24 and 3.04 eV were reported (Cao et al. 2007), which the authors assign to transitions between two different donors with 90 and 290 meV binding energy and a phosphorous (as claimed by the authors) acceptor with $E_A = 122$ meV.

4.3.3.5 Extended Defects

Similar to the situation for gallium nitride, threading dislocations are typically present in heteroepitaxial ZnO layers, with densities of up to some 10^9 cm^{-2}. By off-axis electron holography in a transmission electron microscope, Müller et al. (2006) found charge densities of up to 5×10^{20} cm^{-3} around the dislocation cores. Ohno et al. (2008) created dislocations in ZnO bulk crystals by plastic deformation at elevated temperatures. They report two emission bands at 3.10 and 3.345 eV being introduced. From thermal PL quenching, they estimate two deep levels with 0.3 and 0.05 eV being involved and suggest that point defect complexes involving the dislocations are the origin.

Schirra et al. (2008) identified by cathodoluminescence a frequently reported band at 3.314 eV as a free electron to neutral acceptor (e,A^0) transition emitted from BSFs (see Figure 4.14). The nature of the

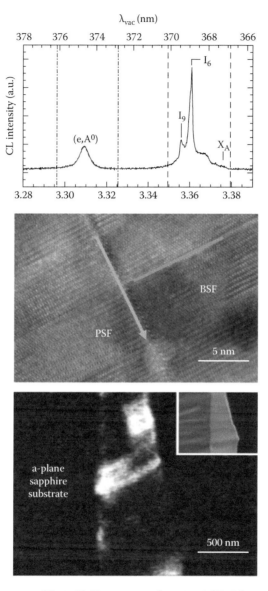

FIGURE 4.14 (a) Low-temperature (T ≈ 15 K) CL spectrum of an epitaxial ZnO layer grown on a-plane sapphire. Besides strong bound exciton luminescence, a band at 3.314 eV due to a free electron to neutral acceptor (e,A⁰) transition is observed. (b) High-resolution TEM image of a (11–20) cross section through a faceted region of the ZnO layer. BSFs that are terminated by prismatic stacking faults (PSF) are found. The downward arrow marks the c direction. (c) CL mapping of the defect emission recorded on the same sample region. Whereas the major part of the sample yields luminescence in the excitonic region only and no 3.314 eV signal, some stripes perpendicular to [0001] inversely emit the (e,A⁰) band exclusively. Inset: SEM picture of the sample. (From Thonke, K., Schirra, M., Schneider, R., Reiser, A., Prinz, G.M., Feneberg, M., Sauer, R., Biskupek, J., and Kaiser, U.: The role of stacking faults and their associated 0.13 eV acceptor state in doped and undoped ZnO layers and nanostructures. *Phys. Stat. Sol. (b)*. 2010. 247. 1464. Copyright Wiley-VCH Verlag GmbH & Co. KGaA. With permission.)

acceptors might be some intrinsic defect such as Zn vacancies especially formed in high concentration around the stacking faults (Travlos et al. 2009).

A very sharp PL line at 3.333 eV presumably related to structural defects was reported to occur in ZnO bulk samples. It is partially superimposed on the TES of the shallow (D^0,X) lines (see Figure 4.8, line marked by "Y"). Alves et al. (2003) conclude from their CL measurements that it must be correlated with structural defects and show features similar to the "Y"-line (excitons bound to structural defects) commonly seen in ZnSe and ZnTe (Meyer et al. 2004). From our own magneto-PL measurements, we find exactly the same splitting behavior as for shallow donor-bound excitons (D^0,X), classifying this transition as bound exciton-like despite of its low energy.

4.3.3.6 Surface-Related Emissions

When discussing luminescence from ZnO, especially from nanocrystalline ZnO samples (see Section 4.1.3 and Figure 4.15), often a strong influence of the semiconductor surface is reported. To unambiguously prove a surface origin of any luminescence band, in principle, a change of surface properties (chemistry or band bending) should be accomplished. Indeed, a chemical change of the ZnO surface can already be found when exposing it to ambient atmosphere after cleaving under controlled conditions in high vacuum. Travnikov et al. (1990) showed the origin of a broad band centered around 3.367 eV— which is higher in energy than the main part of bound exciton emissions—to be surface contamination related. The same band was later investigated in detail by further groups (Grabowska et al. 2005; Wischmeier et al. 2006), presenting elaborate models for the surface decay mechanism. The emission is interpreted to be of excitonic origin, i.e., excitons bound to the ZnO surface itself or to point defects located in close proximity of the surface. It is interesting to note that the same ultraviolet surface emission of ZnO at around 3.367 eV dominates spectra from many different solids cooled down to liquid helium temperature (Savikhin and Freiberg 1993). This suggests that ZnO is a frequent surface contaminant of many solids.

Further luminescence bands reported to be surface related are located at 3.31 eV (Fallert et al. 2007). A correlation of average grain size of ZnO powder with luminescence intensity leads to this conclusion. However, care should be taken in distinguishing the band from the contribution related to BSFs (see Section 4.3.3.4). Therefore, the existence of a surface band centered at 3.31 eV remains questionable.

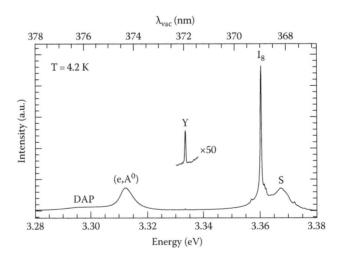

FIGURE 4.15 Typical low-temperature PL spectrum from a sample of ZnO nano-pillars. The dominant donor recombination is the I_8a line, while the surface band labeled (S) centered at 3.367 eV is also visible. The DAP recombination around 3.3 eV and its accompanying acceptor-band transition (e,A^0) at 3.312 eV are related to basal plane stacking faults (see Section 5.3.3.4).

Finally, also the influence of the surface properties on the mid-gap luminescence is discussed in the literature. As examples, two different approaches shall be presented briefly here. Shalish et al. (2004) discuss the size effect of ZnO nanowires on the intensity ratio of near-band-edge and deep luminescence contributions. Interestingly, already in room temperature PL spectra of nanowire ensembles without very uniform geometry or especially low impurity concentrations, they find a linear relationship between nanowire dimension and intensity ratio. This also holds for wires with diameters up to 300 nm, which is orders of magnitudes larger than any upper diameter limit of about 2 nm, for which quantum-size effects in ZnO could be expected. In a different approach, Richters et al. (2008) performed studies where ensembles of nanowires were covered by organic coatings altering again the intensity ratio of mid-gap to band-edge luminescence drastically. They performed their luminescence experiments at 5 K also allowing for the observation of the surface band centered at 3.367 eV as described above.

Further studies are inevitable to uncover the influence of the surface on different luminescence mechanisms in ZnO. Up to now, only empirical relationships are available. But, the emerging attraction of the ZnO surface luminescence bands—monitored by an increasing number of publications on this field—will hopefully soon lead to more specific results like chemical identification of related contaminants or an agreement on recombination models involving the surface. This research will of course take advantage from the improving material quality.

4.3.3.7 Confinement Effects: Quantum Wells, Wires, and Dots

4.3.3.7.1 *Quantum Wells Based on Zn(Cd,Mg)O*

To make ZnO usable for optoelectronic devices, successful bandgap engineering is a prerequisite. The II–VI material system Zn(Cd,Mg)O allows in principle to reach energies from 1.7 eV (CdO) to about 4.5 eV (MgO). However, as already stated in Section 4.1.3, both CdZnO and MgZnO crystallize for higher Cd or Mg content in the rocksalt structure making their wurtzite crystal type incorporation into ZnO a demanding task for growth. To obtain confinement effects and tailored quantized electronic states by variation of the thickness of QWs or the diameter of 1D or 0D objects, the shortest dimension has to be smaller than the exciton radius $r_{B,X}$. This radius can be estimated from a scaled hydrogen model to

$$r_{B,X} = r_{B,H} \varepsilon_r \frac{m_0}{\mu} \qquad (4.5)$$

with the exciton reduced mass $\mu = m_e m_h/(m_e + m_h)$, $r_{B,H}$ the hydrogen Bohr radius, and the relative dielectric constant ε_r. Using the values $m_e = 0.24 m_0$, $m_h = 0.55 m_0$, and for the relative dielectric constant either the high frequency limit $\varepsilon_r = 3.7$ or the static limit $\varepsilon_r = 8$, we estimate an exciton radius of $r_{B,X} \approx 1.2$–2.5 nm. This range of values is much smaller than for any of the commonly used other technologically relevant semiconductors and requires that QWs—especially from ZnO/ZnMgO—have to be extremely thin with very well-defined smooth side walls. Further complication arises from the high piezoelectric response of ZnO (Dal Corso et al. 1994; Malashevich and Vanderbilt 2007), which leads to spontaneous and strain-induced piezo fields (Brandt et al. 2010), similarly to the well-known case of InGaN on GaN (Takeuchi et al. 2000).

Several reports on ZnCdO QW structures are available representing the growing interest in and continuously increasing quality of such heterostructures. Already in 2001, Makino et al. (2001) succeeded in incorporating up to ~7% of Cd into the ZnO matrix as thin layers (Makino et al. 2001). However, they only could show emission for a CdO fraction of 1% consisting of a broad band at around 3.34 eV, which is interestingly close to the emission band related to BSFs. More recently, Matsui and coworkers presented luminescence shifting from 3.035 to 2.73 eV with different QW widths, while maintaining the Cd incorporation at about 8% (Matsui et al. 2010). Lange et al. (2010) showed a [Cd]-dependent energy shift from about 3.056 to 3.19 eV for some 10 nm wide heterostructures. They estimate their CdO content to be in the range from 2.1% to 4.6%. Finally, Sadofev et al. (2007) were able to demonstrate optically pumped lasing from ZnCdO/ZnO QW structures. Their maximum CdO content is about 16% leading to laser emission at around 2.55 eV. These examples show both the great potential as well as the necessity of improvements in ZnCdO QW structures.

QWs made of ZnO as active layer and ZnMgO as barriers are leading to shorter wavelength emission, i.e., higher photon energies. QWs that are too wide lead to emission from the ZnO bandgap energy only—eventually red-shifted by the quantum confined stark effect. Therefore, the QW widths have to be well below 3 nm to have any quantum confinement effect visible in optical spectra. Thus, the design of emitters today is still limited to a narrow window of thin QWs and barriers with low Mg content and concomitantly to a narrow energy range from approximately 3.37 eV (bulk ZnO) to about 3.8 eV. The latter value can be reached by ultrathin QWs of ZnO (1 nm) and barriers containing about 37% of MgO (Gruber et al. 2004; Sadofev et al. 2005; Misra et al. 2006; Gu et al. 2007; Ashrafi and Segawa 2008; Prinz 2008; Lorenz et al. 2009).

A different bandgap engineering approach makes use of the rather toxic element beryllium substituting for Mg leading to an alloy system of BeZnO (Ryu et al. 2006b) with the upper end of BeO having a bandgap energy of about 10.5 eV. The same group has presented controversially discussed results on QW devices including p-type ZnO (Ryu et al. 2006, 2007). No new experimental results were presented to confirm the earlier claims. However, researchers have begun to investigate even mixed quarternary Zn(Be,Mg)O alloys (Yang et al. 2008b).

Confinement effects in ZnO quantum dots were reported very early. Mostly for colloidal particles grown by wet chemistry, a size-dependent shift of the absorption edge was reported for particles with diameters below 2.5 nm in a suspension (Koch et al. 1985) or for particles with diameters up to 7 nm (Meulenkamp 1998). A general problem is that in ZnO particles produced by wet chemistry, different ions used in the growth often get absorbed on the surface together with a hydroxylated layer introducing surface states and band bending. In PL measurements, mostly the green defect band with long lifetime ($\tau \approx 1\,\mu s$) is dominant (Monticone et al. 1998). Wong and Searson find in their room temperature PL measurements a correlated shift of the bandgap-near emission and the green band with changing particle size (Wong and Searson 1999). A shift of the green band with different Na and Li doping was reported by Haranath et al. (2008). In ZnO nanoparticles with diameters from 3.9 to 6.4 nm prepared from metal–organic precursors, Wang et al. (2006b) find a diameter-dependent shift in their absorption, PL, and PL-excitation spectra. Yanhong et al. (2004) report the influence of confinement effects on the surface photovoltage of ZnO dots with 3 nm size.

For quantum wires with 2.2 nm diameter, Gu et al. (2004) observe a slightly shifted excitonic gap in absorption and mainly broad, overlapping bands in the visible region in RT PL. Kim et al. (2008c) observe from short ZnO rods with 5 nm diameter dominant UV PL at room temperature, which is slightly upshifted as compared to the bulk emission. For ZnO wires with 4 nm diameter, Stichtenoth et al. (2007) report a PL peak at 3.327 eV and interpret it as a LO replica of an upshifted exciton, for which the no-phonon transition remains invisible for unknown reasons. Presumably, a defect-related (e, A^0) transition (see Section 4.3.3.4) might be a more natural explanation. Generally, model calculations expect an increase of the exciton binding energy by a factor of up to four for narrow wires, which means a drastic compensation of the confinement effect (Zhang and Mascarenhas 1999; Pedersen 2005).

4.4 Nano-Resonators and Nano-Lasers

Since ZnO nano-pillars and platelets can be grown with high crystalline perfection and well-defined hexagonal side facets, these structures readily form optical resonators for the UV near-bandgap and visible emission. Under higher pumping, these ZnO pillars hereby can act simultaneously as both resonators and active optical gain media giving rise to laser emission.

Some examples for resonator-specific optical features observed in experiments are the so-called whispering gallery modes, for which multiples of the emitted wavelength must fit into a closed path in the

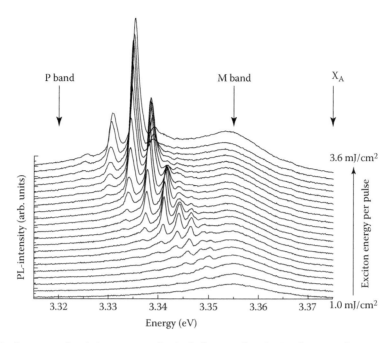

FIGURE 4.16 Sequence of emission spectra of a single free-standing ZnO pillar excited at ~10 K with increasing density by a pulsed frequency doubled Ti:sapphire laser ($\lambda = 355$ nm, $\tau = 150$ fs). First, the biexciton band ("M") shows up, and then lasing modes at lower energy are observed. The number of modes decreases with increasing excitation. "X_A" marks the position of the free exciton and "P" is the exciton–exciton scattering band. (With kind permission from Springer Science+Business Media: *Adv. Solid State Phys.*, ZnO nanostructures: Optical resonators and lasing, 48, 2009, 39, Thonke, K., Reiser, A., Schirra, M., Feneberg, M., Prinz, G.M., Röder, T., Sauer, R., Fallert, J., Stelzl, F., Kalt, H., Gsell, S., Schreck, M., and Stritzker, B., Copyright Springer.)

hexagonal cross section of the ZnO pillar (Nobis and Grundmann 2005; Sun et al. 2009) or platelet (Kim et al. 2009). For standing polariton waves in nanoresonators, very high-Rabi splitting between the polariton branches of 50–100 meV was reported (van Vugt et al. 2006; Sturm et al. 2009). For much smaller ZnO pillar diameters around 130 nm, only a few resonator modes are left. The coupling of two of the fundamental modes was shown to lead to standing waves with spatially resolvable beat frequency in cathodoluminescence measurements (Schirra et al. 2009).

The first example of room temperature lasing in ZnO nanowires was shown in 2001 (Huang et al. 2001). Lasing can only occur when the major part of the optical mode fits into the hexagonal pillar, which can be expected for diameters above ~150 nm (Johnson et al. 2003; Hauschild et al. 2006b). Time-resolved PL studies of singular, free standing, well-faceted ZnO pillars revealed that the main optical gain mechanism is due to an EHP (Fallert et al. 2008) (Figure 4.16). Also, other nanostructures such as nanoribbons (Yan et al. 2003) and tetrapods (Leung et al. 2005; Szarko et al. 2005) were demonstrated to show laser activity. For larger diameters in the range of several micrometers, lasing on whispering gallery modes occurs (Czekalla et al. 2008).

4.5 Applications

One of the most promising applications of the high-bandgap semiconductor ZnO is for blue or UV LEDs. Due to the p-type doping problem, some of the early attempts to produce LEDs were based on the concept of n-ZnO/p-GaN or n-ZnO/p-AlGaN heterostructures, since for these materials the lattice mismatch is only 1.8% (Alivov et al. 2003b). A problem with this concept is that the bands of GaN and ZnO are not aligned, with the ZnO bands being offset to much lower absolute energy, such that only ~1 eV is left as bandgap for spatially indirect band-to-band transitions (van de Walle and Neugebauer 2003). The emitted EL peaking at ~390–430 nm was mainly produced in the p-GaN layer (Alivov et al. 2003a). By adding an intrinsic ZnO interlayer, Liang et al. (2010) could obtain EL from this ZnO region at 387 nm. More complicated triple n-MgZnO/n-ZnO/p-AlGaN/p-GaN heterostructures designed to create high hole density at the p-ZnO/n-AlGaN-interface also showed EL at ~390 nm from the ZnO region and good rectifying characteristics (Osinsky et al. 2004). Another concept uses an intrinsic ZnO layer between n-type ZnO and a metal contact (metal–insulator–semiconductor [MISFET] diode) (Hwang et al. 2007).

During the last few years, homojunction ZnO LEDs were reported (Aoki et al. 2000; Guo et al. 2001; Jiao et al. 2006; Lim et al. 2006). MBE grown homojunction-LEDs on ScAlMgO$_4$ showed high series resistance, high forward voltages, and relatively long wavelength emission at 440 nm (Tsukazaki et al. 2005b). MOCVD grown LEDs showed better I(V) characteristics after annealing and an emission at 384 nm ascribed to DAP transitions on the p-side. Recent MBE grown ZnO:Ga/ZnO:Sb homojunctions on c-plane sapphire show ~32 nW output power at 60 mA driving current with a 384 nm emission peak at room temperature (Yang et al. 2010, Figure 4.17). Generally, so far, the EL observed from hetero- or homojunction devices is relatively weak, and efficiencies are far below any useful level.

Ryu et al. (2006) produced multiple QW structures using BeZnO barriers. The natural choice would be to use MgZnO barriers in heterostructures, e.g., as reported by Wang et al. (2008). Instead of layered structures, ZnO nanorods embedded in polymer were tested as active element in LEDs (Yang et al. 2008, Willander et al. 2009). Room temperature optically pumped lasing was reported in 1997 for bulk ZnO (Bagnall et al. 1997). Optically pumped third-order distributed feedback lasers were made by dry etching from ZnO/SiN$_x$ layers and yielded at the 383 nm center wavelength up to 14 mW output power under 1.1 MW cm^{-2} pumping at 220 K, and the threshold was estimated to be around 120 kW cm^{-2} (Hofstetter et al. 2007, 2008). Also, inverse opals made from ZnO forming photonic crystals showed lasing activity under optical pumping (Scharrer et al. 2006).

The observation of lasing action not only by optical pumping but also by electrical pumping of their multiple QWs embedded in BeZnO barriers was claimed by Ryu et al. (2007), although the EL spectra were extremely noisy. Similar EL spectra were obtained by Zhu et al. (2009) from their columnar

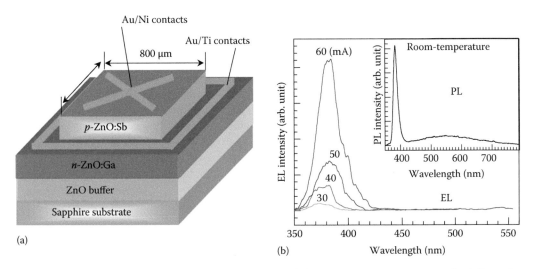

FIGURE 4.17 (a) Schematic of a ZnO:Sb/ZnO:GaP n-LED and (b) room-temperature EL spectra for different forward currents. Inset: PL recorded at room temperature. (From Yang, Z., Chu, S., Chen, W.V., Li, L., Kong, J., Ren, J., Yu, P.K.L., and Liu, J., ZnO:Sb/ZnO:Ga Light emitting diode on c-plane sapphire by molecular beam epitaxy, *Appl. Phys. Exp.*, 3, 032101. With permission of Japanese Society of Applied Physics.)

n-ZnO/p-GaN layers with interfacial MgO electron blocking barrier. Recently, Chu et al. (2008) reported on room temperature lasing of their ZnO:Ga/MgZnO/ZnO/MgZnO/ZnO:Sb pn-Diode with embedded ZnO QW grown on Si. The structure also showed columnar morphology, good rectifying behavior, but no clear mode structure, which was ascribed to random lasing in grains.

The "inverse" optoelectronic application of ZnO is the use in "solar blind" UV detectors. Ryu et al. (2005) produced such pn diode devices with ZnO:As for the p-side and nominally undoped ZnO for the n-side. Several other groups investigated different kinds of hetero-photodiodes, e.g., n-ZnO/p-SiC mesas (Alivov et al. 2005). Alternatively, metal–semiconductor–metal (MSM) photodetectors were also shown to give solar-blind UV response—but sometimes with long response and decay times (Liang et al. 2001; Zheng et al. 2006; Du et al. 2009). Quantum efficiencies around 10% were quoted (Young et al. 2006). Also, ZnO nanowires instead of bulk layers showed a pronounced persistent conductivity effect (Hullavarad et al. 2009), which might be due to oxygen vacancies (Lany and Zunger 2005).

One commercially existing application for ZnO is the use as TCO. Here, typically highly Al-doped ZnO serves as transparent, highly conductive front contact in photovoltaic thin film devices and flat panel displays. It replaces the much more expensive and toxic indium-tin oxide (ITO), although the presently achievable conductivities are typically lower by a factor of two.

Transparent thin-film transistors (TTFTs) are another field of research for applications mainly as drivers in active-matrix liquid-crystal displays, replacing the polysilicon transistors. The ZnO TTFTs can be produced in polycrystalline form on glass substrate even at room temperature and show good switching behavior (Hoffman et al. 2003; Masuda et al. 2003; Fortunato et al. 2004; Lim et al. 2007). For higher channel mobilities, instead of the MISFET concept also metal–semiconductor FETs (MESFETs) are investigated (Frenzel et al. 2008).

Since the ZnO surface easily reacts with different chemical groups like OH resulting in a drastic change of the surface conductivity and Fermi level, applications in gas sensors—especially in the form of nanorods with their large surface—are a broad field of research (Xu et al. 2000; Wan et al. 2004; Li et al. 2010; Park et al. 2010b). The sharp tips of ZnO nanopillars enhance the local electric field and thus can act as cold field emitters for increased electron emission with low turn-on field (Lee et al. 2002; Wan et al. 2003; Zhu et al. 2003; Ye et al. 2007; Chen et al. 2009).

Wang and Song (2006) and Qin et al. (2008) suggested using ZnO nanopillars as converters for mechanical energy into electrical energy: a top metal contact making a Schottky-type diode should rectify the surface charges induced by bending the wires. Meanwhile, ZnO nanostructures were implanted by these authors into a living rat and yielded—driven by the heart movement—a few pA of AC current and mV of AC voltage (Li et al. 2010b). Higher voltages could be obtained by forming arrays of some 300,000 relatively long ZnO nanowires on a substrate (Zhu et al. 2010). Other authors raised some fundamental skepticism about the validity of this concept of "energy harvesting" (Alexe et al. 2008).

Numerous studies have been published on MOSFET devices made from single ZnO nanowires. Mostly, the wires are transferred from the substrate used for growth to some other support with either predefined metal contacts or contacts written selectively after deposition (Arnold et al. 2003; Park et al. 2004; LaRoche et al. 2005). Good pinch-off and saturation characteristics were obtained. These FETs are sensitive to light and sensitive to a change of the oxygen content in the ambient which causes a change in the surface depletion layer thickness (Fan et al. 2004). By coating single-rod FETs with polyimide, Park et al. (2005) got an improved on/off ratio of 10^5, a transconductance of $1.9\,\mu S$, and an electron mobility of $1000\,cm^2/Vs$.

4.6 Further Reading

During the last years, several review articles on ZnO have been published. For a rather comprehensive review on all aspects from growth to applications, see Özgür et al. (2005). Meyer et al. (2004) have collected a lot of optical data. Klingshirn et al. (2006) review excitonic properties and broader aspects (Klingshirn 2007a,b). Look (2005) has published a review on optical and electrical data of p-type ZnO.

For defects, a lot of information was recently collected by McCluskey and Jokela (2009). For a theorist's view on the doping problem, see Janotti and van de Walle (2009). Especially on nanostructures, Wang (2004b) collected valuable information in his review articles. A survey on thin films, growth, doping, and LEDs can be found in (Hwang et al. 2007b). Recent results on the growth of nanowires and devices made from these are collected in Heo et al. (2004) and Willander et al. (2009).

Acknowledgments

The authors thank A. Schildknecht and B. Neuschl for several data sets of optical measurements on ZnO, which entered here in Figures 4.3, 4.6 and 4.7. Thilo Zoberbier contributed Figure 4.1. Some of the ZnO nanostructures in Figure 4.4 were grown by Anton Reiser and Manfred Madel (University of Ulm). We are indebted to R. Helbig for the loan of several high-quality ZnO crystals. The data on lasing of single ZnO pillars were recorded by J. Fallert and H. Kalt (University of Karlsruhe). The authors thank Christopher T. Littler (University of North Texas) for helpful comments on this manuscript.

References

Alexe, M.; Senz, S.; Schubert, M. A.; Hesse, D.; and Gösele, U., 2008, Energy harvesting using nanowires?, *Advanced Materials 20*: 4021–4026.

Alivov, Y. I.; Kalinina, E. V.; Cherenkov, A. E.; Look, D. C.; Ataev, B. M.; Omaev, A. K.; Chukichev, M. V.; and Bagnall, D. M., 2003a, Fabrication and characterization of n-ZnO/p-AlGaN heterojunction light-emitting diodes on 6H-SiC substrates, *Applied Physics Letters, 83*: 4719–4721.

Alivov, Y. I.; Nostrand, J. E. V.; Look, D. C.; Chukichev, M. V.; and Ataev, B. M., 2003b, Observation of 430 nm electroluminescence from ZnO/GaN heterojunction light-emitting diodes, *Applied Physics Letters 83*: 2943–2945.

Alivov, Y. I.; Özgür, U.; Doğan, S.; Johnstone, D.; Avrutin, V.; Onojima, N.; Liu, C.; Xie, J.; Fan, Q.; and Morkoç, H., 2005, Photoresponse of n-ZnO/p-SiC heterojunction diodes grown by plasma-assisted molecular-beam epitaxy, *Applied Physics Letters 86*: 241108.

Allenic, A.; Guo, W.; Chen, Y.; Katz, M.; Zhao, G.; Che, Y.; Hu, Z.; Liu, B.; Zhang, S.; and Pan, X., 2007, Amphoteric phosphorus doping for stable p-type ZnO, *Advanced Materials 19*: 3333–3337.

Alves, H.; Pfisterer, D.; Zeuner, A.; Riemann, T.; Christen, J.; Hofmann, D. M.; and Meyer, B. K., 2003, Optical investigations on excitons bound to impurities and dislocations in ZnO, *Optical Materials, Proceedings of the 8th International Conference on Electronic Materials, IUMRS-ICEM 200 23*: 33–37.

Ando, K.; Saito, H.; Jin, Z.; Fukumura, T.; Kawasaki, M.; Matsumoto, Y.; and Koinuma, H., 2001, Magneto-optical properties of ZnO-based diluted magnetic semiconductors, *Journal of Applied Physics 89*: 7284–7286.

Andress, B. and Mollwo, E., 1959. Zur Kantenlumineszenz des Zinkoxyds, *Naturwissenschaften 46*: 623–624.

Aoki, T.; Hatanaka, Y.; and Look, D. C., 2000, ZnO diode fabricated by excimer-laser doping, *Applied Physics Letters 76*: 3257–3258.

Arnold, M. S.; Avouris, P.; Pan, Z. W.; and Wang, Z. L., 2003, Field-effect transistors based on single semi-conducting oxide nanobelts, *The Journal of Physical Chemistry B 107*: 659–663.

Ashida, A.; Fujita, A.; Shim, Y.; Wakita, K.; and Nakahira, A., 2008, ZnO thin films epitaxially grown by electrochemical deposition method with constant current, *Thin Solid Films 517*: 1461–1464.

Ashrafi, A. and Segawa, Y., 2008, Blueshift in Mg[sub x]Zn[sub 1 − x]O alloys: Nature of bandgap bowing, *Journal of Applied Physics 104*: 123528.

Azamat, D. V.; Debus, J.; Yakovlev, D. R.; Ivanov, V. Y.; Godlewski, M.; Fanciulli, M.; and Bayer, M., 2010, Photo-EPR and magneto-optical spectroscopy of iron centres in ZnO, *Physica Status Solidi (b) 247*: 1517–1520.

Aziz, M.; Otoi, S.; Rusop, M.; and Soga, T. (eds.), 2009, Preparation and luminescence properties of rare earth doped nanostructured zinc oxide thin films by sol gel technique, *Nanoscience and Nanotechnology: International Conference on Nanoscience and Nanotechnology 2008, AIP 1136:* 207–213.

Baer, W. S., 1967, Faraday rotation in ZnO: Determination of the electron effective mass, *Physical Review 154:* 785–789.

Bagnall, D. M.; Chen, Y. F.; Zhu, Z.; Yao, T.; Koyama, S.; Shen, M. Y.; and Goto, T., 1997, Optically pumped lasing of ZnO at room temperature, *Applied Physics Letters 70:* 2230–2232.

Bandyopadhyay, A.; Modak, S.; Acharya, S.; Deb, A.; and Chakrabarti, P., 2010, Microstructural, magnetic and crystal field investigations of nanocrystalline Dy3+ doped zinc oxide, *Solid State Sciences 12:* 448–454.

Barnes, T. M.; Olson, K.; and Wolden, C. A., 2005, On the formation and stability of p-type conductivity in nitrogen-doped zinc oxide, *Applied Physics Letters 86:* 112112.

Bertram, F.; Christen, J.; Dadgar, A.; and Krost, A., 2007, Complex excitonic recombination kinetics in ZnO: Capture, relaxation, and recombination from steady state, *Applied Physics Letters 90:* 041917.

Bhushan, S.; Pandey, A. N.; and Kaza, B. R., 1979, Photo- and electroluminescence of undoped and rare earth doped ZnO electroluminors, *Journal of Luminescence 20:* 29–38.

Blattner, G.; Kurtze, G.; Schmieder, G.; and Klingshirn, C., 1982, Influence of magnetic fields up to 20 T on excitons and polaritons in CdS and ZnO, *Physical Review B 25:* 7413–7427.

Block, D.; Hervé, A.; and Cox, R. T., 1982, Optically detected magnetic resonance and optically detected ENDOR of shallow indium donors in ZnO, *Physical Review B 25:* 6049.

Boemare, C.; Monteiro, T.; Soares, M. J.; Guilherme, J. G.; and Alves, E., 2001, Photoluminescence studies in ZnO samples, *Physica B: Condensed Matter 308–310:* 985–988.

Børseth, T. M.; Svensson, B. G.; Kuznetsov, A. Y.; Klason, P.; Zhao, Q. X.; and Willander, M., 2006, Identification of oxygen and zinc vacancy optical signals in ZnO, *Applied Physics Letters 89:* 262112.

Brandt, M.; Lange, M.; Stölzel, M.; Müller, A.; Benndorf, G.; Zippel, J.; Lenzner, J.; Lorenz, M.; and Grundmann, M., 2010, Control of interface abruptness of polar MgZnO/ZnO quantum wells grown by pulsed laser deposition, *Applied Physics Letters 97:* 052101.

Cao, B. Q.; Lorenz, M.; Rahm, A.; von Wenckstern, H.; Czekalla, C.; Lenzner, J.; Benndorf, G.; and Grundmann, M., 2007, Phosphorus acceptor doped ZnO nanowires prepared by pulsed-laser deposition, *Nanotechnology 18:* 455707.

Che, P.; Meng, J.; and Guo, L., 2007, Oriented growth and luminescence of ZnO: Eu films prepared by sol-gel process, *Journal of Luminescence 122–123:* 168–171.

Chen, Y.; Bagnall, D. M.; Zhu, Z.; Sekiuchi, T.; Park, K.-t.; Hiraga, K.; Yao, T.; Koyama, S.; Shen, M. Y.; and Goto, T., 1997, Growth of ZnO single crystal thin films on c-plane (0 0 0 1) sapphire by plasma enhanced molecular beam epitaxy, *Journal of Crystal Growth 181:* 165–169.

Chen, Z. H.; Tang, Y. B.; Liu, Y.; Yuan, G. D.; Zhang, W. F.; Zapien, J. A.; Bello, I.; Zhang, W. J.; Lee, C. S.; and Lee, S. T., 2009, ZnO nanowire arrays grown on Al:ZnO buffer layers and their enhanced electron field emission, *Journal of Applied Physics 106:* 064303.

Chen, L. L.; Ye, Z. Z.; Lu, J. G.; and Chu, P. K., 2006, Control and improvement of p-type conductivity in indium and nitrogen codoped ZnO thin films, *Applied Physics Letters 89:* 252113.

Cheng, X. M. and Chien, C. L., 2003, Magnetic properties of epitaxial Mn-doped ZnO thin films, *Journal of Applied Physics 93:* 7876–7878.

Cheng, B.; Xiao, Y.; Wu, G.; and Zhang, L., 2004, Controlled growth and properties of one-dimensional ZnO nanostructures with Ce as activator/dopant, *Advanced Functional Materials 14:* 913–919.

Chu, S.; Olmedo, M.; Yang, Z.; Kong, J.; and Liu, J., 2008, Electrically pumped ultraviolet ZnO diode lasers on Si, *Applied Physics Letters 93:* 181106.

Cox, R. T.; Block, D.; Hervé, A.; Picard, R.; Santier, C.; and Helbig, R., 1978, Exchange broadened, optically detected ESR spectra for luminescent donor-acceptor pairs in Li doped ZnO, *Solid State Communications 25:* 77–80.

Cui, J. B.; Thomas, M. A.; Soo, Y. C.; Kandel, H.; and Chen, T. P., 2009, Effects of nitrogen on the growth and optical properties of ZnO thin films grown by pulsed laser deposition, *Journal of Physics D: Applied Physics 42:* 155407.

Czekalla, C.; Sturm, C.; Schmidt-Grund, R.; Cao, B.; Lorenz, M.; and Grundmann, M., 2008, Whispering gallery mode lasing in zinc oxide microwires, *Applied Physics Letters 92:* 241102.

Dal Corso, A.; Posternak, M.; Resta, R.; and Baldereschi, A., 1994, Ab initio study of piezoelectricity and spontaneous polarization in ZnO, *Physical Review B 50:* 10715–10721.

Desgreniers, S., 1998. High-density phases of ZnO: Structural and compressive parameters *Physical Review B 58:* 14102–14105.

Dietz, R. E.; Kamimura, H.; Sturge, M. D.; and Yariv, A., 1963, Electronic structure of copper impurities in ZnO, *Physical Review 132:* 1559–1569.

Ding, L.; Yang, C. L.; He, H. T.; Jiang, F. Y.; Wang, J. N.; Tang, Z. K.; Foreman, B. A.; and Ge, W. K., 2007, Unambiguous symmetry assignment for the top valence band of ZnO by magneto-optical studies of the free A-exciton state, e-print arXiv:0706.3965.

Dinges, R.; Fröhlich, D.; Staginnus, B.; and Staude, W., 1970, Two-photon magnetoabsorption in ZnO, *Physical Review Letters 25:* 922–924.

Dingle, R., 1969, Luminescent transitions associated with divalent copper impurities and the green emission from semiconducting zinc oxide, *Physical Review Letters 23:* 579–581.

Du, X.; Mei, Z.; Liu, Z.; Guo, Y.; Zhang, T.; Hou, Y.; Zhang, Z.; Xue, Q.; and Kuznetsov, A. Y., 2009, Controlled growth of high-quality ZnO-based films and fabrication of visible-blind and solar-blind ultra-violet detectors, *Advanced Materials 21:* 4625–4630.

Ehrentraut, D.; Sato, H.; Miyamoto, M.; Fukuda, T.; Nikl, M.; Maeda, K.; and Niikura, I., 2006, Fabrication of homoepitaxial ZnO films by low-temperature liquid-phase epitaxy, *Journal of Crystal Growth 287:* 367–371.

Fallert, J.; Hauschild, R.; Stelzl, F.; Urban, A.; Wissinger, M.; Zhou, H.; Klingshirn, C.; and Kalt, H., 2007, Surface-state related luminescence in ZnO nanocrystals, *Journal of Applied Physics 101:* 073506.

Fallert, J.; Stelzl, F.; Zhou, H.; Reiser, A.; Thonke, K.; Sauer, R.; Klingshirn, C.; and Kalt, H., 2008, Lasing dynamics in single ZnO nanorods, *Optics Express 16:* 1125–1131.

Fan, Z.; Wang, D.; Chang, P.-C.; Tseng, W.-Y.; and Lu, J. G., 2004, ZnO nanowire field-effect transistor and oxygen sensing property, *Applied Physics Letters 85:* 5923–5925.

Fan, J. C.; Zhu, C. Y.; Fung, S.; Zhong, Y. C.; Wong, K. S.; Xie, Z.; Brauer, G.; Anwand, W.; Skorupa, W.; To, C. K.; Yang, B.; Beling, C. D.; and Ling, C. C., 2009, Arsenic doped p-type zinc oxide films grown by radio frequency magnetron sputtering, *Journal of Applied Physics, 106:* 073709.

Fortunato, E. M. C.; Barquinha, P. M. C.; Pimentel, A. C. M. B. G.; Gonçalves, A. M. F.; Marques, A. J. S.; Martins, R. F. P.; and Pereira, L. M., 2004, Wide-bandgap high-mobility ZnO thin-film transistors produced at room temperature, *Applied Physics Letters 85:* 2541–2543.

Frenzel, H.; Lajn, A.; Brandt, M.; von Wenckstern, H.; Biehne, G.; Hochmuth, H.; Lorenz, M.; and Grundmann, M., 2008, ZnO metal-semiconductor field-effect transistors with Ag-Schottky gates, *Applied Physics Letters 92:* 192108.

Fukumura, T.; Jin, Z.; Ohtomo, A.; Koinuma, H.; and Kawasaki, M., 1999, An oxide-diluted magnetic semiconductor: Mn-doped ZnO, *Applied Physics Letters 75:* 3366–3368.

Fukumura, T.; Yamada, Y.; Toyosaki, H.; Hasegawa, T.; Koinuma, H.; and Kawasaki, M., 2004, Exploration of oxide-based diluted magnetic semiconductors toward transparent spintronics, *Applied Surface Science 223:* 62–67.

Gao, P. and Wang, Z., 2003, Mesoporous polyhedral cages and shells formed by textured self-assembly of ZnO nanocrystals, *Journal of the American Chemical Society 125:* 11299–11305.

Gao, P. X. and Wang, Z. L., 2004, Nanopropeller arrays of zinc oxide, *Applied Physics Letters 84:* 2883–2885.

Grabowska, J.; Meaney, A.; Nanda, K.K.; Mosnier, J.-P.; Henry, M.O.; Duclère, J.-R.; and McGlynn, E., 2005, Surface excitonic emission and quenching effects in ZnO nanowire/nanowall systems: Limiting effects on device potential, *Physical Review B 71:* 115439.

Gruber, T.; Kirchner, C.; Kling, R.; Reuss, F.; and Waag, A., 2004, ZnMgO epilayers and ZnO–ZnMgO quantum wells for optoelectronic applications in the blue and UV spectral region, *Applied Physics Letters 84*: 5359–5361.

Gruber, T.; Kirchner, C.; and Waag, A., 2002, MOCVD Growth of ZnO on different substrate materials, *Physica Status Solidi (b) 229*: 841–844.

Gu, Y., Kuskovsky, I.L., Yin, M., O'Brien, S.; and Neumark, G.F., 2004, Quantum confinement in ZnO nanorods, *Applied Physics Letters 85*: 3833–3835.

Gu, X. Q.; Zhu, L. P.; Ye, Z. Z.; He, H. P.; Zhang, Y. Z.; Huang, F.; Qiu, M. X.; Zeng, Y. J.; Liu, F.; and Jaeger, W., 2007, Room-temperature photoluminescence from ZnO/ZnMgO multiple quantum wells grown on Si(111) substrates, *Applied Physics Letters 91*: 022103.

Guo, X.-L.; Choi, J.-H.; Tabata, H.; and Kawai, T., 2001, Fabrication and optoelectronic properties of a transparentZnO homostructural light-emitting diode, *Japanese Journal of Applied Physics, The Japan Society of Applied Physics 40*: L177–L180.

Gutowski, J., Presser, N.; and Broser, I., 1988, Acceptor-exciton complexes in ZnO: A comprehensive analysis of their electronic states by high-resolution magnetooptics and excitation spectroscopy, *Physical Review B 38*: 9746–9758.

Haranath, D.; Sahai, S.; and Joshi, P., 2008, Tuning of emission colors in zinc oxide quantum dots, *Applied Physics Letters 92*: 233113.

Hauschild, R.; Lange, H.; Priller, H.; Klingshirn, C.; Kling, R.; Waag, A.; Fan, H. J.; Zacharias, M.; and Kalt, H., 2006b, Stimulated emission from ZnO nanorods, *Physica Status Solidi (b) 243*: 853–857.

Hauschild, R.; Priller, H.; Decker, M.; Brückner, J.; Kalt, H.; and Klingshirn, C., 2006, Temperature dependent band gap and homogeneous line broadening of the exciton emission in ZnO, *Physica Status Solidi (c) 3*: 976–979.

Haynes, J.R., 1960, Experimental proof of the existence of a new electronic complex in Silicon, *Physical Review Letters 4*: 361–363.

Heiland, G., 1954, Zum Einfluß von adsorbiertem Sauerstoff auf die elektrische Leitfähigkeit von Zinkoxydkristallen, *Zeitschrift für Physik A 138*: 459–464.

Heinze, S.; Krtschil, A.; Bläsing, J.; Hempel, T.; Veit, P.; Dadgar, A.; Christen, J.; and Krost, A., 2007, Homoepitaxial growth of ZnO by metalorganic vapor phase epitaxy in two-dimensional growth mode, *Journal of Crystal Growth 308*: 170–175.

Heitz, R.; Hoffmann, A.; and Broser, I., 1992, Fe^{3+} center in ZnO, *Physical Review B 45*: 8977–8988.

Heitz, R.; Hoffmann, A.; Hausmann, B.; and Broser, I., 2002, Zeeman spectroscopy of V^{3+}-luminescence in ZnO, *Journal of Luminescence 48–49*: 689–692.

Helbig, R., 1972, Über die Züchtung von grösseren reinen und dotierten ZnO-Kristallen aus der Gasphase, *Journal of Crystal Growth 15*: 25–3.

Heo, Y.; Norton, D.; Tien, L.; Kwon, Y.; Kang, B.; Ren, F.; Pearton, S.; and LaRoche, J., 2004, ZnO nanowire growth and devices, *Materials Science and Engineering: R: Reports 47*: 1–47.

Hofmann, D. M.; Hofstaetter, A.; Leiter, F.; Zhou, H.; Henecker, F.; Meyer, B. K.; Orlinskii, S. B.; Schmidt, J.; and Baranov, P. G., 2002, Hydrogen: A relevant shallow donor in zinc oxide, *Physical Review Letters 88*: 045504.

Hoffman, R. L.; Norris, B. J.; and Wager, J. F., 2003, ZnO-based transparent thin-film transistors, *Applied Physics Letters 82*: 733–735.

Hofmann, D.; Pfisterer, D.; Sann, J.; Meyer, B.; Tena-Zaera, R.; Munoz-Sanjose, V.; Frank, T.; and Pensl, G., 2007, Properties of the oxygen vacancy in ZnO, *Applied Physics A: Materials Science & Processing 88*: 147–151.

Hofstetter, D.; Bonetti, Y.; Giorgetta, F. R.; El-Shaer, A.-H.; Bakin, A.; Waag, A.; Schmidt-Grund, R.; Schubert, M.; and Grundmann, M., 2007, Demonstration of an ultraviolet ZnO-based optically pumped third order distributed feedback laser, *Applied Physics Letters 91*: 111108.

Hofstetter, D.; Théron, R.; El-Shaer, A.-H.; Bakin, A.; and Waag, A., 2008, Demonstration of a ZnO/MgZnO-based one-dimensional photonic crystal multiquantum well laser, *Applied Physics Letters 93*: 101109.

Hopfield, J. J. and Thomas, D. G., 1965, Polariton absorption lines, *Physical Review Letters 15:* 22–25.

Huang, M. H.; Mao, S.; Feick, H.; Yan, H.; Wu, Y.; Kind, H.; Weber, E.; Russo, R.; and Yang, P., 2001, Room-temperature ultraviolet nanowire nanolasers, *Science 292:* 1897–1899.

Hullavarad, S.; Hullavarad, N.; Look, D.; and Claflin, B., 2009, Persistent photoconductivity studies in nanostructured ZnO UV sensors, *Nanoscale Research Letters 4:* 1421–1427.

Hümmer, K., 1973, Interband magnetoreflection of ZnO, *Physica Status Solidi (b) 56:* 249–260.

Hwang, D.-K.; Oh, M.-S.; Choi, Y.-S.; and Park, S.-J., 2008, Effect of pressure on the properties of phosphorus-doped p-type ZnO thin films grown by radio frequency-magnetron sputtering, *Applied Physics Letters 92:* 161109.

Hwang, D.-K.; Oh, M.-S.; Lim, J.-H.; Choi, Y.-S.; and Park, S.-J., 2007, ZnO-based light-emitting metal-insulator-semiconductor diodes, *Applied Physics Letters 91:* 121113.

Hwang, D.-K.; Oh, M.-S.; Lim, J.-H.; and Park, S.-J., 2007b, ZnO thin films and light-emitting diodes, *Journal of Physics D: Applied Physics 40:* R387.

Imanaka, Y.; Oshikiri, M.; Takehana, K.; Takamasu, T.; and Kido, G., 2001. Cyclotron resonance in n-type ZnO, *Physica B: Condensed Matter 298:* 211–215.

Ip, K.; Overberg, M. E.; Heo, Y. W.; Norton, D. P.; Pearton, S. J.; Kucheyev, S. O.; Jagadish, C.; Williams, J. S.; Wilson, R. G.; and Zavada, J. M., 2002, Thermal stability of ion-implanted hydrogen in ZnO, *Applied Physics Letters 81:* 3996–3998.

Ishizumi, A. and Kanemitsu, Y., 2005, Structural and luminescence properties of Eu-doped ZnO nanorods fabricated by a microemulsion method, *Applied Physics Letters 86:* 253106.

Iwata, K.; Fons, P.; Niki, S.; Yamada, A.; Matsubara, K.; Nakahara, K.; Tanabe, T.; and Takasu, H., 2000, ZnO growth on Si by radical source MBE, *Journal of Crystal Growth 214–215:* 50–54.

Janotti, A. and de Walle, C. G. V., 2005, Oxygen vacancies in ZnO, *Applied Physics Letters 87:* 122102.

Janotti, A. and de Walle, C. G. V. 2007, Native point defects in ZnO, *Physical Review B 76:* 165202.

Janotti, A. and van de Walle, C. G. V., 2009, Fundamentals of zinc oxide as a semiconductor, *Reports on Progress in Physics 72:* 126501.

Jiao, S.; Lu, Y.; Zhang, Z.; Li, B.; Yao, B.; Zhang, J.; Zhao, D.; Shen, D.; and Fan, X., 2007, Optical and electrical properties of highly nitrogen-doped ZnO thin films grown by plasma-assisted molecular beam epitaxy, *Journal of Applied Physics 102:* 113509.

Jiao, S. J.; Zhang, Z. Z.; Lu, Y. M.; Shen, D. Z.; Yao, B.; Zhang, J. Y.; Li, B. H.; Zhao, D. X.; Fan, X. W.; and Tang, Z. K., 2006, ZnO p-n junction light-emitting diodes fabricated on sapphire substrates, *Applied Physics Letters 88:* 031911.

Jin, Z.; Fukumura, T.; Kawasaki, M.; Ando, K.; Saito, H.; Sekiguchi, T.; Yoo, Y. Z.; Murakami, M.; Matsumoto, Y.; Hasegawa, T.; and Koinuma, H., 2001, High throughput fabrication of transition-metal-doped epitaxial ZnO thin films: A series of oxide-diluted magnetic semiconductors and their properties, *Applied Physics Letters 78:* 3824–3826.

Jin, Z.-W.; Yoo, Y.-Z.; Sekiguchi, T.; Chikyow, T.; Ofuchi, H.; Fujioka, H.; Oshima, M.; and Koinuma, H., 2003, Blue and ultraviolet cathodoluminescence from Mn-doped epitaxial ZnO thin films, *Applied Physics Letters 83:* 39–41.

Johnson, J.; Yan, H.; Yang, P.; and Saykally, R., 2003, Optical cavity effects in ZnO nanowire lasers and waveguides, *Journal of Physical Chemistry B 107:* 8816–8828.

Joseph, M.; Tabata, H.; and Kawai, T., 1999, p-Type Electrical conduction in ZnO thin films by Ga and N codoping, *Japanese Journal of Applied Physics 38:* L1205–L1207.

Kang, H. S.; Ahn, B. D.; Kim, J. H.; Kim, G. H.; Lim, S. H.; Chang, H. W.; and Lee, S. Y., 2006, Structural, electrical, and optical properties of p-type ZnO thin films with Ag dopant, *Applied Physics Letters 88:* 202108.

Kim, H.; Cepler, A.; Cetina, C.; Knies, D.; Osofsky, M.; Auyeung, R.; and Piqué, A., 2008b, Pulsed laser deposition of Zr–N codoped p -type ZnO thin films, *Applied Physics A: Materials Science & Processing 93:* 593–598.

Kim, K.-K.; Kim H.-S.; Hwang, D.-K.; Lim, J.-H.; and Park, S.-J., 2003, Realization of p-type ZnO thin films via phosphorus doping and thermal activation of the dopant, *Applied Physics Letters 83*, 63–65.

Kim, W. M.; Kim, I. H.; Ko, J. H.; Cheong, B.; Lee, T. S.; Lee, K. S.; Kim, D.; and Seong, T.-Y., 2008, Density-of-state effective mass and non-parabolicity parameter of impurity doped ZnO thin films, *Journal of Physics D: Applied Physics 41:* 195409.

Kim, S. S.; Kim, Y.-J.; Yi, G.-C.; and Cheong, H., 2009, Whispering-gallery-modelike resonance of luminescence from a single hexagonal ZnO microdisk, *Journal of Applied Physics 106:* 094310.

Kim, S. Y.; Yeon, Y. S.; Park, S. M.; Kim, J. H.; and Song, J. K., 2008c, Exciton states of quantum confined ZnO nanorods, *Chemical Physics Letters 462:* 100–103.

Kimpel, B. M. and Schulz, H.-J., 1991, Infrared luminescence of ZnO:Cu^{2+}(d9), *Physical Review B 43:* 9938–9940.

Klingshirn, C., 2007a, ZnO: Material, physics and applications, *ChemPhysChem 8:* 782–803.

Klingshirn, C., 2007b, ZnO: From basics towards applications, *Physica Status Solidi (b) 244:* 3027.

Klingshirn, C.F., 2007c, *Semiconductor Optics*, Springer, Berlin, Germany.

Klingshirn, C.; Hauschild, R.; Fallert, J.; and Kalt, H., 2007, Room-temperature stimulated emission of ZnO: Alternatives to excitonic lasing, *Physical Review B (Condensed Matter and Materials Physics)*, *APS 75:* 115203.

Klingshirn, C.; Priller, H.; Decker, M.; Brückner, J.; Kalt, H.; Hauschild, R.; Zeller, J.; Waag, A.; Bakin, A.; Thonke, H.; Sauer, R.; Kling, R.; Reuss, F.; and Kirchner, C., 2006, Excitonic properties of ZnO, *Advances in Solid State Physics, 45:* 275–287.

Koch, U.; Fojtik, A.; Weller, H.; and Henglein, A., 1985, Photochemistry of semiconductor colloids. Preparation of extremely small ZnO particles, fluorescence phenomena and size quantization effects, *Chemical Physics Letters 122:* 507–510.

Kong, X. Y.; Ding, Y.; Yang, R.; and Wang, Z. L., 2004, Single-crystal nanorings formed by epitaxial self-coiling of polar nanobelts, *Science 303:* 1348–1351.

Kordi Ardakani, H., 1996, Electrical conductivity of in situ "hydrogen-reduced" and structural properties of zinc oxide thin films deposited in different ambients by pulsed excimer laser ablation, *Thin Solid Films 287:* 280–283.

Kossanyi, J.; Kouyate, D.; Pouliquen, J.; Ronfard-Haret, J. C.; Valat, P.; Oelkrug, D.; Mammel, U.; Kelly, G. P.; and Wilkinson, F., 1990, Photoluminescence of semiconducting zinc oxide containing rare earth ions as impurities, *Journal of Luminescence 46:* 17–24.

Kuhnert, R. and Helbig, R., 1981, Vibronic structure of the green photoluminescence due to copper impurities in ZnO, *Journal of Luminescence 26:* 203–206.

Lambrecht, W. R. L.; Rodina, A. V.; Limpijumnong, S.; Segall, B.; and Meyer, B. K., 2002, Valence-band ordering and magneto-optic exciton fine structure in ZnO, *Physical Review B 65:* 075207.

Lange, M.; Dietrich, C.P.; Czekalla, C.; Zippel, J.; Benndorf, G.; Lorenz, M.; Zuniga-Perez, J.; and Grundmann, M., 2010, Luminescence properties of ZnO/ZnCdO/ZnO double heterostructures, *Journal Applied Physics 107:* 093530.

Lany, S. and Zunger, A., 2005, Anion vacancies as a source of persistent photoconductivity in II–VI and chalcopyrite semiconductors, *Physical Review B 72:* 035215.

Lany, S. and Zunger, A., 2010, Generalized Koopmans density functional calculations reveal the deep acceptor state of N_O in ZnO, *Physical Review B 81:* 205209.

LaRoche, J.; Heo, Y.; Kang, B.; Tien, L.; Kwon, Y.; Norton, D.; Gila, B.; Ren, F.; and Pearton, S., 2005, Fabrication approaches to ZnO nanowire devices, *Journal of Electronic Materials 34:* 404–408.

Lavrov, E. V.; Herklotz, F.; and Weber, J., 2009, Identification of two hydrogen donors in ZnO, *Physical Review B 79:* 165210.

Lee, J.; Koteles, E. S.; and Vassell, M. O., 1986, Luminescence linewidths of excitons in GaAs quantum wells below 150 K, *Physical Review B 33:* 5512–5516.

Lee, C. J.; Lee, T. J.; Lyu, S. C.; Zhang, Y.; Ruh, H.; and Lee, H. J., 2002, Field emission from well-aligned zinc oxide nanowires grown at low temperature, *Applied Physics Letters 81:* 3648–3650.

Leiter, F.; Alves, H.; Hofstaetter, A.; Hofmann, D.; and Meyer, B., 2001, The oxygen vacancy as the origin of a green emission in undoped ZnO, *Physica Status Solidi (b) 226:* R4–R5.

Leung, Y. H.; Kwok, W. M.; Djurisic, A. B.; Phillips, D. L.; and Chan, W. K., 2005, Time-resolved study of stimulated emission in ZnO tetrapod nanowires, *Nanotechnology 16:* 579–582.

Li, L.; Yang, H.; Zhao, H.; Yu, J.; Ma, J.; An, L.; and Wang, X., 2010, Hydrothermal synthesis and gas sensing properties of single-crystalline ultralong ZnO nanowires, *Applied Physics A: Materials Science & Processing 98:* 635–641.

Li, Z.; Zhu, G.; Yang, R.; Wang, A. C.; and Wang, Z. L., 2010b, Muscle-driven in vivo nanogenerator, *Advanced Materials 22:* 2534–2537.

Liang, S.; Sheng, H.; Liu, Y.; Huo, Z.; Lu, Y.; and Shen, H., 2001, ZnO Schottky ultraviolet photodetectors, *Journal of Crystal Growth 225:* 110–113.

Liang, H.; Yu, S.; and Yang, H., 2010, Edge-emitting ultraviolet n-ZnO:Al/i-ZnO/p-GaN heterojunction light-emitting diode with a rib waveguide, *Optics Express 18:* 3687–3692.

Lim, S. J.; Ju Kwon, S.; Kim, H.; and Park, J.-S., 2007, High performance thin film transistor with low temperature atomic layer deposition nitrogen-doped ZnO, *Applied Physics Letters 91:* 183517.

Lim, J.-H.; Kang, C.-K.; Kim, K.-K.; Park, I.-K.; Hwang, D.-K.; and Park, S.-J., 2006, UV Electroluminescence emission from ZnO light-emitting diodes grown by high-temperature radiofrequency sputtering, *Advanced Materials 18:* 2720–2724.

Liu, L.; Yu, P. Y.; Ma, Z.; and Mao, S. S., 2008, Ferromagnetism in GaN:Gd: A density functional theory study, *Physical Review Letters 100:* 127203.

Liu, C.; Yun, F.; and Morkoc, H., 2005, Ferromagnetism of ZnO and GaN: A review, *Journal of Materials Science: Materials in Electronics 16:* 555–597.

Look, D. C., 2005, Electrical and optical properties of p-type ZnO, *Semiconductor Science and Technology 20:* S55–S61.

Look, D., 2007, Donors and acceptors in bulk ZnO grown by the hydrothermal, vapor-phase, and melt processes, *Materials Research Society Symposia Proceedings 957:* K08–05.

Look, D. C.; Reynolds, D. C.; Litton, C. W.; Jones, R. L.; Eason, D. B.; and Cantwell, G., 2002, Characterization of homoepitaxial p-type ZnO grown by molecular beam epitaxy, *Applied Physics Letters 81:* 1830–1832.

Look, D. C.; Reynolds, D. C.; Sizelove, J. R.; Jones, R. L.; Litton, C. W.; Cantwell, G.; and Harsch, W. C., 1998, Electrical properties of bulk ZnO, *Solid State Communications 105:* 399–401.

Lorenz, M.; Brandt, M.; Lange, M.; Benndorf, G.; von Wenckstern, H.; Klimm, D.; and Grundmann, M., 2009, Homoepitaxial $MgxZn_{1-x}O$ $(0 = x = 0.22)$ thin films grown by pulsed laser deposition, *Thin Solid Films 518:* 4623–4629.

Lorenz, M.; Hochmuth, H.; Schmidt-Grund, R.; Kaidashev, E.; and Grundmann, M., 2004, Advances of pulsed laser deposition of ZnO thin films, *Annalen der Physik 13:* 59–60.

Lyons, J. L.; Janotti, A.; and de Walle, C. G. V., 2009, Why nitrogen cannot lead to p-type conductivity in ZnO, *Applied Physics Letters 95:* 252105.

Madel, M., 2010, University of Ulm, private communication.

Maeda, K.; Sato, M.; Niikura, I.; and Fukuda, T., 2005, Growth of 2 inch ZnO bulk single crystal by the hydrothermal method, *Semiconductor Science and Technology 20:* S49.

Makino, T.; Segawa, Y.; Kawasaki, M.; Ohtomo, A.; Shiroki, R.; Tamura, K.; Yasuda, T.; and Koinuma, H., 2001, Band gap engineering based on MgZnO and CdZnO ternary alloy films, *Applied Physics Letters 78:* 1237–1239.

Malashevich, A. and Vanderbilt, D., 2007, First-principles study of polarization in $Zn_{1-x}Mg_xO$, *Physical Review B 75:* 045106.

Masuda, S.; Kitamura, K.; Okumura, Y.; Miyatake, S.; Tabata, H.; and Kawai, T., 2003, Transparent thin film transistors using ZnO as an active channel layer and their electrical properties. *Journal of Applied Physics 93:* 1624–1630.

Matsui, H., Osone, T.; and Tabata, H., 2010, Band alignment and excitonic localization of ZnO/CdZnO quantum wells, *Journal of Applied Physics 107:* 093523.

McCluskey, M. D. and Jokela, S. J., 2009, Defects in ZnO, *Journal of Applied Physics 106:* 071101.

McCluskey, M. D.; Jokela, S. J.; Zhuravlev, K. K.; Simpson, P. J.; and Lynn, K. G., 2002, Infrared spectroscopy of hydrogen in ZnO. *Applied Physics Letters 81:* 3807–3809.

Meulenkamp, E. A., 1998, Synthesis and growth of ZnO nanoparticles, *The Journal of Physical Chemistry B 102:* 5566–5572.

Meyer, B. K.; Alves, H.; Hofmann, D. M.; Kriegseis, W.; Forster, D.; Bertram, F.; Christen, J.; Hoffmann, A.; Straßburg, M.; Dworzak, M.; Haboeck, U.; and Rodina, A. V., 2004, Bound exciton and donor-acceptor pair recombinations in ZnO, *Physica Status Solidi (b) 241:* 231–260.

Meyer, B.; Stehr, J.; Hofstaetter, A.; Volbers, N.; Zeuner, A.; and Sann, J., 2007, On the role of group I elements in ZnO, *Applied Physics A: Materials Science & Processing 88:* 119–123.

Minami, T.; Yamamoto, T.; and Miyata, T., 2000, Highly transparent and conductive rare earth-doped ZnO thin films prepared by magnetron sputtering, *Thin Solid Films 366:* 63–68.

Misra, P., Sharma, T.K., Porwal, S.; and Kukreja, L.M., 2006, Room temperature photoluminescence from ZnO quantum wells grown on (0001) sapphire using buffer assisted pulsed laser deposition, *Applied Physics Letters 89:* 161912.

Mollwo, E., 1954, Die Wirkung von Wasserstoff auf die Leitfähigkeit und Lumineszenz von Zinkoxydkristallen, *Zeitschrift für Physik A Hadrons and Nuclei 138:* 478–488.

Monteiro, T.; Neves, A. J.; Carmo, M. C.; Soares, M. J.; Peres, M.; Wang, J.; Alves, E.; Rita, E.; and Wahl, U., 2005, Near-band-edge slow luminescence in nominally undoped bulk ZnO, *Journal of Applied Physics 98:* 013502.

Monticone, S.; Tufeu, R.; and Kanaev, 1998, Complex nature of the UV and visible fluorescence of colloidal ZnO nanoparticles, *The Journal of Physical Chemistry B 102:* 2854–2862.

Müller, E.; Gerthsen, D.; Bruckner, P.; Scholz, F.; Gruber, T.; and Waag, A., 2006, Probing the electrostatic potential of charged dislocations in n-GaN and n-ZnO epilayers by transmission electron holography, *Physical Review B 73:* 245316.

Nause, J. and Nemeth, B., 2005, Pressurized melt growth of ZnO boules, *Semiconductor Science and Technology 20:* S45.

Nickel, N. H. and Fleischer, K., 2003, Hydrogen local vibrational modes in zinc oxide, *Physical Review Letters 90:* 197402.

Nobis, T. and Grundmann, M., 2005, Low-order optical whispering-gallery modes in hexagonal nanocavities, *Physical Review A 72:* 063806.

Oh, M.-S.; Hwang, D.-K.; Choi, Y.-S.; Kang, J.-W.; Park, S.-J.; Hwang, C.-S.; and Cho, K. I., 2008, Microstructural properties of phosphorus-doped p-type ZnO grown by radio-frequency magnetron sputtering, *Applied Physics Letters 93:* 111905.

Ohashi, N.; Ebisawa, N.; Sekiguchi, T.; Sakaguchi, I.; Wada, Y.; Takenaka, T.; and Haneda, H., 2005, Yellowish-white luminescence in codoped zinc oxide, *Applied Physics Letters 86:* 091902.

Ohashi, N.; Ishigaki, T.; Okada, N.; Sekiguchi, T.; Sakaguchi, I.; and Haneda, H., 2002, Effect of hydrogen doping on ultraviolet emission spectra of various types of ZnO, *Applied Physics Letters 80:* 2869–2871.

Ohno, Y.; Koizumi, H.; Taishi, T.; Yonenaga, I.; Fujii, K.; Goto, H.; and Yao, T., 2008, Optical properties of dislocations in wurtzite ZnO single crystals introduced at elevated temperatures, *Journal of Applied Physics 104:* 073515.

Oshikiri, M.; Aryasetiawan, F.; Imanaka, Y.; and Kido, G., 2002, Quasiparticle effective-mass theory in semiconductors, *Physical Review B 66:* 125204.

Osinsky, A.; Dong, J. W.; Kauser, M. Z.; Hertog, B.; Dabiran, A. M.; Chow, P. P.; Pearton, S. J.; Lopatiuk, O.; and Chernyak, L., 2004, MgZnO/AlGaN heterostructure light-emitting diodes, *Applied Physics Letters 85:* 4272–4274.

Özgür, U.; Alivov, Y. I.; Liu, C.; Teke, A.; Reshchikov, M. A.; Dogan, S.; Avrutin, V.; Cho, S.-J.; and Morkoc, H., A, 2005, Comprehensive review of ZnO materials and devices, *Journal of Applied Physics, 98:* 041301.

Pan, Z. W.; Dai, Z. R.; and Wang, Z. L., 2001, Nanobelts of semiconducting oxides, *Science 291:* 1947–1949.

Pan, F.; Song, C.; Liu, X.; Yang, Y.; and Zeng, F., 2008, Ferromagnetism and possible application in spintronics of transition-metal-doped ZnO films, *Materials Science and Engineering: R: Reports: 62,* 1–35.

Park, J.; Choi, S.-W.; and Kim, S., 2010b, Fabrication of a highly sensitive chemical sensor based on ZnO nanorod arrays, *Nanoscale Research Letters 5:* 353–359.

Park, W. I.; Kim, J. S.; Yi, G.-C.; Bae, M. H.; and Lee, H.-J., 2004, Fabrication and electrical characteristics of high-performance ZnO nanorod field-effect transistors, *Applied Physics Letters 85:* 5052–5054.

Park, Y. S.; Litton, C. W.; Collins, T. C.; and Reynolds, D. C., 1966, Exciton spectrum of ZnO, *Physical Review 143:* 512–519.

Park, S.; Minegushi, T.; Oh, D.; Lee, H.; Taishi, T.; Park, J.; Jung, M.; Chang, J.; Im, I.; Ha, J.; Hong, S.; Yonenaga, I.; Chikyow, T.; and Yao, T., 2010, High-quality p-type ZnO films grown by co-doping of N and Te on Zn-face ZnO substrates, *Applied Physics Express 3:* 031103.

Park, C. H.; Zhang, S. B.; and Wei, S.-H., 2002, Origin of p-type doping difficulty in ZnO: The impurity perspective, *Physical Review B 66:* 073202.

Pässler, R., 2002, Dispersion-related description of temperature dependencies of band gaps in semiconductors, *Physical Review B 66:* 085201.

Pedersen, T. G., 2005, Quantum size effects in ZnO nanowires, *Physica Status Solidi (c) 2:* 4026–4030.

Prinz, G.M., 2008, Optische Spektroskopie an Halbleitern mit großer Bandlücke und ZnO-Nanostrukturen (in German), PhD thesis, University of Ulm, Germany.

Qin, Y.; Wang, X.; and Wang, Z. L., 2008, Microfibre-nanowire hybrid structure for energy scavenging, *Nature 451:* 809–813.

Quesada, A.; García, M.; Crespo, P.; and Hernando, A., 2006, Materials for spintronic: Room temperature ferromagnetism in Zn-Mn-O interfaces, *Journal of Magnetism and Magnetic Materials 304:* 75–78.

Reiser, A.; Ladenburger, A.; Prinz, G. M.; Schirra, M.; Feneberg, M.; Zayan, D. O.; Röder, U.; Leute, R.A.R.; Sauer, R.; and Thonke, K., 2011, Optimized growth conditions for well-defined zinc oxide nanopillars grown by the VLS process on different substrate materials, to be published (2011).

Reynolds, D. C.; Litton, C. W.; and Collins, T. C., 1965, Zeeman effects in the edge emission and absorption of ZnO, *Physical Review 140:* A1726–A1734.

Reynolds, D. C.; Litton, C. W.; Look, D. C.; Hoelscher, J. E.; Claflin, B.; Collins, T. C.; Nause, J.; and Nemeth, B., 2004, High-quality, melt-grown ZnO single crystals, *Journal of Applied Physics 95:* 4802–4805.

Reynolds, D. C.; Look, D. C.; and Jogai, B., 2001, Fine structure on the green band in ZnO, *Journal of Applied Physics 89:* 6189–6191.

Reynolds, D. C.; Look, D. C.; Jogai, B.; Litton, C. W.; Cantwell, G.; and Harsch, W. C., 1999, Valence-band ordering in ZnO, *Physical Review B 60:* 2340–2344.

Richters, J.-P.; Voss, T.; Wischmeier, L.; Rückmann, I.; and Gutowski J., 2008, Influence of polymer coating on the low-temperature photoluminescence properties of ZnO nanowires, *Applied Physics Letters 92:* 011103.

Ryu, M. K.; Lee, S. H.; Jang, M. S.; Panin, G. N.; and Kang, T. W., 2002, Postgrowth annealing effect on structural and optical properties of ZnO films grown on GaAs substrates by the radio frequency magnetron sputtering technique, *Journal of Applied Physics 92:* 154–158.

Ryu, Y.; Lee, T.-S.; Lubguban, J. A.; White, H. W.; Kim, B.-J.; Park, Y.-S.; and Youn, C.-J., 2006, Next generation of oxide photonic devices: ZnO-based ultraviolet light emitting diodes, *Applied Physics Letters 88:* 241108.

Ryu, Y.R.; Lee, T.S.; Lubguban, J.A.; Corman, A.B.; White, H.W.; Leem, J.H.; Han, M.S.; Park, Y.S.; Youn, C.J.; and Kim, W.J., 2006b, Wide-band gap oxide alloy: BeZnO, *Applied Physics Letters 88:* 052103.

Ryu, Y. R.; Lee, T. S.; Lubguban, J. A.; White, H. W.; Park, Y. S.; and Youn, C. J., 2005, ZnO devices: Photodiodes and p-type field-effect transistors, *Applied Physics Letters 87:* 153504.

Ryu, Y.R., Lee, T.S.; and White, H.W., 2003, Properties of arsenic-doped p-type ZnO grown by hybrid beam deposition, *Applied Physics Letters 83:* 87–89.

Ryu, Y.R.; Lubguban, J.A.; Lee, T.S.; White, H.W.; Jeong, T.S.; Youn, C.J.; and Kim, B.J., 2007, Excitonic ultraviolet lasing in ZnO-based light emitting devices, *Applied Physics Letters 90*: 131115.

Sadofev, S.; Blumstengel, S.; Cui, J.; Puls, J.; Rogaschewski, S.; Schafer, P.; Sadofyev, Y. G.; and Henneberger, F., 2005, Growth of high-quality ZnMgO epilayers and ZnO/ZnMgO quantum well structures by radical-source molecular-beam epitaxy on sapphire, *Applied Physics Letters 87*: 091903.

Sadofev, S.; Kalusniak, S.; Puls, J.; Schäfer, P.; Blumstengel, S.; and Henneberger, F., 2007, Visible-wavelength laser action of ZnCdO/(Zn,Mg)O multiple quantum well structures, *Applied Physics Letters 91*: 231103.

Saeki, H.; Tabata, H.; and Kawai, T., 2001, Magnetic and electric properties of vanadium doped ZnO films, *Solid State Communication 120*: 439–443.

Sakurai, K.; Kubo, T.; Kajita, D.; Tanabe, T.; Takasu, H.; Fujita, S.; and Fujita, S., 2000, Blue photoluminescence from ZnCdO films grown by molecular beam epitaxy, *Japanese Journal of Applied Physics 39*: L1146–L1148.

Savikhin, S. and Freiberg, A, 1993, Origin of "universal" ultraviolet luminescence from the surfaces of solids at low temperatures, *Journal of Luminescence 55*: 1–3.

Scharrer, M.; Yamilov, A.; Wu, X.; Cao, H.; and Chang, R. P. H., 2006, Ultraviolet lasing in high-order bands of three-dimensional ZnO photonic crystals, *Applied Physics Letters 88*: 201103.

Schildknecht, A., 2003b, Optische spektroskopie von halbleitern mit großer bandlücke: ZnO und AlGaN (in German), Diploma thesis, University of Ulm, Germany.

Schildknecht, A.; Sauer, R.; and Thonke, K, 2003, Donor-related defect states in ZnO substrate material, *Physica B 340*: 205.

Schirmer, O. F. and Zwingel, D., 1970, The yellow luminescence of zinc oxide, *Solid State Communications 8*: 1559–1563.

Schirra, M.; Feneberg, M.; Prinz, G. M.; Reiser, A.; Röder, T.; Thonke, K.; and Sauer, R., 2009, Beating of coupled ultraviolet light modes in zinc oxide nanoresonators, *Physical Review Letters 102*: 073903.

Schirra, M.; Schneider, R.; Reiser, A.; Prinz, G. M.; Feneberg, M.; Biskupek, J.; Kaiser, U.; Krill, C. E.; Thonke, K.; and Sauer, R., 2008, Stacking fault related 3.31-eV luminescence at 130 meV acceptors in zinc oxide, *Physical Review B 77*: 125215.

Schulz, H.-J. and Thiede, M., 1987, Optical spectroscopy of $3d^7$ and $3d^8$ impurity configurations in a wide-gap semiconductor (ZnO:Co,Ni,Cu), *Physical Review B 35*: 18–34.

Segall, B., 1967, Intrinsic absorption "Edge" in II-VI semiconducting compounds with the wurtzite structure, *Physical Review 163*: 769–778.

Sernelius, B. E.; Berggren, K.-F.; Jin, Z.-C.; Hamberg, I.; and Granqvist, C. G., 1988, Band-gap tailoring of ZnO by means of heavy Al doping, *Physical Review B 37*: 10244–10248.

Shalish, I.; Temkin, H.; and Narayanamurti, V., 2004, Size-dependent surface luminescence in ZnO nanowires, *Physical Review B 69*: 245401.

Shi, H.; Zhang, P.; Li, S.-S.; and Xia, J.-B., 2009, Magnetic coupling properties of rare-earth metals (Gd, Nd) doped ZnO: First-principles calculations, *Journal of Applied Physics 106*: 023910.

Singh, S.; Rama, N.; and Rao, M. S. R., 2006, Influence of d-d transition bands on electrical resistivity in Ni doped polycrystalline ZnO, *Applied Physics Letters 88*: 222111.

Stichtenoth, D.; Ronning, C.; Niermann, T.; Wischmeier, L.; Voss, T.; Chien, C.-J.; Chang, P.-C.; and Lu, J. G., 2007, Optical size effects in ultrathin ZnO nanowires, *Nanotechnology 18*: 435701.

Sturm, C.; Hilmer, H.; Schmidt-Grund, R.; and Grundmann, M., 2009, Observation of strong exciton-photon coupling at temperatures up to 410 K, *New Journal of Physics 11*: 073044.

Sun, L.; Chen, Z.; Ren, Q.; Yu, K.; Zhou, W.; Bai, L.; Zhu, Z. Q.; and Shen, X., 2009, Polarized photoluminescence study of whispering gallery mode polaritons in ZnO microcavity, *Physica Status Solidi (c) 6*: 133–136.

Sun, J. W.; Lu, Y. M.; Liu, Y. C.; Shen, D. Z.; Zhang, Z. Z.; Yao, B.; Li, B. H.; Zhang, J. Y.; Zhao, D. X.; and Fan, X. W., 2007, Nitrogen-related recombination mechanisms in p-type ZnO films grown by plasma-assisted molecular beam epitaxy, *Journal of Applied Physics 102*: 043522.

Syrbu, N.; Tiginyanu, I.; Zalamai, V.; Ursaki, V.; and Rusu, E., 2004, Exciton polariton spectra and carrier effective masses in ZnO single crystals, *Physica B: Condensed Matter 353*: 111–115.

Szarko, J. M.; Song, J. K.; Blackledge, C. W.; Swart, I.; Leone, S. R.; Li, S.; and Zhao, Y., 2005, Optical injection probing of single ZnO tetrapod lasers, *Chemical Physics Letters 404*: 171–176.

Takeuchi, T., Amano, H.; and Akasaki, I., 2000, Theoretical study of orientation dependence of piezoelectric effects in wurtzite strained GaInN/GaN heterostructures and quantum wells, *Japanese Journal of Applied Physics 39*: 413–41.

Thomas, D. G., 1960, The exciton spectrum of zinc oxide, *Journal of Physics and Chemistry of Solids 15*: 86–96.

Thonke, K.; Gruber, T.; Teofilov, N.; Schönfelder, R.; Waag, A.; and Sauer, R., 2001, Donor-acceptor pair transitions in ZnO substrate material, *Physica B: Condensed Matter 308–310*: 945–948.

Thonke, K., Kerwien, N., Wysmolek, A., Potemski, M., Waag, A.; and Sauer, R., 2003, Magneto-PL study of donors in ZnO substrates, *Proceedings of the 26th ICPS, Edinburgh (2002)*, ed.: A. R. Long, and J. H. Davies, IOP, Bristol, P22.

Thonke, K.; Reiser, A.; Schirra, M.; Feneberg, M.; Prinz, G. M.; Röder, T.; Sauer, R.; Fallert, J.; Stelzl, F.; Kalt, H.; Gsell, S.; Schreck, M.; and Stritzker, B., 2009, ZnO nanostructures: Optical resonators and lasing, *Advances in Solid State Physics 48: 39*.

Thonke, K.; Schirra, M.; Schneider, R.; Reiser, A.; Prinz, G. M.; Feneberg, M.; Sauer, R.; Biskupek, J.; and Kaiser, U., 2010, The role of stacking faults and their associated 0.13 eV acceptor state in doped and undoped ZnO layers and nanostructures, *Physica Status Solidi (b)247: 1464*.

Travlos, A.; Boukos, N.; Chandrinou, C.; Kwack, H.-S.; and Dang, L. S., 2009, Zinc and oxygen vacancies in ZnO nanorods, *Journal of Applied Physics 106: 104307*.

Travnikov, V.V.; Freiberg, A.; and Savikhin, S.F., 1990, Surface excitons in ZnO crystals, *Journal of Luminescence 47*: 107–112.

Tsukazaki, A.; Kubota, M.; Ohtomo, A.; Onuma, T.; Ohtani, K.; Ohno, H.; Chichibu, S. F.; and Kawasaki, M., 2005, Blue light-emitting diode based on ZnO, *Japanese Journal of Applied Physics 44: L643–L645*

Tsukazaki, A.; Ohtomo, A.; Onuma, T.; Ohtani, M.; Makino, T.; Sumiya, M.; Ohtani, K. et al., 2005b, Repeated temperature modulation epitaxy for p-type doping and light-emitting diode based on ZnO, *Nature Materials 4: 42–46*.

Ungureanu, M.; Schmidt, H.; Xu, Q.; von Wenckstern, H.; Spemann, D.; Hochmuth, H.; Lorenz, M.; and Grundmann, M., 2006, Electrical and magnetic properties of RE-doped ZnO thin films (RE = Gd, Nd), *Superlattices and Microstructures 42*: 231–235.

Van de Walle, C. G., 2000, Hydrogen as a cause of doping in zinc oxide, *Physics Review Letters 85*: 1012–1015.

Van de Walle, C. G. and Neugebauer, J., 2003, Universal alignment of hydrogen levels in semiconductors, insulators and solutions, *Nature 423*: 626–628.

van Vugt, L. K.; Ruhle, S.; Ravindran, P.; Gerritsen, H. C.; Kuipers, L.; and Vanmaekelbergh, D., 2006, Exciton polaritons confined in a ZnO nanowire cavity, *Physical Review Letters 97*: 147401

Venkatesan, M.; Fitzgerald, C. B.; Lunney, J. G.; and Coey, J. M. D., 2004, Anisotropic ferromagnetism in substituted zinc oxide, *Physical Review Letters 93: 177206*.

Viña, L.; Logothetidis, S.; and Cardona, M., 1984, Temperature dependence of the dielectric function of germanium, *Physical Review B 30*: 1979–1991.

Wagner, C. and Schottky, W., 1930, Theorie der geordneten mischphasen, *Zeitschrift für Physikalische Chemie 22: 163*.

Wagner, M. R.; Schulze, J.-H.; Kirste, R.; Cobet, M.; Hoffmann, A.; Rauch, C.; Rodina, A. V.; Meyer, B. K.; Röder, U.; and Thonke, K., 2009, Γ_7 valence band symmetry related hole fine splitting of bound excitons in ZnO observed in magneto-optical studies, *Physical Review B 80: 205203*.

Wahl, U.; Correia, J. G.; Mendonça, T.; and Decoster, S., 2009, Direct evidence for Sb as a Zn site impurity in ZnO, *Applied Physics Letters 94: 261901*.

Wahl, U.; Rita, E.; Correia, J. G.; Marques, A. C.; Alves, E.; and Soares, J. C., 2005, Direct evidence for As as a Zn-site impurity in ZnO, *Physical Review Letters 95: 215503*.

Wakano, T.; Fujimura, N.; Morinaga, Y.; Abe, N.; Ashida, A.; and Ito, T., 2001, Magnetic and magneto-transport properties of ZnO:Ni films, *Physica E: Low-Dimensional Systems and Nanostructures 10*: 260–264.

Wan, Q.; Li, Q. H.; Chen, Y. J.; Wang, T. H.; He, X. L.; Li, J. P.; and Lin, C. L., 2004, Fabrication and ethanol sensing characteristics of ZnO nanowire gas sensors, *Applied Physics Letters, 84*: 3654–3656.

Wan, Q.; Yu, K.; Wang, T. H.; and Lin, C. L., 2003, Low-field electron emission from tetrapod-like ZnO nanostructures synthesized by rapid evaporation, *Applied Physics Letters 83*: 2253–2255.

Wang, Z. L., 2004a, Zinc oxide nanostructures: Growth, properties and applications, *Journal of Physics: Condensed Matter 16*: R829R858.

Wang, Z. L., 2004b, Nanostructures of zinc oxide, *Materials Today 7*: 26–33

Wang, L. and Giles, N. C., 2003, Temperature dependence of the free-exciton transition energy in zinc oxide by photoluminescence excitation spectroscopy, *Journal of Applied Physics 94*: 973–978.

Wang, L. and Giles, N. C., 2004, Determination of the ionization energy of nitrogen acceptors in zinc oxide using photoluminescence spectroscopy, *Applied Physics Letters 84*: 3049–3051.

Wang, Y.S; John Thomas, P.; and O'Brien, P., 2006b, Nanocrystalline ZnO with ultraviolet luminescence, *Journal of Physical Chemistry B 110*: 4099.

Wang, Y.-L.; Ren, F.; Kim, H.; Norton, D.; and Pearton, S., 2008, Materials and process development for ZnMgO/ZnO light-emitting diodes, *Selected Topics in Quantum Electronics 14*: 1048–1052.

Wang, Z. L. and Song, J., 2006, Piezoelectric nanogenerators based on zinc oxide nanowire arrays, *Science 312*: 242–246.

Weber, M.; Parmar, N.; Jones, K.; and Lynn, K., 2010, Oxygen deficiency and hydrogen turn ZnO red, *Journal of Electronic Materials 39*: 573–576.

Willander, M.; Nur, O.; Zhao, Q. X.; Yang, L. L.; Lorenz, M.; Cao, B. Q.; Pérez, J. Z. et al., 2009, Zinc oxide nanorod based photonic devices: Recent progress in growth, light emitting diodes and lasers, *Nanotechnology 20*: 332001.

Wischmeier, L.; Voss, T.; Rückmann, I.; Gutowski, J.; Mofor, A. C.; Bakin, A.; and Waag, A., 2006, Dynamics of surface-excitonic emission in ZnO nanowires, *Physical Review B 74*: 195333.

Wong, E. M. and Searson, P. C., 1999, ZnO quantum particle thin films fabricated by electrophoretic deposition, *Applied Physics Letters 74*: 2939–2941.

Xiao, Z. Y.; Liu, Y. C.; Mu, R.; Zhao, D. X.; and Zhang, J. Y., 2008, Stability of p-type conductivity in nitrogen-doped ZnO thin film, *Applied Physics Letters 92*: 052106.

Xiong, G.; Ucer, K. B.; Williams, R. T.; Lee, J.; Bhattacharyya, D.; Metson, J.; and Evans, P., 2005, Donor-acceptor pair luminescence of nitrogen-implanted ZnO single crystal, *Journal of Applied Physics 97*: 043528.

Xu, J.; Pan, Q.; Shun, Y.; and Tian, Z., 2000, Grain size control and gas sensing properties of ZnO gas sensor, *Sensors and Actuators B: Chemical, 66*: 277–279.

Xu, Q.; Schmidt, H.; Zhou, S.; Potzger, K.; Helm, M.; Hochmuth, H.; Lorenz, M.; Setzer, A.; Esquinazi, P.; Meinecke, C.; and Grundmann, M., 2008, Room temperature ferromagnetism in ZnO films due to defects, *Applied Physics Letters 92*: 082508.

Xue, D.; Zhang, J.; Yang, C.; and Wang, T., 2008, PL and EL characterizations of ZnO:Eu3+, Li+ films derived by sol-gel process, *Journal of Luminescence 128*: 685–689.

Yan, H.; He, R.; Johnson, J.; Law, M.; Saykally, R.; and Yang, P., 2003c, Dendritic nanowire ultraviolet laser array, *Journal of American Chemical Society 125*: 4728–4729.

Yan, H.; He, R.; Pham, J.; and Yang, P., 2003b, Morphogenesis of one-dimensional ZnO nano- and micro-crystals, *Advanced Materials 15*: 402–405.

Yan, H.; Johnson, J.; Law, M.; He, R.; Knutsen, K.; McKinney, J.; Pham, J.; Saykally, R.; and Yang, P., 2003, ZnO nanoribbon microcavity lasers, *Advanced Materials 15*: 1907–1911.

Yan, L.; Ong, C. K.; and Rao, X. S., 2004, Magnetic order in Co-doped and (Mn, Co) codoped ZnO thin films by pulsed laser deposition, *Journal of Applied Physics 96*: 508–511.

Yang, Z.; Chu, S.; Chen, W. V.; Li, L.; Kong, J.; Ren, J.; Yu, P. K. L.; and Liu, J., 2010, ZnO:Sb/ZnO:Ga light emitting diode on c-plane sapphire by molecular beam epitaxy, *Applied Physics Express 3*: 032101.

Yang, C.; Li, X.M.; Gu, Y.F.; Yu, W.D.; Gao, X.D.; and Zhang, Y.W., 2008b, ZnO based oxide system with continuous bandgap modulation from 3.7 to 4.9 eV, *Applied Physics Letters 93*: 112114.

Yang, Y.; Sun, X. W.; Tay, B. K.; You, G. F.; Tan, S. T.; and Teo, K. L., 2008, A p-n homojunction ZnO nanorod light-emitting diode formed by As ion implantation, *Applied Physics Letters 93*: 253107.

Yanhong, L.; Dejun, W.; Qidong, Z.; Min, Y.; and Qinglin, Z., 2004, A study of quantum confinement properties of photogenerated charges in ZnO nanoparticles by surface photovoltage spectroscopy, *The Journal of Physical Chemistry B 108*: 3202–3206.

Ye, Z.; Yang, F.; Lu, Y.; Zhi, M.; Tang, H.; and Zhu, L., 2007, ZnO nanorods with different morphologies and their field emission properties, *Solid State Communications 142*: 425–428.

Yoshikawa, H. and Adachi, S., 1997, Optical constants of ZnO, *Japanese Journal of Applied Physics 36*: 6237–6243.

Young, S. J.; Ji, L. W.; Chuang, R. W.; Chang, S. J.; and Du, X. L., 2006, Characterization of ZnO metal-semi-conductor-metal ultraviolet photodiodes with palladium contact electrodes, *Semiconductor Science and Technology 21*: 1507.

Yu, W.; Yang, L. H.; Teng, X. Y.; Zhang, J. C.; Zhang, Z. C.; Zhang, L.; and Fu, G. S., 2008, Influence of structure characteristics on room temperature ferromagnetism of Ni-doped ZnO thin films, *Journal of Applied Physics 103*: 093901.

Zeng, X.; Yuan, J.; and Zhang, L., 2008, Synthesis and photoluminescent properties of rare earth doped ZnO hierarchical microspheres, *The Journal of Physical Chemistry C 112*: 3503–3508.

Zeuner, A.; Alves, H.; Hofmann, D.; Meyer, B.; Hoffmann, A.; Haboeck, U.; Strassburg, M.; and Dworzak, M., 2002, Optical properties of the nitrogen acceptor in epitaxial ZnO, *Physica Status Solidi (b) 234*: R7–R9.

Zhang, B. P.; Binh, N. T.; Wakatsuki, K.; Liu, C. Y.; Segawa, Y.; and Usami, N., 2005, Growth of ZnO/MgZnO quantum wells on sapphire substrates and observation of the two-dimensional confinement effect, *Applied Physics Letters 86*: 032105.

Zhang, Y.; Lin, B.; Sun, X.; and Fu, Z., 2005b, Temperature-dependent photoluminescence of nanocrystalline ZnO thin films grown on Si (100) substrates by the sol–gel process, *Applied Physics Letters 86*: 131910.

Zhang, Y. and Mascarenhas, A., 1999, Scaling of exciton binding energy and virial theorem in semiconductor quantum wells and wires, *Physical Review B 59*: 2040–2044.

Zheng, X.; Li, Q.; Zhao, J.; Chen, D.; Zhao, B.; Yang, Y.; and Zhang, L., 2006, Photoconductive ultraviolet detectors based on ZnO films, *Applied Surface Science 253*: 2264–2267.

Zhu, H.; Shan, C.-X.; Yao, B.; Li, B.-H.; Zhang, J.-Y.; Zhang, Z.-Z.; Zhao, D.-X.; Shen, D.-Z.; Fan, X.-W.; Lu, Y.-M.; and Tang, Z.-K., 2009, Ultralow-threshold laser realized in zinc oxide, *Advanced Materials 21*: 1613–1617.

Zhu, G.; Yang, R.; Wang, S.; and Wang, Z. L., 2010, Flexible high-output nanogenerator based on lateral ZnO nanowire array, *Nano Letters 10*: 3151–3155.

Zhu, Y. W.; Zhang, H. Z.; Sun, X. C.; Feng, S. Q.; Xu, J.; Zhao, Q.; Xiang, B.; Wang, R. M.; and Yu, D. P., 2003, Efficient field emission from ZnO nanoneedle arrays, *Applied Physics Letters 83*: 144–146.

Zwingel, D. and Gärtner, F., 1974, Paramagnetic and optical properties of Na-doped ZnO single crystals, *Solid State Communications 14*: 45–49.

5

Novel Applications of ZnO: Random Lasing and UV Photonic Light Sources

Hui Cao
Yale University

Robert P.H. Chang
Yale University

5.1 Introduction

5.1.1 Optical Properties of ZnO

ZnO is a II–VI semiconducting compound that forms as ionic crystals of the wurtize structure. ZnO has found a wide range of industrial applications from skin care, phosphors, varistors, to transparent conductive oxides for flat screen displays. Recently, ZnO has received much attention for its potential as an optoelectronic material. It has a wide, direct electronic band gap of approximately 3.37 eV at room temperature. Under photoexcitation, it emits light in the ultraviolet (UV) spectrum at a corresponding band-edge wavelength of approximately 380 nm. This photoluminescence (PL) is due to the radiative recombination of excitons. ZnO possesses a very large exciton-binding energy of about 60 meV, making excitons stable at room temperature. Therefore, ZnO is a promising candidate for blue and UV light-emitting diodes (LEDs) and lasers.

Besides the relatively narrow and short-lifetime UV emission band, a broader emission band is also found in the visible spectrum. The nature of the visible luminescence in ZnO has been the subject of intense research, but is still not fully understood [1]. Oxygen vacancies and impurities have often been assumed to be the most likely recombination centers, based on annealing experiments of macrocrystalline ZnO powders in reducing or oxidizing atmosphere. However, the interpretation of these results is complicated because the emission intensity also depends on the width of the surface depletion layer, which is influenced by the defect concentration as well. Studies on the PL of nanocrystalline ZnO powders suggest that both the UV and visible emission are strongly influenced by surface states. It is suggested that holes can be easily trapped at the surface and that the green luminescence is due to recombination of excited electrons with deeply trapped holes in the bulk [1]. This model is supported by the fact that the intensity of the UV emission depends strongly on the surface chemistry of the ZnO nanoparticles and their size [2–6].

ZnO has also been found to be a promising candidate for application in nonlinear optics. It possesses large second-order and third-order susceptibilities. Efficient second-harmonic [7–9] and third-harmonic [10] generations of near-infrared (IR) laser light have been demonstrated in ZnO thin films. The nonlinear susceptibilities increase strongly with decreasing crystallite size. The wide electronic bandgap of ZnO avoids absorption of the second harmonics or third harmonics in the visible frequencies, allowing efficient frequency conversion.

5.1.2 Random Laser

A laser has two basic components: a gain medium that amplifies light by stimulated emission, and a cavity that traps light and provides feedback. For a long time, optical scattering has been considered detrimental to laser because such scattering removes photons from the lasing modes of a conventional laser cavity. However, in a disordered medium with gain, multiple scattering of light plays a positive role by increasing the path length or dwell time of light in an active medium and enhancing light amplification. In addition, interference of scattered light can provide coherent feedback for lasing oscillation [11].

Since the pioneering work of Letokhov et al. [12], lasing in disordered media has been the subject of intense theoretical and experimental studies [11,13–15]. It represents the process of light amplification by stimulated emission with feedback mediated by optical scattering from random fluctuation of the dielectric constant in space. There are two kinds of feedback: one is intensity or energy feedback; the other is field or amplitude feedback. The field feedback is phase sensitive (i.e., coherent), and therefore frequency dependent (i.e., resonant). The intensity feedback is phase insensitive (i.e., incoherent) and frequency independent (i.e., non-resonant) [13].

In the mid-1980s, intense stimulated radiation was observed in a wide variety of laser crystal powders, for example, neodymium-doped glass powder and titanium-doped sapphire powder [16,17], when pumped by laser pulses. It was shown that the stimulated emission was enhanced by the diffusion of light in the powder, namely, the photons spontaneously emitted by the doped ions of neodymium or titanium bounced from one particle to another many times before leaving the powder [18,19]. In 1994, Lawandy et al. reported laser-like emission from a laser dye solution containing titanium dioxide microparticles [20]. The emission spectrum narrows dramatically above a pumping threshold, illustrating a collapse of the emission linewidth. The peak emission intensity also increases rapidly above the threshold. The strong dependence of the threshold on the density of particles in the solution revealed that the feedback was related to scattering. The smooth emission spectrum above the threshold indicated the feedback was frequency-insensitive, i.e., non-resonant.

In the late 1990s, we reported a different kind of lasing process in highly disordered ZnO powder and polycrystalline films [21,22]. The feedback is resonant and coherent, leading to multiple narrow peaks in the lasing spectra. Interference of scattered light reduces light leakage at certain frequencies, producing laser emission with high first-order coherence (narrow spectral width) and second-order coherence (suppression of photon number fluctuations in single modes) [23]. Vardeny and coworkers also observed similar phenomena in disordered polymers and organic materials [24].

5.1.3 Photonic Crystal

A photonic crystal (PhC) is an artificial multidimensional periodic structure with a spatial period on the order of optical wavelength [25]. Since its invention 20 years ago [26,27], PhC has attracted much attention for the promise of full control of light propagation and emission [28–30]. One unique property of the PhC is a photonic bandgap (PBG) which inhibits or restricts the existence of optical modes. It allows dramatic modification of spontaneous emission [31,32] and realization of strong light localization. One important application is semiconductor laser with low threshold and small modal volume [33,34]. There has been tremendous progress in the development of PhC lasers that operate in the visible and near-IR spectrum range. However, it is much more difficult to fabricate UV PhC laser, because the feature size scales down with the emission wavelength. The reduced feature size demands higher fabrication accuracy, which is technologically challenging for commonly used wide band gap materials such as GaN and ZnO. On the other hand, the demand for blue and UV compact light sources has prompted enormous research effort into wide band gap semiconductors. We recently developed several techniques to fabricate two-dimensional (2D) and three-dimensional (3D) PhC lasers with ZnO [35–38]. We utilized the focused ion beam (FIB) etching to make periodic arrays of air holes in the single-crystalline ZnO films grown on sapphire substrates [36]. We also fabricated ZnO inverse opals of high quality using the atomic layer deposition (ALD) technique [39]. In both systems we realized lasing with optical pumping at room temperature.

Another interesting feature of a PhC is the abnormal dispersion of photonic bands, which can reduce the group velocity of light by orders of magnitude [40]. The slow light can greatly enhance light–matter interactions, for example, light amplification and nonlinear optical processes. We utilized the slow modes in 3D ZnO PhCs to achieve efficient and robust lasing at near UV, and were able to tune the lasing frequency over a significant range with structural parameters. The advantage of employing a higher-order photonic band instead of the fundamental PBG is that the feature size can be larger, thus relaxing the stringent fabrication requirements for UV lasers. Moreover, we demonstrated that the frozen mode at a stationary inflection point of the dispersion curve for a photonic band can strongly modify the intensity, directionality, and polarization of visible luminescence from a ZnO inverse opal. This may greatly enhance the efficiency of LEDs.

5.2 Random Lasing in ZnO Nanostructures

5.2.1 Fabrication of ZnO Nanorods and Nanopowder

ZnO was deposited on c-plane sapphire substrates by a plasma-enhanced metal-organic chemical vapor deposition (MOCVD) method [41]. We were able to achieve a morphological transition of ZnO from crystalline film to columnar-grained nanocrystalline films, and finally to well-aligned nanorods on sapphire substrates as the growth temperature decreases. Diethylzinc (DEZn) and oxygen gas were used as precursors for zinc and oxygen, respectively. A series of ZnO samples were grown at different temperatures while other conditions were fixed. At very high temperature (~750°C), the ZnO films were found to be epitaxial crystalline films. When the growth temperatures were between 500°C and 700°C, the ZnO films exhibited columnar-shaped nanocrystalline grains. These grains typically had diameters around 100 nm, and they were well aligned vertically to the substrate surface. Interestingly, ZnO nanorods were obtained on c-plane sapphire at 300°C–400°C. As shown by the scanning electron microscope (SEM) image in Figure 5.1a, needle-like nanorods were densely grown on sapphire substrates. The diameters of the ZnO nanorods were around 50–100 nm, and the length could be as long as several micrometers, depending on the growth time. The top-view SEM images (not shown in Figure 5.1a) revealed that these ZnO nanorods were uniformly grown on sapphire substrates and they were clearly separated from each other, although the distances between them were quite small [41]. The cross-sectional view SEM image illustrated that these nanorods were very well-aligned along the same direction, perpendicular to the substrate surface. When ZnO was grown at very low temperatures, for example, at 100°C, the morphology of the film on *c*-plane sapphire was found to consist of randomly oriented irregular-shaped grains, which were different from the columnar grains observed in ZnO films deposited at higher temperatures.

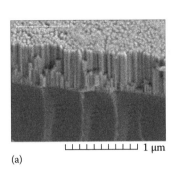
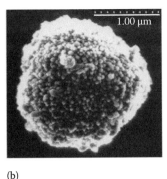

(a)   (b)

FIGURE 5.1 (a) Tilt-view SEM image of two-dimensional random array of ZnO nanorods. (Cao, H., Random lasers: Development, features and applications, *Opt. Photon. News*, 16, 24–30, 2005. With permission of Optical Society of America.) (b) SEM image of a micro-scale cluster of ZnO nanoparticles. (Reprinted with permission from [Cao, H. et al., Microlasers made of disordered medial, *Appl. Phys. Lett.*, 76, 2997] Copyright 2000, American Institute of Physics.)

We also synthesized ZnO colloidal particles by a precipitation reaction [35]. The process involved hydrolysis of zinc salt in a polyol medium. Zinc acetate dihydrate was added to diethylene glycol. The solution was heated to 160°C. As the solution was heated, more zinc acetate was dissociated. When the Zn^{2+} concentration in the solution exceeded the nucleation threshold, ZnO nanocrystallites precipitated and agglomerated to form clusters. The size of the clusters was controlled by varying the rate at which the solution was heated. Figure 5.1b shows the SEM image of a typical ZnO cluster. The ZnO nanocrystallites were polydisperse and had an average size of 50 nm. The size of the clusters varied from submicron to a few microns.

5.2.2 Optical Measurement of Random Lasing

To introduce optical gain to ZnO, the frequency-tripled output of a pulsed Nd:YAG laser ($\lambda = 355$ nm, pulse width = 30 ps, repetition rate = 10 Hz) was focused by a lens onto the ZnO nanorods from the top. The ZnO nanorods served both as gain medium and scattering element. Figure 5.2 shows the spectra of emission from the ZnO nanorod array. At low pumping, the spectrum displayed a single broad spontaneous emission peak of width ~12 nm. As the pump intensity was increased, the emission peak became narrower due to the preferential amplification at frequencies close to the maximum of the gain spectrum. When the pumping intensity exceeded a threshold, discrete narrow peaks emerged on top of the broad spontaneous emission peak. The linewidth of these peaks was less than 0.2 nm, much smaller than the width of the spontaneous emission peak. When the pump intensity was increased further, additional narrow peaks appeared. The frequencies of the discrete peaks depended on the sample position. When the pump position on the sample was fixed, the peak frequencies remained the same, while the peak height varied from pulse to pulse. However, when we pumped a different part of the sample, the peak frequencies changed. This phenomenon suggested that the discrete spectral peaks resulted from spatial resonances for light that were determined by the local configurations of ZnO nanorods.

The ZnO nanorod array is a 2D scattering system, that is, light is scattered by the nanorods in the plane perpendicular to the rods. After multiple scattering, light may return to a position from which it visited before. In fact, light may return to its original position through many different paths. All the backscattered waves interfere; thus the feedback is field or amplitude feedback. Only at certain frequencies the interference is constructive, and light can be effectively confined inside the random system. Lasing occurs at these frequencies, producing discrete peaks in the emission spectrum. Above the lasing

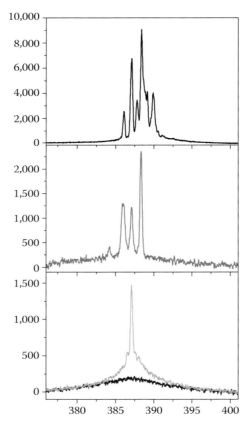

FIGURE 5.2 Emission spectra of the ZnO nanorods when excited by the third harmonics of a pulsed Nd:YAG laser. The incident pump power is (bottom to top) 3.2, 4.5, 6.1, and 11.1 J/cm². (Cao, H., Random lasers: Development, features and applications, *Opt. Photon. News*, 16, 24–30, 2005. With Permission of Optical Society of America.)

threshold, the emission intensity increased much more rapidly with the pumping intensity; meanwhile, the emission pulse length was reduced from 200 ps to nearly 20 ps. In the photon counting experiment, we confirmed that the laser emission is indeed coherent. This result is counterintuitive in the sense that coherent light can be generated from inherently disordered structure.

The abovementioned behaviors of a random laser with coherent feedback are similar to those of a conventional laser. Yet there is a significant difference between these two: a random laser does not have directional output. Because the disorder-induced scattering is random in direction, the random laser output is multidirectional. In fact, the directions of the output for individual lasing modes are different.

Optical scattering can be utilized not only for lasers, but also for microlasers. The key issue for microlaser fabrication is the confinement of light in a small volume with dimensions on the order of optical wavelength. Several types of microlasers have been developed over the past two decades [42]. We have demonstrated that microlasers can be made out of disordered media [43]. Strong scattering in random media provided an alternative way of confining light for microlaser.

Our micro random laser was based on micrometer-sized clusters of closely packed ZnO nanoparticles. Figure 5.1b shows a cluster of size ~1.7 μm. It contains roughly 20,000 ZnO nanocrystallites. The clusters were taken from the reaction solution and placed sparsely on a quartz substrate. The optical experiment was performed on a single cluster. The ZnO nanoparticles were excited by the fourth harmonic of a pulsed Nd:YAG laser ($\lambda = 266$ nm). The pump light was focused by a microscope objective lens to a single cluster. We simultaneously measured the spectrum of emission from the

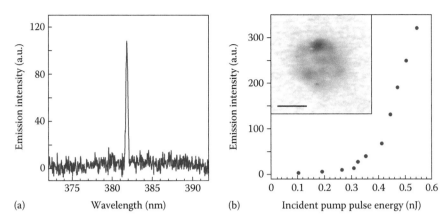

(a) Wavelength (nm)   (b) Incident pump pulse energy (nJ)

FIGURE 5.3 Lasing data for the ZnO microcluster shown in Figure 5.1b, when it is excited by the third harmonics of a pulsed Nd:YAG laser. (a) Spectrum of emission from the cluster at the incident pump pulse energy of 0.35 nJ. (b) Spectrally integrated emission intensity as a function of the incident pump pulse energy. Inset: Optical image of the emitted light distribution across the cluster at the incident pump pulse energy of 0.35 nJ. The scale bar is 1 μm.

cluster with a spectrometer, and imaged the spatial distribution of the emitted light intensity across the cluster with a CCD camera. The pump light was blocked by a bandpass filter placed in front of the camera.

At low pump intensity, the emission spectrum consisted of a single broad spontaneous emission speak. The spatial distribution of the spontaneous emission intensity was uniform across the cluster. When the pump intensity exceeded a threshold, a sharp peak emerged in the emission spectrum (Figure 5.3a). Simultaneously, a few bright spots appeared in the image of the emitted light distribution in the cluster (inset of Figure 5.3b). When the pump intensity was increased further, a second sharp peak emerged in the emission spectrum. Correspondingly, additional bright spots appeared in the image of the emitted light distribution. The curve of the total emission intensity as a function of the pump intensity in Figure 5.3b displayed a distinct change in slope at a threshold. This threshold corresponded to the onset of lasing in the micron-sized cluster. Well above the threshold, the total emission intensity increased almost linearly with the pump intensity. The incident pump pulse energy at the lasing threshold was ~0.3 nJ. Note that less than 1% of the incident pump light was absorbed; the rest was scattered.

Three-dimensional confinement of light in the micrometer-sized cluster was realized through the process of multiple scattering and wave interference. The photons, spontaneously emitted by ZnO inside the cluster, could not easily get out of the cluster. They were scattered by ZnO nanocrystallites many times before finally escaping through the boundary of the cluster. During the process of multiple scattering, light was amplified by stimulated emission from ZnO, and it might return to the position which it had visited before. The interference of all returned light, when it was constructive, minimized light leakage out of the cluster. However, the constructive interference happened only at certain frequencies that were determined by the spatial configuration of the scatterers. Lasing occurred at these special frequencies because of lower lasing threshold. In a different cluster, the spatial arrangement of the ZnO nanocrystallites changed, so did the phase of the scattered light. Hence, the interference of the scattered light was constructive at different frequencies, at which lasing occurred. Therefore, the lasing frequencies were the fingerprint of the unique random structure of a specific cluster. Because optical confinement was not caused by light reflection at the surface of a cluster but by scattering inside the cluster, we were able to achieve lasing in clusters with irregular shapes and rough surfaces.

5.3 Photonic Crystal Light Sources

5.3.1 Two-Dimensional UV Photonic Crystal Laser

In 2004, we realized the first ZnO photonic crystal slab (PhCS) lasers operating in the near-UV regime at room temperature [36]. We developed a procedure to fabricate 2D periodic structures in ZnO films with the FIB etching. Post-thermal annealing was employed to recover the ZnO PL. Lasing was achieved in the strongly localized defect modes near the edge of photonic band gap by optical pumping. We also calculated the band structure of our samples using the 3D plane wave expansion method to explain the experimental results [44].

The 2D triangular lattice PhC structure was fabricated in ZnO. First, 200 nm ZnO films were grown on c-plane sapphire substrate by plasma-enhanced MOCVD at 750°C. The selected area electron diffraction pattern of the ZnO film revealed that single crystalline ZnO film was grown along the *c*-axis. Next, arrays of cylindrical air voids were fabricated in the ZnO films by focused Ga^{3+} ion beam etching at 30 keV. FIB has been widely used for maskless, resistless nano-scale patterning. It allowed us to control the position, size, and density of air cylinders in the ZnO films. However, the ZnO emission was quenched due to structural damages caused by FIB. To recover the ZnO emission, we annealed the patterned films in O_2 at 600°C for 1 h. To overlap the ZnO gain spectrum with the photonic band gap, the lattice constant *a* and the radius of the air cylinders *r* were varied over a wide range. The resolution of our FIB system limited the smallest step of length variation to 15 nm. The size of each pattern was about 8×8 μm with roughly 4000 air cylinders. A top-view SEM image of part of a pattern is shown in Figure 5.4 ($a = 130$ nm and $r/a = 0.25$). Side-view SEM images (not shown) revealed the air cylinders were etched through the ZnO film with nearly vertical walls.

The samples were optically pumped by the third harmonics of a mode-locked Nd:YAG laser at room temperature. A 10× microscope objective lens focused the pump beam to a 4 μm spot on the pattern, and also collected the emission to a spectrometer. Since the sapphire substrate was double-side polished and transparent in both visible and UV regimes, a 20× microscope objective lens was placed on the back side of the sample for simultaneous measurement of the spatial distribution of laser emission. The pump light was blocked by a bandpass filter. The image of lasing mode profile was projected by the objective lens onto a UV-sensitive CCD camera. The sample was also illuminated by a white light source so that we can identify the position of the lasing modes in the photonic lattice.

Among all fabricated patterns with lattice constant *a* varying from 100 to 160 nm, lasing was realized only in the structures of $a = 115$ and 130 nm. Figure 5.5a shows the spectrum of emission from a pattern

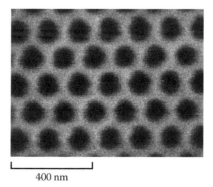

400 nm

FIGURE 5.4   Top-view SEM image of a triangular array of air holes (black) in a 200 nm-thick ZnO crystalline film. The lattice constant $a = 130$ nm, and air cylinder radius $r = 33$ nm. (With permission from Wu, X., Yamilov, A., Liu, X., Li, S., Dravid, V.P., Chang, R.P.H., and Cao, H., Ultraviolet photonic crystal laser, *Appl. Phys. Lett.*, 85, 3657–3659, 2004. Copyright 2004, American Institute of Physics.)

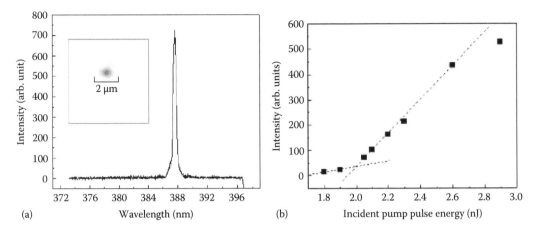

FIGURE 5.5 (a) Emission spectrum of a ZnO photonic crystal film displaying a single lasing peak. $a = 115$ nm, and $r/a = 0.25$. The incident pump pulse energy is 2.3 nJ. The inset is the near-field image of the lasing mode. (b) Emission intensity of this mode as a function of the incident pump pulse energy. (With permission from Wu, X., Yamilov, A., Liu, X., Li, S., Dravid, V.P., Chang, R.P.H., and Cao, H., Ultraviolet photonic crystal laser, *Appl. Phys. Lett.*, 85, 3657–3659, 2004. Copyright 2004, American Institute of Physics.)

of $a = 115$ nm and $r/a = 0.25$. It displayed a single sharp peak at 387.7 nm. Figure 5.5b plots the emission intensity integrated over this peak as a function of the incident pump pulse energy. The threshold behavior was clearly seen. Above the threshold, the spectral width of this lasing peak was merely 0.24 nm. These data indicated that lasing oscillation occurred in this structure. The near-field image of this lasing mode was obtained simultaneously and shown in the inset of Figure 5.5a. The white square represents the boundary of the pattern. The lasing mode was spatially localized in a small region of ~1.0 μm inside the pattern. As we moved the pump spot across the pattern, the lasing modes changed in both frequency and spatial pattern. This behavior suggested that the lasing modes were spatially localized defect states. They were formed by structural disorder, which was unintentionally introduced during the fabrication process. In the SEM image (Figure 5.4), non-uniformity of the size and shape of air cylinders is evident. Lasing was also observed in the patterns of $a = 130$ nm and $r/a = 0.25$. However, the lasing modes had longer wavelength, and their lasing threshold was higher.

To understand our experimental results, we calculated the photonic band structures using the 3D plane wave expansion method [44]. We focused on the transverse-electric (TE) mode, because it was known that the polarization of ZnO exciton emission was predominantly perpendicular to the c-axis. Since the c-axis of our ZnO film was normal to the film/substrate interface, the emission was mainly into the TE modes, as we confirmed experimentally from the measurement of polarization of emission from the side of a ZnO film. The unavoidable disorder introduced into the structure during the fabrication created the spatially localized defect modes near the band edges [45]. By comparing the calculation results to the experimental data, we concluded that the lasing modes in the $a = 115$ nm structure correspond to the defect modes near the dielectric band edge, while the lasing modes in the $a = 130$ nm structure are the defect modes near the air band edge. The lasing threshold of the former structure was lower than that of the latter, because the defects modes at the dielectric band edge were concentrated inside ZnO and experience more gain.

5.3.2 UV Lasing in High-Order Bands of Three-Dimensional ZnO Photonic Crystals

We fabricated the ZnO inverse opals using the ALD technique [39]. Polystyrene opal templates were deposited by self-assembly onto glass substrates, infiltrated with ZnO by ALD, and removed by firing at elevated temperatures. The resulting PhC structures are face-centered cubic (FCC) arrays of air spheres surrounded by ZnO dielectric shells with a thickness of typically ~50 layers. Interstitial porosity between the shells is an inherent feature of the ALD infiltration of an opal lattice, and it leads to beneficial changes in the high-order band structure to be discussed later. The refractive index of ZnO is ~2.0 in the visible spectrum, but increases to over 2.5 as wavelength λ approaches the excitonic emission wavelength. Although this is not sufficient for the formation of an isotropic PBG, the refractive index contrast is large enough for strong directional gaps.

We tuned the position of the first-order gap throughout the visible spectrum by selecting appropriate polystyrene sphere diameters (i.e., PhC lattice constants) for the opal template. To study the high-order band structure we used spheres with diameters ranging from $d = 330$–383 nm. The optical effects of high-order bands were more complicated, as shorter waves experienced simultaneous Bragg diffraction from multiple sets of lattice planes and coupling of many Bloch modes occurred. Even without a complete PBG, stimulated emission was strongly enhanced in weakly dispersive "flat" higher bands [37], because the group velocity was reduced over an extended region of k space (group velocity anomaly).

Figure 5.6 shows reflection spectra for white light incident along the [111] direction (normal to the sample. The first-order Bragg peaks were observed at wavelengths from 655–765 nm (peak R1). We also observed two strong reflection peaks (R2 and R3) and several fine features at wavelengths approaching the ZnO absorption edge. In the lasing experiments, the sample surfaces were illuminated by a white-light source and imaged by an objective lens onto a CCD camera. Highly reflective areas free of cracks were selected for measurements. The samples were pumped at 10 Hz with short pulses of $\lambda = 355$ nm from a mode-locked Nd:YAG laser, with a spot diameter of ~20 μm and light incident along the [111] direction. At low pump level the emission spectra of all the samples featured a broad spontaneous emission peak with a maximum at $\lambda \approx 385$ nm. With increasing pump intensity a single lasing peak appeared (labeled L in Figure 5.6), with λ at the short-wavelength shoulder of the third reflection peak. The lasing peak initially had a width of 0.6–1.0 nm, but became narrower with increasing pump intensity (Figure 5.7a), reaching a full width at half maximum of 0.1–0.3 nm. The emission increased rapidly and

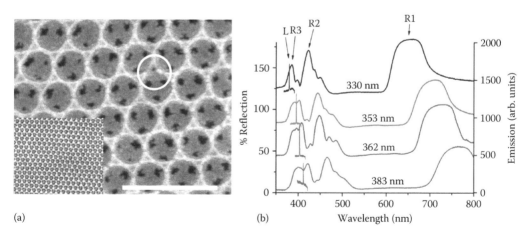

(a)   (b)

FIGURE 5.6   (a) SEM image of exposed (111) top surface of 356 nm ZnO inverted opals. (Scale bar equals 1 μm.) The interstitial pores inherent in the infiltration process remain open after firing of the structure (circle). Insert: Lower magnification image of (111) surface. (b) Reflection and lasing spectra of ZnO inverse opals of four different sphere diameters (written next to the curves). The reflection spectrum of a white light, incident along the [111] direction normal to the sample surface, contains three major peaks labeled as R1, R2 and R3. They shift to shorter wavelength as the sphere diameters decrease. Lasing (peak L) occurs at the high-frequency shoulder of the third peak (R3) in all samples. (With permission from Scharrer, M., Wu, X., Yamilov, A., Cao, H., and Chang, R.P.H., Fabrication of inverted opal ZnO photonic crystals by atomic layer deposition, *Appl. Phys. Lett.*, 86, 151113, 2005. Copyright 2005, American Institute of Physics.)

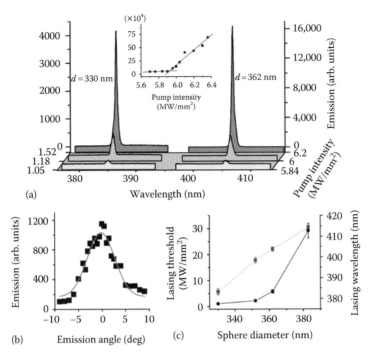

FIGURE 5.7 (a) Emission spectra of ZnO inverse opal with sphere diameter $d = 330$ and 362 nm for increasing pump intensity. The peak positions blueshift slightly with increasing pump level. Inset: input-output curve for the 362 nm sample (with a clear threshold at ~5.9 MW/mm²). (b) Angular distribution of lasing emission. 0° corresponds to the [111] direction normal to the sample surface. The emission is strongly directional with a full width at half-maximum ~6°. (Squares represent measured data, the red line is a guide for the eye.) (c) Lasing threshold (circle) and wavelength (square) versus sphere diameter d. Lasing occurs at longer wavelengths for increasing lattice constants. As the lasing wavelength shifts away from the ZnO gain maximum, the lasing threshold increases sharply. (With permission from Scharrer, M., Wu, X., Yamilov, A., Cao, H., and Chang, R.P.H., Ultraviolet using in high-order bands of three-dimensional ZnO photonic crystals, *Appl. Phys. Lett.*, 88, 2006. Copyright 2005. American Institute of Physics.)

displayed a pronounced threshold behavior (Figure 5.7a inset). Lasing output was highly directional, with a divergence angle of approximately 6° (Figure 5.7b). The directionality was preserved if the pump light was incident at an angle slightly off the surface normal. This indicated that lasing was strongly confined in the [111] direction. Changing the PhC lattice constant led to a shift in lasing wavelength λ_{las}, from $\lambda_{las} \approx 383$ nm for $d = 330$ nm to $\lambda_{las} \approx 415$ nm for $d = 383$ nm (Figure 5.7c), closely matching the shift in the high-order band structure. The gain spectrum of ZnO is centered at ~385 nm. Accordingly, the threshold increases with growing distance of λ_{las} from the gain maximum, for example, by a factor of 30 from $\lambda_{las} = 383$ nm ($d = 330$ nm) to $\lambda_{las} = 415$ nm ($d = 383$ nm). The fact that lasing occurred at frequencies so far from the gain maximum revealed that the confinement effect of the high-order band structure is remarkably strong in these PhCs. Well above threshold, secondary lasing peaks appeared within a wavelength range of approximately ±3 nm from the main peak. These peaks had more isotropic output and might result from defects or lasing in other crystal directions. However, the directional nature of the primary lasing mode, the reproducibility and stability of λ_{las}, and its dependence on the PhC lattice constant suggested that the primary lasing was related to the photonic band structure.

To understand our experimental results, we correlated them to the photonic band structure. Three directional PBGs match up closely with the prominent reflection peaks in the experimental data (Figure 5.6). At the high-frequency edge of the third gap, there is an isolated degenerate band which coincided with the lasing wavelength for all samples of different lattice constants. This photonic band has a group velocity v_g less than 0.09 c, where c is the speed of light in vacuum. Light amplification in this band is strongly enhanced due to the extremely low v_g over the whole Γ–L range. This suggests that lasing occurs due to distributed feedback of slow propagating modes in this band. Such a flat band is a feature of the inverse opal structure with interstitial porosity and that the corresponding band in a fully filled inverse opal structure shows considerably more curvature (thus larger v_g). The increase of the refractive index in the vicinity of the electronic band edge leads to additional flattening of the band. Our calculations showed that 25%–32% of the energy of a mode in this band is concentrated in the dielectric structure and thus can be efficiently amplified by the gain material. Therefore, even without a full PBG, the flat band can lead to robust and directional lasing.

5.3.3 Photoluminescence Modification by a High-Order Photonic Band with Abnormal Dispersion in ZnO Inverse Opal

In addition to lasing, we also demonstrated that a high-order photonic band with abnormal dispersion can significantly modify the PL intensity, directionality, and polarization in an inverse opal [46]. The dispersion of photons in a PhC can be dramatically different from that in free space. Let us denote the dispersion of a photonic mode by $\omega(k)$, where ω is photon frequency and k is wave vector. A mode with $d\omega/dk \simeq 0$ is called a slow mode, because the group velocity is nearly zero. There have been many proposals of utilizing the slow modes of PhC to reduce the speed of light by orders of magnitude for a wide range of applications. However, a serious problem that hinders the slow light application is that a typical slow mode with $d^2\omega/dk^2 \neq 0$ has large impedance mismatch at the PhC/air interface; thus, the conversion efficiency of incident light into the slow mode is very low. A solution to this problem is to use the photonic mode at the stationary inflection point of dispersion curve of a photonic band [47]. Such a mode has both $d\omega/dk \simeq 0$ and $d^2\omega/dk^2 \simeq 0$. It is called a frozen mode. When the incident light is in resonance with a frozen mode, the vanishing group velocity is offset by the diverging electromagnetic energy density. The energy flux inside the PhC is finite and comparable to the incident flux. Hence, the incident light can be completely converted to the frozen mode instead of being reflected. Our aim is to employ the unique properties of a frozen mode in a 3D PhC to tailor spontaneous emission. The vanishing group velocity enhances emission into the frozen mode, while the perfect impedance match at the PhC/air interface leads to efficient extraction of emission from PhC.

When pumped by a He:Cd laser at $\lambda = 325$ nm, the ZnO inverse opal has PL in both the UV and visible frequencies. The UV emission is ascribed to electron transition from the conduction band to valence band, and the visible emission is via the defect states within the electronic bandgap. At room temperature the defect emission, which results from various material defects such as oxygen vacancy, zinc interstitial, and oxygen interstitial, covers over 300 nm wavelength range. Here we concentrated on the defect emission whose broad spectral range allowed us to observe the effects of many different-order photonic bands on emission.

We measured the angle- and polarization-resolved PL spectra of ZnO inverse opals. The sample was mounted on a goniometer stage. Spectra of emission into different angles θ were measured as the detector was scanned in the ΓLK plane of the FCC lattice. A linear polarizer was placed in front of the detector to select s- or p-polarized light with electric field perpendicular or parallel to the detection plane (made of the detection arm and the normal of the sample surface). Figure 5.8 shows the PL spectra of a ZnO inverse opal of $d = 400$ nm and a reference sample of random structure. The spectral shape of PL from the random sample did not change with observation angle θ. The PL of ZnO inverse opal was strongly modified, and the modification was angle-dependent. For comparison, the emission spectra taken at identical θ were scaled so that they overlapped at $\lambda = 880$ nm, well below the lowest-order PBG. Suppression of emission at longer wavelength is evident in the ZnO inverse opal. At shorter wavelength there is significant enhancement of s-polarized emission. This enhancement is not related to stimulated emission, as the emission intensity is confirmed to vary linearly with pump intensity.

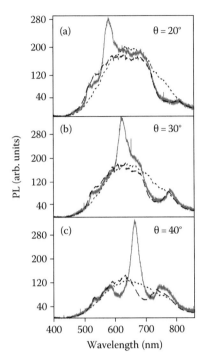

FIGURE 5.8 Measured PL spectra of a ZnO inverse opal (sphere diameter $d = 400\,nm$) and a random sample (dotted line). The emission angle θ from the surface normal is 20° (a), 30° (b), and 40° (c). The solid curve and dashed curve represent the s and p-polarized emission from the ZnO inverse opal, respectively. (With permission from Noh, H., Scharrer, M., Anderson, M.A., Chang, R.P.H., and Cao, H., Photoluminescence modification by a high-order photonic band with abnormal dispersion in ZnO inverse opal, *Phys. Rev. B*, 77, 115136, 2008. Copyright 2008 by the American Physical Society.)

We extracted the PL enhancement factor by dividing the PL spectrum of ZnO inverse opal by that of the reference sample for same θ. If the normalized PL intensity was less (or more) than unity, the luminescence was suppressed (or enhanced). Figure 5.9 shows the normalized PL spectra for both s- and p-polarizations with θ varying from 0° to 50°. The dip at λ ~800 nm for θ = 0° blueshift with increasing θ. Its wavelength coincided with a primary peak in the reflection spectra taken at the incident angle θ (not shown). Our earlier studies of angle-resolved reflection spectra confirmed that the primary reflection peak resulted from the lowest-order PBG in [111] direction. This partial gap suppressed the emission due to depletion of density of states (DOS) within certain angle. The peaks at higher frequencies in the normalized PL spectra illustrated the emission enhancement by higher-order photonic band structures. Most enhancement peaks for both polarizations were weakly dispersive with angle, except one for s-polarized PL. This peak redshifted dramatically with increasing θ. Its amplitude reached a maximal value of 2.3 at θ = 40°, exceeding all other peaks. It was responsible for strong enhancement of s-polarized PL in Figure 5.8. This PL peak had similar dispersion as a reflection peak that existed only for s-polarization. To compare their frequencies, we overlay the normalized PL spectra and reflection spectra in Figure 5.10 (1st row) for θ = 20°–50°. It is evident that the two peaks do not overlap spectrally; instead the PL peak is always at the low-frequency shoulder of the reflection peak.

To interpret the experimental results, we performed numerical simulations of ZnO inverse opals using the experimental values. To account for the angular dependence of reflection and PL, we calculated photonic bands and DOS corresponding to a specified angle of incidence/exit in air. These bands and DOS are called reduced bands and reduced DOS [46], as opposed to angle-integrated bands and DOS. We calculated the reduced photonic bands of s polarization in Figure 5.10 (second row) and

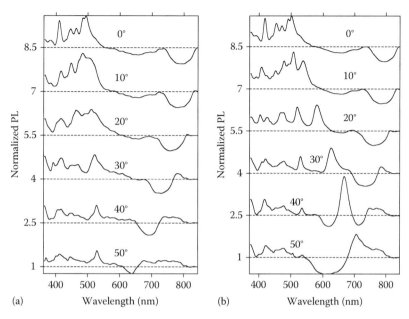

FIGURE 5.9 Normalized PL spectra of the ZnO inverse opal at various emission angle θ. The values of θ are written in the graph. The spectra are vertically shifted with a constant offset of 1.5. The reference line of unity for each spectrum is plotted as a horizontal dashed line. (a) *p*-polarized emission, (b) *s*-polarized emission. (With permission from Noh, H., Scharrer, M., Anderson, M.A., Chang, R.P.H., and Cao, H., Photoluminescence modification by a high-order photonic band with abnormal dispersion in ZnO inverse opal, *Phys. Rev. B*, 77, 115136, 2008. Copyright 2008 by the American Physical Society.)

p polarization bands in Figure 5.11 (second row). The projection of *k* vector in the *z*-axis that is parallel to [111] direction and normal to the surface of ZnO inverse opal is represented by k_z. Comparing the reduced *s*- and *p*-polarized bands revealed their significant differences. For example, at θ = 30°, the second and third *s*-polarized bands (labeled as 2*s* and 3*s* in Figure 5.10) exhibit frequency anti-crossing around $k_z a/2\pi = 0.7$, while 2*p* and 3*p* bands nearly cross in Figure 5.11. For both polarizations, the second and third reduced bands originate from Bragg diffraction of light by (111) and ($\bar{1}$11) planes for $k_z > 0$, and by (111) and (200) planes for $k_z < 0$. At their crossing point near $k_z a/2\pi \sim 0.7$, simultaneous Bragg diffraction by (111) and ($\bar{1}$11) planes results in band repulsion [48]. The anti-crossing of 2*s* and 3*s* bands indicates strong band coupling via multiple diffraction. The interaction of 2*p* and 3*p* bands, however, is much weaker. Such difference can be explained by the dependence of diffraction efficiency on polarization. It is well known for x-ray diffraction that when the Bragg angle is close to 45°, the intensity of diffracted beam is extremely weak for *p*-polarized wave. The suppression of Bragg diffraction has formal analogy to the Brewster effect on reflection of *p*-polarized light by homogeneous medium. The diffraction efficiency for *p*-polarized light in a PhC can be greatly reduced if the incident angle approaches the critical angle [49,50]. The weak coupling of 2*p* and 3*p* bands is attributed to low efficiency of multiple diffraction of *p*-polarized light at the band crossing point because one of the Bragg angles is close to the critical angle.

Comparison of the calculated band structure to the measured reflection spectra in Figures 5.10 and 5.11 confirmed that the primary reflection peak corresponded to the lowest-order gap between the first and second bands for both polarizations. The additional reflection peak observed only for *s*-polarization overlaps with the gap opened by anti-crossing of 2*s* and 3*s* bands. As θ increases, the gap moves toward lower frequency, and the reflection peak follows. This reflection peak is not observed for *p*-polarization because 2*p* and 3*p* bands have little repulsion. In fact the calculated reflection spectrum for θ = 40°

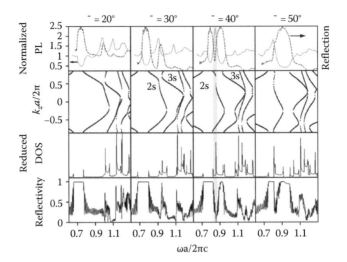

FIGURE 5.10 First row: angle-resolved *s*-polarized reflection spectra (dashed line) of the ZnO inverse opal over-laid with the normalized PL spectra (solid line) of same polarization and angle θ. Second row: calculated *s*-polarized reduced band structure of the ZnO inverse opal for a fixed angle of incidence/exit in air. Lattice constant $a = 566$ nm, the dielectric constant of ZnO is $\varepsilon = 3.8$. Third row: calculated reduced density of *s*-polarized states of the ZnO inverse opal. Fourth row: calculated reflectivity of *s*-polarized light from a ZnO inverse opal whose thickness is 34 layers of air spheres. $\phi = 0$, and $\theta = 20°$ (first column), 30° (second column), 40° (third column), 50° (fourth column). For $\theta = 40°$, a stationary inflection point is developed for 2*s* band at $\omega a/2\pi c = 0.856$. (With permission from Noh, H., Scharrer, M., Anderson, M.A., Chang, R.P.H., and Cao, H., Photoluminescence modification by a high-order photonic band with abnormal dispersion in ZnO inverse opal, *Phys. Rev. B*, 77, 115136, 2008. Copyright 2008 by the American Physical Society.)

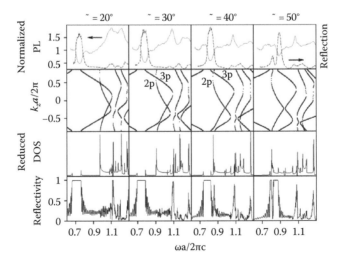

FIGURE 5.11 Same as Figure 5.10 for *p*-polarization. (With permission from Noh, H., Scharrer, M., Anderson, M.A., Chang, R.P.H., and Cao, H., Photoluminescence modification by a high-order photonic band with abnormal dispersion in ZnO inverse opal, *Phys. Rev. B*, 77, 115136, 2008. Copyright 2008 by the American Physical Society.)

exhibits a very narrow peak corresponding to the small gap between $2p$ and $3p$ bands at $k_za/2\pi \sim 0.7$ (Figure 5.11). Such narrow peak is smeared out experimentally by averaging over finite collection angle.

The dispersion of $2s$ band is nearly flat in the vicinity of its avoided crossing with $3s$ band. It produces a peak in the reduced DOS, as shown in Figure 5.10. The enhanced PL peak, which is observed only for s-polarization, coincides with this DOS peak. It follows the DOS peak as it moves to lower frequency at higher θ. At $\theta = 40°$, the dispersion curve for $2s$ band has a stationary inflection point at $k_za/2\pi \simeq 0.6$ where $d\omega/dk_z \simeq 0$ and $d^2\omega/dk_z^2 \simeq 0$.

The existence of stationary inflection point was verified by tracing the evolution of $2s$ band with θ. For $\theta = 50°$, the dispersion curve for $2s$ band exhibits a local minima at $k_za/2\pi \sim 0.5$ and a local maxima at $k_za/2\pi \sim 0.7$. As θ decreases, the local minima and local maxima approach each other, and eventually they merge at $\theta \simeq 40°$. With a further decrease of θ, for example, at $\theta = 30°$, the dispersion curve has neither local minima nor local maxima; its slope does not change sign throughout the region of interest. Such band evolution confirms that $2s$ band has a stationary inflection point at $\theta = 40°$ where the merging of a local minima and a local maxima gives not only $d\omega/dk_z = 0$ but also $d^2\omega/dk_z^2 = 0$. The evolution of $2p$ band with θ, shown in the second row of Figure 5.11, is completely different from that of $2s$ band. It reveals that $2p$ band does not have a stationary inflection point near $\theta = 40°$, possibly due to its tiny anti-crossing with $3p$ band.

The DOS at the stationary inflection point diverges in an infinite large PhC. In a real sample such divergence is avoided because of finite sample size. Nevertheless, the DOS peak has the maximal amplitude at $\theta \simeq 40°$ where the stationary inflection point is developed. The large DOS enhances the spontaneous emission process. Although $2p$ band does not have a stationary inflection point, its dispersion is relatively flat in the neighborhood of $k_za/2\pi = 0.6$ (Figure 5.11). It produces a peak in the reduced DOS, which should enhance emission. Experimentally the p-polarized PL is not enhanced.

The question arised why the DOS peak leads to enhanced PL for s-polarization but not for p-polarization. The answer lied in the emission extraction efficiency. One unique property of the frozen mode at the stationary inflection point is its efficient coupling to free photon mode outside the PhC. It leads to vanishing reflectivity at the sample/air interface, which is confirmed by our calculation and measurement of reflectivity from ZnO inverse opal. The calculated reflection spectrum for $\theta = 40°$ (Figure 5.10) reveals that the reflectivity at the stationary inflection point $\omega a/2\pi c = 0.856$ is almost zero. Experimentally, the measured reflectivity is low but not zero due to averaging over finite collection angle. We verified that the zero reflectivity was not caused by the Fabry–Perot resonance in a finite PhCS, as its frequency did not vary with the slab thickness. Hence, the vanishing reflectivity was an intrinsic property of the photonic band, more specifically, the dispersion of $2s$ band. This result demonstrated perfect impedance match for the frozen mode at the interface of ZnO inverse opal and air. Light that is emitted to the frozen mode of the PhC had little reflection when reaching the sample surface, namely it mostly transmitted through the sample surface to air. Hence, s-polarized PL at the stationary inflection point of $2s$ band was efficiently extracted from the sample. The reflection of p-polarized light did not vanish due to the absence of frozen mode; thus, p-polarized PL could not escape easily from the sample.

Although the data presented earlier were taken from the ZnO inverse opal with $d = 400\,\text{nm}$, we repeated the experiments with several samples of different sphere diameter and obtained similar results. The enhancement of PL by the frozen mode is a common phenomenon because many high-order bands of ZnO inverse opal have stationary inflection points in their dispersion curves. In addition to scanning in the ΓLK plane, we also scanned in the ΓLW plane and observed PL enhancement at a different angle θ.

5.4 Conclusion

In summary, we have demonstrated the applications of nanostructured ZnO to UV lasers and light emitting diodes (LEDs). The UV lasers have potential applications in high-density optical storage, high-resolution laser printing, and sharper displays, as the shorter lasing wavelength gives better spatial resolution. The combination of a near-UV LED with red and green emitting phosphors becomes one of the most promising methods of generating white light for applications such as display backlighting, accent lighting, and general illumination.

Both the top-down and bottom-up approaches have been developed to fabricate ZnO nanostructures. We demonstrated random lasing in 2D arrays of ZnO nanorods. Optical scattering in the disordered structures traps light inside the gain material for a longer time, and enhances light amplification. The interference of scattered light also provides feedback for lasing action. Moreover, light can be localized in micrometer-sized cluster of ZnO nanoparticles through the process of multiple scattering and wave interference. Utilizing this mechanism, we fabricated a new type of microlaser with disordered materials.

In addition to random structures, we also utilized periodic structures to enhance light emission and lasing in ZnO. We fabricated 2D PhC structures in crystalline ZnO films with FIB etching. Lasing was realized in the near-UV frequency at room temperature under optical pumping. The lasing modes were found to be the strongly localized defect modes near the edges of photonic band gap. These defect modes originated from the structure disorder unintentionally introduced during the fabrication process.

We also realized UV lasing in 3D ZnO PhCs at room temperature. The PhCs were inverse opals with high refractive index contrast that simultaneously confined light and provided gain under optical excitation. The laser output was highly directional and the lasing frequency was tuned by the structural parameters. Comparison of the experimental results to the calculated band structure showed that lasing occurred in high-order bands with abnormally low group velocity. This demonstrated that the high-order band structure of 3D PhCs could be used to effectively confine light and enhance lasing action.

Finally, we investigated the modification of broadband defect emission in the ZnO inverse opals by photonic bands with abnormal dispersion. The angle- and polarization-resolved PL spectra revealed a significant enhancement of spontaneous emission in particular directions for a specific polarization. A detailed theoretical analysis and numerical simulation illustrated that the frozen mode at a stationary inflection point of a dispersion curve for a high-order photonic band can strongly modify the intensity, directionality, and polarization of spontaneous emission and enhance the extraction efficiency. In the future we will utilize the frozen modes to control the exciton emission of ZnO. It can be an effective mechanism for improving the efficiency of UV LEDs.

Acknowledgments

We wish to thank our former and current students, postdocs, and collaborators for their contributions to the work described in this chapter. In alphabetical order, they are as follows: Mark Anderson, Yong Ling, Xiang Liu, Heeso Noh, Michael Scharrer, Eric W. Seelig, Xiaohua Wu, Junying Xu, and Alexey Yamilov. Our research program has been sponsored by the National Science Foundation, the David and Lucille Packard Foundation, the Alfred P. Sloan Foundation, and Northwestern University Materials Research Center.

References

1. A. van Dijken, E. A. Meulenkamp, D. Vanmaekelbergh, and A. Meijerink. The kinetics of the radiative and nonradiative processes in nanocrystalline ZnO particles upon photoexcitation. *J. Phys. Chem. B*, 104:1715–1723, 2000.
2. H. Zhou, H. Alves, D. M. Hofmann, W. Kriegseis, B. K. Meyer, G. Kaczmarczyk, and A. Hoffmann. Behind the weak excitonic emission of ZnO quantum dots: ZnO/Zn(OH)(2) core-shell structure. *Appl. Phys. Lett.*, 80:210–212, 2002.
3. S. Sakohara, M. Ishida, and M. A. Anderson. Visible luminescence and surface properties of nano-sized ZnO colloids prepared by hydrolyzing zinc acetate. *J. Phys. Chem. B*, 102:10169–10175, 1998.

4. S. Mahamuni, K. Borgohain, B. S. Bendre, V. J. Leppert, and S. H. Risbud. Spectroscopic and structural characterization of electrochemically grown ZnO quantum dots. *J. Appl. Phys.*, 85:2861–2865, 1999.

5. C. L. Yang, J. N. Wang, W. K. Ge, L. Guo, S. H. Yang, and D. Z. Shen. Enhanced ultraviolet emission and optical properties in polyvinyl pyrrolidone surface modified ZnO quantum dots. *J. Appl. Phys.*, 90:4489–4493, 2001.

6. P. D. Cozzoli, M. L. Curri, A. Agostiano, G. Leo, and M. Lomascolo. ZnO nanocrystals by a non-hydrolytic route: Synthesis and characterization. *J. Phys. Chem. B*, 107:4756–4762, 2003.

7. H. Cao, J. Y. Wu, H. C. Ong, J. Y. Dai, and R. P. H. Chang. Second harmonic generation in laser ablated zinc oxide thin films. *Appl. Phys. Lett.*, 73:572–574, 1998.

8. G. Wang, G. T. Kiehne, G. K. L. Wong, J. B. Ketterson, X. Liu, and R. P. H. Chang. Large second harmonic response in ZnO thin films. *Appl. Phys. Lett.*, 80:401–403, 2002.

9. X. Q. Zhang, Z. K. Tang, M. Kawasaki, A. Ohtomo, and H. Koinuma. Resonant exciton second-harmonic generation in self-assembled ZnO microcrystallite thin films. *J. Phys. Cond. Matter*, 15:5191–5196, 2003.

10. G. I. Petrov, V. Shcheslavskiy, V. V. Yakovlev, I. Ozerov, E. Chelnokov, and W. Marine. Efficient third-harmonic generation in a thin nanocrystalline film of ZnO. *Appl. Phys. Lett.*, 83:3993–3995, 2003.

11. H. Cao. Lasing in disordered media. In Wolf, E., Ed., *Progress in Optics*, vol. 45 of *Progress in Optics*, pp. 317–370. North-Holland, Amsterdam, the Netherlands, 2003.

12. R. V. Ambartsumyan, N. G. Basov, P. G. Kryukov, and V. S. Letokhov. Non-resonant feedback in lasers. In J. H. Sanders and K. W. H. Stevens, Eds., *Progress in Quantum Electronics*, Vol. 1 of *Progress in Optics*, pp. 105–193. Pergamon Press, London, U.K., 1970.

13. H. Cao. Review on latest developments in random lasers with coherent feedback. *J. Phys. A*, 38:10497–10535, 2005.

14. M. Noginov. *Solid-State Random Lasers*, vol. 105 of *Springer Series in Optical Sciences*. Springer, New York, 2005.

15. D. S. Wiersma. The physics and applications of random lasers. *Nat. Phys.*, 4:359–367, 2008.

16. V. M. Markushev, V. F. Zolin, and C. M. Briskina. Luminescence and induced emission from neodymium in powders of double sodium-lanthanum molybdate. *Kvantovaya Elektronika*, 13:427–430, 1986.

17. V. M. Markushev, N. E. Tergabrielyan, C. M. Briskina, V. R. Belan, and V. F. Zolin. Kinetics of the neodymium powder laser action. *Kvantovaya Elektronika*, 17:854–858, 1990.

18. M. A. Noginov, N. E. Noginova, H. J. Caulfield, P. Venkateswarlu, T. Thompson, M. Mahdi, and V. Ostroumov. Short-pulsed stimulated emission in the powders of NdAl3(BO3)(4) NdSc3(BO3)(4), and Nd:Sr-5(PO4)(3)F laser crystals. *J. Opt. Soc. Am. B*, 13:2024–2033, 1996.

19. D. S. Wiersma and A. Lagendijk. Light diffusion with gain and random lasers. *Phys. Rev. E*, 54:4256–4265, 1996.

20. N. M. Lawandy, R. M. Balachandran, A. S. L. Gomes, and E. Sauvain. Laser action in strongly scattering media. *Nature*, 368:436–438, 1994.

21. H. Cao, Y. G. Zhao, H. C. Ong, S. T. Ho, J. Y. Dai, J. Y. Wu, and R. P. H. Chang. Ultraviolet lasing in resonators formed by scattering in semiconductor polycrystalline films. *Appl. Phys. Lett.*, 73:3656–3658, 1998.

22. H. Cao, Y. G. Zhao, S. T. Ho, E. W. Seelig, Q. H. Wang, and R. P. H. Chang. Random laser action in semiconductor powder. *Phys. Rev. Lett.*, 82:2278–2281, 1999.

23. H. Cao, Y. Ling, J. Y. Xu, C. Q. Cao, and P. Kumar. Photon statistics of random lasers with resonant feedback. *Phys. Rev. Lett.*, 86:4524–4527, 2001.

24. S. V. Frolov, Z. V. Vardeny, K. Yoshino, A. Zakhidov, and R. H. Baughman. Stimulated emission in high-gain organic media. *Phys. Rev. B*, 59:R5284–R5287, 1999.

25. S. G. Johnson and J. D. Joannopoulos. *Photonic Crystals: The Road from Theory to Practice*, 2nd edn. Springer, New York, 2002.

26. E. Yablonovitch. Inhibited spontaneous emission in solid-state physics and electronics. *Phys. Rev. Lett.*, 58:2059–2062, 1987.

27. S. John. Strong localization of photons in certain disordered dielectric superlattices. *Phys. Rev. Lett.*, 58:2486–2489, 1987.

28. J. D. Joannopoulos, S. G. Johnson, J. N. Winn, and R. D. Meade. *Photonic Crystals: Molding the Flow of Light*, 2nd edn. Princeton University Press, Princeton, NJ, 2008.

29. C. M. Soukoulis, Ed. *Photonic Crystals and Light Localization in the 21st Century*, vol. 563 of *NATO Science Series C*. Springer, New York, 2001.

30. S. Noda and T. Baba. *Roadmap on Photonic Crystals*. Springer, New York, 2010.

31. S. Noda, M. Fujita, and T. Asano. Spontaneous-emission control by photonic crystals and nanocavities. *Nat. Photon.*, 1:449–458, 2007.

32. P. Lodahl, A. F. van Driel, I. S. Nikolaev, A. Irman, K. Overgaag, D. L. Vanmaekelbergh, and W. L. Vos. Controlling the dynamics of spontaneous emission from quantum dots by photonic crystals. *Nature*, 430:654–657, 2004.

33. O. Painter, R. K. Lee, A. Scherer, A. Yariv, J. D. O'Brien, P. D. Dapkus, and I. Kim. Two-dimensional photonic band-gap defect mode laser. *Science*, 284:1819–1821, 1999.

34. H. Altug, D. Englund, and J. Vuckovic. Ultrafast photonic crystal nanocavity laser. *Nat. Phys.*, 2:484–488, 2006.

35. E. W. Seelig, B. Tang, A. Yamilov, H. Cao, and R. P. H. Chang. Self-assembled 3D photonic crystals from ZnO colloidal spheres. *Mater. Chem. Phys.*, 80:257–263, 2003.

36. X. Wu, A. Yamilov, X. Liu, S. Li, V. P. Dravid, R. P. H. Chang, and H. Cao. Ultraviolet photonic crystal laser. *Appl. Phys. Lett.*, 85:3657–3659, 2004.

37. M. Scharrer, A. Yamilov, X. H. Wu, H. Cao, and R. P. H. Chang. Ultraviolet lasing in high-order bands of three-dimensional ZnO photonic crystals. *Appl. Phys. Lett.*, 88:201103, 2006.

38. M. Scharrer, H. Noh, X. Wu, M. A. Anderson, A. Yamilov, H. Cao, and R. P. H. Chang. Five-fold reduction of lasing threshold near the first Gamma L-pseudogap of ZnO inverse opals. *J. Opt.*, 12, 024007, 2010.

39. M. Scharrer, X. Wu, A. Yamilov, H. Cao, and R. P. H. Chang. Fabrication of inverted opal ZnO photonic crystals by atomic layer deposition. *Appl. Phys. Lett.*, 86, 151113, 2005.

40. T. Baba. Slow light in photonic crystals. *Nat. Photon.*, 2:465–473, 2008.

41. X. Liu, A. Yamilov, X. H. Wu, J. G. Zheng, H. Cao, and R. P. H. Chang. Effect of ZnO nanostructures on 2-dimensional random lasing properties. *Chem. Mater.*, 16:5414–5419, 2004.

42. K. J. Vahala. Optical microcavities. *Nature*, 424:839–846, 2003.

43. H. Cao, J. Y. Xu, S. H. Chang, and S. T. Ho. Transition from amplified spontaneous emission to laser action in strongly scattering media. *Phys. Rev. E*, 61:1985–1989, 2000.

44. A. Yamilov, X. Wu, and H. Cao. Photonic band structure of ZnO photonic crystal slab laser. *J. Appl. Phys.*, 98, 103102, 2005.

45. A. Yamilov, X. Wu, X. Liu, R. P. H. Chang, and H. Cao. Self-optimization of optical confinement in an ultraviolet photonic crystal slab laser. *Phys. Rev. Lett.*, 96, 083905, 2006.

46. H. Noh, M. Scharrer, M. A. Anderson, R. P. H. Chang, and H. Cao. Photoluminescence modification by a high-order photonic band with abnormal dispersion in ZnO inverse opal. *Phys. Rev. B*, 77, 115136, 2008.

47. A. Figotin and I. Vitebskiy. Slow light in photonic crystals. *Waves in Random and Complex Media*, 16:293–382, 2006.

48. H. M. van Driel and W. L. Vos. Multiple Bragg wave coupling in photonic band-gap crystals. *Phys. Rev. B*, 62:9872, 2000.

49. A. V. Baryshev, A. B. Khanikaev, H. Uchida, M. Inoue, and M. F. Limonov. Interaction of polarized light with three-dimensional opal-based photonic crystals. *Phys. Rev. B*, 73:033103–033104, 2006.

50. A. A. Dukin, N. A. Feoktistov, A. V. Medvedev, A. B. Pevtsov, V. G. Golubev, and A. V. Sel'kin. Polarization inhibition of the stop-band in distributed Bragg reflectors. *J. Opt. A*, 8:625–629, 2006.
51. H. Cao, J. Y. Xu, E. W. Seelig, and R. P. H. Chang. Microlasers made of disordered medial. *Appl. Phys Lett.*, 76:2997, 2000.
52. H. Cao, Random lasers: Development, features, and applications. *Opt. Photon. News*, 16:24–29, 2005.

6

Luminescent ZnO and MgZnO

Leah Bergman
University of Idaho

Jesse Huso
University of Idaho

John L. Morrison
University of Idaho

M. Grant Norton
Washington State University

6.1 Introduction

Zinc oxide (ZnO) is a direct bandgap semiconductor with a bandgap of ~3.37 eV at room-temperature and relatively deep excitonic binding energy of 60 meV, both attributes of which make ZnO an efficient UV optical material at and above room temperature [1–6]. Due to their environmentally friendly chemical nature, resistivity to harsh environments, and deep excitonic level, ZnO as well as $Mg_xZn_{1-x}O$ (where $0 \leq x \leq 1$ is the composition) are emerging as promising materials capable of high-efficiency luminescence in a wide range of the ultraviolet (UV) spectrum [2,7]. The exceptional optical properties of ZnO have led to the realization of various optoelectronic devices. Random lasing in microcrystalline films and optically pumped stimulated emission in films and nanostructures have been demonstrated, and room temperature polariton lasers as well as light-emitting diodes and solar cells have also been realized [8–15]. Although the topic of ZnO has been extensively researched, less is known concerning the properties of the $Mg_xZn_{1-x}O$ alloy system. $Mg_xZn_{1-x}O$ thin films and nanostructures that include nanocrystals as well as core–shell structures have recently been studied with the objective of achieving a viable alloy family with tunable bandgap and luminescence at the UV range [16–26].

ZnO has the hexagonal wurtzite structure, while MgO has the NaCl cubic structure of a direct bandgap ~7.5 eV and excitonic binding energy ~140 meV [27,28]. Figure 6.1 presents the crystal structures of those materials; upon alloying, Zn and Mg atoms replace each other in their sublattice. Although the two oxides do not show complete solid solubility, an alloy system between the two oxides, $Mg_xZn_{1-x}O$ over a wide composition range, has been reported to take place in thin films made by pulsed-laser deposition (PLD) and by molecular beam epitaxy (MBE) techniques [16–21]. Other research groups have extended research efforts into the field of $Mg_xZn_{1-x}O$ in its nanocrystalline and nanostructure forms [22–26]. The $Mg_xZn_{1-x}O$ alloy system may provide an optically tunable family of wide bandgap materials

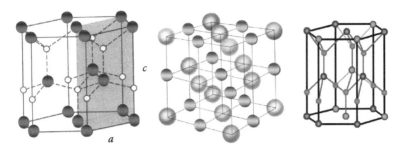

FIGURE 6.1 The hexagonal wurtzite crystal structure of ZnO (left), the cubic rocksalt structure of MgO (middle), and the wurtzite structure of MgZnO alloy, where the Mg and Zn atoms substitute for each other (right). For the wurtzite structure, the lattice parameters are $a = 3.24$ Å and $c = 5.20$ Å. The covalent radius of Zn = 1.25 Å. For the rocksalt structure, the lattice parameter is $a = 4.24$ Å. The covalent radius of Mg = 1.36 Å.

that can be used in various UV luminescences, absorption, lighting, and display applications in the range of 3.4–7.5 eV.

Although the covalent radii of Mg and Zn are comparable, due to their different crystal structures a very low solubility limit of the ZnO–MgO alloy system has been predicted. According to the phase diagram of the ZnO–MgO alloy system, the thermodynamic solubility limit of MgO in ZnO has been reported to be less than 4%, while the solubility of ZnO in MgO is ~56% (at ~1600°C) [29,30]. As was mentioned above, a few groups have grown $Mg_xZn_{1-x}O$ thin films via PLD and MBE methods. Overcoming the solubility limit in these methods of growth has been attributed to the lack of thermodynamic equilibrium conditions. In those studies, at up to approximately 30% of Mg concentration, the alloys were found to have the wurtzite structure, while at a composition range of ~30%–50% a phase separation occurs in which mixed phases of hexagonal and cubic structures as well as MgO precipitates have been observed [17]. At above ~50%, the alloy was found to exhibit the cubic structure. Figure 6.2 presents the bandgap of $Mg_xZn_{1-x}O$ over the entire composition range of the PLD thin films [19]. In general, each growth method and associated parameters should exhibit its particular solubility limit. For comparison purposes, the bandgap as a function of composition for the $Al_xGa_{1-x}N$ alloy system is

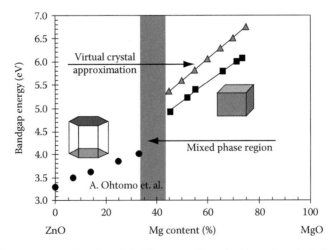

FIGURE 6.2 Bandgap energy as a function of the Mg content for pulsed laser deposited $Mg_xZn_{1-x}O$ thin films as obtained from transmission experiments. (With permission from Choopun, S., Vispute, R.D., Yang, W., Sharma, R.P., Venkatesan, T., and Shen, H., *Appl. Phys. Lett.*, 80, 1529, 2002. Copyright 2002, American Institute of Physics.)

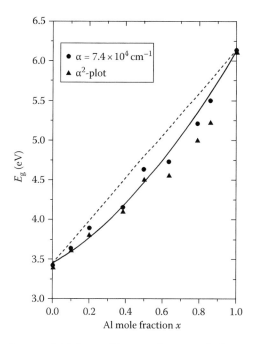

FIGURE 6.3 The bandgap of epitaxial $Al_xGa_{1-x}N$ films as a function of Al composition. The bandgap energy was determined via the absorption coefficient α. (With permission from Angerer, H., Brunner, D., Freudenberg, F., Ambacher, O., Stutzmann, M., Hopler, R., Metzger, T., Born, E., Dollinger, G., Bergmaier, A., Karsch, S., and Korner, H.-J., *Appl. Phys. Lett.*, 71, 1504, 1997. Copyright 1997, American Institute of Physics.)

presented in Figure 6.3 [31]. GaN and AlN both have the wurtzite structure, and solid solutions at the entire composition range have been previously achieved.

6.2 Experimental

The photoluminescence experiments utilized a JY-Horiba micro-Raman/PL system consisting of a high-resolution T-64000 triple monochromator and a UV confocal microscope capable of focusing on a spot size of ~1 micron diameter. Two lasers were used: a CW-Kimmon laser with a wavelength of 325 nm (3.8 eV), and a Lexel laser of 244 nm (5.1 eV) line. The cold temperature measurements were conducted in an INSTEC UV-compatible micro-cell at temperatures ranging from 77 to 900 K. The absorption experiments utilized the Varian–Cary 300 UV-visible system.

6.3 Photoluminescence Properties of $Mg_xZn_{1-x}O$ Nanocrystals

In this section, studies of the optical properties of $Mg_xZn_{1-x}O$ nanocrystals below and at the phase transition range are presented. The $Mg_xZn_{1-x}O$ nanocrystals were synthesized via a thermal decomposition method using zinc and magnesium acetates in a quartz tube furnace [32]. The compositions of the $Mg_xZn_{1-x}O$ nanocrystals were determined by x-ray photoelectron spectroscopy (XPS) and energy dispersive x-ray spectroscopy (EDS). The Mg content in each of the nanoalloy samples was found to be on the average ~3%, 12%, 23%, 30%, 40%, and 50% respectively. The x-ray diffraction (XRD) indicated that for samples of up to ~30% Mg composition, the nanocrystals retain mainly their wurtzite structure, while the samples of a higher Mg composition exhibit both the wurtzite as well as the cubic structure.

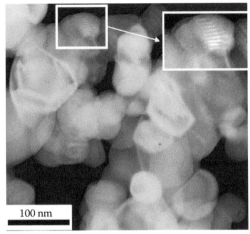

FIGURE 6.4 TEM images of ZnO (left) and MgZnO (right) nanocrystals. The average crystal size is ~40 nm. Most of the nanocrystals have surface imperfections. (With permission from Morrison, J.L., Huso, J., Hoeck, H., Casey, E., Mitchelle, J., Bergman, L., and Norton, M.G., *J. Appl. Phys.*, 104, 123519, 2008. Copyright 2008, American Institute of Physics.)

The first part of this section focuses on the study of the properties of $Mg_xZn_{1-x}O$ nanocrystals at the composition range $x = 0$–0.3. Figure 6.4 shows a representative transmission electron microscope (TEM) image of the ZnO and $Mg_{0.23}Zn_{0.77}O$ nanocrystal samples. As can be seen in the image, the samples consist of crystallites of an average size, ~40 nm. The crystallites are faceted and many display a banded contrast that is consistent with the presence of twins as well as surface imperfections. The PL and the transmission spectra of the $Mg_xZn_{1-x}O$ are presented in Figures 6.5 and 6.6, respectively, while Figure 6.7 depicts the functional behavior of the PL and the absorption edge. The optical spectra for that part of the study were acquired at room temperature.

The results presented in the above figures indicate that at a composition range of up to 30% Mg, the PL energy blueshifted by ~0.3 eV, and a similar blueshift is also evident in the absorption edge. Thus, in this

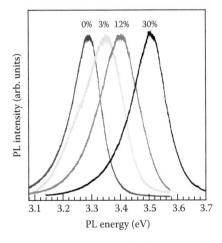

FIGURE 6.5 PL spectra of the $Mg_xZn_{1-x}O$ nanocrystals. The PL was acquired at room temperature; the probing laser spot size is ~1 μm. (With permission from Morrison, J.L., Huso, J., Hoeck, H., Casey, E., Mitchelle, J., Bergman, L., and Norton, M.G., *J. Appl. Phys.*, 104, 123519, 2008. Copyright 2008, American Institute of Physics.)

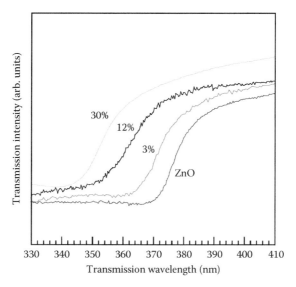

FIGURE 6.6 Transmission spectra of the $Mg_xZn_{1-x}O$ nanocrystals. The spectra were acquired at room temperature. (From Morrison, J.L. et al., *J. Appl. Phys.*, 104, 123519, 2008.)

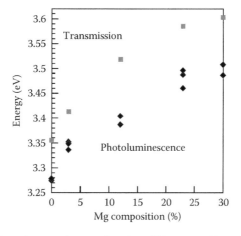

FIGURE 6.7 PL and absorption edge energies, as a function of Mg composition, of the $Mg_xZn_{1-x}O$ nanocrystals. Data points were compiled from the experiments presented in Figures 6.5 and 6.6. The multiple data points of the PL reflect the spatial fluctuation of the composition across each sample. The probing size of the PL is ~1 μm. (From Morrison, J.L. et al., *J. Appl. Phys.*, 104, 123519, 2008.)

regard, bandgap engineered nanoalloys of tunable luminescence properties were achieved [22,32]. In epitaxial $Mg_xZn_{1-x}O$ thin film grown PLD, a blueshift of ~0.5 eV was observed in the PL for the composition range of ~30% [17]. The optical properties of the PLD samples are depicted in Figures 6.8 and 6.9. The smaller blueshift in the optical spectra of the nanocrystals relative to that of the PLD thin films may indicate that a density of in-gap defect states exists, which in turn minimizes the bandgap, an implication that is reasonable given the large surface-to-volume ratio of the crystallites and the existence of surface defects as can be seen in the TEM images. In addition, it is plausible that the phase separation of the nanocrystals has an onset ~25% Mg, as can be inferred from the deviation from linearity of the PL trend presented in Figure 6.7.

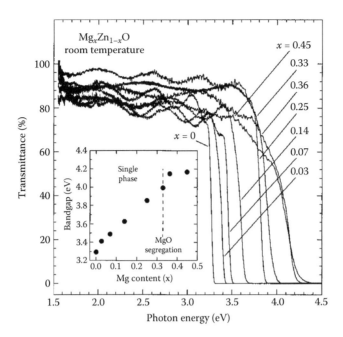

FIGURE 6.8 Transmission spectra of $Mg_xZn_{1-x}O$ films grown via PLD measured at room temperature. The inset shows the bandgap energy as a function of Mg compositions. The bandgap was determined from the transmission spectra via calculating the absorption coefficient. (With permission from Ohtomo, A., Kawasaki, M., Koida, T., Masubuchi, K., Koinuma, H., Sakurai, Y., Yoshida, Y., Yasuda, T., and Segawa, Y., *Appl. Phys. Lett.*, 72, 2466, 1998. Copyright 1998, American Institute of Physics.)

The determination of the origins of the PL in a semiconductor is important from two perspectives: One is the basic knowledge gained on the optical interactions of the semiconductor, and the other is the application of that material to devices. In particular, a comprehensive knowledge of the dominant PL emissions is needed in order to achieve a viable optoelectronic device. In most of the ZnO nanocrystals synthesized for the above research, the PL was consistently lower by a few tens of meV relative to the excitons of the ZnO. A detailed discussion of the excitons in ZnO is given in Chapter 4. In order to gain insight into the origins of the light emission of the nanocrystals, the PL was acquired at a temperature range of 77 K up to room temperature. Figure 6.10 presents the PL spectra at 77 K for the ZnO and the $Mg_{0.23}Zn_{0.77}O$ nanocrystals along with that of a reference ZnO bulk. As can be seen in Figure 6.10, the PL of the bulk ZnO consists of two main PL spectrallines at 3.362 and 3.376 eV. These two PL emissions have been previously observed and studied by various groups, and have been assigned to a neutral donor-bound exciton D^0X and to the free A-exciton (denoted as X), respectively [6]. As is shown in Figure 6.10 in the spectrum of the ZnO nanocrystals, the D^0X and the A-exciton are present in the same energies as those of the bulk. Additionally, phonon replicas are also present at the lower energy range of the PL spectra of the bulk and the nanocrystals. The key point is that in addition to the two excitonic emission lines, the PL of the ZnO nanocrystals exhibits a strong line at 3.32 eV, referred to as the ε-PL, which is redshifted by ∼56 meV from the A-exciton.

In general, in a semiconductor, various forms of impurities and defects may give rise to relatively deep radiative centers. The growth of the nanocrystals studied here took place under ambient conditions, and the incorporation of unintentional dopants from the synthesis environment is expected. Additionally, due to the large surface-to-volume ratio, structural surface defects are unavoidable. The origin of the PL at ∼3.31 eV is still under investigation by various research groups [33–36]; however, straightforward experimental approaches using imaging and luminescence have found compelling evidence that it is due to stacking faults [33]. Specifically, optical studies of ZnO thin films utilizing

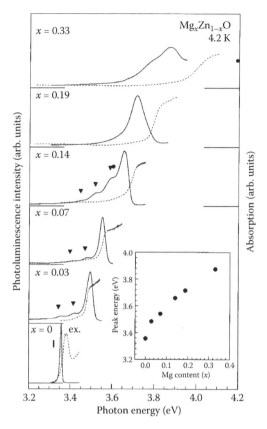

FIGURE 6.9   PL (solid lines) and absorption (dotted lines) spectra of $Mg_xZn_{1-x}O$ films grown via PLD. The spectra were taken at 4.2 K. The inset to the figure shows the PL peak position as a function of the Mg content. (With permission from Ohtomo, A., Kawasaki, M., Koida, T., Masubuchi, K., Koinuma, H., Sakurai, Y., Yoshida, Y., Yasuda, T., and Segawa, Y., *Appl. Phys. Lett.*, 72, 2466, 1998. Copyright 1998, American Institute of Physics.)

cathodoluminescence (CL) in conjunction with high resolution imaging techniques have indicated that the 3.31 eV PL in the ZnO thin film originates predominantly from distinct lines on sample surfaces due to stacking faults [33]. In that research, the CL was found to have the characteristics of a conduction band to acceptor-like states transition, and it was suggested that electronic states due to dangling bonds are the most likely source of the acceptor states [33]. The ZnO nanocrystals are nominally undoped and XPS, and EDS did not show significant traces of impurities; so due to the large surface area of the nanocrystals and their surface imperfections, it is reasonable to suggest that the ε-PL originates from surface related defects.

In contrast to the relatively sharp multiple emissions of the ZnO nanocrystals, the PL spectrallines of the $Mg_xZn_{1-x}O$ nanoalloys were all found to be very broad, even at 77 K, and the sharp excitonic lines could not be resolved. The PL spectrum acquired at 77 K of the $Mg_{0.23}Zn_{0.77}O$ nanocrystals is depicted in Figure 6.10, and as can be seen in the figure, the line broadening masks the phonon replica at the lower energy range and the excitons at the higher range. The featureless lineshape inhibits the straightforward identification of the transition mechanism involved in the PL of the nanoalloys; however, due to the continuous and consistent behavior of the PL as a function of Mg composition (Figure 6.7), it is reasonable to assume that it is of similar origin to that found for the ZnO nanocrystals. In general, the linewidths of ternary semiconductor alloys are expected to be significantly wide due to the inherent spatial variation of the alloy constituents [37]. In addition, alloy clustering can also contribute to the linewidth of the PL. PL line broadening mechanisms have been previously investigated for $Al_xGa_{1-x}N$ and $Al_xGa_{1-x}As$

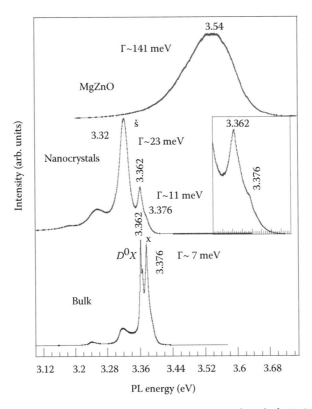

FIGURE 6.10 The PL spectra at 77 K of $Mg_{0.23}Zn_{0.77}O$, of ZnO nanocrystals and of a ZnO bulk reference sample. The inset to the figure is the focused spectral range of the excitons of the ZnO nanocrystals. The PL of the nano-crystals exhibits an emission at ~3.32 eV, in addition to the excitons at 3.362 and 3.376 eV. The PL of the MgZnO nanoalloys has a very broad linewidth $\Gamma = 141$ meV, indicative of inhomogeneous broadening mechanisms. The PL linewidths of the ZnO nanocrystals are relatively narrower. (With permission from Morrison, J.L., Huso, J., Hoeck, H., Casey, E., Mitchell, J., Bergman, L., and Norton, M.G., *J. Appl. Phys.*, 104, 123519, 2008. Copyright 2008, American Institute of Physics.)

thin films, and were attributed to the compositional fluctuation in the films [38,39]. The PL broadening mechanism of the $Mg_xZn_{1-x}O$ nanocrystals can be similarly attributed to the expected compositional fluctuation in the nanocrystal ensemble; moreover, due to the incomplete nature of the solubility of the MgO–ZnO system clustering even at low Mg composition range is expected that in turn will result in a broad lineshape.

The second part of this section addresses the properties of the nanoalloys at higher composition range. As was discussed in the introduction, the ZnO–MgO is not a complete soluble alloy system. In the $Mg_xZn_{1-x}O$ thin films grown via the PLD technique, a mixed phase region due to the solubility limit was observed in samples that contained ~30%–50% Mg [17,19]. The characteristics of the mixed phase region of the nanoalloys of composition $Mg_{0.4}Zn_{0.6}O$ and $Mg_{0.5}Zn_{0.5}O$ was studied as follows. Figure 6.11 depicts a representative room-temperature PL spectrum of $Mg_{0.5}Zn_{0.5}O$, and Figure 6.12 that of its XRD; for comparison purposes an XRD spectrum of one of the lower Mg composition samples is included. As is seen in Figure 6.11, the PL of the $Mg_{0.5}Zn_{0.5}O$ sample has three principal pronounced emission lines: 3.22, 3.5, and 3.93 eV. The XRD, depicted in Figure 6.12, indicates that in $Mg_{0.5}Zn_{0.5}O$ the cubic structure of MgO is present along that of the wurtzite structure. Thus, the multiple PL emissions of the $Mg_{0.5}Zn_{0.5}O$ nanoalloy sample may be assigned to a compositional segregation effect: The low energy peak at 3.22 eV is due to the ZnO-rich phase corresponding to the wurtzite structure, while the higher

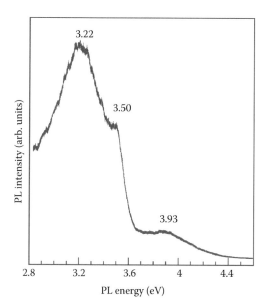

FIGURE 6.11 The PL spectra, acquired at room temperature, of the $Mg_{0.5}Zn_{0.5}O$ nanoalloys exhibiting a phase separation phenomenon. A PL due to the ZnO-rich phase is at 3.22 eV; the ones at 3.5 and 3.93 eV are attributed to regions of the sample that contain higher Mg compositions. (With permission from Morrison, J.L., Huso, J., Hoeck, H., Casey, E., Mitchell, J., Bergman, L., and Norton, M.G., *J. Appl. Phys.*, 104, 123519, 2008. Copyright 2008, American Institute of Physics.)

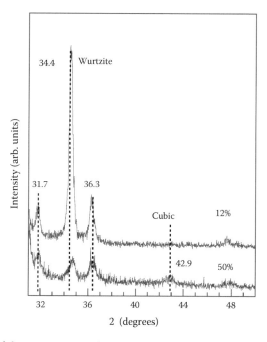

FIGURE 6.12 The XRD of the $Mg_{0.5}Zn_{0.5}O$ and $Mg_{0.12}Zn_{0.88}O$ samples. The powder diffraction reference of the wurtzite ZnO and that of the cubic MgO are included (lines). (With permission from Morrison, J.L., Huso, J., Hoeck, H., Casey, E., Mitchell, J., Bergman, L., and Grant Norton, M., *J. Appl. Phys.*, 104, 123519, 2008. Copyright 2008, American Institute of Physics.)

energy PL lines correspond to phases of higher Mg composition, some of which have the cubic structure. The $Mg_{0.4}Zn_{0.6}O$ sample was found to exhibit optical properties similar to that of $Mg_{0.5}Zn_{0.5}O$, implying that the phase separation for the nanoalloys is very significant in those samples.

6.4 Raman Scattering and Phonons in MgZnO

A very powerful technique that is widely used in the study of material properties is Raman scattering [40–46]. In general, Raman scattering experiments utilize the same optical system that is used for photoluminescence studies, conditional on the system's spectral resolution and throughput. There are several books that discuss aspects of the optical instrumentation for PL and Raman experiments; the reader is referred, for example, to *Solid-State Spectroscopy* by H. Kuzmany (Springer, 1998), *Spectrochemical Analysis* by J. D. Ingle and S. R. Crouch (Prentice Hall, 1988), *Analytical Raman Spectroscopy* edited by J. G. Grasselli and B. J. Bulkin (John Wiley & Sons, 1991), *Scattering of Light By Crystals* by W. Hayes and R. Loudon (John Wiley & Sons, 1978), and *Semiconductor Optics* by C. F. Klingshirn (Springer, 1997). In many studies of optical properties of semiconductors, photoluminescence as well as Raman spectroscopy are employed as complementary techniques. For completeness, a short overview of Raman scattering is given in the following paragraphs.

The Raman effect in solids is an inelastic scattering of the incident photons by the crystal vibrational modes, the latter which are known as phonons [40–46]. In Raman scattering, the incoming photons exchange a quantum of energy with the crystal via the creation or annihilation of phonons. As a result, the scattered photons lose or gain an energy quantum depending on whether a phonon was created (Stokes process) or annihilated (anti-Stokes process). In Raman spectroscopy, the energy of the scattered photons is measured; thus, a characteristic value of the vibration energy of a specific material may be obtained.

Figure 6.13 illustrates the Raman scattering process. The sample is irradiated by a monochromatic laser beam of energy E_L. The incident photons create or annihilate crystal vibrations of energy E_{phonon}; thus, the observed scattered beam consists of light of energy $E_L - E_{phonon}$ and $E_L + E_{phonon}$ due to the Stokes and anti-Stokes mechanisms, respectively. The scattered beam has an additional light component of the same frequency as the incident light (E_L) that arises from the elastic Rayleigh scattering. A typical Raman spectrum of a crystal exhibits sharp bands at frequencies $\pm E_{phonon}$, which are measured as a shift, called the Raman shift, from the incident beam frequency E_L. Usually, Raman scattering due to Stokes mechanism is studied. The spectral unit of wavenumber cm^{-1}, known also as the Raman frequency or the Raman shift, is conventionally employed in Raman spectroscopy. The unit of the Raman wavenumber can be converted to the unit of energy since $E = hc/\lambda$, where $1/\lambda$ is the Raman unit [46]. Detailed descriptions of the Raman effect can be found in [40–46] and references contained therein.

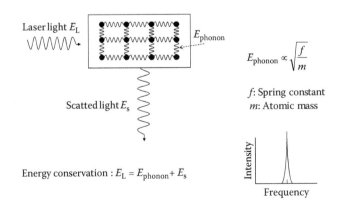

FIGURE 6.13 A schematic representation of Raman scattering in solids.

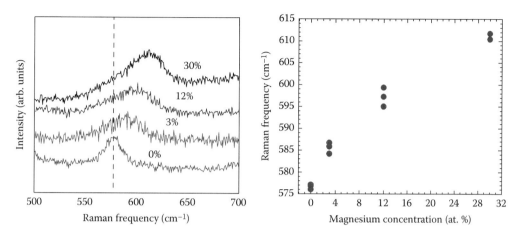

FIGURE 6.14 The Raman spectra of the LO mode of the MgZnO nanocrystals (left), and the Raman frequency as a function of Mg composition (right). The Raman spectra were acquired at room temperature utilizing the 325 nm laser line; the probing laser spot size is ~1 μm. The multiple data points are due to the spatial fluctuation of the alloy constituents. (With permission from Bergman, L., Morrison, J.L., Chen, X.-B., Huso, J. and Hoeck, H., *Appl. Phys. Lett.*, 88, 023103, 2006. Copyright 2006. American Institute of Physics.)

Representative Raman spectra from $Mg_xZn_{1-x}O$ nanocrystals and its analysis are presented in Figure 6.14. The spectra show the longitudinal optical (LO) mode of the crystallites [22]. As can be seen in Figure 6.14, for an Mg composition range of 0%–30%, the LO mode exhibits a significant shift of ~33 cm⁻¹. The Raman selection rules dictate that the wurtzite structure has two symmetry types of Raman active LO modes; A1(LO) and E1(LO), the observation of which depends on the crystal orientation and the geometry of the Raman scattering experiment [41,47]. In crystallite of tilted orientation, the two LO modes are expected to interact and create one mode of mixed A1-E1 symmetry known as a quasi-LO mode [41]. Studies concerning the phonon modes of $Mg_xZn_{1-x}O$ thin films (of well-defined crystal orientation) grown via a PLD method found that the A1(LO) and the E1(LO) exhibit a significant blueshift [48]. Such a shift is expected and indicates that an alloy was created: The LO Raman mode of ZnO is in the neighborhood of 575 cm⁻¹ and that of MgO is ~710 cm⁻¹, and therefore, upon alloying ZnO with Mg, the LO mode is expected to have a blueshift trend. So in this regard, the Raman study is consistent with the PL investigation, implying that a ZnO–MgO solid solution of nanocrystals was created, overcoming the predicted (~4%) low solubility limit of that alloy system.

6.5 Photoluminescence Properties of $Mg_xZn_{1-x}O$ Flexible Nanocrystalline Films and Ceramics

The topic of flexible electronic structures has been extensively studied [49–52]. Flexible devices have distinctive advantageous properties, such as conforming to irregular surfaces, sustaining bending and deformation, and coating of large areas for ultra-light-weight applications. Flexible devices have promising potential for applications in displays, solar cells, and large scale sensor arrays.

In the following, a research direction that focuses on ZnO polycrystalline films grown on a flexible and UV-compatible substrate is presented. While there have been a number of studies on the optical properties of ZnO films grown on flexible substrates, most have used flexible substrates that are transparent in the visible region [53–59]. In those studies, flexible ZnO transistors, light-emitting diodes, and circuits were fabricated, and thin films and nanostructured ZnO were deposited for various optical applications [53–59]. However, there have been relatively few studies on UV-compatible flexible films. The substrate chosen for the experiments is commercially available fluorinated ethylene propylene (FEP,

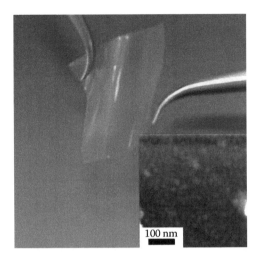

FIGURE 6.15 A photograph of ZnO/FEP film, and an SEM image showing the film morphology. The average crystallite size of the film is ~30 nm.

also known as Teflon FEP) due to its high UV transmission and flexibility. FEP is an ideal candidate for ZnO and $Mg_xZn_{1-x}O$ flexible thin films due to its inert chemistry, high UV and visible transparency, and its thermoplastic properties. The optical properties of the substrate are discussed in detail in the following paragraphs.

Samples were grown on prepared FEP substrates utilizing magnetron sputtering of Mg–Zn targets under an argon plasma. The samples were then oxidized for 2 h at a temperature of 275°C under an atmosphere of 99.99% pure oxygen. Figure 6.15 shows a representative photograph of the resulting films and the corresponding scanning electron microscopy (SEM) image. The photograph shows that the film is highly flexible, while the polycrystalline features are visible in the SEM image. The individual crystallites have an average size of approximately 30 nm. Due to the chemical structure of the FEP, it is unlikely that the crystallites are chemically bonded to the substrate. In addition, it was observed that even after repeated flexing and deformation, the ZnO film remained attached to the substrate, suggesting that the crystallites become partially embedded in the FEP substrate during the oxidation step so as to create a stable film/substrate system. Figure 6.16 shows the transmission spectrum of the FEP

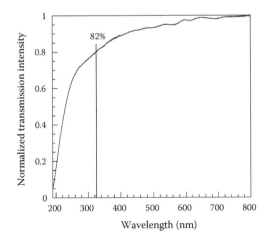

FIGURE 6.16 The transmission spectrum of the FEP substrate. The substrate exhibits substantial transparency in the UV spectral range above ~250 nm.

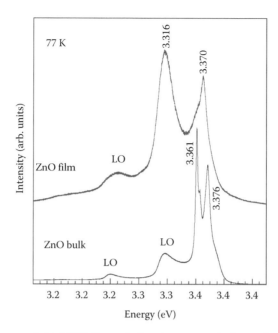

FIGURE 6.17 The PL spectra of a ZnO/FEP film and of a reference ZnO bulk sample. The PL of the film exhibits the ε-emission, similar to that of ZnO nanocrystals and the excitons. The PL spectra were acquired at 77 K.

substrate, demonstrating the high transparency of the substrate well into the UV range. In a manner similar to that discussed for the nanocrystals, the origin of the PL of the flexible ZnO samples merits closer investigation. Figure 6.17 depicts the PL of that sample at 77 K along with a reference bulk ZnO. It is evident from Figure 6.17 that the dominant PL corresponds to the ε-emission, similar to that of the nanocrystals (see Figure 6.10).

For practical applications, knowledge concerning the properties of the films under bending conditions may prove useful. A powerful tool that may be used to gain insight into the stress state of a semiconductor is Raman scattering. In general terms, the phonon vibrations can be approximated in the framework of the harmonic oscillator model, i.e., Hooke's law, for which the Raman frequency is related to the square root of the ratio of the force constant to the atomic masses involved. Under tensile stress, for example, the force constant may be weakened (due to weaker interaction among the nearest-neighbor crystal planes), and as a result the Raman frequency is expected to shift to a smaller value relative to that of an unstrained solid.

In any flexible device, the effect of mechanical deformation is a major concern. To study the effect of flexing, the films were bent at 180° and clamped in a hemostat, and then the resulting highly strained edge was studied via Raman spectroscopy. The results of the experiments are presented in Figure 6.18a and b. The figure depicts the Raman spectra of the bare FEP substrate and the ZnO/FEP film. Both these spectra were acquired without the deformation condition. Along with these spectra, a spectrum of the ZnO/FEP film under the bending condition is presented. From the shift of the Raman modes toward lower frequencies, the following conclusions can be drawn. First, looking at the two lower traces in Figure 6.18a (no deformation condition), it is evident that in the ZnO/FEP film, the underlying FEP substrate is under tensile stress. The existence of tensile strain in the ZnO/FEP film may stem from the very large differences in thermal expansion between the ZnO at 4.75×10^{-6}/K and that of the transparent substrate, in the order of 10^{-4}/K [60–62]. During growth, the FEP is very soft and near its melting point, thus allowing for the ZnO crystallites to become embedded in it. During cooling, the FEP begins to contract significantly, and each of the ZnO crystallites acts as a mechanical barrier to the contraction, thus resulting in a net tensile strain in the underlying FEP substrate. The second observation is that the

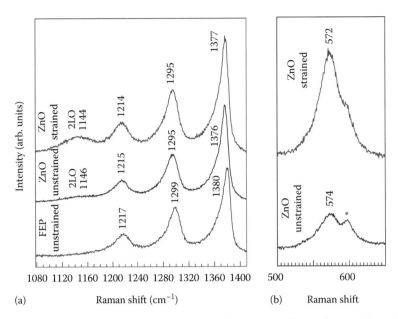

FIGURE 6.18 The Raman spectra of the FEP substrate and the ZnO/FEP film: Both non-deformed [in (a) lower two traces], and of the ZnO/FEP film under deformation [in (a) the upper trace]. The Raman spectrum of the non-deformed ZnO/FEP film indicates that the underlying FEP substrate is under tensile stress. The spectrum ZnO/FEP film under deformation indicates that the deformation has no impact on the FEP substrate. The Raman spectra of the ZnO/FEP film in the spectral range of the LO mode of ZnO are shown in (b). The * denotes a Raman mode of the underlying FEP. Under the deformation condition, the LO mode of ZnO has a small redshift $\sim$2 cm^{-1}, indicative that the ZnO in under minimal tensile stress.

deformation condition (upper trace in Figure 6.18a) did not impact the FEP substrate. However, there is a minor impact on the ZnO crystallite: The LO mode of the flexed sample (Figure 6.18b) has a $\sim$2 cm^{-1} redshift relative to that of the unstrained sample, indicating that the ZnO itself is under small tensile stress due to the deformation.

So far two material forms of the $Mg_xZn_{1-x}O$ samples have been discussed: The nanocrystals that were synthesized via a chemical approach, and thin films that were grown via a sputtering technique. In the following section, the properties of $Mg_xZn_{1-x}O$ ceramics are presented. Ceramics, in general, are relatively easy to achieve and are cost effective. ZnO ceramics, in particular, have been widely used as varistors, have potential applications in optoelectronics, and, moreover, are easy to handle in experimental setups [63–66].

For the ceramic synthesis, the desired mole fraction of 99.99% pure magnesium oxide and 99.99% pure zinc oxide powders was ball milled in isopropyl alcohol, then heated to evaporate the excess remaining alcohol. A small amount of the dried mixture was cold-pressed into a 3 mm disc. The sample was then sintered in a quartz tube furnace at 1100°C for 24 h, where alloying of the precursors occurred, and then allowed to slowly cool to room temperature. Figure 6.19 shows SEM images of the $Mg_xZn_{1-x}O$ ceramics, along with their photograph. As can be inferred from the SEM images, the morphology of ceramics degrades as the Mg composition increases. In previous sections, it was discussed that the PL of the ZnO nanocrystals and film is primarily due to the ε-emission rather than of a purely excitonic origin. Figure 6.20 shows the resulting PL spectra acquired at 77 K for the ZnO ceramic sample along with that of bulk ZnO. As is apparent in the spectra, the peak positions of the bulk and ceramic are identical indicating that the origin of the PL in the ceramic material is of excitonic origin. The large grains, of micron-scale, of the ZnO ceramic sample is comprised of material having quality comparable

MgZnO ceramic alloys

FIGURE 6.19 SEM images of the Mg$_x$Zn$_{1-x}$O ceramics and their photograph. The ceramics are ~3 mm diameter in size. The morphology of the ceramics degrades as a function of Mg composition indicative of the low solubility of the alloyed ceramics.

to that of bulk ZnO, unlike that of the nanoscale grains of the crystallites and the film. The morphology of the ceramics is conducive to excitonic emission rather than to PL whose origin stems from surface crystal defects.

The compositional dependence of the PL is presented in Figure 6.21; the straight line is an approximation to the PL behavior of PLD thin films [17]. As can be inferred from the figure, a phase segregation has an onset at ~15% Mg in comparison to that of nanocrystals at ~25% (Figure 6.7). The implication is that the process of achieving the ceramics is more one of thermal equilibrium (more so than that of the nanocrystals), which derives the phase segregation. Further studies are needed in order to investigate growth conditions that may result in a better solubility of the alloy constituents. Yet, the solubility limit for the ceramics is greater than the one previously predicted (~4% Mg) for the MgO-ZnO solid solution under thermal equilibrium conditions [29–30].

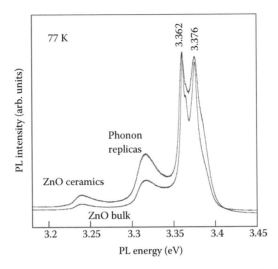

FIGURE 6.20 The PL spectra of the ZnO ceramic sample and of a reference ZnO bulk. The PL of both materials exhibits the ZnO excitonic emission. The PL due to surface defects, ~3.32 eV, is not present.

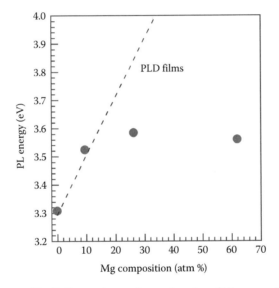

FIGURE 6.21 The PL energy of the ZnO ceramic samples as a function of Mg composition. The dotted line represents the trend of the PLD film discussed in the chapter. Solubility limit at ~15% Mg, is predicted for the ceramics grown with the specific conditions discussed in the text.

6.6 Photoluminescence Properties of MgZnO under Extreme Environments

Nanomaterials under high pressure may have different properties than bulk materials. For example, previous x-ray studies on the structural stability of the nanocrystalline ZnO found that a phase transition from the wurtzite to the rocksalt structure takes place at a pressure of 15.1 GPa in comparison to 9.9 GPa for the bulk [67]. A similar trend of increasing phase transition pressure as a function of size was also found for CdSe nanocrystallites [68–70]. The optical properties of bulk and

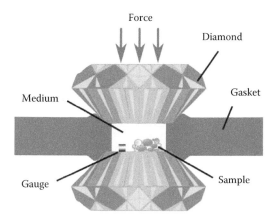

FIGURE 6.22 A diamond anvil cell used for the high-pressure experiments.

nanoscale ZnO at ambient conditions have been extensively investigated; in contrast, less is known about their properties under the influence of applied pressure and still less so for MgZnO nanomaterials [66,71–74].

For the hydrostatic pressure measurements a D'Anvils diamond anvil cell with 0.6 mm culets and INCONEL gaskets was employed. The standard 4:1 methanol–ethanol mix was used as the pressure medium. The pressure was measured using the shift in the TO-Raman mode of cubic boron nitride as a pressure gauge calibrated previously by Datchi and Canny [75]. Figure 6.22 presents a schematic of the pressure cell. Figure 6.23a and b depicts the room temperature UV-PL spectra of the ZnO and $Mg_{0.15}Zn_{0.85}O$ crystallites for various pressures, and Figure 6.24 presents the dependence of PL energy on the applied pressure for both samples [66]. The reason for the blueshifted PL behavior stems from the functional response of the bandgap to the shrinking of the lattice spacing; a general scheme is presented in Figure 6.25. In particular, as can be seen in Figure 6.24, the PL energy of the ZnO nanocrystallites exhibits a blueshift of ~160 meV as the pressure increases from ambient pressure up to ~7 GPa. For the same pressure range, the PL energy of the $Mg_{0.15}Zn_{0.85}O$ crystallites shifts by ~190 meV. To

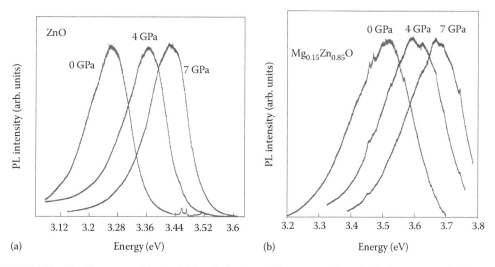

FIGURE 6.23 The PL spectra of the ZnO (a) and of MgZnO (b) nanocrystals under high-pressure. The data were acquired at room temperature. (From Huso, J. et al., *Appl. Phys. Lett.*, 94, 061919, 2009.)

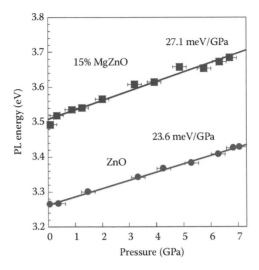

FIGURE 6.24 The PL energy as a function of pressure for ZnO and MgZnO nanocrystals. (From Huso, J. et al., *Appl. Phys. Lett.*, 94, 061919, 2009.)

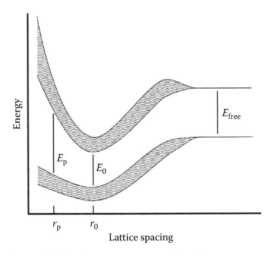

FIGURE 6.25 A general scheme of the bandgap energy in a semiconductor.

obtain the pressure coefficient, $a_P = \partial E/\partial P$, the data in Figure 6.24 have been fitted to a linear function of the form [66]:

$$E(P) = E_0 + a_P P \qquad (6.1)$$

where E_0 is the PL energy at atmospheric pressure. It was found that $a_P = 23.6 \pm 0.4$ meV/GPa for ZnO and 27.1 ± 1.3 meV/GPa for the $Mg_{0.15}Zn_{0.85}O$ crystallites.

The pressure coefficients of several-micron-size ZnO particles, which were obtained via absorption and transmission measurements, were previously reported to be in the range $\sim$23.3–24.5 meV/GPa [72,73]. In those studies, the pressure dependence of the fundamental absorption edge was investigated. A more detailed absorption study concerning the pressure response of the excitonic levels in micro-ZnO particles found that the pressure coefficient is a function of the exciton type and its energy levels; in particular the *A*- and *B*-excitons in their ground state were found to have pressure coefficients of 23.6

and 24.4 meV/GPa, respectively [71]. Thus, in that regard, the value obtained for the ZnO nanocrystals, 23.6 meV/GPa, is in close proximity to that of the bulk.

The variation in values of the pressure coefficients of $Mg_{0.15}Zn_{0.85}O$ (27.1 meV/GPa) and ZnO (23.6 meV/GPa) nanocrystals may be explained within the framework of the theoretical study by Wei and Zunger on the pressure dynamics of tetrahedral-bonded semiconductors [76]. According to the theory, the d-orbitals reduce the pressure coefficient via p-d coupling; for example, the calculated pressure coefficient of MgO (of the zinc-blende structure) is ~41.2 meV/GPa [77], which is higher than that of ZnO. The higher value for MgO stems from the lack of occupied d-orbitals in that material. In the nanoalloy, there is ~15% Mg in the ZnO matrix, thus effectively reducing the density of occupied d-orbitals in the alloy, and resulting in the increase of the pressure coefficient.

In subsequent work, the bulk modulus for $Mg_{0.15}Zn_{0.85}O$ ZnO and nanocrystals were found utilizing XRD under high-pressure; the experiments were done at the Advanced Light Source (ALS) in Lawrence Berkeley National Laboratory [78]. It was found that between pressures of 9.45 and 10.7 GPa, $Mg_{0.15}Zn_{0.85}O$ transformed into the rocksalt structure. The bulk modulus of $Mg_{0.15}Zn_{0.85}O$ was reported to be 136.5 ± 2.1 GPa for the wurtzite phase, and 172 ± 6 GPa for the rocksalt phase. The bulk modulus of ZnO nanocrystals of the wurtzite structure was found to be 144 GPa. The value for ZnO nanocrystals is similar to that of a bulk material, while the bulk modulus for $Mg_xZn_{1-x}O$ alloy had not been previously reported.

The temperature response of the photoluminescence may be a useful technique to probe the nature of the emission of a semiconductor at a given temperature range. Figure 6.26 presents the PL of ZnO bulk sample as a function of temperature; the inset to the figure is the temperature response of a GaN thin film. For both materials, the data was fitted to the electron–phonon interaction model [79]. The key point is that for both GaN and ZnO, there exists a critical temperature for which the PL energy increases: $T \sim 400$ K for GaN and $T \sim 700$ K for ZnO. These temperatures correspond approximately to the binding energies of the A-excitons of the two materials, i.e., ~30 meV for GaN and ~60 meV for ZnO [80,81]. The implication of such behavior is that at the critical temperature, the PL undergoes a characteristic change: the excitonic emission undergoes an onset to become a bandgap emission that

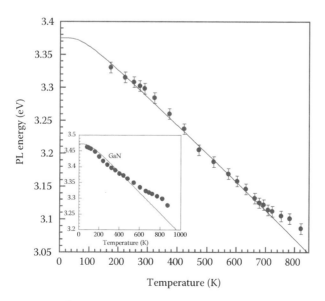

FIGURE 6.26 The PL energy as a function of temperature for the ZnO bulk crystal. The inset to the figure is the temperature response of a GaN thin film. For both materials, a critical temperature that corresponds to a change of the emission character is evident.

involves an electron-hole recombination mechanism [80,81]. The high-temperature response of ZnO implies that ZnO is a thermally stable material suitable for optical applications at and above room temperature.

6.7 Concluding Remarks

Optical and material properties of $Mg_xZn_{1-x}O$ were studied using photoluminescence. The materials were grown via various approaches so as to create nanocrystals, nanocrystalline films, and ceramics of micron-size grains. The photoluminescence of the ZnO ceramics was found to be due to the inherent excitons of ZnO. On the other hand, the photoluminescence of the nano-size ZnO was found to exhibit a dominant PL of lower energy relative to that of the excitons, which was attributed to surface defect states. The solubility limit of the $Mg_xZn_{1-x}O$ alloys depends on the growth technique. For a specific technique, a key issue is to explore the growth parameters that enable an optimal solubility. For the study of the optical and material properties of semiconductors, photoluminescence has been established as a powerful and versatile spectroscopic technique [82–84]. Specifically, this chapter presented the application of photoluminescence to the study of the various aspects of alloys. A comprehensive knowledge of the alloy's properties may be obtained via photoluminescence, along with complementary techniques such as Raman scattering [85], absorption [86], and XRD [87].

Acknowledgments

Leah Bergman and her group at the University of Idaho greatly acknowledge the National Science Foundation, Department of Energy, and the Petroleum Research Fund for supporting the research presented in this chapter.

References

1. Y. S. Park, C. W. Litton, T. C. Collins, and D. C. Reynolds, *Phys. Rev.* **143**, 512 (1966).
2. U. Ozgur, Ya I. Alivov, C. Liu, A. Teke, M. A. Reshchikov, S. Dogan, V. Avrutin, S.-J. Cho, and H. Morkoc, *J. Appl. Phys.* **98**, 041301 (2005).
3. W. Y. Liang and A. D. Yoffe, *Phys. Rev. Lett.* **20**, 59 (1968).
4. A. Teke, U. Ozgur, S. Dogan, X. Gu, H. Morkoc, B. Nemeth, J. Nause, and H. O. Everitt, *Phys. Rev. B* **70**, 195207 (2004).
5. R. Laskowski and N. E. Christensen, *Phys. Rev. B* **73**, 045201 (2006).
6. B. K. Meyer, H. Alves, D. M. Hofmann, W. Kriegseis, D. Forster, F. Bertram, J. Christen, A. Hoffmann, M. Strasburg, M. Dworzak, U. Haboeck, and A. V. Rodina, *Phys. Stat. Sol. (b)* **241**(2), 231 (2004).
7. S. Aslam, F. Yan, D. E. Pugel, D. Franz, L. Miko, F. Herrero, M. Matsumara, S. Babu, and C.M. Stahle, *Proc. of SPIE* **5901**, 59011J-1 (2005).
8. A. Nakamura, T. Ohashi, K. Yamamoto, J. Ishihara, T. Aoki, J. Temmyo, and H. Gotoh, *Appl. Phys. Lett.* **90**, 093512 (2007).
9. C. Yuen, S. F. Yu, E. S. P. Leong, H. Y. Yang, S. P. Lau, N. S. Chen, and H. H. Hng, *Appl. Phys. Lett.* **86**, 031112 (2005).
10. H. Cao, Y. G. Zhao, X. Liu, E. W. Seelig, and R. P. H. Chang, *Appl. Phys. Lett.* **75**, 1213 (1999).
11. M. H. Huang, S. Mao, H. Feick, H. Yan, Y. Wu, H. Kind, E. Weber, R. Russo, and P. Yang, *Science* **292**, 1897 (2001).
12. M. Zamfirescu, A. Kavokin, B. Gil, G. Malpuech, and M. Kaliteevski, *Phys. Rev. B* **65**, 161205 (2002).
13. C. Bayram, F. H. Teherani, D.J. Rogers, and M. Razeghi, *Appl. Phys. Lett* **93**, 08111 (2008).

14. J.W. Dong, A. Osinsky, B. Hertog, A.M. Dabiran, P.P Chow, Y.W. Heo, D.P. Norton, and S.J. Pearton, *J. Electron. Mat.* **34**, 416 (2005).

15. M. Law, L. E. Greene, J. C. Johnson, R. Saykally, and P. Yang, *Nat. Mater.* **4**, 455 (2005).

16. T. A. Wassner, B. Laumer, S. Maier, A. Laufer, B. K. Meyer, M. Stutzmann, and M. Eickhoff, *J. Appl. Phys.* **105**, 023505 (2009).

17. A. Ohtomo, M. Kawasaki, T. Koida, K. Masubuchi, H. Koinuma, Y. Sakurai, Y. Yoshida, T. Yasuda, and Y. Segawa, *Appl. Phys. Lett.* **72**, 2466 (1998).

18. A. K. Sharma, J. Narayan, J. F. Muth, C. W. Teng, C. Jin, A. Kvit, R. M. Kolbas, and O. W. Holland, *Appl. Phys. Lett.* **75**, 3327 (1999).

19. S. Choopun, R. D. Vispute, W. Yang, R. P. Sharma, T. Venkatesan, and H. Shen, *Appl. Phys. Lett.* **80**, 1529 (2002).

20. Z. Vashaei, T. Minegishi, H. Suzuki, T. Hanada, M. W. Cho, T. Yao, and A. Setiawan, *J. Appl. Phys.* **98**, 054911 (2005).

21. F. K. Shan, B. I. Kim, G. X. Liu, Z. F. Liu, J. Y. Sohn, W. J. Lee, B. C. Shin, and Y. S. Yu, *J. Appl. Phys.* **95**, 4772 (2004).

22. L. Bergman, J. L. Morrison, X-B. Chen, J. Huso, and H. Hoeck, *Appl. Phys. Lett.* **88**, 023103 (2006).

23. J. Huso, J. L. Morrison, H. Hoeck, E. Casey, L. Bergman, T. D. Pounds, and M. G. Norton *Appl. Phys. Lett.* **91**, 111906 (2007).

24. J. Huso, J. L. Morrison, H. Hoeck, X.-B. Chen, L. Bergman, S. J. Jokela, M. D. McCluskey, and T. Zheleva, *Appl. Phys. Lett.* **89**, 171909 (2006).

25. H. C. Hsu, C. Y. Wu, H. M. Cheng, and W. F. Hsieh, *Appl. Phys. Lett.* **89**, 013101 (2006).

26. M. Ghosh and A. K. Raychaudhuri, *J. Appl. Phys.* **100**, 034315 (2006).

27. R. C. Whited, C. J. Flaten, and W. C. Walker, *Solid State Commun.* **13**, 1903 (1973).

28. P. D. Johnson, *Phys. Rev.* **94**, 845 (1954).

29. E. R. Segnit and A. E. Holland, *J. Am. Ceramic Soc.* **48**, 409 (1965).

30. J. F. Sarver, F. L. Katnack, and F. A. Hummel, *J. Electrochemical Soc.* **106**, 960 (1959).

31. H. Angerer, D. Brunner, F. Freudenberg, O. Ambacher, M. Stutzmann, R. Hopler, T. Metzger, E. Born, G. Dollinger, A. Bergmaier, S. Karsch, and H.-J. Korner, *Appl. Phys. Lett.* **71**, 1504 (1997).

32. J. L. Morrison, J. Huso, H. Hoeck, E. Casey, J. Mitchell, L. Bergman, and M. G. Norton, *J. Appl. Phys.* **104**, 123519 (2008).

33. M. Schirra, R. Schneider, A. Reiser, G. M. Prinz, M. Feneberg, J. Biskupek, U. Kaiser, C. E. Krill, K. Thonke, and R. Sauer, *Phys. Rev. B* **77**, 125215 (2008).

34. Y. Li, M. Feneberg, A. Reiser, M. Schirra, R. Enchelmaier, A. Ladenburger, A. Langlois, R. Sauer, K. Thonke, J. Cai, and H. Rauscher, *J. Appl. Phys.* **99**, 054307 (2006).

35. D.-K. Hwang, H.-S. Kim, J.-H. Lim, J.-Y. Oh, J.-H. Yang, S.-J. Park, K.-K. Kim, D. C. Look, and Y. S. Park, *Appl. Phys. Lett.* **86**, 151917 (2005).

36. Y. R. Ryu, T. S. Lee, and H. W. White, *Appl. Phys. Lett.* **83**, 87 (2003).

37. E. F. Schubert, *Doping in III–V Semiconductors* (Cambridge University Press, New York, 1993).

38. G. Coli, K. K. Bajaj, J. Li, J. Y. Lin, and H. X. Jiang, *Appl. Phys. Lett.* **78**, 1829 (2001).

39. E. F. Schubert, E. O. Göbel, Y. Horikoshi, K. Ploog, and H. J. Queisser, *Phys. Rev. B* **30**, 813 (1984).

40. M. Cardona (ed.) *Topics in Applied Physics; Light Scattering in Solids I* (Springer-Verlag, New York, 1983).

41. C. A. Arguello, D. L. Rousseau, and S. P. S. Porto, *Phys. Rev.* **181**, 1351 (1969).

42. G. Turrell (ed.) *Infrared and Raman Spectra of Crystals* (Academic Press, New York, 1972).

43. W. H. Weber and R. Merlin (eds.) *Raman Scattering in Materials Science* (Springer, New York, 2000).

44. J. G. Grasselli and B. J. Bulkin (eds.) *Analytical Raman Spectroscopy* (John Wiley & Sons, Inc., New York 1991).

45. W. Hayes and R. Loudon (eds.) *Scattering of Light by Crystals* (John Wiley & Sons, New York 1978).

46. J. M. Hollas (ed.) *Modern Spectroscopy* (John Wiley & Sons, New York 1992).

47. L. Bergman, X.-B. Chen, J. Huso, J. L. Morrison, and H. Hoeck, *J. Appl. Phys.* **98**, 93507 (2005).

48. C. Bundesmann, A. Rahm, M. Lorenz, M. Grundmann, and M. Schubert, *J. Appl. Phys.* **99**, 113504 (2006).

49. W.-Y. Chang, T.-H. Fang, Y.-T. Shen, and Y.-C. Lin, *Rev. Sci. Instrum.* **80**, 84701 (2009).

50. T. Sekitani, T. Yokota, U. Zchieschang, H. Klauk, S. Bauer, K. Takeuchi, M. Takamiya, T. Sakurai, and T. Someya, *Science* **326**, 1516 (2009).

51. Z. Fan, H. Razavi, J.-W. Do, A. Moriwaki, O. Ergen, Y.-L. Chueh, P. W. Leu, J. C. Ho, T. Takahashi, L. A. Reichertz, S. Neale, K. Yu, M. Wu, J. W. Ager, and A. Javey, *Nat. Mater.* **8**, 648 (2009).

52. P. Heremans, *Nature* **444**, 828 (2006).

53. E. M. C. Fortunato, P. M. C. Barquinha, A. C. M. B. G. Pimentel, A. M. F. Goncalves, A. J. S. Marques, L. M. N. Pereira, and R. F. P. Martins, *Adv. Mater.* **17**, 590 (2005).

54. K. Nomura, H. Ohta, A. Takagi, T. Kamiya, M. Hirano, and H. Hosono, *Nature* **432**, 488 (2004).

55. D. Zhao, D. A. Mourey, and T. N. Jackson, *IEEE Electron Dev. Lett.* **31**, 323 (2010).

56. R. Konenkamp, R. C. Word, and C. Schlegel, *Appl. Phys. Lett.* **85**, 6004 (2004).

57. E. S. P. Leong, S. F. Yu, S. P. Lau, and A. P. Abiyasa, *IEEE Photon. Technol. Lett.* **19**, 1792 (2007).

58. A. N. Banerjee, C. K. Ghosh, K. K. Chattopadhyay, H. Minoura, A. K. Sarkar, A. Akiba, A. Kamiya, and T. Endo, *Thin Solid Films* **496**, 112 (2006).

59. C.-C. Lin, H.-P. Chen, H.-C. Liao, and S.-Y. Chen, *Appl. Phys. Lett.* **86**, 183103 (2005).

60. Y. Araki, *J. Appl. Polym. Sci.* **9**, 421 (1965).

61. H. Ibach, *Phys. Stat. Sol.* **33**, 257 (1969).

62. R. D. Vispute, V. Talyansky, S. Choopun, R. P. Sharma, T. Venkatesan, M. He, X. Tang, J. B. Halpern, M. G. Spencer, Y. X. Li, L. G. Salamanca-Riba, A. A. Iliadis, and K. A. Jones, *Appl. Phys. Lett.* **73**, 348 (1998).

63. I. V. Markevich, and V. I. Krushnirenko, *Solid State Commun.* **149**, 866 (2009).

64. A. T. Santhanam, T. K. Gupta, and W. G. Carlson, *J. Appl. Phys.* **50**, 852 (1979).

65. F. A. Modine, R. W. Major, S. I. Choi, L. Bergman, and M. N. Silver, *J. Appl. Phys.* **68**, 339 (1990).

66. J. Huso, J. L. Morrison, J. Mitchell, E. Casey, H. Hoeck, C. Walker, L. Bergman, W. M. Hlaing Oo, and M. D. McCluskey, *Appl. Phys. Lett.* **94**, 061919 (2009).

67. J. Z. Jiang, J. S. Olsen, L. Gerward, D. Frost, D. Rubie, and J. Peyronneau, *Europhys. Lett.* **50**, 48 (2000).

68. S. H. Tolbert and A. P. Alivisatos, *Science* **265**, 373 (1994).

69. R. W. Meulenberg and G. F. Strouse, *Phys. Rev. B* **66**, 03517 (2002).

70. S. H. Tolbert and A. P. Alivisatos, *J. Chem. Phys.* **102**, 4642 (1995).

71. A. Mang, K. Reimann, and St. Rubenacke, *Solid State Commun.* **94**, 251 (1995).

72. D. R. Huffmann, L. A. Schwalbe, and D. Schiferl, *Solid State Commun.* **44**, 521 (1982).

73. A. Segura, J. A. Sans, F. J. Manjón, A. Muñoz, and M. J. Herrera-Cabrera, *Appl. Phys. Lett.* **83**, 278 (2003).

74. W. Shan, W. Walukiewicz, J. W. Ager, III, K. M. Yu, Y. Zhang, S. S. Mao, R. Kling, C. Kirchner, and A. Waag, *Appl. Phys. Lett.* **86**, 153117 (2005).

75. F. Datchi and B. Canny, *Phys. Rev. B* **69** 144106 (2004).

76. S.-H. Wei and A. Zunger, *Phys. Rev. B* **60**, 5404 (1999).

77. S.-H. Wei, National Renewable Energy Laboratory, Private communication.

78. K. K. Zhuravlev, W. M. Hlaing Oo, M. D. McCluskey, J. Huso, J. L. Morrison, and L. Bergman, *J. Appl. Phys.* **106**, 13511 (2009).

79. L. Vina, S. Logothetidis, and M. Cardona, *Phys. Rev. B* **30**, 1979 (1984).

80. X. B. Chen, J. Huso, J. L. Morrison, and L. Bergman, *J. Appl. Phys.* **99**, 046105 (2006).

81. X. B. Chen, J. Huso, J. L. Morrison, and L. Bergman, *J. Appl. Phys.* **102**, 116105 (2007).

82. H. B. Bebb and E. W. Williams, Photoluminescence theory, in *Semiconductors and Semimetals*, Vol. 8, R. K. Willardson and A. C. Beer (eds.) (Academic Press, New York, 1972).

83. J. I. Pankove, *Optical Processes in Semiconductors* (Dover Publications Inc., New York, 1971).

84. P. C. Schmidt, K. C. Mishra, B. Di Bartolo, J. McKittrick, and A. M. Srivastava (eds.) *Luminescence and Luminescence Materials*, Materials Research Society vol. 667, 2001.

85. D. A. Long, *Raman Spectroscopy* (McGraw-Hill, New York, 1977).

86. J. D. Ingle Jr. and S. R. Crouch, *Spectrochemical Analysis* (Prentice Hall, Englewood Cliffs, NJ, 1988).

87. C. Suryanarayana and M. G. Norton, *X-Ray Diffraction a Practical Approach* (Plenum Publishing Corporation, New York, 1998).

7

Luminescence Studies of Impurities and Defects in III-Nitride Semiconductors

Bo Monemar
Linköping University

Plamen P. Paskov
Linköping University

7.1 Introduction

Although studies of the optical properties of single crystalline III-nitrides have been conducted during the last four decades, knowledge in the areas of defects and impurities is still being developed. This is related to the quality of the materials, which still lags far behind what has been demonstrated for the lower bandgap III–V compounds. The gradual development of growth procedures for III-nitrides has been focused on thin films grown on foreign substrates for the development of inexpensive device structures. On the other hand, the materials suitable for accurate studies of optical spectra related to dopants and most other defects are strain-free bulk samples or layers grown homoepitaxially on such bulk substrates [1]. The availability of such samples has so far been very limited for GaN and AlN, while for InN, only heteroepitaxial samples grown on foreign substrates such as sapphire exist to date. Bulk material of GaN and AlN now exists with a density of structural defects (dislocations) $<10^6$ cm^{-2} in GaN and even lower in AlN. Such material is suitable for optical studies of defects, provided the density of impurities and other point defects are also low (a spectroscopic line width of <1 meV is desirable). The concentration of residual impurities in nominally undoped bulk GaN (mainly donors like O and Si) is at best about 10^{16} cm^{-3} for GaN, and considerably higher ($>10^{18}$ cm^{-3}) for bulk AlN. For samples grown on foreign substrates, the linewidth in photoluminescence (PL) spectra is typically several meV due to the inhomogeneous strain. For InN, this problem continues to exist, and in addition, InN samples so far are degenerate n-type that cause an additional broadening of the PL spectra. Alloys of III-nitrides are now only available as heteroepitaxial layers, and such samples suffer an additional broadening due to the alloy composition fluctuation potentials.

7.2 Donors in GaN

The shallow donors that have been studied most extensively in GaN are substitutional O on the nitrogen site (O_N) and Si on the gallium site (Si_{Ga}). Of these, only the Si donor is suitable for applications in *n*-doping. The main reason is that the complexes between Si donors and gallium vacancy (V_{Ga}) acceptors are unstable at the common growth temperatures of GaN (1050°C–1150°C), i.e., a high fraction of the Si atoms will remain as isolated shallow donors in the grown material upon cooldown [2]. For O donors, this situation is different: V_{Ga}-O complexes are stable up to about 1250°C, and act as deep acceptors in the GaN matrix [3,4]. Other O-related deep levels, such as interstitial O acceptors are also expected to occur [5] making the use of O donors for the purpose of n-doping less convenient.

The shallow donors are easily detected and studied via neutral donor bound exciton (typical short notations in the literature are DBE or D^0X) PL spectra. The schematic coupling of electrons and holes in the DBE complex is shown in Figure 7.1a. An analogous scheme for neutral acceptor bound excitons (ABE) is shown in Figure 7.1b. In Figure 7.1c is shown a PL spectrum at 2 K in the near bandgap region for a nominally undoped bulk GaN sample with a total shallow donor concentration of about 10^{16} cm^{-3}. Two distinct peaks are shown at 3.4714 eV (A exciton bound to neutral O donors, O^0X_A) and 3.4722 eV (A exciton bound to neutral Si donors, Si^0X_A) [6]. Very weak excited states [O^0X_A (a), Si^0X_A (a), Si^0X_A (b)] are also present in the spectrum at higher energies. These are better studied with selective excitation [7]. These excited states correspond to rotational states of the outer hole bound in the DBE complex [8], and their population is more pronounced at temperatures >2 K. The emission line at 3.466 eV (A^0X_A) is related to an exciton bound to a neutral acceptor and will be discussed subsequently. Since the free A exciton energy is about 3.4784 eV at 2 K [6], the bound exciton (BE) binding energies for Si and O donors in GaN are 6.2 and 7.0 meV, respectively. It has been suggested that the so-called Haynes'rule is valid for shallow donors in GaN [9], meaning that the donor BE binding energy $E_{BE} \approx 0.2E_d$, where E_d is the donor binding energy. For a more complete discussion of the BE properties in GaN, see Ref. [6].

A PL spectrum of the same sample in a lower energy region, showing a rich structure of phonon replicas, together with the so-called two-electron spectra (TES), is shown in Figure 7.2. The latter occur by recombination of the DBE with the simultaneous excitation of the donor to an excited state in the final state of the transition [10]. The selection rules for these transitions, which as initial states involve both the DBE ground state and the rotational DBE excited states and as final states both *s*- and *p*-symmetry

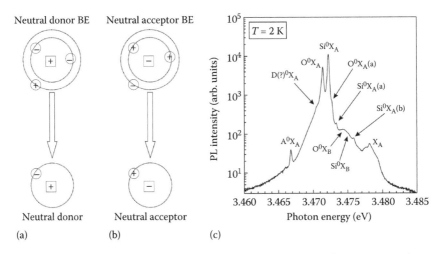

FIGURE 7.1 (a) Schematic picture of a neutral donor DBE with a core two-electron state and an outer hole; (b) corresponding picture for a neutral acceptor ABE; (c) low-temperature near bandgap edge PL spectrum of a bulk GaN sample showing the main excitonic transitions.

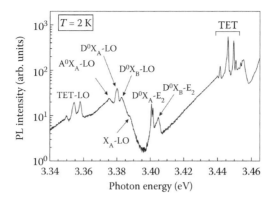

FIGURE 7.2   Low-temperature PL spectrum of the same sample as in Figure 7.1 but at lower energies showing two-electron transition (TET) lines as well as phonon replicas of the excitonic transitions.

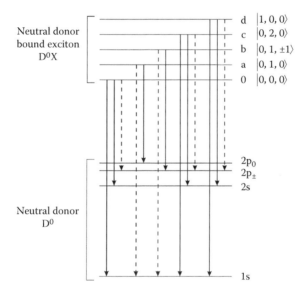

FIGURE 7.3   Schematic diagram showing all possible optical transitions of a neutral donor bound exciton (DBE) recombination. Dipole-allowed and dipole-forbidden (but quadrupole-allowed) transitions are indicated by solid and dashed lines, respectively.

excited donor states, were recently discussed [8], and are schematically illustrated in Figure 7.3. Allowed transitions from the DBE ground state are to s-like final donor states, while transitions to donor p-states require involvement of the excited DBE states. This scheme accounts for the drastic changes observed in the TES with a variation of sample temperature in the range 2–15 K (Figure 7.4).

From spectra like those in Figure 7.4, an accurate mapping of the neutral donor excited states is obtained. Since the $2p$ states, and to a higher degree the $3p$ states, are expected to have a very small central cell shift, these data allow an accurate determination of the donor binding energies. In the effective mass picture, small corrections have to be made for the anisotropy of the dielectric constant and the electron effective mass, and for the polaron effects [11]. As a result, a binding energy E_d of 30.2 ± 0.3 meV for the Si donor and $E_d = 33.5 \pm 0.4$ for the O donor are deduced [11]; the error bars are due to some scatter between different samples. These data agree well with previous infrared magneto-absorption data [12].

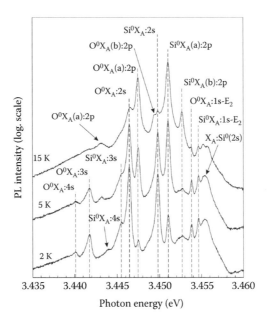

FIGURE 7.4 Two-electron transition spectra of the same sample as in Figure 7.1 measured at different temperatures. The emission lines are labeled with the initial DBE state and the final states of neutral donors. For further details see Ref. [11].

Although O and Si appear to be the dominant residual donors present in nominally undoped GaN grown by hydride vapor phase epitaxy (HVPE) and metalorganic vapor phase epitaxy (MOVPE), in some samples, weak evidence of additional residual donors can be seen via their BE lines in PL spectra. In Figure 7.5 is shown near bandgap PL spectra from such nominally undoped samples, where a weak DBE line at 3.4704 eV is observed, but the corresponding donor is not identified.

An interesting impurity in GaN (as well as in most other semiconductors) is hydrogen. Hydrogen is believed to be introduced as a (deep) donor in p-type GaN, and as a deep acceptor in n-GaN [13]. These energy levels are expected to be important in controlling the Fermi level during growth (and the doping)

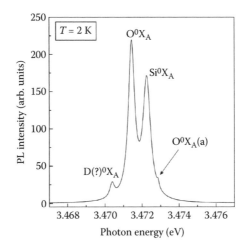

FIGURE 7.5 Low-temperature near bandgap PL spectrum of another bulk GaN sample showing an additional unidentified DBE line.

of GaN, e.g., it has been suggested that the incorporation of Mg is enhanced by the presence of H donors during growth [14]. Hydrogen in p-GaN can passivate the acceptors by a complex formation (in addition to compensation), but can be efficiently removed by a post growth thermal annealing procedure [15]. At room temperature, the influence of the H content on carrier concentration from doping can therefore be kept small; the H concentration in optimized p-GaN can be in the low 10^{17} cm^{-3} range after proper annealing [16]. While H is expected to mainly create deep levels in *n*-type as well as *p*-type materials, the possibility of observing shallow donor states can be invoked from muon spin resonance spectroscopy where very shallow donor states have been identified in GaN [17]. In some GaN materials, a very shallow donor with binding energy about 20 meV has been observed via the BE PL line with a BE binding energy of just about 4 meV [18]. This could be evidence of a shallow metastable H_0 neutral donor state, but this suggestion so far remains speculative [18].

7.3 Acceptors in GaN

Doping with acceptors was studied early for GaN, since *p*-type material was obviously needed for efficient light emitters based on GaN. About four decades ago, Be, Mg, Zn, and Cd were introduced during HVPE growth, and the doping effects were evaluated in electrical and optical experiments [19–22]. Approximate binding energies were deduced from optical data, but no *p*-type conductivity was observed at the time. In retrospect, this was mainly due to the lack of understanding of the role of H in the material, i.e., H as a passivation agent for acceptors. This property of H was first understood in Si and III-V materials during the 1980s [23] and was further studied during the 1990s [24].

An intense development period for III-nitrides was initiated when a method to obtain low resistive GaN by Mg doping in MOVPE growth was discovered by Amano et al. [25]. Activation could be done by electron irradiation, and it was later shown by Nakamura et al. that a more convenient method with thermal annealing could also be used [15], in a similar way as with other semiconductor materials. Remarkably, only Mg was found to have the property to offer low resistivity *p*-conduction after proper annealing. Other dopants, such as C, Zn, Be, and Cd have never been reported to be useful for *p*-doping so far. We shall briefly discuss the optical properties of these different acceptor dopants, with an emphasis on the Mg acceptor.

In Figure 7.6 is shown a synopsis of the approximate energy positions of the above mentioned acceptors in GaN, as obtained from PL data on the donor–acceptor pair (DAP) emissions. The shallowest acceptor observed is Be related, with a binding energy of just about 100 meV [26]. However, no *p*-conductivity has been reported with Be-doping during growth. This has been attributed to the simultaneous creation

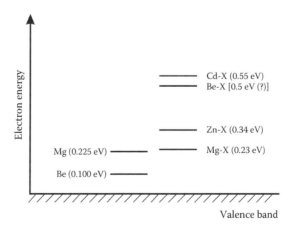

FIGURE 7.6 Schematic diagram of the binding energies of common acceptors in GaN. The acceptors that are deep and show strong phonon coupling are denoted A-X, and are believed to have a relaxed acceptor configuration.

of compensating Be-related donors in GaN [27]. It is also possible that a deep acceptor state exists for Be-doping, in analogy to the Li acceptor in ZnO [28,29]. The near bandgap PL spectrum shows a shallow BE peak with a binding energy <10 meV, presumably related to a shallow ABE transition [26].

Interestingly, a high hole concentration has been reported from Mg and Be co-doping via Be implantation in Mg-doped GaN [30]. No confirmation of this result has been presented so far, to our knowledge.

Zn-doping gives rise to a strong violet DAP emission with a no-phonon line at 3.10 eV [31], consistent with a binding energy of about 0.34 eV [32,33]. The strong phonon coupling present in the optical transition is consistent with the deep acceptor state. The Zn acceptor, accordingly, is known to cause semi-insulating behavior in GaN. At high Zn-doping, additional deeper states (presumably deeper acceptors) are manifested via emissions in the visible and red part of the spectrum [32,34]. The identities of these deeper states are not known, and no theoretical prediction of such states has been provided, to our knowledge. The near bandgap PL spectrum for Zn-doped material demonstrates an ABE spectrum at about 3.452 eV in strain-free GaN [35]. This spectrum consists of three closely spaced lines, and the temperature dependence shows that these lines are derived from a split ABE ground state. This fact shows that the traditional wisdom about the electronic structure of ABEs in wurtzite wide bandgap materials [36] has to be revised. In Ref. [36], it is assumed that the ABE ground state in these materials is a singlet, as obtained if only the top valence band is considered in discussing the ABE wave function. A proper treatment including all three top valence bands removes this problem, and predicts that the ABE in these materials is likely to have a split ground state [37], as shown experimentally (Figure 7.7).

Carbon on the nitrogen site (C_N) has been suggested to be a shallow acceptor in GaN from early theoretical predictions [38,39]. The experimental results seem not to give any clear confirmation of this situation [40]. C doping typically produces highly resistive material due to a deep acceptor state related to a PL emission in the yellow region [31]. Clear evidence for any shallow DAP emission spectrum related to C acceptors seems to be lacking. A recent theoretical calculation results in a C_N acceptor binding energy about 0.9 eV in GaN, in good agreement with the abovementioned experimental data [41].

Cd doping in GaN causes a deep broad DAP PL band in the blue spectral region, interpreted as related to a deep acceptor with a binding energy about 0.5 eV [20,22]. Obviously, this acceptor is too deep to be of interest for *p*-doping.

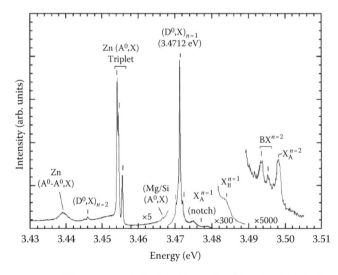

FIGURE 7.7 Low-temperature PL spectrum of a bulk GaN sample with residual Zn doping grown by the Ga-Na melt technique, clearly showing the splitting of the Zn-related ABE [35]. (From *J. Cryst. Growth*, 246, Skromme, B.J., Palle, K.C., Poweleit, C.D. et al., Optical characterization of bulk GaN crystals grown from a Na-Ga melt, 299, Copyright 2002, with permission from Elsevier.)

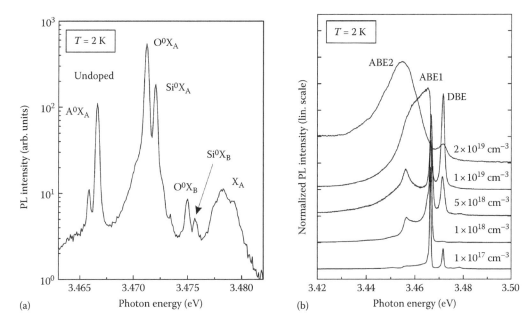

FIGURE 7.8 Low-temperature PL spectra of (a) an undoped homoepitaxial GaN layer showing the splitting of the Mg-related ABE and (b) Mg-doped homoepitaxial GaN layers evidencing the presence of two Mg-related ABEs. (From *Phys. Rev. Lett.*, 102, Monemar, B., Paskov, P.P., Pozina, G. et al., Evidence for two Mg related acceptors in GaN, 235501, 2009, Copyright 2002, from Elsevier.)

The practical alternative for *p*-doping in GaN is thus based on introducing the Mg acceptor. As mentioned above, Mg doping has successfully been employed in GaN-based light emitting devices [42].The physical properties of the Mg acceptor have not been clarified in detail in the literature. It was recently established that an ABE spectrum at about 3.466 eV at 2 K in unstrained GaN is in fact Mg-related [43]. As shown in Figure 7.8a, it consists of two lines about 0.7 meV apart, i.e., the ABE ground state is split in this case as well [37] (similar to the Zn ABE in Figure 7.7). This interpretation differs from the one of Stepniewski et al. in Ref. [44]. The common presence of this ABE spectrum in nominally undoped GaN shows that Mg is a typical contaminant in GaN, independent of the growth technique. It was recently discovered that Mg doping causes an additional ABE state at about 3.454 eV in GaN (Figure 7.8b) [43]. This peak is much broader, and increases in relative strength with increased doping as well as with increased excitation intensity. The two acceptors are labeled A1 and A2, respectively, in the following discussion.

The presence of two ABE signatures in Mg-doped GaN is naturally interpreted to mean that there are in fact two Mg-related acceptors present [43]. In Ref. [43], this situation was tentatively described with one regular Mg_{Ga} state, and in addition, an Mg-related complex acceptor. Very recently, a new theoretical development suggests that the Mg_{Ga} acceptor can actually exist in two different configurations, one regular Mg_{Ga} acceptor state where the hole wave function is distributed over the neighboring N ligands (this state is suggested to have an effective mass-like character) and another state where the hole wave function is localized on one neighboring N ligand only; this N atom is then relaxed from its ideal position [29]. The latter configuration is a much more localized state, and large lattice relaxation means that a strong phonon coupling is expected in the corresponding optical spectra. This type of dual acceptor character is shown to occur in ZnO:Li [28], and is predicted to be typical for acceptors in wide bandgap oxides [45].

The ABE peaks described above for Mg-doped GaN (Figure 7.8b) cannot be used as an indication for the difference in binding energies for the two acceptors, since it is well known that the so-called

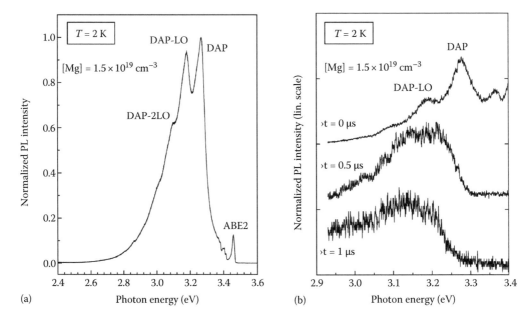

FIGURE 7.9 Continuous wave (a) and time-resolved (b) spectra of a GaN:Mg homoepitaxial layer with [Mg] = 1.5×10^{19} cm^{-3}. The spectra in (b) are taken after annealing of the sample. (From Monemar, B., Paskov, P.P., Pozina, G. et al., Evidence for two Mg related acceptors in GaN, *Phys. Rev. Lett.*, 102, 235501, 2009. Copyright 2009 by the American Physical Society.)

Haynes' rule does not apply for acceptors in direct bandgap materials [36]. However, the corresponding binding energies can be obtained from an evaluation of the shallow DAP spectra for the two acceptors A problem is that the A1 DAP spectrum usually dominates by far in intensity (see Figure 7.9a). This spectrum has a no-phonon line at about 3.27 eV, and well resolved longitudinal-optical (LO) phonon replicas toward lower energies. The DAP spectrum for the A2 acceptor is much weaker, but can be retrieved in time-resolved spectra at long delay times (in the μs range, see Figure 7.9b). A comparison of the estimated position of the no-phonon lines in both spectra shows that the acceptor binding energies are very similar. The value E_a for A1 has been given as 225 ± 5 meV [31], for A2 a less accurate value $E_a = 240 \pm 10$ meV may be deduced.

In the framework of the two-acceptor model of Refs. [29,43], A1 is the shallow effective mass like neutral acceptor state (i.e., occupied with a hole), while A2 corresponds to the "deep" distorted configuration for the bound hole with stronger phonon coupling. The negatively charged state is assumed to be the same in both cases (unrelaxed). Accidentally, in the case of the Mg acceptors, the hole binding energies turn out to be very similar in the two acceptor states. An illustration of the development of the occupancy of the two states is given in Figure 7.10. The PL spectra for a Mg-doped sample before annealing are shown as a function of time during prolonged continuous excitation at 2 K. Before annealing, the equilibrium Fermi level is well above both acceptor levels. Mg acceptors are supposed to be at least partly passivated by H before anneal [43]. The photoexcited holes are initially captured mainly into the A1 acceptor level, but given time, these acceptors gradually relax and transform into the A2 configuration. In this A2 state, the recombination with electrons at shallow donors is much slower, allowing a gradual buildup of hole occupancy in A2. This is reflected in the gradual transformation of the PL spectrum from the A1-related DAP to the broader A2-related DAP with a long continuous excitation time at 2 K (Figure 7.10) [43].

In addition to the spectra related to the two acceptors A1 and A2, there seem to be more spectral features related to the Mg doping. At moderate Mg doping, there is a forest of rather sharp line spectra

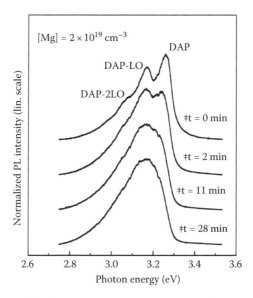

FIGURE 7.10 Transformation of the low-temperature DAP spectrum of an as-grown homoepitaxial GaN:Mg layer during UV laser excitation. (From Monemar, B., Paskov, P.P., Pozina, G. et al., Evidence for two Mg related acceptors in GaN, *Phys. Rev. Lett.*, 102, 235501, 2009. Copyright 2009 by the American Physical Society.)

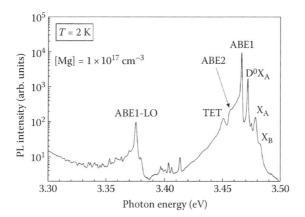

FIGURE 7.11 Low-temperature PL spectra of a homoepitaxial Mg-doped GaN layer showing a number of sharp emission lines tentatively attributed to discrete DAP recombination related to the A1 acceptor.

present, as previously described by Stepniewski et al. [46]. These spectra are most prominent at a rather low Mg doping, and show a multitude of lines in the range 3.34–3.43 eV (Figure 7.11). The lines persist up to high doping, although broadened above [Mg] $>10^{19}$ cm^{-3}. They were interpreted as discrete DAP lines in Ref. [46], and an analysis of the line energies led to an acceptor binding energy of 260 meV [46]. This value is clearly inconsistent with the A1 binding energy of 225 meV, so the analysis in Ref. [46] needs to be reconsidered. (The A1 acceptor would be the only candidate for discrete DAP lines, since the large phonon coupling of A2 would prohibit sharp line observations for that DAP). The many lines present up to higher energies might well be connected to the A1 DAP spectrum, but a detailed line assignment is difficult. More future work is needed to clarify the detailed origin of these features.

At high Mg doping levels (about 10^{20} cm^{-3}), additional broad PL features occur at lower energies (Figure 7.12). The PL background in the range 3.0–3.2 eV is broadened, and a new broad peak is observed

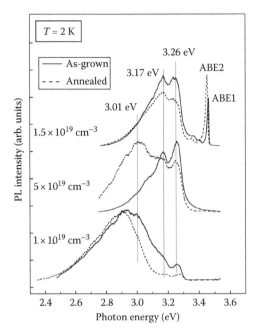

FIGURE 7.12   Low-temperature PL spectra of three homoepitaxial GaN layers with different Mg-doping concentrations before and after annealing. (Reprinted with permission from Pozina, G. et al., *Appl. Phys. Lett.*, 92, 151904, 2008. Copyright 2008, American Institute of Physics.)

at about 2.9 eV in the studied MOVPE samples. This peak was discussed in the literature some time ago, and interpreted as evidence for a deep donor Mg-V_N complex, i.e., the transition would be of DAP nature involving a deep ~0.26 eV donor and the ~0.2 eV Mg acceptors [47,48]. This model could explain the optical data, provided the deep donor concentration is high, >10^{19} cm^{-3}. More recent data show that this interpretation is in doubt. Positron annihilation experiments have indeed observed the Mg-V_N complex, but its concentration should be at the most 10^{17} cm^{-3} [49], inconsistent with the strong up-shift in energy observed for the 2.9 eV peak with excitation intensity [47]. Magnetic resonance data for Mg-doped GaN (with [Mg] >10^{19} cm^{-3}) shows that the dominant donors in the material are residual shallow donors [50]. This calls for a reinterpretation of the 2.9 eV DAP as a shallow-donor to the Mg acceptor DAP transition, which should have a much stronger oscillator strength as compared to the case of a deep donor DAP with a localized donor wave function. The strong spectral up-shift typically observed with excitation intensity in these spectra [51] could be interpreted as due to a gradual saturation of deeper DAP transitions favoring the more shallow ones with increased excitation [16]. Electrical measurements of deep hole traps in Mg-doped GaN show evidence of several acceptor defects in the range 0.2–0.6 eV from the valence band top [16,52]. Such acceptors may be involved in the strong PL background typically observed in the range 3.0–3.1 eV in MOVPE GaN:Mg.

7.4 Donor and Acceptor Spectra in AlN

Recently, the purity of AlN samples has improved rather drastically, so that near bandgap optical spectra can give detailed information on the intrinsic excitons as well as excitons bound at shallow impurities. The free exciton (FE) energies are obtained via a combination of reflectance data and PL or cathodoluminescence (CL) data [53,54]. It turns out that even for virtually strain-free bulk AlN samples, the value of the FE energy varies between samples of seemingly similar quality, presumably reflecting unknown concentrations of various defects in the material, which affect the lattice parameters, and consequently, the bandgap [55]. In fact, a similar situation has been observed for bulk GaN [56]. The FE position at ~5 K in bulk AlN has been stated as 6.03 eV by Silveira et al. [57], and about 6.04 eV by Prinz et al. [55], denoting the present spectral accuracy. In fact, the FE peak is composed of several components, possibly explaining some scatter in the data [58]. Spectra obtained for AlN layers on SiC have a lower energy for the FE, about 5.985 eV and a low T is reported for growth on c-plane 6H-SiC [59], while the FE energy for AlN layers grown on sapphire varies between different authors, depending on the orientation of the used sapphire substrate, but is typically higher than for the bulk AlN. For layers grown on the c-plane sapphire, an FE energy of 6.14 eV at 12 K is reported for a sample with the lowest defect density [60], while samples with higher dislocation density were found to have a lower FE energy, by as much as ~100 meV in some cases [61].

The presence of up to four narrow BE lines at energies about 14, 23, 30, and 34 meV below the FE peak for the AlN layers is suggested to be related to four different shallow donors in AlN [60]. The only such donor known to commonly exist is the Si donor. In a separate study, the binding energy of the Si DBE is given as 24 meV [59]. In a recent paper on PL, a higher accuracy of the spectral energies is claimed from strain free bulk AlN [58]. In this work, the DBE localization energies for the four DBE peaks are given as 28.2, 22.1, 12.8, and 7.2 meV, respectively; the corresponding donors still remain unidentified [58]. The Si donor binding energy has been given in a rather wide range from previous works, from about 180 meV to about 300 meV [62–64]. O is a very common contaminant in AlN, but is supposed to be a deep level (a so-called DX state) [65] and may therefore not show any shallow DBE peak. More work is needed to identify the shallow donors present in AlN.

A recent electron paramagnetic resonance (EPR) study has shed new light on the question of meta-stability of the shallow Si donor. The effective mass binding energy of a donor in AlN was recently estimated as 65 meV [66]. Studies of EPR data as a function of temperature give evidence that the Si donor actually shows DX behavior just like the O donor, but the two-electron level of Si is sufficiently shallow (about 0.14 eV below the conduction band) to depopulate this state at room temperature, with the consequence that Si actually works as a shallow donor with a binding energy of just 0.065 eV in AlN [67].

Acceptors in AlN are known to be deep, but the binding energies are not so well known. Mg doping has been applied in MOVPE growth, and a BE with a binding energy of about 40 meV has been claimed to be the Mg ABE [68]. A corresponding broad band at a lower energy has been assigned to an Mg acceptor state with a binding energy of about 0.5 eV [68] (see Figure 7.13). Measuring the hole concentration versus temperature in Mg-doped AlN gives an activation energy of 630 meV [64]. Doping with Be in MOVPE growth produces a BE with a binding energy of 33 meV, suggesting a slightly shallower binding energy to that of the Mg acceptor [69]. We note, however, that the so-called Haynes' rule is known not to apply to acceptors in direct bandgap semiconductors. While these deep acceptors seem to make p-doping very difficult, there is a report of C-doped AlN that claims respectable hole conductivity [70]. These interesting results seem not to have been confirmed by other work, though. Early theoretical work predicted that C_N is more shallow than the Al site acceptors in AlN [38]. In a more recent calculation, the C acceptor is found to be much deeper in AlN, of the order 1 eV, and should therefore not be useful for p-doping [41].

7.5 Shallow Donors and Acceptors in InN

Donors in InN are expected to be very shallow according to the effective mass model, since the effective mass for electrons is only $(0.04–0.05)m_0$ [71]. Unfortunately, so far, no detailed studies of optical properties of shallow donors could be done for InN since the (bulk) electron concentration in samples studied so far has been in the range $>10^{17}$ cm^{-3}, which is above the limit for degenerate n-doping [72]. The very high electron affinity of InN typically places the Fermi level at the surface of about 1.6 eV higher than the conduction band edge, leading to an electron accumulation layer at the surface. The degenerate doping situation even in the bulk has the consequence that defects that normally are deep levels in other

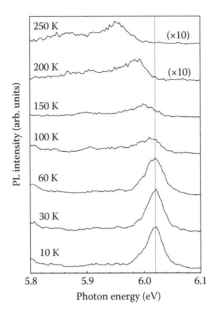

FIGURE 7.13   PL spectra of an Mg-doped AlN layer measured from 10 to 250 K. (With permission from Nepal, N., Nakarmi, L.M., Nam, K.B. et al., Acceptor-bound exciton transition in Mg-doped AlN epilayer, *Appl. Phys. Lett.*, 85, 2271, 2004. Copyright 2004, American Institute of Physics.)

materials have states resonant with the conduction band in InN, and thus they have a tendency to act as shallow donors and deliver electrons to the conduction band edge.

The origin of the dominant *n*-type conductivity in bulk InN has been vigorously debated recently. One model supported by experimental as well as theoretical results was nitrogen vacancies associated with dislocations as the main electron source, with resonant levels in the conduction band [73,74]. More recent results found no real correlations with dislocation density and *n*-type doping, but instead a strong indication that H could be the donors involved [75,76]. H can act as single as well as double donors in InN [77]. Obviously, H can be incorporated in high concentration by MBE during growth, and also during room temperature storage after growth. The obvious way to reduce the bulk electron concentration in InN would then be to avoid H incorporation.

Optical spectra related to donor doping in InN do not reveal any spectral features of individual donors, as long as the material is degenerately doped. As shown in Figure 7.14, rather broad near band-gap emission is interpreted as transitions from filled conduction band states to localized hole states close to the valence band. At somewhat lower energies, transitions from the conduction band states to the acceptor states are seen [78,79]. In Mg-doped InN, the acceptor-related conduction band-to-acceptor peak is clearly seen (Figure 7.15), and the binding energy is evaluated at about 61 meV [80]. As noted above, BEs have so far not been observed in InN due to excessive *n*-doping.

7.6 Point Defect–Related PL in GaN and AlN

Intrinsic point defects in III-nitrides are mainly vacancies and interstitials. These give rise to deep levels in the bandgap, which have been studied theoretically [81], and to a lesser extent, experimentally. It appears that the optically active defects are mainly pairs of point defects and impurities. An example is the V_{Ga}-O_N pair defect in GaN, which is a deep level involved in the so-called yellow emission typically observed in n-GaN. The proof that this PL spectrum is related to a V_{Ga} defect was the correlation between PL and the positron annihilation data published by Saarinen et al. [82]. It was later shown that the V_{Ga}-O_N pair is stable up to very high temperatures (>1200°C) [3], while the isolated V_{Ga} defect is

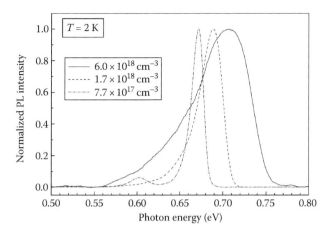

FIGURE 7.14 Low-temperature near bandgap PL spectra of molecular beam epitaxy grown InN layers with different free electron concentrations.

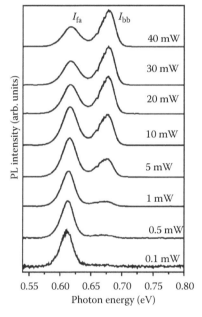

FIGURE 7.15 PL spectra ($T = 16\,K$) of an Mg-doped InN layer measured at different excitation powers. (With permission from Wang, X.Q., Che, S.B., Ishitani, Y. et al., Growth and properties of Mg-doped In-polar InN films, *Appl. Phys. Lett.*, 90, 201913, 2007. Copyright 2007, American Institute of Physics.)

unstable already at much lower temperatures [83]. Thus, the V_{Ga}-O_N pair is the defect present after cool down from growth. Recently, the magnetic resonance signature of this defect has also been demonstrated in bulk n-GaN [84]. There are also reports on metastability involving the transfer between two different charge states of this defect in GaN, optically observed via another PL spectrum in the blue spectral range [85]. In recent room temperature PL work, deep level emissions around 1.2–1.3 eV were studied [86]. This infrared emission was suggested to be related to the V_{Ga}-O_N pair defect (valence band to acceptor transition), and thus complementary to the previously studied yellow luminescence (YL) band [82]. Similar complexes between V_{Ga} and Si donors may also be formed, but are unstable at the growth temperature, and therefore, not expected to be present after cooldown from growth [2].

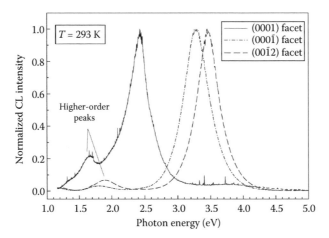

FIGURE 7.16 Room-temperature CL spectra taken from different facets of a bulk AlN sample grown by sublimation.

In p-type GaN, defects related to N vacancies (V_N) are expected [81]. The isolated V_N is known to be a moderately shallow donor [87], suggesting that a previously studied shallower donor with a binding energy of only 25 meV [88] is not V_N related [87]. Evidence for the existence of Mg-V_N deep donors are presented by positron annihilation data [49]. The correlation with a blue PL emission in Mg doped GaN [47] is questionable, however; it seems more likely that the blue PL band in Mg doped GaN is related to a deep Mg-related acceptor state and shallow donors [16].

Spectra related to Ga interstitials (Ga_i) and related complexes seem to be present in the infrared part of the spectrum, and broad PL bands centered around 0.95 eV have been studied in optically detected magnetic resonance (ODMR) experiments [89,90]. However, there is no clear picture of the donor energy level related to isolated Ga_i. The interstitial N is suggested to be a deep acceptor [91], and no optical activities for this defect have been reported, to our knowledge.

In the case of AlN, and according to theoretical calculations [81,92], the intrinsic defects are expected to be similar to the corresponding ones in GaN. The V_{Al}-O_N pair defect has very recently been identified by EPR in bulk AlN [93]; it is often associated with deep broad PL bands at photon energies between 3 and 4 eV (Figure 7.16) [94]. Some support of these suggestions comes from correlation with positron annihilation data [95]. Indeed, O is a very common contaminant in bulk AlN, so the dominant association of V_{Al} with O is natural [94]. It is difficult to keep down the O concentration even in epitaxial material, so for epilayers, the deep PL emissions are also believed to be dominated by the V_{Al}-O_N complexes [61]. There may also be other complexes involving V_{Al} that are nonradiative [61]. The nitrogen vacancy has also recently been identified as a deep donor center in EPR data for bulk AlN [96]. There are suggestions in the literature that a PL emission band peaking at about 5.87 eV at low T in implanted AlN could be related to this deep donor, indicating a binding energy of the order <300 meV [97]. However, these results need a definite confirmation. AlGaN alloys are suggested to behave in a way similar to AlN concerning vacancy defects, although in this case, both V_{Al}-O_N complexes and V_{Ga}-O_N pairs would be present. It should be mentioned that there are ideas that the strong nonradiative recombination associated with dislocations in AlInGaN-based LEDs could be due to the nonradiative cation vacancy complexes present preferentially around dislocations [61,98].

In the case of the primary alloy material for visible LEDs, i.e., InGaN, most of the optical data in this case are for narrow (say 3 nm) InGaN quantum wells (QW), and very little work has been done on thicker bulk-like layers. Therefore, no detailed point defect–related optical spectra are available so far for thicker InGaN layers. Similarly, for InN, very little work has been done to explore optical spectra related, for example, to the predicted V_{In} complexes in n-InN [99,100].

7.7 PL and CL Related to Structural Defects in III-Nitrides

Structural defects like dislocations and stacking faults (SFs) are very common in III-nitride structures due to the widespread habit of growing the materials on foreign substrates. Many studies have been made on the structural properties of these defects; less detailed knowledge is present concerning their electronic structure and optical properties. Concerning dislocations of edge or screw type in GaN (and other III-nitrides), they are not known to cause any radiative spectra in PL or CL; rather, numerous CL studies have shown dark contrast for these defects, indicating that they are nonradiative recombination sites (Figure 7.17) Whether this nonradiative recombination is caused by intrinsic dislocation electronic states, or by point defects and/or impurities that aggregate around these line defects is still less clear [61]. It is well known that in some other semiconductors, like Si and II–VI materials, dislocations are known to produce specific radiative recombination spectra [101,102]; such results are still lacking for GaN.

Stacking faults are interesting objects that are known to cause luminescence spectra in semiconductors. Good examples are recently shown for various SiC polytypes [103,104]. Similar effects occur in GaN, i.e., SF-related luminescence spectra occur in the spectral region 3.29–3.45 eV. Such spectra may be dominant in case of nonpolar epitaxial layers where the SF density is often very high (Figure 7.18)

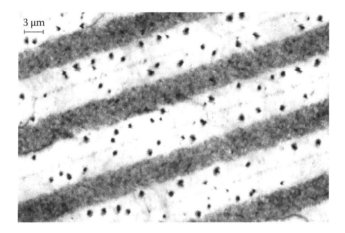

FIGURE 7.17   Panchromatic CL image of an $Al_{0.22}Ga_{0.78}N$ layer grown by MOVPE on a patterned GaN/sapphire template. The black spots in the image reveal the surface pits terminating the treading dislocations.

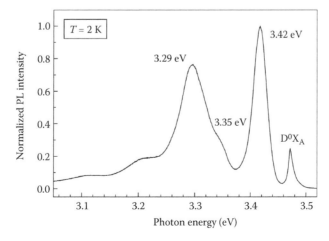

FIGURE 7.18   Low-temperature PL spectra of an *a*-plane GaN layer grown by MOVPE on an *r*-plane sapphire.

andbook of Luminescent Semiconductor Materials

[105]. Recently, different spectra observed in CL have been correlated with different types of SFs, e.g., the commonly observed features around 3.41–3.42 eV is related to the basal-plane SFs (BFSs) of type I_1 [106]. The type I_1 BSF consists of one violation of the stacking sequence of atomic planes along the [0001] direction. The BSFs in GaN do not introduce localized states in the bandgap, but can be considered as a thin zinc-blende (ZB) layer embedded in the wurtzite (W) matrix. The theory predicts that this polytypic QW has a type-II band alignment with a conduction band offset of about 220–270 meV [107]. In a simplified wurtzite GaN band diagram, the SF defect can then be described as associated with a corresponding bound electron state about 0.2 eV below the conduction band edge. The electrons are confined in the ZB QW-like region, while the holes residing in the QW barrier are trapped in a shallow potential caused by strain and polarization fields (at low temperatures) or attracted to bound electrons via Coulomb interaction. Therefore, the 3.41–3.42 eV emission can be described as a recombination of type-II (indirect) excitons. The large linewidth of this emission has been explained by the presence of bundles of BSFs [105] or by the donors localized in the vicinity of a single BSF [108]. The origin of the emissions in the 3.29–3.36 eV region is less understood. BSFs of type I_2 (in which the thickness of the ZB layer is doubled comparing with the type I_1 BSF) are invoked to explain the emission at 3.35 eV [109]. The emission at 3.29–3.30 eV has been attributed to the prismatic SFs (PSFs) intersecting BSFs [110] or to the partial dislocations (PDs) terminating the BSFs [106]. This emission behaves more like as a DAP emission with pronounced LO replicas [111] implying an impurity decoration of SFs or PDs. Obviously, more work has to be done in order to clarify the exact origin of the structural defect–related emissions in GaN. The weak binding of holes to SF defects will result in such holes becoming free at room temperature, and may affect the hole conductivity in p-GaN. The high hole concentration reported in Ref. [112] for hetero-epitaxial m-plane GaN:Mg might well be influenced by the SF density in such layers.

Concerning AlN and InN, there are no reports on the structural defect-related emissions in these materials so far.

7.8 Conclusions

The studies of defects in the III-nitrides are in a stage of steady development, with detailed optical data present mainly for donors in GaN. Acceptor doping remains a bottleneck in light emitting devices; therefore, it is likely that acceptor studies will be a priority also in the near future, so that more detailed optical spectra can be obtained. This is still of interest for GaN, but will be particularly important for AlN and InN, as well as for related alloys. The role of intrinsic point defects for III-nitride devices is still unclear, and more work is needed to verify their interplay with structural defects. A combination of optical and magnetic resonance work would be useful in this area. The case of dislocations and SFs needs a more careful analysis to develop an understanding of the electronic properties and how these in turn affect the optical properties.

Acknowledgments

We would like to thank Prof. B. J. Skromme, Prof. H. X. Jiang, Dr. M. Bickermann, and Dr. G. Pozina for providing electronic versions of Figures 7.7, 7.13, 7.16, and 7.17.

References

1. H. Morkoç, *Handbook of Nitride Semiconductors and Devices*, Wiley, Wienheim, Germany (2008).
2. J. Oila, V. Ranki, J. Kivioja et al., Influence of dopants and substrate material on the formation of Ga vacancies in epitaxial GaN layers, *Phys. Rev. B*. **63**, 045205 (2001).
3. F. Toumisto, K. Saarinen, T. Paskova et al., Thermal stability of in-grown vacancy defects in GaN grown by hydride vapor phase epitaxy, *J. Appl. Phys.* **99**, 066105 (2006).

4. J. Neugebauer and C. G. Van de Walle, Gallium vacancies and the yellow luminescence in GaN, *Appl. Phys. Lett.* **69**, 503 (1996).

5. A. F. Wright, Substitutional and interstitial oxygen in wurtzite GaN, *J. Appl. Phys.* **98**, 103531 (2005).

6. B. Monemar, P. P. Paskov, J. P. Bergman et al., Recombination of free and bound excitons in GaN, *Phys. Status Solidi (b)* **245**, 1723 (2008).

7. G. Neu, M. Teisseire, B. Beaumont et al., Near-band gap selective photoluminescence in wurtzite GaN, *Phys. Status Solidi (b)* **216**, 79 (1999).

8. B. Gil, P. Bigenwald, M. Leroux et al., Internal structure of the neutral donor-bound exciton complex in cubic zinc-blende and wurtzite semiconductors, *Phys. Rev. B.* **75**, 085204 (2007).

9. B. K. Meyer, Free and bound excitons in GaN epitaxial films, *Mater. Res. Symp. Proc.* **449**, 497 (1997).

10. P. J. Dean, J. D. Cuthbert, D. G. Thomas et al., Two-electron transitions in the luminescence of excitons bound to neutral donors in gallium phosphide, *Phys. Rev. Lett.* **18**, 122 (1967).

11. P. P. Paskov, B. Monemar, A. Toropov et al., Two-electron transition spectroscopy of shallow donors in bulk GaN, *Phys. Status Solidi (c)* **4**, 2601 (2007).

12. W. J. Moore, J. A. Freitas Jr., G. C. B. Braga et al., Identification of Si and O donors in hydride-vapor-phase epitaxial y GaN, *Appl. Phys. Lett.* **79**, 2570 (2001).

13. J. Neugebauer and C. G. Van de Walle, Hydrogen in GaN: Novel aspects of a common impurity, *Phys. Rev. Lett.* **75**, 4452 (1995).

14. J. Neugebauer and C. G. Van de Walle, Role of hydrogen in doping of GaN, *Appl. Phys. Lett.* **68**, 1829 (1996).

15. S. Nakamura, T. Mukai, M. Sehoh et al., Thermal annealing effect on p-type Mg-doped GaN films, *Jpn. J. Appl. Phys.* **31**, L139 (1992).

16. B. Monemar, P. P. Paskov, G. Pozina et al., Mg-related acceptors in GaN, *Phys. Status Solidi (c)* **7**, 1850–1852 (2010).

17. K. Shimomura, R. Kadono, K. Ohishi et al., Muonium as a shallow center in GaN, *Phys. Rev. Lett.* **92**, 135505 (2004).

18. B. Monemar, P. P. Paskov, J. P. Bergman et al., A hydrogen-related shallow donor in GaN?, *Physica B* **376–377**, 460 (2006).

19. H. G. Grimmeis and B. Monemar, Low-temperature luminescence of GaN, *J. Appl. Phys.* **41**, 4054 (1970).

20. M. Ilegems, R. Dingle, and R. A. Logan, Luminescence of Zn- and Cd-doped GaN, *J. Appl. Phys.* **43**, 3797 (1972).

21. M. Ilegems and R. Dingle, Luminescence of Be- and Mg-doped GaN, *J. Appl. Phys.* **44**, 4234 (1973).

22. O. Lagerstedt and B. Monemar, Luminescence in epitaxial GaN:Cd, *J. Appl. Phys.* **45**, 2266 (1974).

23. J. I. Pankove, D. E. Carlson, J. E. Berkeyheiser et al., Neutralization of shallow acceptor levels in silicon by atomic hydrogen, *Phys. Rev. Lett.* **51**, 2224 (1983).

24. S. J. Pearton, J. W. Corbett, and T. S. Shi, Hydrogen in crystalline semiconductors, *Appl. Phys. A* **43**, 153 (1987).

25. H. Amano, M. Kirro, K. Hiramatsu et al., P-type conduction in Mg-doped GaN treated with low-energy electron beam irradiation (LEEBI), *Jpn. J. Appl. Phys.* **28**, L2112 (1989).

26. T. H. Myers, A. J. Ptak, L. Wang et al., Magnesium and beryllium doping during rf-plasma MBE growth of GaN, *IPAP Conf. Series* **1**, 451 (2000).

27. C. G. Van de Walle and S. Limpijumnong, First-principle studies of beryllium doping of GaN, *Phys. Rev. B* **63**, 245205 (2001).

28. B. K. Meyer, J. Sann, and A. Zeuner, Lithium and sodium acceptors in ZnO, *Superlatt. Microstruct* **38**, 344 (2005).

29. S. Lany and A. Zunger, Dual nature of acceptors in GaN and ZnO: The curious case of the shallow Mg_{Ga} deep state, *Appl. Phys. Lett.* **96**, 142114 (2010).

30. C. C. Yu, C. F. Chu, J. Y. Tsai et al., Beryllium-implanted p-type GaN with high carrier concentration, *Jpn. J. Appl. Phys.* **40**, L417 (2001).

31. M. A. Reshchikov and H. Morkoç, Luminescence properties of defect in GaN, *J. Appl. Phys.* **97**, 061301 (2005).

32. B. Monemar, O. Lagerstedt, and H. P. Gislason, Properties of Zn-doped VPE-drown GaN. I. Luminescence data in relation to doping conditions, *J. Appl. Phys.* **51**, 625 (1980).

33. B. Monemar, H. P. Gislason, and O. Lagerstedt, Properties of Zn-doped VPE-drown GaN. II. Optical cross sections, *J. Appl. Phys.* **51**, 640 (1980).

34. J. I. Pankove, J. E. Berkeyheiser, and E. A. Miller, Properties of Zn-doped GaN. I. Photoluminescence, *J. Appl. Phys.* **45**, 1280 (1974).

35. B. J. Skromme, K. C. Palle, C. D. Poweleit et al., Optical characterization of bulk GaN crystals grown from a Na-Ga melt, *J. Cryst. Growth* **246**, 299 (2002).

36. P. J. Dean and D. C. Herbert, Bound excitons in semiconductors, in K. Cho (Ed.) *Excitons*, Springer, Berlin, Germany, 1979, p. 55.

37. B. Gil, P. Bigenwald, P. P. Paskov et al., Internal structure of acceptor-bound excitons in wide-band-gap wurtzite semiconductors, *Phys. Rev. B* **81**, 085211 (2010).

38. P. Boguslawski and J. Bernholc, Doping properties of C, Si, and Ge impurities in GaN and AlN, *Phys. Rev. B* **56**, 9496 (1997).

39. A. F. Wright, Substitutional and interstitial carbon in wurtzite GaN, *J. Appl. Phys.* **92**, 2575 (2002).

40. M. A. Reshchikov, R. H. Patillo, and K. C. Travis, Photoluminescence in wurtzite GaN containing carbon, *Mater. Res. Soc. Proc.* **892**, FE23.12 (2006).

41. J. L. Lyons, A. Janotli, and L. G. Vande Walle, Carbon impurities and the yellow luminescence in GaN, *Appl. Phys. Lett.* **97**, 152108 (2010).

42. S. Nakamura and G. Fasol, *The Blue Laser Diode*, Springer-Verlag, Berlin, Germany, 1997.

43. B. Monemar, P. P. Paskov, G. Pozina et al., Evidence for two Mg related acceptors in GaN, *Phys. Rev. Lett.* **102**, 235501 (2009).

44. R. Stepniewski, A. Wysmolek, M. Potemski et al., Fine structure of effective mass acceptors in gallium nitride, *Phys. Rev. Lett.* **91**, 226404 (2003).

45. S. Lany and A. Zunger, Polaronic hole localization and multiple hole binding of acceptors in oxide wide-gap semiconductors, *Phys. Rev. B* **80**, 085202 (2009).

46. R. Stepniewski, A. Wysmolek, M. Potemski et al., Impurity-related luminescence of homoepitaxial GaN studied with high magnetic fields, *Phys. Status Solidi* (*b*) **210**, 373 (1998).

47. U. Kaufmann, M. Kunzer, H. Obloh et al., Origin of defect-related photoluminescence bands in doped and nominally undoped GaN, *Phys. Rev. B* **59**, 5561 (1999).

48. C. G. Van de Walle, C. Stampfl, and J. Neugebauer, Theory of doping and defects in III–V nitrides, *J. Cryst. Growth* **189/190** 505 (1998).

49. S. Hautakangas, K. Saarinen, L. Liszkay et al., Role of open volume defects in Mg-doped GaN films studied by positron annihilation spectroscopy, *Phys. Rev. B* **72**, 165303 (2005).

50. E. R. Glaser, W. E. Carlos, G. C. B. Braga et al., Magnetic resonance studies of Mg-doped GaN epi-taxial layers grown by organometallic chemical vapor deposition, *Phys. Rev. B* **65**, 085312 (2002).

51. M. A. Reshchikov, G. C. Yi, and B. W. Wessels, Behavior of 2.8- and 3.2-eV photoluminescence bands in Mg-doped GaN at different temperatures and excitation densities, *Phys. Rev. B* **59**, 13176 (1999).

52. P. Hacke, H. Nakayama, T. Detchprohm et al., Deep levels in the upper band-gap region of lightly Mg-doped GaN, *Appl. Phys. Lett.* **68**, 1362 (1996).

53. L. Chen, B. J. Skromme, R. F. Dalmau et al., Band-edge exciton states in AlN single crystals and epi-taxial layers, *Appl. Phys. Lett.* **95**, 4334 (2004).

54. E. Silveira, J. A. Freitas Jr., O. J. Glembocki et al., Excitonic structure of bulk AlN from optical reflec-tivity and cathodoluminescence measurements, *Phys. Rev. B* **71**, 041201(R) (2005).

55. G. M. Prinz, A. Ladenburger, M. Feneberg et al., Photoluminescence, cathodoluminescence, and reflectance study of AlN layers and AlN single crystals, *Superlatt. Microstruct.* **40**, 513 (2006).

56. V. Darakchieva, B. Monemar, and A. Usui, On lattice parameters of GaN, *Appl. Phys. Lett.* **91**, 031911 (2007).

57. E. Silveira, J. A. Freitas Jr., S. B. Schujman et al., AlN bandgap temperature dependence from its optical properties, *J. Cryst. Growth* **310**, 4007 (2008).
58. M. Feneberg, R. A. R. Leute, B. Neuschl et al., High-excitation and high-resolution photoluminescence spectra of bulk AlN, *Phys. Rev. B* 82, 075208 (2010)
59. G. M. Prinz, A. Ladenburger, M. Schirra et al., Cathodoluminescence, photoluminescence, and reflectance of an aluminum nitride layer grown on silicon carbide substrate, *J. Appl. Phys.* **101**, 023511 (2007).
60. T. Onuma, T. Shibata, K. Kosaka et al., Free and bound exciton fine structures in AlN epilayers grown by low-pressure metalorganic vapor phase epitaxy, *J. Appl. Phys.* **101**, 023529 (2009).
61. S. F. Chichibu, A. Uedono, T. Onuma et al., Impact of point defects on the luminescence properties of (Al, Ga)N, *Mater. Sci. Forum* **590**, 233 (2008).
62. M. L. Nakarmi, K. H. Kim, K. Zhu, et al., Transport properties of highly conductive n-type Al-rich $Al_xGa_{1-x}N$ ($x \geq 0.7$), *Appl. Phys. Lett.* **85**, 3769 (2004).
63. R. Zeisel, M. W. Bayerl, S. T. B. Goennenwein et al., DX-behavior of Si in AlN, *Phys. Rev B* **61**, R16283 (2000).
64. Y. Taniyasu, M. Kasu, and T. Makimoto, An aluminium nitride light-emitting diode with a wavelength of 210 nanometers, *Nature* **441**, 325 (2006).
65. C. G. Van de Walle, DX-center formation in wurtzite and zinc-blende $Al_xGa_{1-x}N$, *Phys. Rev. B* **57**, R2033 (1998).
66. I.G. Ivanov, Private communication (2010).
67. N. T. Son, M. Bickermann, and E. Janzén, Shallow donor and DX states of Si in AlN, *Appl. Phys. Lett.* 98, 092104 (2011).
68. N. Nepal, L. M. Nakarmi, K. B. Nam et al., Acceptor-bound exciton transition in Mg-doped AlN epilayer, *Appl. Phys. Lett.* **85**, 2271 (2004).
69. A. Sedhain, T. M. Al Tahtamouni, J. Li et al., Beryllium acceptor binding energy in AlN, *Appl. Phys. Lett.* **93**, 141104 (2008).
70. K. Wongchotigul, N. Chen, D. P. Zhang et al., Low resistivity aluminum nitride: Carbon (AlN:C) films grown by metal organic chemical vapor deposition, *Mater. Lett.* **26**, 223 (1996).
71. T. Hofmann, V. Darakchieva, B. Monemar et al., Optical Hall effect in hexagonal InN, *J. Electron. Mater.* **37**, 611 (2008).
72. B. Monemar, P. P. Paskov, and A. Kasic, Optical properties of InN—The bandgap question, *Superlatt. Microstruct.* **38**, 38 (2005).
73. V. Cimalla, V. Lebedev, F. M. Morales et al., Model for the thickness dependence of electron concentration in InN film, *Appl. Phys. Lett.* **89**, 172109 (2006).
74. L. F. J. Piper, T. D. Veal, C. F. McConville et al., Origin of the n-type conductivity of InN: The role of positively charged dislocations, *Appl. Phys. Lett.* **88**, 252109 (2006).
75. V. Darakchieva, K. Lorenz, N. P. Barradas et al., Hydrogen in InN: A ubiquitous phenomenon in molecular beam epitaxy grown material, *Appl. Phys. Lett.* **96**, 081907 (2010).
76. C. S. Gallinat, G. Koblmüller, and J. S. Speck, The role of threading dislocations and unintentionally incorporated impurities on the bulk electron conductivity of In-face InN, *Appl. Phys. Lett.* 95, 022103 (2009).
77. A. Janotti and C. G. Van de Walle, Source of unintentional conductivity in InN, *Appl. Phys. Lett.* **92**, 032104 (2008).
78. B. Arnaudov, T. Paskova, P. P. Paskov et al., Energy position of near-band-edge emission spectra of InN epitaxial layers with different doping levels, *Phys. Rev. B* **69**, 115216 (2004).
79. A. A. Klochikhin, V. Yu. Davydov, V. V. Emtsev et al., Acceptor states in the photoluminescence spectra of n-InN, *Phys. Rev. B* **71**, 195207 (2005).
80. X. Q. Wang, S. B. Che, Y. Ishitani et al., Growth and properties of Mg-doped In-polar InN films, *Appl. Phys. Lett.* 90, 201913 (2007).
81. C. G. Van de Walle and J. Neugebauer, First-principle calculations for defects and impurities: Applications to III-nitrides, *J. Appl. Phys.* **95**, 3851 (2004).

82. K. Saarinen, T. Laine, S. Kuisma et al., Observation of native Ga vacancies in GaN by positron anni-hilation, *Phys. Rev. Lett.* **79**, 3030 (1997).

83. K. Saarinen, T. Suski, I. Grzegory et al., Thermal stability of isolated and complexed Ga vacancies in GaN bulk crystals, *Phys. Rev. B* **64**, 233201 (2001).

84. N. T. Son, C. G. Hemmingsson, T. Paskova et al., Identification of the gallium vacancy-oxygen pair defect in GaN, *Phys. Rev. B* **80**, 153202 (2009).

85. S. Dhar and S. Ghosh, Optical metastability in undoped GaN grown on Ga-rich GaN buffer layers, *Appl. Phys. Lett.* **80**, 4519 (2002).

86. A. Sedhain, J. Li, J. Y. Lin et al., Nature of deep center emissions in GaN, *Appl. Phys. Lett.* **96**, 151902 (2010).

87. D. C. Look, G. C. Farlow, P. J. Drevinski et al., On the nitrogen vacancy in GaN, *Appl. Phys. Lett.* **83**, 3525 (2003).

88. Q. Yang, H. Feick, and E. R. Weber, Observation of a hydrogen donor in the luminescence of elec-tron-irradiated GaN, *Appl. Phys. Lett.* **82**, 3002 (2003).

89. M. Linde, S. J. Uftring, G. D. Watkins et al., Optical detection of resonance in electron-irradiated GaN, *Phys. Rev. B* **55**, R10177 (1997).

90. K. H. Chow, G. D. Watkins, A. Usui et al., Detection of interstitial Ga in GaN, *Phys. Rev. Lett.* **85**, 2761 (2000).

91. D. C. Look, D. C. Reynolds, J. W. Hemsky et al., Defect donor and acceptor in GaN, *Phys. Rev. Lett.* **79**, 2273 (1997).

92. C. Stampfl and C. G. Van de Walle, Theoretical investigation of native defects, impurities, and com-plexes in aluminum nitride, *Phys. Rev. B* **65**, 155212 (2002).

93. N. T. Son, private communication, 2010.

94. M. Bickermann, B. M. Epelbaum, O. Filip et al., Point defect content and optical transitions in bulk aluminum nitride crystals, *Phys. Status Solidi* (*b*) **246**, 1181 (2009).

95. A. Uedono, S. Ishibashi, S. Keller et al., Vacancy-oxygen complexes and their optical properties in AlN epitaxial film studied by positron annihilation, *J. Appl. Phys.* **105**, 054501 (2009).

96. S. M. Evans, N. C. Giles, L. E. Halliburton et al., Electron paramagnetic resonance of a donor in alu-minum nitride crystals, *Appl. Phys. Lett.* **88**, 062112 (2006).

97. N. Nepal, K. B. Nam, M. L. Nakarmi et al., Optical properties of the nitrogen vacancy in AlN epilay-ers, *Appl. Phys. Lett.* **84**, 1090 (2004).

98. S. F. Chichibu, A. Uedono, T. Onuma et al., Origin of defect-insensitive emission probability in In-containing (Al, In, Ga) N alloy semiconductors, *Nat. Mater.* **5**, 810 (2006).

99. C. Stampfl, C. G. Van de Walle, D. Vogel et al., Native defects and impurities in InN: First-principle studies using the local-density approximation and self-interaction and relaxation-corrected pseudo-potentials, *Phys. Rev. B* **61**, R7846 (2000).

100. J. Oila, A. Kemppinen, A. Laakso et al., Influence of layer thickness on the formation of In vacancies in InN grown by molecular beam epitaxy, *Appl. Phys. Lett.* **84**, 1486 (2004).

101. R. Sauer, Ch. Kisielowski-Kemmerich, and H. Alexander, Dislocation-width-dependent radiative recombination of electrons and holes at widely split dislocations in silicon, *Phys. Rev. Lett.* **57**, 1472 (1986).

102. K. Wolf, S. Jilka, H. Sahin et al., Relaxation process and luminescence of lattice defects in epitaxially grown ZnSe/GaAs layers, *J. Cryst. Growth* **152**, 34 (1995).

103. S. G. Sridhara, F. H. C. Carlsson, J. P. Bergman et al., Luminescence from stacking faults in 4H SiC, *Appl. Phys. Lett.* **79**, 3944 (2001).

104. H. Iwata, U. Lindefelt, S. Öberg et al., Localized states around stacking faults in silicon carbide, *Phys. Rev. B* **65**, 033203 (2001).

105. P. P. Paskov, R. Schifano, B. Monemar et al., Emission properties of a-plane GaN grown by metal-organic chemical vapor deposition, *J. Appl. Phys.* **98**, 093519 (2005).

106. R. Liu, A. Bell, F. A. Ponce et al., Luminescence from stacking faults in gallium nitride, *Appl. Phys. Lett.* **86**, 021908 (2005).
107. C. Stampfl and C. G. Van de Walle, Energetics and electronic structure of stacking faults in AlN, GaN, and InN, *Phys. Rev. B* **57**, R15052 (1998).
108. P. Corfdir, P. Lefebvre, J. Ristic et al., Electron localization by a donor in the vicinity of a basal stacking fault in GaN, *Phys. Rev. B* **80**, 153309 (2009).
109. Y. J. Sun, O. Brandt, U. Jahn et al., Impact of nucleation conditions on the structural and optical properties of m-plane GaN(1–100) grown on γ-LiAlO$_2$, *J. Appl. Phys.* **92**, 5714 (2002).
110. J. Mei, S. Srinivasan, R. Liu et al., Prismatic stacking faults in epitaxially laterally overgrown GaN, *Appl. Phys. Lett.* **88**, 141912 (2006).
111. P. P. Paskov, R. Schifano, T. Paskova et al., Structural defect-related emissions in nonpolar a-plane GaN, *Physica B* **376–377**, 473 (2006).
112. M. McLaurin, T. E. Mates, F. Wu et al., Growth of p-type and n-type m-plane GaN by molecular beam epitaxy, *J. Appl. Phys.* **100**, 063707 (2006).
113. G. Pozina, C. Hemmingsson, P. P. Paskov et al., Effect of annealing on metastable shallow acceptors in Mg-doped GaN layers grown on GaN substrates, *Appl. Phys. Lett.* **92**, 151904 (2008).

8

Narrow-Gap Semiconductors for Infrared Detectors

Antoni Rogalski
*Military University
of Technology*

8.1 Introduction

The years during World War II saw the origins of modern infrared (IR) detector technology. Recent success in applying IR technology to remote sensing problems has been made possible by the successful development of high-performance IR detectors over the last six decades. Many materials have been investigated in the IR field. Spectral detectivity curves for a number of commercially available IR detectors are shown in Figure 8.1. Interest has centered mainly on the wavelengths of the two atmospheric windows from 3–5 to 8–14 m, though in recent years there has been increasing interest in longer wavelengths stimulated by space applications.

During the 1950s, IR detectors were built using single-element-cooled lead salt detectors, primarily for anti-air-missile seekers. Usually lead salt detectors were polycrystalline and were produced by vacuum evaporation and chemical deposition from a solution, followed by a post-growth sensitization process. The first extrinsic photoconductive detectors were reported in the early 1950s after the discovery of the transistor, which stimulated a considerable improvement in the growth and material purification techniques. Since the techniques for controlled impurity introduction became available for germanium at an earlier date, the first high-performance extrinsic detectors were based on the use of germanium. The extrinsic

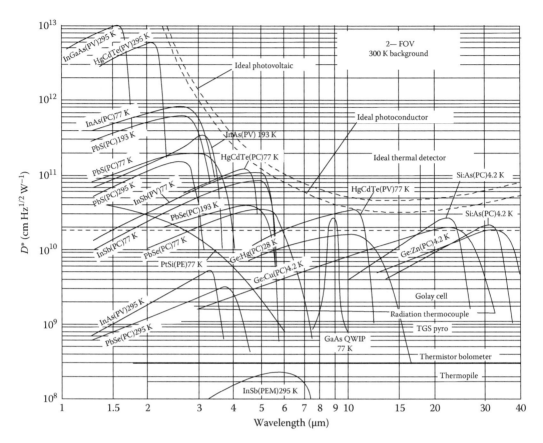

FIGURE 8.1 Comparison of the D^* of various commercially available IR detectors when operated at the indicated temperature. Chopping frequency is 1000 Hz for all detectors except the thermopile (10 Hz), thermocouple (10 Hz), thermistor bolometer (10 Hz), Golay cell (10 Hz), and pyroelectric detector (10 Hz). Each detector is assumed to view a hemispherical surrounding at a temperature of 300 K. Theoretical curves for the background-limited D^* (dashed lines) for ideal photovoltaic and photoconductive detectors and thermal detectors are also shown. PC, photoconductive detector; PV, photovoltaic detector; and PEM, photoelectromagnetic detector.

photoconductors were widely used at wavelengths beyond 10 m, prior to the development of the intrinsic detectors. They must be operated at lower temperatures to achieve performance similar to that of intrinsic detectors, and a sacrifice in quantum efficiency is required to avoid impractically thick detectors.

At the same time, rapid advances were being made in narrow bandgap semiconductors that would later prove useful in extending wavelength capabilities and improving sensitivity. The first such material was InSb, a member of the newly discovered III–V compound semiconductor family. The interest in InSb stemmed not only from its small energy gap, but also from the fact that it could be prepared in single crystal form using a conventional technique. The end of the 1950s and the beginning of the 1960s saw the introduction of narrow-gap semiconductor alloys in III–V ($InAs_{1-x}Sb_x$), IV–VI ($Pb_{1-x}Sn_xTe$), and II–VI ($Hg_{1-x}Cd_xTe$) material systems. These alloys allowed the bandgap of the semiconductor and hence the spectral response of the detector to be custom tailored for specific applications. In 1959, research by Lawson and coworkers [1] triggered development of variable bandgap $Hg_{1-x}Cd_xTe$ (HgCdTe) alloys, providing an unprecedented degree of freedom in IR detector design. This first paper reported both photoconductive and photovoltaic response at the wavelength extending out to 12 μm. Soon thereafter, working under a U.S. Air Force contract with the objective of devising an 8–12 μm background-limited semiconductor IR detector that would operate at temperatures as high as 77 K, the group lead by Kruse at the Honeywell Corporate Research Center in Hopkins, Minnesota developed a modified Bridgman crystal growth technique for HgCdTe. They soon reported both photoconductive and photovoltaic detection in rudimentary HgCdTe devices [2].

The fundamental properties of HgCdTe ternary alloys (high optical absorption coefficient, high electron mobility, and low thermal generation rate), together with the capability for bandgap engineering, make these alloy systems almost ideal for a wide range of IR detectors. The difficulties in growing HgCdTe material, significantly due to the high vapor pressure of Hg, encouraged the development of alternative detector technologies over the past 40 years. One of these was PbSnTe, which was vigorously pursued in parallel with HgCdTe in the late 1960s, and early 1970s [3–5]. PbSnTe was comparatively easy to grow and good-quality long-wavelength IR (LWIR) photodiodes were readily demonstrated. However, in the late of 1970s two factors led to the abandonment of PbSnTe detector work: high dielectric constant and large temperature expansion coefficient (TEC) mismatch with Si. Scanned IR imaging systems of the 1970s last century required relatively fast response times so that the scanned image is not smeared in the scan direction. With the trend today toward staring arrays, this consideration might be less important than it was when first generation systems were being designed. The second drawback, large TCE, can lead to failure of the indium bonds in hybrid structure (between silicon readout and the detector array) after repeated thermal cycling from room temperature to the cryogenic temperature of operation.

During the 1970s, a new generation of materials, so called low-dimensional solids, for IR detectors have been proposed. The HgTe/CdTe superlattice (SL) system was proposed in 1979 [6], only a few years after the first GaAs/AlGaAs quantum heterostructures were fabricated by MBE. Next, two additional structures were introduced: InAs/InAsSb SLs with strain-induced bandgap reduction [7] and InAs/GaInSb SL with superlattice-induced band inversion [8,9]. Recently InAs/GaInSb strained-layer superlattices (SLSs) are considered as an alternative to HgCdTe material system.

The operation of the intrinsic IR photodetector is based on band-to-band transitions caused by absorption of the photons in a narrow-gap semiconductor. Among the many types of IR photodetectors, intrinsic photodetectors offer the best performance at high temperatures.

8.2 Narrow-Gap Semiconductor Materials

The properties of narrow-gap semiconductors that are used as the material systems for IR detectors result from the direct energy bandgap structure: a high density of states in the valence and conduction bands, which results in strong absorption of IR radiation and a relatively low rate of thermal generation. Table 8.1 compares important parameters of narrow-gap semiconductors used in IR detector fabrication.

TABLE 8.1 Some Physical Properties of Narrow-Gap Semiconductors

Material	E_g (eV) 77 K	E_g (eV) 300 K	n_i (cm^{-3}) 77 K	n_i (cm^{-3}) 300 K	ε	μ_e (10^4 cm^2/V s) 77 K	μ_e (10^4 cm^2/V s) 300 K	μ_h (10^4 cm^2/V s) 77 K	μ_h (10^4 cm^2/V s) 300 K
InAs	0.414	0.359	6.5×10^3	9.3×10^{14}	14.5	8	3	0.07	0.02
InSb	0.228	0.18	2.6×10^9	1.9×10^{16}	17.9	100	8	1	0.08
In$_{0.53}$Ga$_{0.47}$As	0.66	0.75		5.4×10^{11}	14.6	7	1.38		0.05
PbS	0.31	0.42	3×10^7	1.0×10^{15}	172	1.5	0.05	1.5	0.06
PbSe	0.17	0.28	6×10^{11}	2.0×10^{16}	227	3	0.10	3	0.10
PbTe	0.22	0.31	1.5×10^{10}	1.5×10^{16}	428	3	0.17	2	0.08
Pb$_{1-x}$Sn$_x$Te	0.1	0.1	3.0×10^{13}	2.0×10^{16}	400	3	0.12	2	0.08
Hg$_{1-x}$Cd$_x$Te	0.1	0.1	3.2×10^{13}	2.3×10^{16}	18.0	20	1	0.044	0.01
Hg$_{1-x}$Cd$_x$Te	0.25	0.25	7.2×10^8	2.3×10^{15}	16.7	8	0.6	0.044	0.01

The wide body of information now available concerning different methods of crystal growth and physical properties of materials used for IR photodetectors makes it difficult to review all aspects in detail. As a result, only selected topics are reviewed in this chapter. More information can be found in many comprehensive reviews and monographs (see, for example, Refs. [3–5,10–18]).

8.2.1 Some Physical Properties of III–V Narrow-Gap Semiconductors

Development of crystal growth techniques in the early 1950s led to InSb and InAs bulk single crystal detectors. Since then the quality of single crystal growth has improved immensely.

Two methods of obtaining these crystals have been used commercially: Czochralski or horizontal Bridgman techniques. Single crystals can be grown with relatively high purity, low dislocation density, and ingot sizes that permit wafer diameters up to 100 mm range, suitable for convenient handling and photolithography. The growth of InSb single crystals is reviewed by Hulme, Mullin, Liang, Micklethwaite, and Johnson [19–23]. A wide spectrum of topics in materials that today's engineers, material scientists, and physicists need is included in a comprehensive treatises on electronic and photonic materials gathered in the *Springer Handbook of Electronic and Photonic Materials* [24].

Unlike InSb, reaction of the elements to form an InAs compound is not a simple matter. To keep the arsenic from disappearing because of its high vapor pressure near the melting point, it is necessary to let the constituents react in a sealed quartz ampoule. Purification of InAs is also more difficult than InSb.

$In_xGa_{1-x}As$ ternary alloy has been of great interest for the short-wavelength infrared (SWIR), low-cost detector applications. The $x=0.53$ alloy is lattice-matched to InP substrates, has a bandgap of 0.73 eV, and covers the wavelength range from 0.9 to 1.7 μm. High-quality InP substrates are available with diameters as large as 100 mm. $In_xGa_{1-x}As$ with 53% InAs is often called "standard InGaAs" without bothering to note the values of "x" or "$1-x$." This is mature material, driven by the mass-production of fiber-optic receivers at 1.3 and 1.55 μm. At present, InGaAs is also becoming the choice for high-temperature operation in the 1–3 μm spectrum. By increasing the indium content to $x=0.82$, the wavelength response of $In_xGa_{1-x}As$ can be extended out to 2.6 μm. Single-element InGaAs detectors have been made with up to 2.6 μm cutoffs while linear arrays and cameras have been demonstrated to 2.2 μm.

The energy bandgap of InGaAsP quaternary system range from 0.35 eV (InAs) to 2.25 eV (GaP), with InP (1.29 eV) and GaAs (1.43 eV) falling between [25,26]. InGaAsP alloys have been epitaxially grown by hydride and chloride vapor-phase epitaxy (VPE); liquid-phase epitaxy (LPE); molecular-beam epitaxy (MBE), and metalorganic chemical vapor deposition (MOCVD) [27]. A brief comparison of the four techniques is given by Olsen and Ban [28]. While each of these techniques has certain advantages, hydride VPE is well suited for InGaAsP/InP optoelectronic devices. Epitaxy methods are also used for more sophisticated structures of modern InSb, InAs, InGaAs, $InAs_{1-x}Sb_x$ (InAsSb) and $Ga_xIn_{1-x}Sb$ (GaInSb) devices [12,29,30].

Figure 8.2 shows the composition dependence of the energy gap and the electron effective mass at the Γ-conduction bands of $Ga_xIn_{1-x}As$, $InAs_xSb_{1-x}$ and $Ga_xIn_{1-x}Sb$ ternaries.

InAsSb is an attractive semiconductor material for detectors covering the 3–5 and 8–14 μm spectral ranges [29]. However, progress in this ternary system has been limited by crystal synthesis problems. The large separation between the liquidus and solidus (Figure 8.3) and the lattice mismatch (6.9% between InAs and InSb) place stringent demands upon the method of crystal growth [31]. These difficulties are being overcome systematically using MBE and MOCVD.

The III–V detector materials have a zinc-blende structure and direct energy gap at the Brillouin zone center. The shape of the electron band and the light mass hole band is determined by the $\mathbf{k} \cdot \mathbf{p}$ theory. The momentum matrix element varies only slightly for different materials and has an approximate value of 9.0×10^{-8} eV cm. Then, the electron effective masses and conduction band densities of states are similar for materials with the same energy gap.

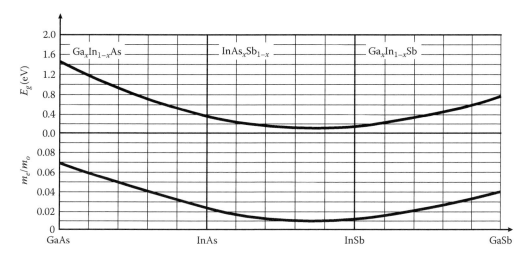

FIGURE 8.2 Variation of the bandgap energy and electron effective mass at the Γ-conduction bands of $Ga_xIn_{1-x}As$, $InAs_xSb_{1-x}$, and $Ga_xIn_{1-x}Sb$ ternary alloys at room temperature.

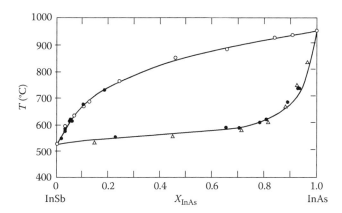

FIGURE 8.3 Pseudobinary phase diagram for the InAs-InSb system. (After Stringfellow, G.B. and Greene, P.R., *J. Electrochem. Soc.*, 118, 805, 1971. With permission.)

The III–V materials have a conventional negative temperature coefficient of the energy gap which is well described by the Varshni relation [32]

$$E_g(T) = E_o - \frac{\alpha T^2}{T + \beta} \tag{8.1}$$

where α and β are fitting parameter characteristics of a given material.

Table 8.2 contains some physical parameters of the InAs, InSb, GaSb, $InAs_{0.35}Sb_{0.65}$, and $In_{0.53}Ga_{0.47}As$ semiconductors [17].

The III–V detector material which has been investigated most broadly is InSb. The temperature independent portions of the Hall curves indicate that most of the electrically active impurity atoms in InSb have shallow activation energies and above 77 K are thermally ionized. The Hall coefficient for p-type samples is positive in the low-temperature extrinsic range, and reverses sign to become negative in the intrinsic range because of the higher mobility of the electrons (the mobility ratio $b = \mu_e/\mu_h$ of the order of

TABLE 8.2   Physical Properties of Narrow Gap III–V Compounds

	T (K)	InAs	InSb	GaSb	InAs$_{0.35}$Sb$_{0.65}$	In$_{0.53}$Ga$_{0.47}$As
Lattice structure		Cubic (ZnS)	Cubic (ZnS)	Cubic (ZnS)	Cubic (ZnS)	Cubic (ZnS)
Lattice constant a (nm)	300	0.60584	0.647877	0.6094	0.636	0.58438
Thermal expansion	300	5.02	5.04	6.02		
coefficient α (10^{-6} K^{-1})	80		6.50			
Density ρ (g/cm^3)	300	5.68	5.7751	5.61		5.498
Melting point T_m (K)		1210	803	985		
Energy gap E_g (eV)	4.2	0.42	0.2357	0.822	0.138	0.627
	80	0.414	0.228		0.136	
	300	0.359	0.180	0.725	0.100	0.75
Thermal coefficient of E_g	100–300	-2.810^{-4}	-2.8×10^{-4}			-3.0×10^{-4}
Effective masses						
m_e^*/m	4.2	0.023	0.0145	0.042		0.041
	300	0.022	0.0116		0.0101	
m_{lh}^*/m	4.2	0.026	0.0149			0.0503
m_{hh}^*/m	4.2	0.43	0.41	0.28	0.41	0.60
Momentum matrix		9.2×10^{-8}	9.4×10^{-8}			
element P (eV cm)						
Mobilities						
μ_e (cm^2/V s)	77	8×10^4	10^6		5×10^5	70,000
	300	3×10^4	8×10^4	5×10^3	5×10^4	13,800
μ_h (cm^2/V s)	77		1×10^4	2.4×10^3		
	300	500	800	880		
Intrinsic carrier	77	6.5×10^3	2.6×10^9		2.0×10^{12}	
concentration n_i (cm^{-3})	200	7.8×10^{12}	9.1×10^{14}		8.6×10^{15}	
	300	9.3×10^{14}	1.9×10^{16}		4.1×10^{16}	5.4×10^{11}
Refractive index n_r		3.44	3.96	3.8		
Static dielectric constant ε_s		14.5	17.9	15.7		14.6
High frequency dielectric		11.6	16.8	14.4		
constant ε_∞						
Optical phonons						
LO (cm^{-1})		242	193		≈ 210	
TO (cm^{-1})		220	185		≈ 200	

10^2 is observed). The transition temperature for the p-type samples, at which R_H changes sign, depends on the purity. The samples become intrinsic above certain temperature (above 150 K for pure n-type samples) and below these temperatures (below 100 K for pure n-type samples) there is little variation of Hall coefficients.

There are various carrier scattering mechanisms in semiconductors. Reasonably pure n-type and p-type InSb samples exhibit an increase in mobility up to approximately 20–60 K after which the mobility decreases due to polar and electron–hole scattering. Carrier mobility systematically increases with a decrease in impurity concentration both in temperature 77 as well as in 300 K.

Optical properties of InSb have been reviewed by Kruse [33]. Because of the very small effective mass of electrons, the conduction band density of states is small and it is possible to fill the available band states by doping, thereby appreciably shifting the absorption edge to shorter wavelengths. This has been referred as the Burstein–Moss effect.

The physical properties of InAs are the same as InSb. InAs material with cutoff wavelength of 3.5 μm makes it of limited utility for the middle-wavelength band even though it could theoretically be operated near 190 K. Development work has been limited by growth and passivation problems.

Although effort has been devoted to the development of InAsSb as an alternative to HgCdTe for IR applications, there is relatively little information available on its physical properties. A III–V detector technology would benefit from superior bond strengths and material stability (compared to HgCdTe), well-behaved dopants, and high-quality III–V substrates. The review of the development of InAsSb crystal growth techniques, physical properties and detector fabrication procedures is presented in Rogalski's papers [29,34].

Conventional InAsSb (although it is the smallest bandgap for any bulk III–V materials) does not have a sufficiently small gap at 77 K for operation in the 8–14 μm wavelength range. To build detectors that span the 8–14 μm atmospheric window, Osbourn proposed the use of InAsSb strain layer superlattices [7]. This theoretical prediction reinforced interest in investigation of InAsSb ternary alloy as a material for IR detector applications.

The ternary alloy $Ga_xIn_{1-x}Sb$ is important material for the fabrication of detectors designed for middle-wavelength IR applications. The long wavelength limit of $Ga_xIn_{1-x}Sb$ detectors has been tuned compositionally from 1.52 μm ($x = 1.0$ at 77 K) to 6.8 μm ($x = 0.0$ at room temperature). The bandgap energy of $In_xGa_{1-x}As_ySb_{1-y}$ at room temperature can be fitted by the relationship [35]:

$$E_g(x) = 0.726 - 0.961x - 0.501y + 0.08xy + 0.451x^2 - 1.2y^2 + 0.021x^2y + 0.62xy^2 \qquad (8.2)$$

The lattice matching condition to GaSb imposes the additional constraint that x and y are related as $y = 0.867/(1 - 0.048x)$.

8.2.2 HgCdTe Ternary Alloys

HgCdTe ternary alloy is nearly ideal IR detector material system. Its position is conditioned by three key features:

- Tailorable energy bandgap over the 1–30 μm range
- Large optical coefficients that enable high quantum efficiency
- Favorable inherent recombination mechanisms that lead to high operating temperature

These properties are direct consequence of the energy band structure of this zinc-blende semiconductor. Moreover, the specific advantages of HgCdTe are ability to obtain both low and high carrier concentrations, high mobility of electrons, and low dielectric constant. The extremely small change of lattice constant with composition makes it possible to grow high quality layered and graded gap structures. As a result, HgCdTe can be used for detectors operated at various modes (photoconductor, photodiode, or metal-insulator-semiconductor [MIS] detector).

Table 8.3 summarizes various material properties of $Hg_{1-x}Cd_xTe$ [36].

8.2.2.1 Outlook on Crystal Growth

The time line for the evolution of growth technologies is illustrated in Figure 8.4 [37]. Historically, crystal growth of HgCdTe has been a major problem mainly because a relatively high Hg pressure is present during growth, which makes it difficult to control the stoichiometry and composition of the grown material. The wide separation between the liquidus and solidus (see Figure 8.5), leading to marked segregation between CdTe and HgTe, was instrumental in the development of all the bulk growth techniques to this system. In addition to solidus–liquidus separation, high Hg partial pressures are also influential both during growth and post-growth heat treatments.

TABLE 8.3 Summary of the Material Properties for the $Hg_{1-x}Cd_xTe$ Ternary Alloy, Listed for the Binary Components HgTe and CdTe, and for Several Technologically Important Alloy Compositions

Property	HgTe	$Hg_{1-x}Cd_xTe$						CdTe
x	0	0.194	0.205	0.225	0.31	0.44	0.62	1.0
a (Å)	6.461	6.464	6.464	6.464	6.465	6.468	6.472	6.481
	77 K	77 K	77 K	77 K	140 K	200 K	250 K	300 K
E_g (eV)	−0.261	0.073	0.091	0.123	0.272	0.474	0.749	1.490
λ_c (μm)	—	16.9	13.6	10.1	4.6	2.6	1.7	0.8
n_i (cm^{-3})	—	1.9×10^{14}	5.8×10^{13}	6.3×10^{12}	3.7×10^{12}	7.1×10^{11}	3.1×10^{10}	4.1×10^5
m_c/m_o	—	0.006	0.007	0.010	0.021	0.035	0.053	0.102
g_c	—	−150	−118	−84	−33	−15	−7	−1.2
$\varepsilon_s/\varepsilon_o$	20.0	18.2	18.1	17.9	17.1	15.9	14.2	10.6
$\varepsilon_\infty/\varepsilon_o$	14.4	12.8	12.7	12.5	11.9	10.8	9.3	6.2
n_r	3.79	3.58	3.57	3.54	3.44	3.29	3.06	2.50
μ_e (cm^2/V s)	—	4.5×10^5	3.0×10^5	1.0×10^5	—	—	—	—
μ_{hh} (cm^2/V s)	—	450	450	450	—	—	—	—
$b = \mu_e/\mu_\eta$	—	1000	667	222	—	—	—	—
τ_R (μs)	—	16.5	13.9	10.4	11.3	11.2	10.6	2
τ_{A1} (μs)	—	0.45	0.85	1.8	39.6	453	4.75×10^3	
$\tau_{typical}$ (μs)	—	0.4	0.8	1	7	—	—	—
E_p (eV)	19							
Δ (eV)	0.93							
m_{hh}/m_o	0.40–0.53							
ΔE_v (eV)	0.35–0.55							

Source: After Reine, M.B., Mercury cadmium telluride, in R.D. Guenther, D.G. Steel, and L. Bayvel (eds.), *Encyclopedia of Modern Optics*, Elsevier, Oxford, U.K., 2004, pp. 392–402. With permission.

τ_R and τ_{A1} calculated for n-type HgCdTe with $N_d = 1 \times 10^{15}$ cm^{-3}. The last four material properties are independent of or relatively insensitive to alloy composition.

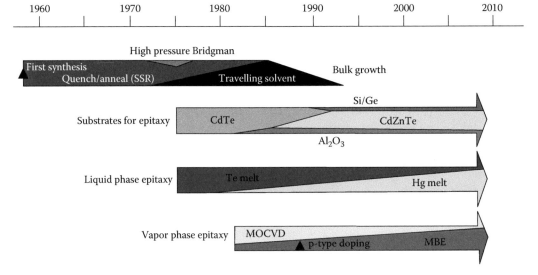

FIGURE 8.4 Evolution of HgCdTe crystal growth technology from 1958 to present. (With permission from Norton, P., HgCdTe infrared detectors, *Opto-Electron. Rev.*, 10, 159–174, 2002. Copyright 2002, American Institute of Physics.)

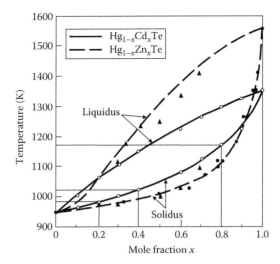

FIGURE 8.5   Liquidus and solidus lines in the HgTe-CdTe and HgTe-ZnTe pseudobinary systems.

Early experiments and a significant fraction of early production were done using a quench-anneal or solid-state recrystallization process. In this method, the charge of a required composition was synthesized, melted, and quenched. Then, the fine dendritic mass (highly polycrystalline solid) obtained in a process was annealed below the liquidus temperature for a few weeks to recrystallize and homogenize the crystals. The material usually requires low-temperature annealing for adjusting the concentration of native defects. The crystals can be also uniformly doped by the introduction of dopants to the charge.

Bridgman growth was attempted for several years near the mid-1970s. At the same time, solvent growth methods from Te-rich melts were initiated to reduce the grown temperature. One successful implementation was the traveling heater method up to 5 cm diameter. The perfect quality of crystals grown by this method is occupied by a low growth rate.

The bulk HgCdTe crystals were initially used for any types of IR photodetectors. At present they are still used for some IR applications such as n-type single element photoconductors, signal processing in the element (SPRITE) detectors, and linear arrays. Bulk growth produced thin rods, generally to 15 mm in diameter, about 20 cm in length, and with nonuniform distribution of composition. Large two-dimensional (2D) arrays could not be realized with bulk crystals. Another drawback to bulk material was need to thin the bulk wafers, generally cut about 500 μm thick, down to final device thickness of about 10 μm. Also, further fabrication stapes (polishing the wafers, mounting them to suitable substrates, and polishing to the final device thickness) was labor intensive.

The epitaxial techniques offer, in comparison with bulk growth techniques, the possibility to grow large area epilayers and sophisticated device structures with good lateral homogeneity, abrupt and complex composition and doping profiles, which can be configured to improve the performance of photodetectors. The growth is performed at low temperatures, which makes it possible to reduce the native defects density.

Among the various epitaxial techniques, the LPE is the most matured method. LPE is a single crystal growth process in which growth from a cooling solution occurs onto a substrate. At present, the VPE growth of HgCdTe is typically done by nonequilibrium methods; MOCVD, MBE, and their derivatives. The great potential benefit of MBE and MOCVD over the equilibrium methods is the ability to modify the growth conditions dynamically during growth to tailor band gaps, add and remove dopants, prepare surfaces and interfaces, add passivations, perform anneals, and even grow on selected areas of a substrate. The growth control is exercised with great precision to obtain basic materials properties comparable to those routinely obtained from equilibrium growth.

Epitaxial growth of HgCdTe layers requires a suitable substrate. CdTe was used initially, since it was available from commercial sources in reasonable large sizes. The main drawback to CdTe is that it has a few percent lattice mismatch with LWIR and middle-wavelength infrared (MWIR) HgCdTe. By the mid-1980s, it was demonstrated that the addition of a few percent ZnTe to CdTe (typically 4%) could create a lattice-matched substrate. CdTe and closely lattice matched CdZnTe substrates are typically grown by the modified vertical and horizontal unseeded Bridgman technique. Most commonly, the (111) and (100) orientations have been used, although others have been tried. Twinning which occurs in (111) layers can be prevented by a suitable disorientation of the substrate. Growth conditions found to be nearly optimal for the (112)B orientation were selected. The limited size, purity problems, Te precipitates, dislocation density (routinely in the low 10^4 cm^{-2} range), nonuniformity of lattice match, and high price ($50–$500 per 1 cm^2, polished) are remaining problems to be solved. It is believed that these substrates will continue to be important for a long time, particularly for the highest performance devices.

LPE growth of thin layer on CdTe substrates began in the early-to-mid-1970s. Initially, Te solutions with dissolved Cd (Cd has a high solubility in Te) and saturated with Hg vapor were used to efficiently grow HgCdTe in temperature range 420°C–600°C. This allowed small-volume melts to be used with slider techniques which did not appreciably deplete during the growth run. Experiments with Hg-solvent LPE began in the late-1970s. Because of the limited solubility of Cd in Hg, the volume of the Hg melts had to be much larger then Te melts (typically about 20 kg) in order to minimize melt depletion during layer growth in temperature range 380°C–500°C. This precluded the slider growth approach and Hg-melt epitaxy has been developed using large dipping vessels.

In the early 1990s, bulk growth has been replaced by LPE and is now very mature for production of first- and second-generation detectors. However, LPE technology is limited for a variety of advanced HgCdTe structures required for third-generation detectors. LPE typically melts off a thin layer of the underlying material and each time an additional layer is grown as a result of relatively high growth temperature. Additionally, the gradient in x-value in the base layer of p$^+$-on-n junction can generate a barrier transport in certain cases due to interdiffusion. These limitations have provided an opportunity for VPE; MBE and MOCVD.

The era of MBE and MOCVD began in the early 1980s by adopting both methods well established in the III–V semiconductor materials. Through the next decade a variety of metalorganic compounds were developed along with a number of reaction-chamber designs. In the case of MBE, a specially designed Hg-source oven was successfully designed to overcome the low sticking coefficient of Hg at the growth temperature. The growth temperature is less than 200°C for MBE but around 350°C for MOCVD, making it more difficult to control the p-type doping in the MOCVD due to the formation of Hg vacancies at higher growth temperatures. At present, MBE is the dominant vapor phase method for HgCdTe. It offers low-temperature growth under an ultrahigh vacuum environment, *in situ* n- and p-type doping, and control of composition, doping, and interfacial profiles. MBE is now the preferred method for growing complex layer structures for multicolor detectors and for avalanche photodiodes (APDs). Although the quality of MBE material is not yet on a par with LPE, it has made tremendous progress in the past decade. Keys of this success have been the doping ability and the reduction of etch pit densities (EPDs) to below 10^5 cm^{-2}.

Near lattice matched CdZnTe substrates severe drawbacks such as lack of large area, high production cost, and more importantly, the difference of TEC in CdZnTe substrates and silicon readout integrated circuits (ROICs) as well as interest in large area based IR FPAs (1024 × 1024 and larger) have resulted in limitations of CdZnTe substrate application. Currently, readily producible CdZnTe substrates are limited to areas of approximately 50 cm^2. At this size, the wafers are unable to accommodate more than two 1024 × 1024 FPAs. Not even a single die can be printed for very large FPA formats (2048 × 2048 and larger) on substrates of this size.

The use of Si substrates is very attractive in IR FPA technology not only because it is less expensive and available in large area wafers but also because the coupling of the Si substrates with Si readout circuitry in an FPA structure allows fabrication of very large arrays exhibiting long-term thermal cycle reliability. The $7 \times 7 \, cm^2$ bulk CdZnTe substrate is the largest commercially available, and it is unlikely to increase much larger than its present size. With the cost of 6 in. Si substrate being $\approx$\$100 versus \$10,000 for the $7 \times 7 \, cm^2$, significant advantages of HgCdTe/Si are evident [38]. Despite the large lattice mismatch ($\approx$19%) between CdTe and Si, MBE has been successfully used for the heteroepitaxial growth of CdTe on Si. Using optimized growth condition for Si(211)B substrates and CdTe/ZnTe buffer system, epitaxial layers with EPD about $10^6 \, cm^{-2}$ range have been obtained. This value of EPD has little effect on both MWIR and LWIR HgCdTe/Si detectors [38,39]. By comparison, HgCdTe epitaxial layers grown by MBE or LPE on bulk CdZnTe have typical EPD value in the 10^4 to mid-$10^5 \, cm^{-2}$ range where there is a negligible effect of dislocation density on detector performance. At 77 K, diode performance with cutoff wavelength in LWIR region is comparable to that on bulk CdZnTe substrates [39].

Sapphire has also been widely used as a substrate for HgCdTe epitaxy. In this case, a CdTe (CdZnTe) film is deposited on the sapphire prior to the growth of HgCdTe. This substrate has excellent physical properties and can be purchased in large wafer sizes. Large lattice mismatch with HgCdTe is accommodated by CdTe buffer layer. Sapphire is transparent from the UV to about 6 µm and has been used to backside-illuminated SWIR and MWIR detectors (it is not acceptable for backside-illuminated LWIR arrays because of the opacity beyond 6 µm).

8.2.2.2 Physical Properties

8.2.2.2.1 Band Structure

The electrical and optical properties of $Hg_{1-x}Cd_xTe$ are determined by the energy gap structure in the vicinity of the Γ-point of the Brillouin zone, in essentially the same way as InSb. The shape of the electron band and the light-mass hole band are determined by the $\mathbf{k} \cdot \mathbf{p}$ interaction, and hence, by the energy gap and the momentum matrix element. The energy gap of this compound at 4.2 K ranges from $-0.300 \, eV$ for semimetallic HgTe, goes through zero at about $x = 0.15$ and opens up to 1.648 eV for CdTe.

Figure 8.6 plots the energy bandgap $E_g(x, T)$ for $Hg_{1-x}Cd_xTe$ versus alloy composition parameter x at temperature 77 K and 300 K. Also plotted is the cutoff wavelength $\lambda_c(x, T)$, defined as that wavelength at which the response has dropped to 50% of its peak value.

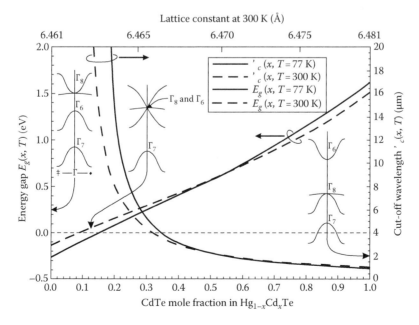

FIGURE 8.6 The bandgap structure of $Hg_{1-x}Cd_xTe$ near the Γ-point for three different values of the forbidden energy gap. The energy bandgap is defined at the difference between the Γ_6 and Γ_8 band extreme at $\Gamma = 0$.

A number of expressions approximating $E_g(x, T)$ are available at present. The most widely used expression is due to Hansen et al. [40]

$$E_g = -0.302 + 1.93x - 0.81x^2 + 0.832x^3 + 5.35 \times 10^{-4}(1 - 2x)T \tag{8.3}$$

where E_g is in eV and T is in K.

The expression which has become the most widely used for intrinsic carrier concentration is that of Hansen and Schmit [41] who used their own $E_g(x, T)$ relationship of (8.3), the $\mathbf{k \cdot p}$ method and a value of $0.443m_o$ for heavy hole effective mass ratio

$$n_i = (5.585 - 3.82x + 0.001753T - 0.001364xT) \times 10^{14} E_g^{3/4} T^{3/2} \exp\left(-\frac{E_g}{2kT}\right) \tag{8.4}$$

The electron m_e^* and light hole m_{lh}^* effective masses in the narrow gap mercury compounds are close and they can be established according to the Kane band model. Here, we used Weiler's expression [42]

$$\frac{m_o}{m_e^*} = 1 + 2F + \frac{E_p}{3}\left(\frac{2}{E_g} + \frac{1}{E_g + \Delta}\right) \tag{8.5}$$

where $E_p = 19\,\text{eV}$, $\Delta = 1\,\text{eV}$, and $F = -0.8$. This relationship can be approximated by $m_e^*/m \approx 0.071E_g$, where E_g is in eV. The effective mass of heavy hole m_{hh}^* is high; the measured values range between 0.3 and $0.7m_o$. The value of $m_{hh}^* = 0.55m_o$ is frequently used in modeling of IR detectors.

8.2.2.2.2 Mobilities

Due to small effective masses the electron mobilities in $Hg_{1-x}Cd_xTe$ are remarkably high, while heavy-hole mobilities are two orders of magnitude lower. A number of scattering mechanisms influence the electron mobility including ionized impurities, alloy disorder, electron–electron and hole–hole events, acoustic phonon modes, and polar phonon modes. Nonpolar optical phonon scattering is significant in p-type and semimetallic n-type materials.

The electron mobility in $Hg_{1-x}Cd_xTe$ (expressed in $cm^2/V\,s$), in composition range $0.2 \leq x \leq 0.6$ and temperature range $T > 50\,\text{K}$, can be approximated as [43]

$$\mu_e = \frac{9 \times 10^8 s}{T^{2r}} \quad \text{where } r = (0.2/x)^{0.6}$$
$$s = (0.2/x)^{7.5} \tag{8.6}$$

The hole mobilities at room temperature range from 40 to 80 $cm^2/V\,s$, and the temperature dependence is relatively weak. A 77 K hole mobility is by one order of magnitude higher. According to Ref. [44], the hole mobility measured at 77 K falls as the acceptor concentration is increased and in the composition range 0.20–0.30 yields the following empirical expression:

$$\mu_h = \mu_o\left[1 + \left(\frac{p}{1.8 \times 10^{17}}\right)^2\right]^{-1/4} \tag{8.7}$$

where $\mu_o = 440\,\text{cm}^2/V\,s$.

For modeling IR photodetectors, the hole mobility is usually calculated assuming that the electron-to-hole mobility ratio $b = \mu_e/\mu_h$ is constant and equal to 100.

The minority carrier mobility is one of the fundamental material properties affecting performance of HgCdTe along with carrier concentration, composition, and minority carrier lifetime. For materials having acceptor concentrations < 10^{15} cm^{-3}, literature results give comparable electron mobilities to those found in n-type HgCdTe [$\mu_e(n)$]. As the acceptor concentration increases, the deviation from n-type electron mobilities increases, resulting in lower electron mobilities for p-type material [$\mu_e(p)$]. Typically, for $x = 0.2$ and $N_a = 10^{16}$ cm^{-3}, $\mu_e(p)/\mu_e(n) = 0.5$–0.7, while for $x = 0.2$ and $N_a = 10^{17}$ cm^{-3}, $\mu_e(p)/\mu_e(n) = 0.25$–0.33. For $x = 0.3$, however, $\mu_e(p)/\mu_e(n)$ ranges from 0.8 for $N_a = 10^{16}$ cm^{-3} to 0.9 for $N_a = 10^{17}$ cm^{-3} [45]. It was found that at temperatures above 200 K there was very little difference between the electron mobility in epitaxial p-type HgCdTe layers and that measured directly in n-type layers.

8.2.2.2.3 Optical Properties

Optical properties of HgCdTe have been investigated mainly at energies near the bandgap. There still appears to be considerable disagreement among the reported results concerning absorption coefficients. This is caused by different concentrations of native defects and impurities, nonuniform composition and doping, thickness inhomogeneities of samples, mechanical strains, and different surface treatments.

In most compound semiconductors, the band structure closely resembles the parabolic energy versus the momentum dispersion relation. The optical absorption coefficient would then have a square-root dependence on energy that follows the electronic density of states, often referred to as the Kane model [46]. The above bandgap absorption coefficient can be calculated for InSb-like band structure semiconductors such as Hg$_{1-x}$Cd$_x$Te, including the Moss–Burstein shift effect. Corresponding expressions were derived by Anderson [47]. Beattie and White proposed an analytic approximation with a wide range of applicability for band-to-band radiative transition rates in direct, narrow-bandgap semiconductors [48].

In high-quality samples, the measured absorption in the short-wavelength region is in good agreement with the Kane model calculation while the situation appears to be complicated in the long-wavelength edge by the appearance of an absorption tail extending at energies lower than the energy gap. This tail has been attributed to the composition-induced disorder. According to Finkman and Schacham [49], the absorption tail obeys a modified Urbach's rule:

$$\alpha = \alpha_o \exp\left[\frac{\sigma(E - E_o)}{T + T_o}\right] \text{ in cm}^{-1} \tag{8.8}$$

where T is in K, E is in eV, and $\alpha_o = exp(53.61x - 18.88)$, $E_o = -0.3424 + 1.838x + 0.148x^2$ (in eV), $T_o = 81.9$ (in K), $\sigma = 3.267 \times 10^4 (1 + x)$ (in K/eV) are fitting parameters which vary smoothly with composition. The fit was performed with data at $x = 0.215$ and $x = 1$ and for temperatures between 80 and 300 K.

Assuming that the absorption coefficient for large energies can be expressed as

$$\alpha(h\nu) = \beta(h\nu - E_g)^{1/2}, \tag{8.9}$$

many researchers assume that this rule can be applied to HgCdTe. For example, Schacham and Finkman used the following fitting parameter $\beta = 2.109 \times 10^5 [(1 + x)/(81.9 + T)]^{1/2}$, which is a function of composition and temperature [50]. The conventional procedure used to locate the energy gap is to use the point inflection, that is, exploit the large change in the slope of $\alpha(h\nu)$ that is expected when the band-to-band transition overtakes the weaker Urbach contribution. To overcome the difficulty in locating the onset of the band-to-band transition, the bandgap was defined as that energy value where $\alpha(h\nu) = 500$ cm^{-1} [49]. Schacham and Finkman analyzed the crossover point and suggested $\alpha = 800$ cm^{-1} was a better choice [50]. Hougen analyzed absorption data of n-type LPE layers and suggested that the best formula was $\alpha = 100 + 5000x$ [51].

8.2.2.2.4 *Generation–Recombination Processes*

Generation–recombination processes in semiconductors are widely discussed in literature (see, for example, Refs. [17,52]). We reproduce here only some dependencies directly related to the performance of photodetectors. Assuming bulk processes only, there are three main thermal generation–recombination processes to be considered in the narrow bandgap semiconductors, namely: Shockley–Read (SR), radiative, and Auger.

The SR mechanism is not an intrinsic and fundamental process as it occurs via levels in the forbidden energy gap. The reported position of SR centers for both n- and p-type materials range anywhere from near the valence to near the conduction band.

The SR mechanism is probably responsible for lifetimes in lightly doped n- and p-type $Hg_{1-x}Cd_xTe$. The possible factors are SR centers associated with native defects and residual impurities. In n-type material ($x = 0.20$–0.21, 80 K) with carrier concentrations less than 10^{15} cm^{-3}, the lifetimes exhibit a broad range of values (0.4–8 μs) for material prepared by various techniques [52]. Dislocations may also influence the recombination time for dislocation densities $> 5 \times 10^5$ cm^{-2} [53,54].

In p-type HgCdTe, SR mechanism is usually blamed for reduction of lifetime with decreasing temperature. The steady-state low-temperature photoconductive lifetimes are usually much shorter than the transient lifetimes. The low-temperature lifetimes exhibit very different temperature dependencies with a broad range of values over three orders of magnitude, from 1 ns to 1 μs ($p \approx 10^{16}$ cm^{-3}, $x \approx 0.2$, $T \approx 77$ K, vacancy doping) [52,55]. This is due to many factors, which may affect the measured lifetime including inhomogeneities, inclusions, surface and contact phenomena. Typically, Cu- or Au-doped materials exhibit lifetimes one order of magnitude larger compared to vacancy doped ones of the same hole concentration [55]. It is believed that the increase of lifetime in impurity doped $Hg_{1-x}Cd_xTe$ arises from a reduction of SR centers.

The SR centers seem not to be the vacancies themselves and thus may be removable. Vacancy doped material with the same carrier concentration, but created under different annealing temperatures may produce different lifetimes. One possible candidate for recombination centers is Hg interstitials. Vacancy doped $Hg_{1-x}Cd_xTe$ exhibits SR recombination center densities roughly proportional to the vacancy concentration.

Measurements at DSR [56] give lifetimes values for extrinsic p-type material

$$\tau_{ext} = 9 \times 10^9 \frac{p_1 + p}{p N_a} \tag{8.10}$$

where

$$p_1 = N_v \exp\left(\frac{q(E_r - E_g)}{kT}\right) \tag{8.11}$$

and E_r is the SR center energy relative to the conduction band. Experimentally, E_r was found to lie at the intrinsic level for As, Cu, and Au dopants, giving $p_1 = n_i$.

For vacancy doped p-type $Hg_{1-x}Cd_xTe$

$$\tau_{vac} = 5 \times 10^9 \frac{n_1}{p N_{vac}} \tag{8.12}$$

where

$$n_1 = N_c \exp\left(\frac{q E_r}{kT}\right) \tag{8.13}$$

E_r is ≈ 30 mV from the conduction band ($x = 0.22$–0.30).

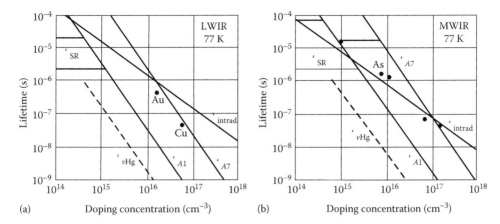

FIGURE 8.7 Measured lifetimes for n- and p-type LWIR (a) and MWIR (b) at 77 K, compared to theory for Auger 1, Auger 7, SR, and internal radiative recombination, as a function of doping concentration. (With kind permission from Springer Science+Business Media: *J. Electron. Mater.*, Minority carrier lifetime in p-HgCdTe, 34, 2005, 880–884, Kinch, M.A., Aqariden, F., Chandra, D., Liao, P.-K., Schaake, H.F., and Shih, H.D.)

As follows from these expressions and Figure 8.7 [57], doping with the foreign impurities (Au, Cu, and As for p-type material) gives lifetimes significantly increased compared to native doping of the same level.

The radiative recombination is an inversed process of annihilation of electron–hole pairs with emission of photons. The radiative recombination rates were calculated for conduction-to-heavy-hole-band and conduction-to-light-hole-band transitions using an accurate analytical form [48].

For a long time, internal radiative processes have been considered to be the main fundamental limit to detector performance and the performance of practical devices has been compared that limit. The role of radiative mechanism in the detection of IR radiation has been critically reexamined by Humpreys [58,59]. He indicated that most of the photons emitted in photodetectors as a result of radiative decay are immediately reabsorbed, so that the observed radiative lifetime is only a measure of how well photons can escape from the body of the detector. Due to reabsorption the radiative lifetime is highly extended, and dependent on the semiconductor geometry. Therefore, internal combined recombination–generation processes in one detector are essentially noiseless. In contrast, the recombination act with cognate escape of a photon from the detector, or generation of photons by thermal radiation from outside the active body of the detector, is a noise producing process. This may readily happen for a case of detector array, where an element may absorb photons emitted by another detectors or a passive part of the structure [60,61]. Deposition of the reflective layers (mirrors) on the back and side of the detector may significantly improve optical insulation preventing noisy emission and absorption of thermal photons.

It should be noted that internal radiative generation could be suppressed in detectors operated under reverse bias where the electron density in the active layer is reduced to well below its equilibrium level [62].

Auger mechanisms dominate generation and recombination processes in high-quality narrow-gap semiconductors such as $Hg_{1-x}Cd_xTe$ and InSb at near room temperatures [63,64]. The Auger generation is essentially the impact ionization by electrons of holes in the high energy tail of Fermi–Dirac distribution. The band-to-band Auger mechanisms in InSb-like band structure semiconductors are classified in 10 photonless mechanisms. Two of them have the smallest threshold energies ($E_T \approx E_g$) and are denoted as Auger 1 (A1) and Auger 7 (A7). In some wider bandgap materials (e.g., InAs and low x $InAs_{1-x}Sb_x$) in which the split-off band energy Δ is comparable to E_g, and the Auger process involving split-off band (AS process) may also play an important role.

The A1 generation is the impact ionization by an electron, generating an electron–hole pair, so this process involves two electrons and one heavy hole. It is well known that the Auger 1 process is an important recombination mechanism in n-type $Hg_{1-x}Cd_xTe$, particularly for x around 0.2 and at higher temperatures [17,52,65].

Auger 7 generation is the impact generation of electron hole pair by a light hole, involving one heavy hole, one light hole, and one electron [66–68]. This process may dominate in p-type material.

The net generation rate due to the Auger 1 and Auger 7 processes can be described as [69]

$$G_A - R_A = \frac{n_i^2 - np}{2n_i^2}\left[\frac{n}{(1+an)\tau_{A1}^i} + \frac{p}{\tau_{A7}^i}\right], \tag{8.14}$$

where

τ_{A1}^i and τ_{A7}^i are the intrinsic Auger 1 and Auger 7 recombination times
n_i is the intrinsic concentration

The last equation is valid for a wide range of concentrations, including degeneration, which easily occurs in n-type materials. This is expressed by the finite value of a. According to Ref. [69], $a = 5.26 \times 10^{-18}$ cm^3. Due to the shape of the valence band, the degeneracy in p-type material occurs only at very high doping levels, which is not achievable in practice.

The Auger 1 intrinsic recombination time is equal [63]

$$\tau_{A1}^i = \frac{h^3\varepsilon_o^2}{2^{3/2}\pi^{1/2}q^4m_o}\frac{\varepsilon^2(1+\mu)^{1/2}(1+2\mu)\exp\left[\left((1+2\mu)/(1+\mu)\right)E_g/kT\right]}{\left(m_e^*/m\right)\left|F_1F_2\right|^2\left(kT/E_g\right)^{3/2}} \tag{8.15}$$

where

μ is the ratio of the conduction to the heavy-hole valence-band effective mass
ε_s is the static-frequency dielectric constant
$|F_1F_2|$ are the overlap integrals of the periodic part of the electron wave functions

The overlap integrals cause the biggest uncertainly in the Auger 1 lifetime. Values ranging from 0.1 to 0.3 have been obtained by various authors. In practice, it is taken as a constant equal to anywhere between 0.1 and 0.3 leading to changes by almost an order of magnitude in the lifetime.

The ratio of Auger 7 and Auger 1 intrinsic times

$$\gamma = \frac{\tau_{A7}^i}{\tau_{A1}^i} \tag{8.16}$$

is another term of high uncertainty. According to Casselman et al. [66,67], for $Hg_{1-x}Cd_xTe$ over the range $0.16 \le x \le 0.40$ and $50\,K \le T \le 300\,K$, $3 \le \gamma \le 6$. Direct measurements of carrier recombination show the ratio γ larger than expected from previous calculations. More recently published theoretical [70,71] and experimental [57,71] results indicate that this ratio is several tens. As γ is higher than unity, higher recombination lifetimes are expected in p-type materials compared to n-type materials of the same doping.

The Auger generation and recombination rates are strongly dependent on temperature via dependence of carrier concentration and intrinsic time on temperature. Therefore, cooling is a natural and a very effective way to suppress Auger processes.

8.2.2.3 Hg-Based Alternatives to HgCdTe

Among the small gap II–VI semiconductors for IR detectors, only Cd, Zn, Mn, and Mg have been shown to open the bandgap of the Hg-based binary semimetals HgTe and HgSe to match the IR wavelength range. It appears that the amount of Mg to introduce in HgTe to match $10\,\mu m$ range is insufficient to introduce a significant reinforcement of the Hg-Te bond [72]. The main obstacles to the technological development of $Hg_{1-x}Cd_x$Se are the difficulties in obtaining type conversion. From the above alloy systems, $Hg_{1-x}Zn_x$Te (HgZnTe) and $Hg_{1-x}Mn_x$Te (HgMnTe) occupy a privileged position.

Both HgZnTe and HgMnTe have never been systematically explored in the device context. The reasons for this are several. Preliminary investigations of these alloy systems came on the scene when development of HgCdTe detectors was well on its way. Moreover, the HgZnTe alloy is a more serious technological problem material than HgCdTe (see Figure 8.5). In the case of HgMnTe, Mn is not a group II element, so that HgMnTe is not a truly II–VI alloy. This ternary compound was viewed with some suspicion by those not directly familiar with its crystallographic, electrical, and optical behavior. In such a situation, proponents of parallel development of HgZnTe and HgMnTe for IR detector fabrication encountered considerable difficulty in selling the idea to industry and to funding agencies.

In 1985, Sher et al. [73] showed from theoretical consideration that the weak HgTe bond is destabilized by alloying it with CdTe, but stabilized by ZnTe. This prediction has stimulated an interest by many groups worldwide in the growth and properties of the HgZnTe alloy system as the material for photodetection application in the IR spectral region. The question of lattice stability in the case of HgMnTe compound is rather ambiguous. According to Wall et al. [74] the Hg-Te bond stability of this alloy is similar to that observed in the binary narrow-gap parent compound. This conclusion is in contradiction with results published in Ref. [75]. It has been established that the incorporation of Mn in CdTe destabilizes its lattice because of the Mn 3d orbital hybridizing into the tetrahedral bonds [76].

The selected topics that concentrate on the growth process, physical properties, and IR detectors of HgZnTe and HgMnTe ternary alloys are reviewed in two comprehensive reviews [77,78] and books [17,79,80].

8.2.3 Lead Salts

The properties of the lead salt binary and ternary alloys have been extensively reviewed [3–5,13,15,17, 81–83]. Therefore, only some of their most important properties will be mentioned here.

Lead salt detectors were developed during World War II by the German military for use as heat-seeking sensors to find weapons. Immediately after the war, communications, fire control, and search system applications began to stimulate a strong development effort that continues to the present day. Sidewinder heat-seeking IR-guided missiles received a great deal of public attention. Some of the commercial applications include spectrometry, protein analysis, fire detection systems, combustion control, and moisture detection and control.

8.2.3.1 Deposition of Polycrystalline PbS and PbSe Films

Although the fabrication methods developed for these photoconductors are not completely understood, their properties are well established. Unlike most other semiconductor IR detectors, lead salt materials are used in the form of polycrystalline films approximately $1\,\mu m$ thick and with individual crystallites ranging in size from approximately 0.1–$1.0\,\mu m$. They are usually prepared by chemical deposition using empirical recipes, which generally yields better uniformity of response and more stable results than the evaporative methods [84,85].

The PbSe and PbS films used in commercial IR detectors are made by chemical bath deposition (CBD), the oldest and most-studied PbSe and PbS thin-film deposition method. It was used to deposit PbS in 1910 [86]. The basis of CBD is a precipitation reaction between a slowly produced anion (S^{2-} or Se^{2-}) and a complexed metal cation. The commonly used precursors are lead salts, $Pb(CH_3COO)_2$ or $Pb(NO_3)_2$, thiourea [$(NH_2)_2CS$] for PbS, and selenourea [$(NH_2)_2CSe$] for PbSe, all in alkaline solutions. Lead may be complexed with citrate, ammonia, triethanolamine, or with selenosulfate itself. Most often, however, the deposition is carried out in a highly alkaline solution where OH^- acts as the complexing agent for Pb^{2+}.

In CBD, the film is formed when the product of the concentrations of the free ions is larger than the solubility product of the compound. Thus, CBD demands very strict control over the reaction temperature, pH, and precursor concentrations. In addition, the thickness of the film is limited, the terminal thickness usually being 300–500 nm. Therefore, in order to get a film with a sufficient thickness (approximately 1 μm in IR detectors, for example), several successive depositions must be done. The benefit of CBD compared to gas phase techniques is that CBD is a low-cost temperature method and the substrate may be temperature-sensitive with the various shapes.

As-deposited PbS films exhibit significant photoconductivity. However, a post-deposition baking process is used to achieve final sensitization. In order to obtain high-performance detectors, lead chalcogenide films need to be sensitized by oxidation. The oxidation may be carried out by using additives in the deposition bath, by post-deposition heat treatment in the presence of oxygen, or by chemical oxidation of the film. The effect of the oxidant is to introduce sensitizing centers and additional states into the bandgap and thereby increase the lifetime of the photoexcited holes in the p-type material

The backing process changes the initial n-type films to p-type films and optimizes performance through the manipulation of resistance. The best material is obtained using a specific level of oxygen and a specific bake time. Only a small percentage (3%–9%) of oxygen influences the absorption properties and response of the detector. Temperatures ranging from 100°C to 120°C and time periods from a few hours to in excess of 24 h are commonly employed to achieve final detector performance optimized for a particular application. Other impurities added to the chemical-deposition solution for PbS have a considerable effect on the photosensitivity characteristics of the films. $SbCl_2$, $SbCl_3$, and As_2O_3 prolong the induction period and increase the photosensitivity by up to 10 times that of films prepared without these impurities. The increase is thought to be caused by the increased absorption of CO_2 during the prolonged induction period. This increases $PbCO_3$ formation, and thus photosensitivity. Arsine sulfide also changes the oxidation states on the surface. Moreover, it has been found that essentially the same performance characteristics can be achieved by baking in an air or a nitrogen atmosphere. Therefore, all of the constituents necessary for sensitization are contained in the raw PbS films as deposited.

The preparation of PbSe photoconductors is similar to PbS ones. The post-deposition baking process for PbSe detectors operating at 77 K is carried out at a higher temperature (>400°C) in an oxygen atmosphere. However, for detectors to be used at ambient and/or intermediate temperatures, the oxygen or air bake is immediately followed by baking in a halogen gas atmosphere at temperatures in the range of 300°C–400°C [87]. A variety of materials can be used as substrates, but the best detector performance is achieved using single-crystal quartz material. PbSe detectors are often matched with Si to obtain higher collection efficiency.

Photoconductors also have been fabricated from epitaxial layers without backing that resulted in devices with uniform sensitivity, uniform response time, and no aging effects. However, these devices do not offset the increased difficulty and cost of fabrication.

8.2.3.2 Physical Properties

Lead chalcogenide semiconductors have a face-centered cubic (rock salt) crystal structure and hence obtained the name "lead salts." Thus, they have [100] cleavage planes and tend to grow in the (100) orientation although they can also be grown in the [111] orientation. Only SnSe possesses the orthorhombic-B29 structure.

Numerous techniques to prepare lead salt single crystals and epitaxial layers have been investigated, and several excellent review articles devoted to this topic have been published [83,88,89]. Bridgman-type or Czochralski methods give crystals with increased size and variable composition. Crystals are mainly used as substrates for the subsequent growth of epilayers. Growth from the solution and the traveling solvent method offer interesting advantages such as higher homogeneity in composition and lower temperatures, which lead to lower concentrations of lattice defects and impurities. The best results have been achieved with the sublimation growth technique since lead salts sublime as molecules.

Thin single-crystal films of IV–VI compounds have found broad application in both fundamental research and applications. The epitaxial layers are usually grown by either VPE or LPE techniques. Recently, the best-quality devices have been obtained using MBE.

Lead salts can exist with very large deviations from stoichiometry, and it is difficult to prepare material with carrier concentrations below approximately 10^{17} cm^{-3}. The width of the solidus field is large in IV–VI compounds ($\approx 0.1\%$) making the doping by native defects very efficient. Deviations from stoichiometry create n- or p-type conduction. Native defects associated with excess metal (nonmetal vacancies or possibly metal interstitials) yield acceptor levels, while those that result from excess nonmetal (metal vacancies or possibly nonmetal interstitials) yield donor levels.

In crystals grown from high-purity elements, the effects of foreign impurities are usually negligible when the carrier concentration, due to lattice defects, is above 10^{17} cm^{-3}. Below this concentration, foreign impurities can play a role by compensating for lattice defects and other foreign impurities.

Table 8.4 contains a list of material parameters for different types of binary lead salts.

TABLE 8.4 Physical Properties of Lead Salts

	T (K)	PbTe	PbSe	PbS
Lattice structure		Cubic (NaCl)	Cubic (NaCl)	Cubic (NaCl)
Lattice constant a (nm)	300	0.6460	0.61265	0.59356
Thermal expansion coefficient α (10^{-6} K^{-1})	300	19.8	19.4	20.3
	77	15.9	16.0	
Heat capacity C_p (J/mol/K)	300	50.7	50.3	47.8
Density γ (g/cm^3)	300	8.242	8.274	7.596
Melting point T_m (K)		1197	1354	1400
Bandgap E_g (eV)	300	0.31	0.28	0.42
	77	0.22	0.17	0.31
	4.2	0.19	0.15	0.29
Thermal coefficient of E_g (10^{-4} eV/K)	80–300	4.2	4.5	4.5
Effective masses				
m_{et}^*/m	4.2	0.022	0.040	0.080
m_{ht}^*/m		0.025	0.034	0.075
m_{el}^*/m		0.19	0.070	0.105
m_{hl}^*/m		0.24	0.068	0.105
Mobilities				
μ_e (cm^2/V s)	77	3×10^4	3×10^4	1.5×10^4
μh (cm^2/V s)		2×10^4	3×10^4	1.5×10^4
Intrinsic carrier concentration ni (cm^{-3})	77	1.5×10^{10}	6×10^{11}	3×10^7
Static dielectric constant ε_s	300	380	206	172
	77	428	227	184
High-frequency dielectric constant ε_∞	300	32.8	22.9	17.2
	77	36.9	25.2	18.4
Optical phonons				
LO (cm^{-1})	300	114	133	212
TO (cm^{-1})	77	32	44	67

Lead salts have direct energy gaps, which occur at the Brillouin zone edge at the L point. The effective masses are therefore higher and the mobilities lower than for a zinc-blende structure with the same energy gap at the Γ point (the zone center). The constant-energy surfaces are ellipsoids characterized by the longitudinal and transverse effective masses m_l^* and m_t^*, respectively. The anisotropy factor for PbTe is of the order of 10. The factor is much less, approximately 2, for PbS and PbSe.

As a consequence of the similar valence and conduction bands of lead salts, the electron and hole mobilities are approximately equal for the same temperatures and doping concentrations. Room-temperature mobilities in lead salts are 500–2000 cm^2/V s. In many high-quality single-crystal samples, the mobility due to lattice scattering varies as $T^{-5/2}$ [82]. This behavior has been attributed to a combination of polar-optical and acoustical lattice scattering and achieves the limiting values in the range of 10^5–10^6 cm^2/V s due to defect scattering.

The interband absorption of the lead salts is more complicated compared to the standard case due to the anisotropic multivalley structure of both conduction and valence bands, nonparabolic Kane-type energy dispersion, and k-dependent matrix elements. Analytical expressions for the absorption coefficient for energies near the absorption edge have been given by several researchers [47,90,91].

8.2.4 InAs/GaInSb Type II Strained Layer Superlattices

The InAs$_{1-x}$Sb$_x$ ternary alloy has the lowest band-gap of all III–V semiconductors but this gap is not suitable ($\lambda_c \approx 9\,\mu$m) for operation in 8–14 μm atmospheric window at 77 K. It was theoretically shown that strain effects in InAsSb superlattices, called also SLSs, were sufficient to achieve wavelength cutoffs of 12 μm at 77 K independent of the band offset which was unknown at that time [7]. Progress in the growth of InAsSb SLSs by both MBE and MOCVD has been observed since Osbourn's proposal. The first decade efforts in development of epitaxial layers are presented in Rogalski's monograph [29]. Difficulties have been encountered in finding the proper growth conditions especially for SLSs in the middle region of composition. This ternary alloy tends to be unstable at low temperatures, exhibiting miscibility gaps, and this can generate phase separation or clustering. Control of alloy composition has been problematic especially for MBE. Due to the spontaneous nature of CuPt-orderings, which result in substantial bandgap shrinkage, it is difficult to accurately and reproducibly control the desired bandgap for optoelectronic device applications.

InAs/Ga$_{1-x}$In$_x$Sb (InAs/GaInSb) SLSs can be considered as an alternative to HgCdTe material systems as a candidate for third generation IR detectors. Due to strong absorption, the SLS structures provide high responsivity, as already reached with HgCdTe, without any need for gratings. Further advantages are a photovoltaic operation mode, operation at elevated temperatures and well established III–V process technology.

InAs/GaInSb material system is however in a very early stage of development. Problems exist in material growth, processing, substrate preparation, and device passivation [92]. Optimization of SL growth is a trade-off between interfaces roughness, with smoother interfaces at higher temperature, and residual background carrier concentrations, which are minimized on the low end of this range. The thin nature of InAs and GaInSb layers (<8 nm) necessitate low growth rates for control of each layer thickness to within 1 (or ½) monolayer (ML). Typical growth rates are less than 1 ML/s for each layer.

8.2.4.1 Material Properties

InAs and GaInSb form an ideal material system for the growth of semiconductor heterostructure because of their small difference in lattice constant. For example, Ga$_{1-x}$In$_x$Sb with an indium concentration of 15% grows compressively strained on a GaSb substrate with a lattice mismatch of $\Delta a/a = 0.94\%$, while InAs is under tensile with a lattice mismatch of $\Delta a/a = -0.62\%$. In an InAs/Ga$_{1-x}$In$_x$Sb superlattice, the compressively strained Ga$_{1-x}$In$_x$Sb layers can be compensate for the tensile strain in the InAs layers.

InAs/Ga$_{1-x}$In$_x$Sb superlattices was proposed for IR detector applications in the 8–14 μm region in 1987 [8]. LWIR response in these SLs arises due to a type II band alignment and internal strain which lowers the conduction band minimum of InAs and raises the heavy-hole band in $Ga_{1-x}In_xSb$ by the deformation potential effect. However, unlike InAsSb materials, effects due to strain are combined with a substantial valence band offset, which for InAs/GaSb SLs exceeds 500 meV. The GaSb valence band edge lies approximately 150 meV above the InAs conduction band edge at low temperature. LWIR absorption however can be achieved in InAs/GaSb SLs for an InAs layer thickness greater than approximately 100 Å, resulting in comparatively poor optical absorption due to weak wave function overlap as the barrier heights are large. Substituting a $Ga_{1-x}In_xSb$ alloy for GaSb it is possible to reach the important 12 µm IR region for thin SLs to obtain large optical absorption. A small lattice mismatch (<5%) between the InGaSb and InAs layers causes the tetragonal distortions which shift the bulk energy levels and split the valence band degeneracies of the light and heavy hole energy levels. The presence of coherent strain shifts the band edges such that the SL energy gap is reduced. This reduced bandgap is advantageous because longer cutoff wavelengths can be obtained with reduced layer thickness in the strained SL.

Summarizing, the type-II superlattice has staggered band alignment such that the conduction band of the InAs layer is lower than the valence band of InGaSb layer. This creates a situation in which the energy bandgap of the superlattice can be adjusted to form either a semimetal (for wide InAs and GaInSb layers) or a narrow bandgap (for narrow layers) semiconductor material. In the SL, the electrons are mainly located in the InAs layers, whereas holes are confined to the GaInSb layers, as shown in Figure 8.8a. This suppresses Auger recombination mechanisms and thereby enhances carrier lifetime but optical transitions occur spatially indirectly and, thus, the optical matrix element for such transitions is relatively small. The bandgap of the SL is determined by the energy difference between the electron miniband E_1 and the first heavy hole state HH_1 at the Brillouin zone center and can be varied continuously in a range between 0 and about 250 meV. An example of the wide tunability of the SL is shown in Figure 8.8b.

In comparison with HgCdTe material system, type II superlattice is mechanically robust and has fairly weak dependence of bandgap on composition (see Figure 8.9). Using InAs/GaInSb SLS we have the ability to fix one component of the material and vary the other to tune the wavelength. As shown in this Figure 8.8b, by fixing the GaSb layer thickness (at 40 Å) and varying the thickness of the InAs (from 40 to 66 Å), the cutoff wavelength of SLS can be tuned from 5 to 25 µm. There is no material composition change needed, what is serious problem in the case of LWIR HgCdTe focal plane array (FPA) material to fulfill requirement of high uniformity.

It has been suggested that $InAs/Ga_{1-x}In_xSb$ SLSs material system can have some advantages over bulk HgCdTe, including lower leakage currents and greater uniformity. Electronic properties of SLSs may be superior to those of the HgCdTe alloy [94]. The effective masses are not directly dependent on the

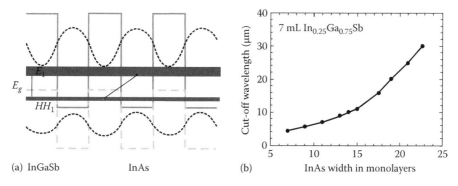

FIGURE 8.8 InAs/GaInSb SLS: (a) band-edge diagram illustrating the confined electron and hole minibands which form the energy bandgap; (b) change in cutoff wavelength with change in one superlattice parameter—InAs layer width. (After Brown, G.J. et al., *Proc. SPIE*, 4999, 457, 2003. With permission from International Society for Optical Engineering.)

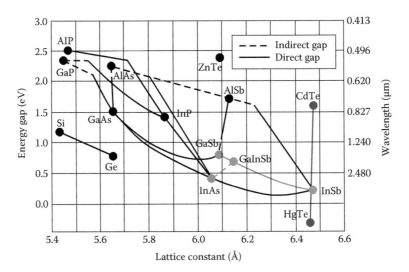

FIGURE 8.9 Composition and wavelength diagram of Sb-based III–V material systems.

bandgap energy, as it is the case in a bulk semiconductor. The electron effective mass of InAs/GaInSb SLS is larger ($m^*/m_o \approx 0.02$–0.03, compared to $m^*/m_o = 0.009$ in HgCdTe alloy with the same bandgap $E_g \approx 0.1\,eV$). Thus, diode tunneling currents in the SL can be reduced compared to the HgCdTe alloy [95]. Although in-plane mobilities drop precipitously for thin wells, electron mobilities approaching 10^4 cm^2/V s have been observed in InAs/GaInSb superlattices with the layers less than 40 Å thick. While mobilities in these SLs are found to be limited by the same interface roughness scattering mechanism, detailed band structure calculations reveal a much weaker dependence on layer thickness, in reasonable agreement with experiment.

Theoretical analysis of band-to-band Auger and radiative recombination lifetimes for InAs/GaInSb SLSs showed that in these objects the p-type Auger recombination rates are suppressed by several orders, compared to those of bulk HgCdTe with similar bandgap [96], but n-type materials are less advantageous. Comparison of theoretically calculated and experimentally observed lifetimes at 77 K for 10 μm InAs/GaInSb SLS and 10 μm HgCdTe indicates on good agreement for carrier densities above 2×10^{17} cm^{-3} [97]. The discrepancy between both types of results for lower carrier densities is due to SR recombination. More recently published upper experimental data [98,99] coincide well with HgCdTe trend-line in the range of lower carrier concentration. In general, however, the SL carrier lifetime is limited by influence of trap centers located in the energy gap. Empirical fitting data gave a minority carrier lifetimes from 35 to 200 ns, with similar absorption layers but different device structure. There is no clear understanding why the minority carrier lifetime varies within the device structure [100].

Narrow bandgap materials require the doping to be controlled to at least 1×10^{15} cm^{-3} or below to avoid deleterious high-field tunneling currents across reduced depletion widths at temperature below 77 K. Lifetimes must be increased to enhance carrier diffusion and reduce related dark currents. At the present stage of development, the residual doping concentration (both n-type as well as p-type) is typically about 5×10^{15} cm^{-3} in superlattices grown at substrate temperature ranging from 360°C to 440°C. Low to mid 10^{15} cm^{-3} residual carrier concentrations are the best that have been achieved so far.

8.3 Photoconductive Detectors

Photoconductive detector (also named as the photoresistor or photoconductor) is essentially a radiation-sensitive resistor. In this device, the radiation changes the electrical conductivity of the material upon which it is incident. The change in conductivity is measured by means of electrodes attached to the

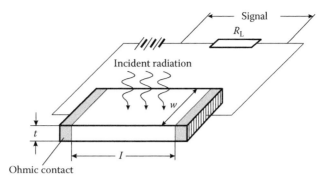

FIGURE 8.10 Geometry and bias of a bulk photoconductive detector.

sample. The sample geometry and circuit employed for detecting the photoconductive effect is shown in Figure 8.10. In almost all cases, the transverse geometry is used, in which the direction of the incident radiation is perpendicular to the direction in which the change in current is measured. The photosignal is detected either as a change in voltage developed across a load resistor in series with the detector as illustrated, or as a change in current through the sample. For low-resistance material, the photoresistor is usually operated in a constant current circuit as showed in Figure 8.10. The series load resistance is large compared to the sample resistance, and the signal is detected as a change in voltage developed across the sample. For high-resistance photoconductors, a constant voltage circuit is preferred and the signal is detected as a change in current in the bias circuit.

8.3.1 Performance of Photoconductive Detectors

The performance of a photoconductor is measured in terms of the following parameters: current (or voltage) responsivity, photoconductive gain, detectivity, and response time.

The basic expression describing photoconductivity in semiconductors under equilibrium excitation (i.e., steady state) is

$$I_{ph} = q\eta A\Phi g \tag{8.17}$$

where

I_{ph} is the short circuit photocurrent at zero frequency (dc)
Φ is the incident photon flux density

The photoconductive gain, g, is the number of carriers passing contacts per one generated pair.

The photoconductive gain can be defined as

$$g = \frac{\tau}{t_t} \tag{8.18}$$

Equation 8.18 indicates that the photoconductive gain is given by the ratio of free carrier lifetime, τ, to transit time, t_t, between the sample electrodes. Carrier lifetime is a strong function of the semiconductor materials used in the photoconductor. The photoconductive gain can be less than or greater than unity depending upon whether the drift length, $L_d = v_d\tau$, is less than or greater than interelectrode spacing, l. The carrier drift velocities, v_d, are given by $v_e = \mu_e E$ for electrons and $v_h = \mu_h E$ for holes (μ is the carrier mobility).

In many semiconductors, the mobility of an electron is greater than that of a hole (see Table 8.1). At high field-strengths E, in excess of approximately 10^5 V/cm, a saturation velocity between 6×10^6 and 10^7 cm/s is observed. In most semiconductors, the saturation velocities for both holes and electrons are similar, with the hole generally being slightly slower, and a single value for saturation velocity v_s is used for both carriers. Since the velocity of the faster carrier determines the transit-time in a photoconductor, high-gain photoconductor are sometimes operated at relatively low voltages.

The value of $L_d > l$ implies that a free charge carrier swept out at one electrode is immediately replaced by injection of an equivalent free charge carrier at the opposite electrode. This corresponds to a photo-generated carrier looping through the circuit many times before recombining. A photoconductive gain of intrinsic photoresistor as large as 10^6 has been measured.

It can be shown that the current responsivity is equal [15,101]

$$R_i = \frac{\lambda \eta}{hc} qg \tag{8.19}$$

instead the frequency dependent responsivity can be determined by the equation [101]

$$R_v = \frac{\eta}{lwt} \frac{\lambda \tau_{ef}}{hc} \frac{V_b}{n_o} \frac{1}{\left(1 + \omega^2 \tau_{ef}^2\right)^{1/2}} \tag{8.20}$$

where τ_{ef} is the effective carrier lifetime, the detector area $A = wl$, and t is the detector thickness.

The expression (8.20) shows clearly the basic requirements for high photoconductive responsivity at a given wavelength λ: one must have high quantum efficiency η, long excess-carrier lifetime τ_{ef}, the smallest possible piece of crystal, low thermal equilibrium carrier concentrations n_o, and the highest possible bias voltage V_b.

The above simple model takes no account of additional limitations related to the practical conditions of photoconductor operation such as sweep-out effects or surface recombination. These are specified in the following.

Equation 8.20 shows that voltage responsivity increases monotonically with the increase of bias voltage. However, there are two limits on applied bias voltage, namely: thermal conditions (Joule heating of the detector element) and sweep-out of minority carriers. The thermal conductance of the detector depends on the device fabrication procedure. If the excess carrier lifetime is long, we cannot ignore the effects of contacts and of drift and diffusion on the device performance. Present-day material technology is such that at moderate bias fields minority carriers can drift to the ohmic contacts in a time short compared to the recombination time in the material. Removal of carriers at an ohmic contact in this way is referred to as "sweep-out." Minority carrier sweep-out limits the maximum applied voltage of V_b. The effective carrier lifetime can be reduced considerably in detectors where the minority carrier diffusion length exceeds the detector length. At low bias, the average drift length of the minority carriers is very much less than the detector length, l, and the minority carrier lifetime is determined by the bulk recombination modified by diffusion to surface and contacts. The carrier densities are uniform along the length of the detector. At higher values of the applied field, the drift length of the minority carriers is comparable to or greater than l. Some of the excess minority carriers are lost at an electrode, and to maintain space charge equilibrium, a drop in excess majority carrier density is necessary. This way the majority carrier lifetime is reduced. It should be pointed out that the loss of the majority carriers at one ohmic contact is replenished by injection at the other, but the minority carriers are not replaced. At high bias, the excess carrier density is nonuniformly distributed along the length of the sample.

To achieve high photoelectric gain, low resistance and low surface recombination velocity contacts are required. The metallic contacts are usually far from expectation. The contacts are characterized by a recombination velocity, which can be varied from infinity (ohmic contacts) to zero (perfectly blocking contacts). In the latter case, a more intensely doped region at the contact (e.g., n$^+$ for n-type devices or heterojunction contact) causes a built-in electric field that repels minority carriers, thereby reducing recombination and increasing the effective lifetime and the responsivity.

In a photoconductor, we can distinguish two groups of noise; the radiation noise and the noise internal to the detector. The radiation noise includes signal fluctuation noise and background fluctuation noise. Under most operating conditions the signal fluctuation limit is operative for ultraviolet and visible detectors.

The random processes occurring in semiconductors give rise to internal noise in detectors even in the absence of illumination. There are two fundamental processes responsible for the noise: fluctuations in the velocities of free carriers due to their random thermal motion, and fluctuations in the densities of free carriers due to randomness in the rates of thermal generation and recombination.

The total noise voltage of photoconductor is

$$V_n = \left(V_j^2 + V_{gr}^2 + V_{1/f}^2 + V_{pa}^2 \right)^{1/2} \tag{8.21}$$

where

V_j is the Johnson–Nyquist noise (the thermal noise)
V_{gr} is the generation–recombination (*gr*) noise
$V_{1/f}$ is the 1/*f* noise

Normally, the preamplifier is carefully chosen such that its noise, V_{pa}, is negligible compared to the other sources of noise.

The most used figure of merit is the detectivity, which is defined as

$$D^\star = \frac{R_v (A\Delta f)^{1/2}}{V_n} = \frac{R_i (A\Delta f)^{1/2}}{V_i} \tag{8.22}$$

where V_n (V_i) is the noise voltage (current).

For mid-IR to far-IR wavelengths the photoconductors are cooled to lower temperature. The lower temperatures reduce thermal effect and increase the photoconductive gain and detection efficiency. For low-level detection at microwave frequencies, however, a photodiode will provide more speed and considerably higher signal-to-noise ratio. Thus photoconductors have limited use in high-frequency optical demodulators, such as in optical mixing. They were, however, extensively used for IR detectors.

Note that as carrier lifetime increases, the bandwidth decreases. This is the opposite of what occurred with photoconductive gain, where an increase in carrier lifetime was beneficial. This allows us to draw an important conclusion. Photoconductive gain cannot be increased arbitrarily by increasing carrier lifetime: there is an inherent gain-bandwidth limitation. Given this constraint, one way to achieve high performance is to first engineer the semiconductor so that the carrier lifetime is as long as possible while still maintaining adequate bandwidth. The photoconductive gain is then increased by decreasing the transit-time as much as possible.

8.3.2 Lead Salt Photoconductors

Around 1920, Case investigated the thallium sulfide photoconductor—one of the first photoconductors to give a response in the near IR region to approximately 1.1 μm [102]. The next group of materials to be studied was the lead salts (PbS, PbSe, and PbTe), which extended the wavelength response to 7 μm. PbS photoconductors from natural galena found in Sardinia were originally fabricated by Kutzscher at the University of Berlin in the 1930s [103]. However, for any practical applications, it was necessary to develop a technique for producing synthetic crystals. PbS thin-film photoconductors were first produced in Germany, next in the United States at Northwestern University in 1944, and then in England at the Admiralty Research Laboratory in 1945 [104]. During World War II, the Germans produced systems that used PbS detectors to detect hot aircraft engines. Immediately after the war, communications, fire control, and search systems began to stimulate a strong development effort that has extended to the present day. After 60 years, low-cost, versatile PbS and PbSe polycrystalline thin films remain the photoconductive detectors of choice for many applications in the 1–3 and 3–5 μm spectral range. Current development with lead salts is in the focal plane arrays (FPAs) configuration.

8.3.2.1 PbS and PbSe Photoconductors

The PbS and PbSe polycrystalline films are deposited, either over or under plated gold electrodes, and on fused quartz, crystal quartz, single crystal sapphire, glass, various ceramics, single crystal strontium titanate, Irtran II (ZnS), Si, and Ge. The most commonly used substrate materials are fused quartz for ambient operation and single crystal sapphire for detectors used at temperatures below 230 K. The very low TEC of fused quartz relative to PbS films results in poorer detector performance at lower operating temperatures. Different shapes of substrates are used: flat, cylindrical, or spherical. To obtain higher collection efficiency, detectors may be deposited directly by immersion onto optical materials with high indices of refraction (e.g., into strontium titanate). Lead salts cannot be immersed directly; special optical cements must be used between the film and the optical element.

As was mentioned in Section 8.2.3.1, in order to obtain high-performance detectors, lead chalcogenide films must be sensitized by oxidation, which may be carried out by using additives in the deposition bath, by postdeposition heat treatment in the presence of oxygen, or by chemical oxidation of the film. Unfortunately, in the older literature the additives are seldom identified and are often referred to only as "an oxidant" [105,106]. A more recent paper deals with the effects of H_2O_2 and $K_2S_2O_8$, both in the deposition bath and in the postdeposition treatment [107]. It was found that both treatments increase the resistivity of PbS films. Although the resistivity usually increases during the oxidation, a different behavior also has been observed [108]. A sensitized PbS film may significantly degrade in air without an overcoating. Possible overcoating materials are As_2S_3, CdTe, ZnSe, Al_2O_3, MgF_2, and SiO_2. Vacuum-deposited As_2S_3 has been found to have the best optical, thermal, and mechanical properties, and it has improved the detector performance. The drawback of the As_2S_3 coating is the toxicity of As and its precursors. Overall, however, the electric properties (resistivity, and in particular, detectivity) of lead salt thin films are rather poorly reported, although there are many papers on PbS and PbSe film growth. The effects of annealing and oxidation treatments on detectivity have not been reported accurately either.

PbS and PbSe materials are peculiar because they have a relatively long response time that affects the significant photoconductive gain. It has been suggested that during the sensitization process the films are oxidized, converting the outer surfaces of exposed PbS and PbSe films to PbO or a mixture of $PbO_xS(Se)_{1-x}$, and forming a heterojunction at the surface [109]. Oxide heterointerfaces create conditions for trapping minority carriers or separating majority carriers and thereby extending the lifetime of the material. As was mentioned above, without the sensitization (oxidation) step, lead salt materials have very short lifetimes and a low response.

The basic detector fabrication steps are electrode deposition and delineation, active-layer deposition and delineation, passivation overcoating, mounting (cover/window), and lead wire attachment. Photolithographic delineation methods are used for complex, high-density patterns of small element sizes. The outer electrodes are produced with vacuum deposition of bimetallic films such as TiAu. To passivate and optimize transmission of radiation into the detector, usually a quarter-wavelength thick overcoating of As_2S_3 is used. Normally, detectors are sealed between a cover plate and the substrate with epoxy cement. The cover-plate material is ordinarily quartz, but other materials such as sapphire may be used to transmit longer wavelengths. This technique seals the detector reasonably well against humid environments. The top surface of the cover is normally made antireflective with a material such as MgF_2.

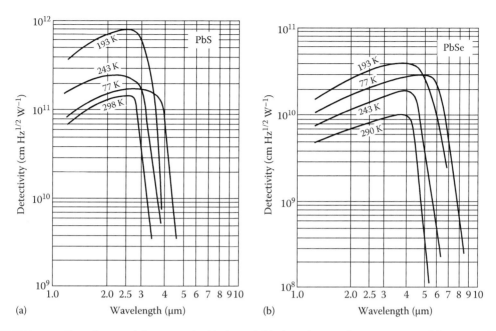

FIGURE 8.11 Typical spectral detectivity for (a) PbS and (b) PbSe photoconductors. (Reprinted from NEP data sheets.)

Standard active-area sizes are typically 1, 2, or 3 mm². However, most manufacturers offer sizes ranging from 0.08×0.08 to 10×10 mm² for PbSe detectors, and 0.025×0.025 to 10×10 mm² for PbS detectors. Active areas are generally square or rectangular; detectors with more exotic geometries have sometimes not performed up to expectations.

When a lead salt detector is operated below −20°C and exposed to UV radiation, semipermanent changes in responsivity, resistance, and detectivity occur [110]. This is the so-called "flash effect." The amount of change and the degree of permanence depend on the intensity of the UV exposure and the length of exposure time. Lead salt detectors should be protected from fluorescent lighting. They are usually stored in a dark enclosure or overcoated with an appropriate UV-opaque material. They are also hermetically sealed so their long-term stability is not compromised by humidity and corrosion.

The spectral distribution of detectivity of lead salt detectors is presented in Figure 8.11. Usually the operating temperatures of detectors are between −196°C and 100°C; it is possible to operate at a temperature higher than recommended, but 150°C should never be exceeded. Table 8.5 contains the performance range of detectors fabricated by various manufacturers [86].

Below 230 K, background radiation begins to limit the detectivity of PbS detectors. This effect becomes more pronounced at 77 K, and peak detectivity is no greater than the value obtained at 193 K. Quantum efficiency of approximately 30% is limited by incomplete absorption of the incident flux in the relatively thin (1–2 μm) detector material. The responsivity uniformity of PbS and PbSe detectors is generally of the order of 3%–10%.

8.3.2.2 HgCdTe Photoconductors

For longer wavelengths (i.e., $\lambda > 8\,\mu m$), prevailing position is occupied by HgCdTe. The operating temperature for HgCdTe detectors is higher than for other types of photon detectors.

The main part of HgCdTe photoconductor structure is a 3–20 μm flake of HgCdTe, supplied with electrodes. The optimum thickness of the active element depends upon the temperature of operation and is smaller in uncooled devices. It is chosen typically to be of order of α^{-1}, where α is the optical absorption

TABLE 8.5   Performance of Lead Salt Detectors (2π FOV, 300 K Background)

	T (K)	Spectral Response (μm)	λ_p (μm)	D^* (λ_p, 1000 Hz, 1) (cm Hz$^{1/2}$W^{-1})	$R/\square$ (MΩ)	τ (μs)
PbS	298	1–3	2.5	$(0.1–1.5) \times 10^{11}$	0.1–10	30–1000
	243	1–3.2	2.7	$(0.3–3) \times 10^{11}$	0.2–35	75–3000
	195	1–4	2.9	$(1–3.5) \times 10^{11}$	0.4–100	100–10,000
	77	1–4.5	3.4	$(0.5–2.5) \times 10^{11}$	1–1000	500–50000
PbSe	298	1–4.8	4.3	$(0.05–0.8) \times 10^{10}$	0.05–20	0.5–10
	243	1–5	4.5	$(0.15–3) \times 10^{10}$	0.25–120	5–60
	195	1–5.6	4.7	$(0.8–6) \times 10^{10}$	0.4–150	10–100
	77	1–7	5.2	$(0.7–5) \times 10^{10}$	0.5–200	15–150

Source: After Harris, R.H., Lead-salt detectors, *Laser Focus/Electro-Optics*, pp. 87–96, December 1983. With permission.

coefficient. In order to obtain bulk-limited lifetimes, it is necessary to reduce the surface recombination velocity by treating the surface in such a way as to leave it slightly accumulated. The frontside surface is usually covered with a passivation layer (native oxide on n-type material) and antireflection coating.

HgCdTe photoconductive detectors operating at 77 K in the 8–14 μm range were widely used in the first-generation thermal imaging systems in linear arrays of up to 200 elements. The production processes of these devices are well established. The material used is n-type with an extrinsic carrier density of about 5×10^{14} cm^{-3}. Commercially available n-type HgCdTe photoconductive detectors are typically manufactured in a square configuration with active size from 25 μm to 4 mm. The length of the photoconductors being used in high-resolution thermal imaging systems ($\approx$50 μm) is typically less than the minority carrier diffusion and drift length in cooled HgCdTe, resulting in reduction of photoelectric gain due to diffusion and drift of photogenerated carriers to the contact regions, called sweep-out effect. This causes the "saturation" of response with increasing electric field. The behavior of a typical device, showing the saturation in responsivity (at about 10^5 V/W), is shown in Figure 8.12a.

The performance of HgCdTe photoconductors at higher temperatures is reduced. The carrier lifetimes at higher temperature are short being fundamentally limited by Auger processes, and the g-r noise limited performance is obtained. Since $\gamma > 1$ (see Equation 8.16), there is in principle an advantage in using p-type material. In practice, however, p-type photoconductors are difficult to passivate and low $1/f$ noise contacts are difficult to form. For these reasons, the majority of device for higher-temperature operation are n-type. Figure 8.12b shows examples of the detectivity as a function of cutoff wavelength, obtained from 230 μm square n-type devices operated at different temperatures. For comparison, theoretical limiting detectivity is shown assuming an extrinsic concentration of 5×10^{14} cm^{-3}, thickness of 7 μm, reflection coefficient at front and back surface of 30%, and $f/1$ optics.

Spectral detectivity curves of HgCdTe Judson's photoconductors are shown in Figure 8.13; their other parameters are included in Table 8.6. The J15D5 series operated at 77 K with peak at 5 μm are recommended for thermal imaging or IR tracking applications. The J15D12 series offer near BLIP performance and fast response time. Applications include thermography, CO_2 laser detection and missile guidance. The J15D22 series are the detectors of choice for general "wide band" spectroscopy.

8.3.2.3 SPRITE Detectors

The SPRITE detector is a type of photoresistor for use in scanned thermal imaging systems. It was originally invented by T. C. Elliott and developed further almost exclusively by British workers [113,114]. The important benefit this device achieves is that the time delay and integration (TDI) required in serial scan thermal imaging systems is performed within a single detector element. In a conventional serial scan system, the image of the scene is scanned across a series of discrete detectors, the output for each device is then amplified and delayed by the correct amount so that all the detector

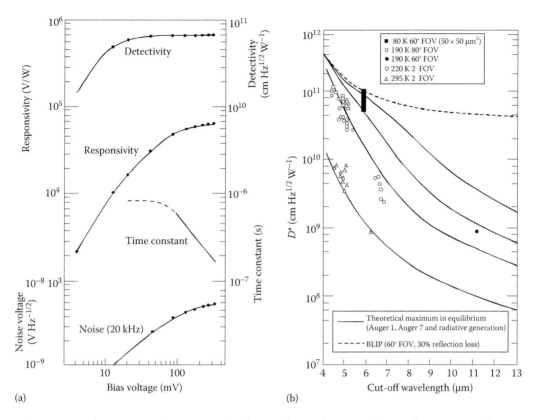

FIGURE 8.12 Characteristics of n-type HgCdTe photoconductive detectors: (a) 50 μm detector operated at 80 K as a function of voltage (the measurements were made in 30° FOV and the responsivity values refer to the peak wavelength response at 12 μm); (b) spectral detectivity (the theoretical curves are calculated including Auger generation and radiative generation only); the experimental points are for 230 μm square n-type detectors, except where indicated. (After Elliott, C.T. and Gordon, N.T., Infrared detectors, in C. Hilsum (ed.), *Handbook on Semiconductors*, Vol. 4, North-Holland, Amsterdam, the Netherlands, 1993, pp. 841–936. With permission.)

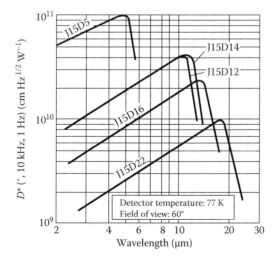

FIGURE 8.13 Spectral dependence of detectivity of Judson's J15D series HgCdTe photoconductive detectors. (After http://www.judsontechnologies.com/files/pdf/MCT_shortform_Dec2002.pdf)

TABLE 8.6   Typical Specifications J15D Series HgCdTe Photoconductors at $T = 77$ K and FOV $= 60°$

Model Number	Active Size (mm)	λ_c (μm)	λ_p (μm)	D^* (cm Hz$^{1/2}$/W) for 10 kHz	R_v (V/W)	R_d (Ω)	I_B (mA)	τ (μs)
J15D5-M204-SO50U	0.05	~5.5	~5	1×10^{11}	2×10^5	100–800	~0.8	1
J15D5-M204-SO1M	1	~5.5	~5	8×10^{10}	2×10^3	100–800	~10	5
J15D12-M204-SO50U	0.05	>12	11 ± 1	5×10^{10}	8×10^4	20–120	~12	0.2
J15D12-M-204 SO1M	1	>12	11 ± 1	4×10^{10}	3×10^3	20–120	~30	0.5
J15D14-M204-SO1M	1	>13.5	~13	4×10^{10}	1×10^3	20–100	~30	0.5

Source: After http://www.judsontechnologies.com/files/pdf/MCT_shortform_Dec2002.pdf

outputs can be integrated in phase. In the SPRITE detectors, these functions are actually performed in the element.

The operating principle of the SPRITE detector is shown in Figure 8.14. The device is essentially an ≈1 mm long, 62.5 μm wide, and 10 μm thick n-type photoresistor with two bias contacts and a readout potential probe. The device is constant current biased with the bias field E set such that the ambipolar drift velocity v_a, which approximates to the minority hole drift velocity v_d, is equal to the image scan velocity v_s along the device. These conditions are fulfilled for HgCdTe and the SPRITE detectors are fabricated only from this material. The length of the device L is typically close to or larger than the drift length $v_d\tau$, where τ is the recombination time. Consider now an element of the image scanned along the device. The excess carrier concentration in the material increases during scan, as illustrated in Figure 8.14. When the illuminated region enters the readout zone, the increased conductivity modulates the output contacts and provides an output signal. Thus, the signal integration, which, for a conventional array is done by external delay line and summation circuitry, is done in the SPRITE detector in the element itself.

The integration time approximates the recombination time τ for long devices. It becomes much longer than the dwell time τ_{pixel} on a conventional element in a fast-scanned serial system. Thus, a proportionally larger ($\propto \tau/\tau_{pixel}$) output signal is observed. When Johnson noise or amplifier noise dominates, this leads to a proportional increase in the signal-to-noise ratio with respect to a discrete element. In the background-limited detector, the excess carrier concentration due to background

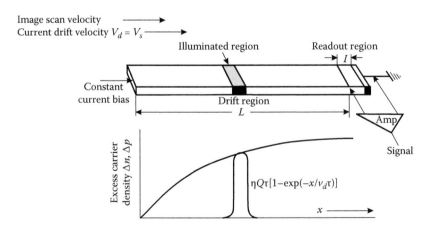

FIGURE 8.14   Operating principle of a SPRITE detector. The upper part of the figure shows a HgCdTe filament with three ohmic contacts. The lower part shows the build-up of excess carrier density in the device as a point in the image is scanned along it. (After Elliott, C.T., Infrared detectors with integrated signal processing, in A. Goetzberger and M. Zerbst (eds.), *Solid State Devices*, Verlag Chemie, Weinheim, Germany, 1983, pp. 175–201. With permission.)

also increases by the same factor, but corresponding noise is proportional only to integrated flux. As a result, the net gain in the signal-to-noise ratio with respect to a discrete element is increased by a factor $(\tau/\tau_{pixel})^{1/2}$.

It can be shown, that the detectivity of SPRITE detector is equal [114]

$$D^\star = (2\eta)^{1/2} D^\star_{BLIP}(s\tau)^{1/2} \tag{8.23}$$

From this equation results that long lifetimes are required to achieve large gains in the signal to noise ratio. Useful improvement in detectivity relative to a BLIP-limited discrete device $D^\star_{BLIP}$ can be achieved when the value of $s\tau$ exceeds unity. The performance of the device can be described in terms of the number of BLIP-limited elements in a serial array giving the same signal-to-noise ratio, $N_{eq}(BLIP) = 2s\tau$. For example, a $60\,\mu$m wide element scanned at a speed of 2×10^4 cm/s and with τ equal to $2\,\mu$s gives $N_{eq}(BLIP) = 13$.

As shown in Table 8.7, to achieve usable performance in the 8–14 μm band the SPRITE devices require liquid nitrogen cooling while 3- or 4-stage Peltier coolers are sufficient for effective operation in the 3–5 μm range. The performance achieved in the 3–5 μm band is illustrated in Figure 8.15. The detectivity increases with the bias field except at the higher fields where the element temperature is raised by Joule heating. Useful performance in this band can be obtained at temperatures up to about 240 K.

The SPRITE detectors are fabricated from lightly doped ($\approx 5 \times 10^{14}$ cm^{-3}) n-type HgCdTe. Single and 2, 4, 8, 16, and 24 element arrays have been demonstrated; the eight-element arrays are the most common at present (Figure 8.16). In order to manufacture the devices in line, it is necessary to reduce the width of the readout zone and corresponding contacts to bring them out parallel to the length of the element within the width of the element as shown in Figure 8.16. Various modifications of the device geometry have been proposed to improve both the detectivity and spatial resolution. The modifications

TABLE 8.7 Performance of SPRITE Detectors

Material	HgCdTe	
Number of elements	8	
Filament length (μm)	700	
Nominal sensitive area (μm)	62.5 × 62.5	
Operating band (μm)	8–14	3–5
Operating temperatures (K)	77	190
Cooling method	Joule–Thompson or heat engine	Thermoelectric
Bias field (V/cm)	30	30
Field of view	f/2.5	f/2.0
Ambipolar mobility (cm^2/V s)	390	140
Pixel rate per element (pixel/s)	1.8×10^6	7×10^5
Typical element resistance (Ω)	500	4.5×10^3
Power dissipation (per element) (mW)	9	1
(total) (mW)	<80	<10
Mean $D^\star$ (500 K, 20 kHz, 1 Hz), 62.5 × 62.5 μm (10^{10} cm Hz$^{1/2}$/W)	>11	4–7
Responsivity (500 K), 62.5 × 62.5 μm (10^4 V/W)	6	1

Source: After *Infrared Phys.* 22, Blackburn, A., Blackman, M.V., Charlton, D.E., Dunn, W.A.E., Jenner, M.D., Oliver, K.J., and Wotherspoon, J.T.M., The practical realization and performance of SPRITE detectors, 57–64, Copyright 1982, with permission from Elsevier.

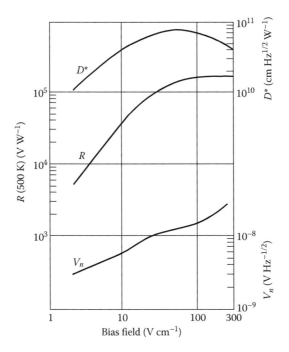

FIGURE 8.15 Performance of a 3–5 μm SPRITE operating at 190 K. (After *Infrared Phys.*, 22, Blackburn, A., Blackman, M.V., Charlton, D.E., Dunn, W.A.E., Jenner, M.D., Oliver, K.J., and Wotherspoon, J.T.M., The practical realization and performance of SPRITE detectors, 57–64, Copyright 1982, with permission from Elsevier.)

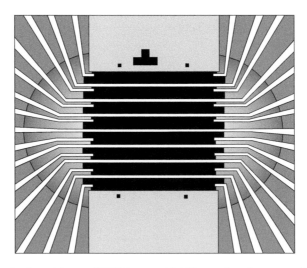

FIGURE 8.16 Schematic of an eight-row SPRITE array with bifurcated readout zones. (After *Infrared Phys.*, 22, Blackburn, A., Blackman, M.V., Charlton, D.E., Dunn, W.A.E., Jenner, M.D., Oliver, K.J., and Wotherspoon, J.T.M., The practical realization and performance of SPRITE detectors, 57–64, Copyright 1982, with permission from Elsevier.)

have included horn geometry of the readout zone to reduce the transit time spread, and slight taper of the drift region to compensate a slight change of drift velocity due to background radiation.

Despite remarkable successes, SPRITE detectors have important limitations such as limited size, stringent cooling requirements, and the necessity to use fast mechanical scanning. The ultimate size of SPRITE arrays is limited by the significant heat load imposed by Joule heating.

8.4 Photovoltaic Detectors

Photoeffects, which occur in structures with built-in potential barriers, are essentially photovoltaic and result when excess carriers are injected optically into the vicinity of such barriers. The role of the built-in electric field is to cause the charge carriers of opposite sign to move in opposite directions depending upon the external circuit. Several structures are possible to observe the photovoltaic effect. These include p–n junctions, heterojunctions, Schottky barriers, metal-semiconductor-metal (MSM) devices, and MIS photocapacitors. Each of these different types of devices has certain advantages depending on the particular applications. The most common example of a photovoltaic detector is the abrupt p–n junction prepared in the semiconductor, which is often referred to simply as a photodiode.

8.4.1 Principle of Operation

The operation of the p–n junction photodiode is illustrated in Figure 8.17. Photons with energy greater than the energy gap, incident on the front surface of the device, create electron–hole pairs in the material on both sides of the junction. By diffusion, the electrons and holes generated within a diffusion length from the junction reach the space-charge region. Then electron–hole pairs are separated by the strong electric field; minority carriers are readily accelerated to become majority carriers on the other side. This way a photocurrent is generated which shifts the current–voltage characteristic in the direction of negative or reverse current. The total current density in the p–n junction is usually written as

$$J(V,\Phi) = J_d(V) - J_{ph}(\Phi) \tag{8.24}$$

where the dark current density, J_d, depends only on bias voltage, V, and the photocurrent depends only on the photon flux density, Φ.

Generally, the current gain in a simple photovoltaic detector (e.g., not an APD) is equal to 1, and then according to Equation 8.17, the magnitude of photocurrent is equal

$$I_{ph} = \eta q A \Phi \tag{8.25}$$

The dark current and photocurrent are linearly independent, which occurs even when these currents are significant.

If the p–n diode is open-circuited, the accumulation of electrons and holes on the two sides of the junction produces an open-circuit voltage (Figure 8.17d). If a load is connected to the diode, a current

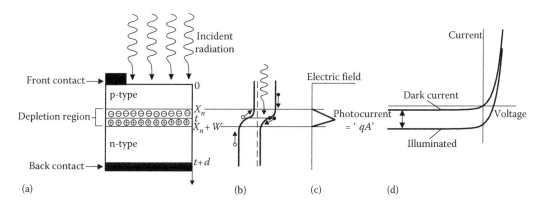

FIGURE 8.17 p–n junction photodiode: (a) structure of abrupt junction, (b) energy band diagram, (c) electric field, and (c) current–voltage characteristics.

will conduct in the circuit. The maximum current is realized when an electrical short is placed across the diode terminals, and this is called the short-circuit current.

The open-circuit voltage can be obtained by multiplying the short-circuit current by the incremental diode resistance $R = (\partial I / \partial V)^{-1}$ at $V = V_b$:

$$V_{ph} = \eta q A \Phi R \tag{8.26}$$

where
    V_b is the bias voltage
    $I = f(V)$ is the current–voltage characteristic of the diode

In many direct applications, the photodiode is operated at zero-bias voltage:

$$R_0 = \left(\frac{\partial I}{\partial V} \right)^{-1}_{\bigg| V_b = 0} \tag{8.27}$$

A frequently encountered figure of merit for IR photodiode is the $R_0 A$ product

$$R_0 A = \left(\frac{\partial J}{\partial V} \right)^{-1}_{\bigg| V_b = 0} \tag{8.28}$$

where $J = I/A$ is the current density.

Assuming diffusion limited dark current of photodiode equal $I_d = I_s[exp(qV/kT) - 1]$, the $R_0 A$ product is equal

$$(R_0 A)_d = \left(\frac{dJ_d}{dV} \right)^{-1}_{\bigg| V_b = 0} = \frac{kT}{qJ_s} \tag{8.29}$$

Different behavior of saturation current depending on diffusion length is explained in the following. For an n-on-p diode structure, the junction resistance is limited by diffusion of minority carriers from the p side into the depletion region. In the case of conventional bulk diodes, where thickness of active region $t \gg L_e$ (L_e is the minority diffusion length), the $R_0 A$ product is equal

$$(R_0 A)_d = \frac{kT}{qJ_s} = \frac{(kT)^{1/2}}{q^{3/2} n_i^2} N_a \left(\frac{\tau_e}{\mu_e} \right)^{1/2} \tag{8.30}$$

where
    J_s is the current density
    n_i is the intrinsic carrier concentration
    N_a is the doping concentration
    μ_e and τ_e are the minority carrier mobility and lifetime

The $R_0 A$ product (saturation current density) is proportional (inversely proportional) to $N_a \tau_e^{1/2}$. By thinning the substrate to a thickness smaller than the minority-carrier diffusion length (thus reducing the volume in which diffusion current is generated) the corresponding dark current decreases, provided that the back surface is properly passivated to reduce surface recombination. As result, the current density can decrease by a factor of L_e/t. If the thickness of the p-type region is such that $t \ll L_e$, we obtain:

$$(R_0 A)_d = \frac{kT}{q^2} \frac{N_a}{n_i^2} \frac{\tau_e}{d} \tag{8.31}$$

On the contrary, if $t \ll L$, the $R_0 A$ product (saturation dark current) is proportional (inversely proportional) to $N\tau$. Of course, analogical formulas can be obtained for p-on-n junctions.

In the detection of radiation, the photodiode is operated at any point of the *I–V* characteristic. High reverse bias may or may not shorten charge collection time, but it generally reduces capacitance. Both result in faster response. On the other hand, increased reverse bias causes increased noise, so that a tradeoff exists between speed and sensitivity.

The intrinsic noise mechanism of a photodiode is shot noise in the current passing through the diode. It is generally accepted that the noise in an ideal diode is given by

$$I_n^2 = \left[2q(I_d + 2I_s) + 4kT(G_J - G_0)\right]\Delta f \tag{8.32}$$

where
G_J is the conductance of the junction
G_0 is the low-frequency value of G_J

In the low-frequency region, the second term on the right-hand side is zero. For a diode in thermal equilibrium (i.e., without applied bias voltage and external photon flux) the mean square noise current is just the Johnson–Nyquist noise of the photodiode zero bias resistance ($R_0^{-1} = qI_s/kT$):

$$I_n^2 = \frac{4kT}{R_0}\Delta f \tag{8.33}$$

Hitherto, photodiodes were analyzed in which the dark current was limited by diffusion. However, several additional excess mechanisms are involved in determining the dark current–voltage characteristics of the photodiode. The dark current is the superposition of current contributions from three diode regions: bulk, depletion region, and surface. In the following, we will be concerned with the current contribution of high-quality photodiodes with high $R_0 A$ products limited by [17].

- Generation–recombination within the depletion region
- Tunneling through the depletion region
- Surface current

The first two mechanisms are schematically illustrated in Figure 8.18.

The generation–recombination current of the depletion region can be described as

$$J_{GR} = \frac{qwn_i}{2\tau_o} \tag{8.34}$$

if a trap is near the intrinsic level of the bandgap energy. In this equation, τ_o is the carrier lifetime in depletion region. The surface leakage current can be described in the same way.

The space-charge region generation–recombination current varies with temperature as n_i, that is, less rapidly than diffusion current which varies as n_i^2.

The third type of dark current component that can exist is a tunneling current caused by electrons directly tunneling across the junction from the valence band to the conduction band (direct tunneling)

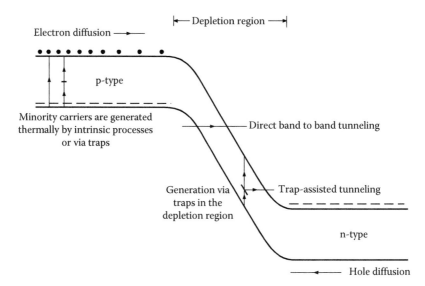

FIGURE 8.18 Schematic representation of some of the mechanisms by which dark current is generated in a reverse biased p–n junction.

or by electrons indirectly tunneling across the junction by way of intermediate trap sites in the junction region (indirect tunneling or trap-assisted tunneling—see Figure 8.18).

8.4.2 InGaAs Photodiodes

The need for high-speed, low-noise $In_xGa_{1-x}As$ (InGaAs) photodetectors for use in lightwave communication systems operating in the 1–1.7 μm wavelength region ($x = 0.53$) is well established. $In_{0.53}Ga_{0.47}As$ alloy is lattice matched to the InP substrate. Having lower dark current and noise than indirect-bandgap germanium, the competing near-IR material, the material is addressing both entrenched applications including low light level night vision and new applications such as remote sensing, eye-safe range finding, and process control [9]. By changing the alloy composition of the InGaAs absorption layer, the photodetector responsivity can be maximized at the desired wavelength of the end user to enhance the signal to noise ratio.

InGaAs-detector processing technology is similar to that used with silicon, but the detector fabrication is different. The InGaAs detector's active material is deposited onto a substrate using chloride VPE or MOCVD techniques adjusted for thickness, background doping, and other requirements. Planar technology evolved from the older mesa technology and at present is widely used due to its simple structure and processing as well as the high reliability and low cost (see Figure 8.19).

Both p-i-n and avalanche InGaAs photodiode structures with total layer thicknesses of 4–7 μm are fabricated. In comparison to APDs operating in the same wavelength region, p-i-n photodiodes offer the advantages of lower dark current, larger frequency bandwidth, and simpler driving circuitry. Thus, although p-i-n diodes do not have internal gain, an optimal combination of a p-i-n diode with a low-noise, large-bandwidth transistor has led to high sensitivity optical receivers operating up to 2.5 Gb/s.

Standard $In_{0.53}Ga_{0.47}As$ photodiodes have detector-limited room temperature detectivity of ~10^{13} cm $Hz^{1/2}$/W. With increasing cutoff wavelength, detectivity decreases. Figure 8.20 shows the spectral response of four such InGaAs detectors at room temperature, whose peak responsivity is optimized at 1.6, 1.9, 2.2, and 2.6 μm, respectively. Next figure (Figure 8.21) shows that the highest quality InGaAs photodiodes have been grown by MOCVD. Their performance is comparable with HgCdTe photodiodes.

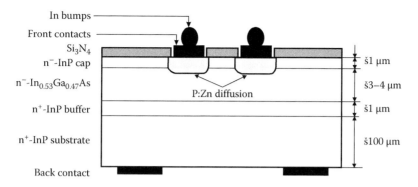

FIGURE 8.19 Cross section of a planar backside illuminated p-i-n InGaAs photodiode.

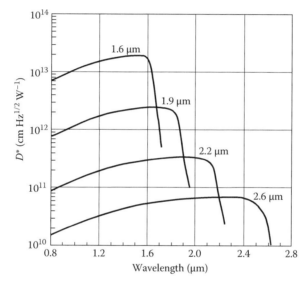

FIGURE 8.20 Room temperature detectivity of InGaAs photodiodes with cutoff wavelength at 1.6, 1.9, 2.2, and 2.6 μm, respectively.

An example of APD is shown in Figure 8.22 together with the band diagram of the entire heterostructure. Light is absorbed in the InGaAs and holes (with a higher impact ionization coefficient than electrons; this ensures low-noise operation) are swept to an InP junction, where the avalanche multiplication takes place. This structure combines low leakage, due to the junction being placed in the high-bandgap material (InP), which sensitivity at longer wavelength provided by the lower-bandgap InGaAs absorption region. Dark currents of the order pA can be obtained. However, there is a potential problem with the operation of a SAM-ADP. Holes can accumulate at the valence band discontinuity at the InGaAs/InP heterojunction and thereby increase the response time. To alleviate this problem, a graded bandgap InGaAsP layer can be inserted between the InP and InGaAs. This modified structure is known as a separate-absorption-graded-multiplication APD (SAGM-APD). APD can achieve 5–10 dB better sensitivity than p-i-n, provided that the multiplication noise is low and the gain-bandwidth product is sufficiently high.

For commercially available InGaAs/InP APDs, it is fairly difficult to respond reliably to 10 Gb/s range signals with moderate multiplication gain, due to their limited gain-bandwidth (BG) products. The GB products in these APDs are typically 20–40 GHz.

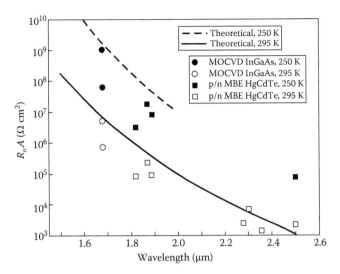

FIGURE 8.21   Short-wavelength IR detector R_0A versus wavelength at 295 and 250 K. (After Kozlowski, L.J. *Proc. SPIE*, 3182, 2, 1997. With permission from International Society for Optical Engineering.)

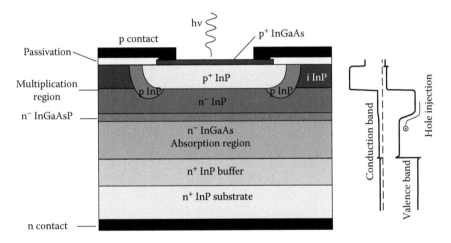

FIGURE 8.22   Cross section of SAGM-APD InP-based photodiode.

8.4.3 InSb Photodiodes

InSb photodiodes have been available since the late 1950s and are generally fabricated by impurity diffusion and ion implantation. Epitaxy is not used; instead, the standard manufacturing technique begins with bulk n-type single crystal wafers with donor concentration about 10^{15} cm^{-3}. Wimmers et al. [116,117] and Rogalski [101] have presented the status of InSb photodiode technology for a wide variety of linear and FPAs.

Typical InSb photodiode RA product at 77 K is 2×10^6 Ω cm^2 at zero bias and 5×10^6 Ω cm^2 at slight reverse biases of approximately 100 mV. As element size decreases below 10^{-5} cm^{-2}, some slight degradation in resistance due to surface leakage occurs.

Figure 8.23 compares the dependence of dark current on temperature between InSb and HgCdTe photodiodes used in the current high performance large FPAs [118]. This comparison suggests that MWIR HgCdTe photodiodes have significant higher performance in the 30–120 K temperature range.

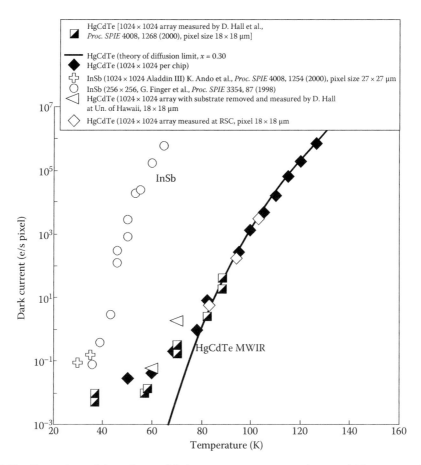

FIGURE 8.23 Comparison of dependence of dark current on temperature between highest reported value for InSb arrays and MBE-grown HgCdTe MWIR FPAs (with $18 \times 18\,\mu m$ pixels). (With kind permission from Springer Science+Business Media: *J. Electron. Mater.*, Mid-wavelength infrared p-on-on $Hg_{1-x}Cd_xTe$ heterostructure detectors: 30–120 Kelvin state-of-the-art performance, 32, 2003, 803–809, Zandian, M., Garnett, J.D., DeWames, R.E., Carmody, M., Pasko, J.G., Farris, M., Cabelli, C.A., Cooper, D.E., Hildebrandt, G., Chow, J., Arias, J.M., Vural, K., and Hall, D.N.B.)

The InSb devices are dominated by generation–recombination currents in the 60–120 K temperature range because of a defect center in the energy gap, whereas MWIR HgCdTe detectors do not exhibit g–r currents in this temperature range and are limited by diffusion currents. In addition, wavelength tunability has made HgCdTe the preferred material.

InSb photodiodes can also be operated in the temperature range above 77 K. Of course, the RA products degrade in this region. At 120 K, RA products of $10^4\,\Omega\,cm^2$ are still achieved with slight reverse bias, making BLIP operation possible. The quantum efficiency in InSb photodiodes optimized for this temperature range remains unaffected up to 160 K. Detectivity increases with reduced background flux (narrow field of view and/or cold filtering) as illustrated in Figure 8.24.

InSb photovoltaic detectors are widely used for ground-based IR astronomy and for applications aboard the Space Infrared Telescope Facility. Recently, impressive progress has been made in the performance of InSb hybrid FPAs. An array size of 2048×2048 is possible because the InSb detector material is thinned to less than $10\,\mu m$ (after surface passivation and hybridization to a readout chip) which allows it to accommodate the InSb/silicon thermal mismatch.

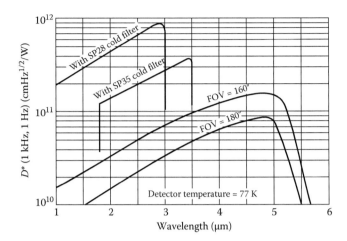

FIGURE 8.24 Detectivity as a function of wavelength for a InSb photodiode operating at 77 K. (After http://www.judsontechnologies.com/files/pdf/MCT_shortform_Dec2002.pdf)

8.4.4 HgCdTe Photodiodes

HgCdTe photodiodes are available to cover the spectral range from 1 to 20 μm. Spectral cutoff can be tailored by adjusting the HgCdTe alloy composition.

Epitaxy is the preferable technique to obtain device-quality HgCdTe epilayers for IR devices. Among the various epitaxial techniques, the LPE is the most matured method. The recent efforts are mostly on low growth temperature techniques; MOCVD and MBE. MOCVD appears to be the most promising for large-scale and low-cost production of epilayers. MBE offers unique capabilities in material and device engineering including the lowest growth temperature, superlattice growth and potential for the most sophisticated composition and doping profiles. The growth temperature is less than 200°C for MBE but around 350°C for MOCVD, making it more difficult to control the p-type doping in the MOCVD due to the formation of Hg vacancies.

Different HgCdTe photodiode architectures have been fabricated that are compatible with backside and frontside illuminated hybrid FPA technology [17,101,119]. Figure 8.25 shows the schematic band profiles of the most commonly used unbiased homo- (n^+-on-p) and heterojunction (p-on-n) photodiodes. To avoid contribution of the tunneling current, the doping concentration in the base region below 10^{16} cm^{-3} is required. In both photodiodes, the lightly doped narrow gap absorbing region ["base" of the photodiode: p(n)-type carrier concentration of about 5×10^{15} cm^{-3} (5×10^{14} cm^{-3})], determines the dark current and photocurrent. The internal electric fields at interfaces are "blocking" for minority carriers

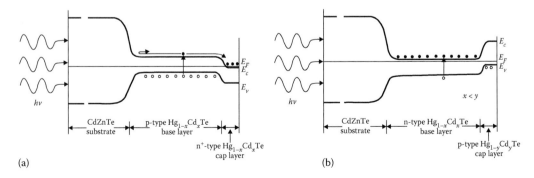

FIGURE 8.25 Schematic band diagrams of n^+-on-p (a) and p-on-n heterojunction (b) photodiodes.

and influence of surface recombination is eliminated. Also suitable passivation prevents the influence of surface recombination. Indium is most frequently used as a well-controlled dopant for n-type doping due to its high solubility and moderately high diffusion. Elements of the VB group are acceptors substituting Te sites. They are very useful for fabrication of stable junctions due to very low diffusivity. Arsenic proved to be the most successful p-type dopant to date. The main advantages are stability in lattice, low activation energy and possibility to control concentration over 10^{15}–10^{18} cm^{-3} range. Intensive efforts are currently underway to reduce a high temperature (400°C) required to activate As as an acceptor.

n-on-p junctions are fabricated in two different manners using Hg vacancy doping and extrinsic doping. Hg vacancies (V_{Hg}) provide intrinsic p-type doping in HgCdTe. In this case, the doping level depends on only one annealing temperature. However, the use of Hg vacancy as p-type doping is known to kill the electron lifetime, and the resulting detector exhibits a higher current than in the case of extrinsic doping using. However, for very low doping (<10^{15} cm^{-3}), the hole lifetime becomes SR limited and does not depend on doping anymore. V_{Hg} technology leads to low minority diffusion length of the order of 10–15 μm, depending on doping level. Generally, n-on-p vacancy doped diodes give rather high diffusion currents but lead to a robust technology as its performance weakly depend on doping level and absorbing layer thickness. Simple modeling manages to describe dark current behavior of V_{Hg} doped n-on-p junctions over a range of at least eight orders of magnitude (see Figure 8.26). In the case of extrinsic doping, Cu, Au, and As are often used. Due to higher minority carrier lifetime, extrinsic doping is used for low dark current (low flux) applications. The extrinsic doping usually leads to larger diffusion length and allows lower diffusion current but might exhibit performance fluctuations, thus affecting yield and uniformity.

In the case of p-on-n configuration, the typical diffusion length, L, is up to 30–50 μm for low doping levels such as 10^{15} cm^{-3} typically reached with indium doping and the dark currier generation is volume limited by the absorbing layer volume itself [120]. Different behavior of saturation current depending on diffusion length is explained by Equations 8.30 and 8.31. If $t \gg L$, the saturation current is inversely proportional to $N\tau^{1/2}$. On the contrary, if $t \ll L$, the saturation dark current is inversely proportional to $N\tau$.

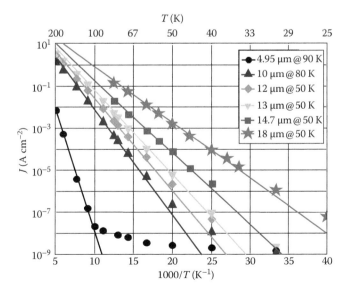

FIGURE 8.26 Mercury vacancy dark current modeling for n-on-p HgCdTe photodiodes with different cutoff wavelengths. (With kind permission from Springer Science+Business Media: *J. Electron. Mater.*, Study of LWIR and VLWIR focal plane array developments: Comparison between p-on-n and different n-on-p technologies on LPE HgCdTe, 38, 2009, 1733–1740, Gravrand, O., Mollard, L., Largeron, C., Baier, N., Deborniol, E., and Chorier, Ph.)

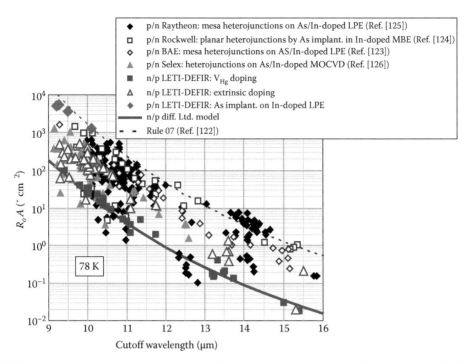

FIGURE 8.27 R_oA product versus cutoff wavelength at 78 K, summarized with bibliographic data. (With kind permission from Springer Science+Business Media: *J. Electron. Mater.*, Study of LWIR and VLWIR focal plane array developments: Comparison between p-on-n and different n-on-p technologies on LPE HgCdTe, 38, 2009, 1733–1740, Gravrand, O., Mollard, L., Largeron, C., Baier, N., Deborniol, E., and Chorier, Ph.)

Figure 8.27 is an accumulation of R_oA-data with different cutoff wavelength, taken from LETI LIR using both V_{Hg} and extrinsic doping n-on-p HgCdTe photodiodes compared with P-on-n data from other laboratories, extracted from various recent literature reports, using various techniques and different diode structures [121–125]. It is clearly shown, that extrinsic doping of n-on-p photodiodes leads to higher R_oA product (lower dark current) whereas mercury vacancy doping remains on the bottom part of the plot. p-on-n structures are characterized by the lowest dark current (highest R_oA product)—see trend line calculated with Teledyne (formerly Rockwell Scientific) empirical model [121].

Up to the present, photovoltaic HgCdTe FPAs have been mainly based on p-type material. Linear (240, 288, 480, and 960 elements), 2D scanning arrays with TDI, and 2D staring formats up to 2048×2048 pixels have been made [101]. Pixel sizes ranging from 18 μm² to over 1 mm have been demonstrated. The best results have been obtained using hybrid architecture.

HgCdTe as an attractive material for room-temperature APDs operate at 1.3–1.6 μm wavelengths for fiber-optical communication applications was recognized in the 1980s. The resonant enhancement occurs when the spin-orbit splitting energy in the valence band, Δ, is equal to the fundamental energy gap E_g. This has the beneficial effect, first pointed out by Verie et al. of making the electron and hole impact ionization rates quite different, which is highly desirable for low-noise APDs [126]. The band structure of HgCdTe gives k-values close to 0—a highly favorable ratio of hole to electron multiplication during avalanche conditions, resulting in very little noise gain. These properties give HgCdTe APDs a figure of merit better than InGaAs APDs, where k is about 0.45. Silicon, in comparison, has an ionization ratio of 0.02, and, therefore, much lower excess noise, however Si is not sensitive to wavelengths greater than 1.1 μm.

Another APD application is LADAR (laser radar) whereby a pulsed laser system is obtained 3D imagery. 3D imagery has been has traditionally obtained with scanned laser systems that operate in the

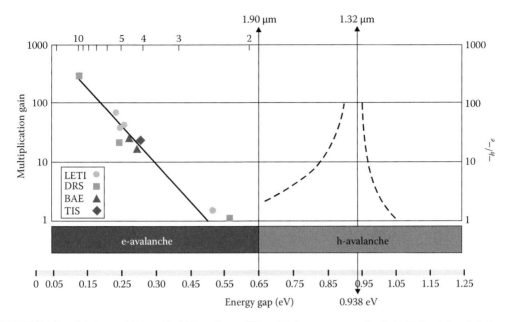

FIGURE 8.28 Distinct e-APD and h-APD regimes of $Hg_{1-x}Cd_xTe$ cross over at $E_g \approx 0.65\,eV$ ($\lambda_c \approx 1.9\,\mu m$). At lower band gaps, the e-APD gain increases exponentially (material for four manufactures shows remarkably consistent results). (From Hall, D.N.B. et al., HgCdTe optical and infrared focal plane array development in the next decade, *Astro2010: The Astronomy and Astrophysics Decadal Survey*, Technology Development Papers, no. 28. With permission.)

visible and near IR (NIR) out to 1 μm. However, eye-safe lasers are required if there is a chance humans will be in the scene. The "eye-safe" range is around 1.55 μm in the NIR region. In addition, the advantages of a single pulse, flood illuminated, and laser imaging system caused researchers to consider 2D arrays of APD sensors. The gated active/passive system promises target detection and identification at longer ranges compared to conventional passive only imaging systems. Because of the desired operating ranges (up to 10 km, typically) and the fact that laser power is limited, there is a need for a NIR solid state detector with a sensitivity approaching single photon [127].

As shown in Figure 8.28 [128], the avalanche properties of $Hg_{1-x}Cd_xTe$ vary dramatically with bandgap. Leveque et al. [129] described two regimes in which the ratio $k = \alpha_h/\alpha_e$ of the hole ionization coefficient to the electron ionization coefficient is either much greater or much less than unity. For cutoff wavelengths shorter than approximately 1.9 μm ($x = 0.65$ at 300 K), they predict $\alpha_h \gg \alpha_e$ because of resonant enhancement of the hole ionization coefficient when $E_g \cong \Delta = 0.938\,eV$. The situation, when $k = \alpha_h/\alpha_e \gg 1$, is favorable for low-noise APDs with hole-initiated avalanche. For more details see Kinch's monograph [130].

The above described performance of HgCdTe e-APDs has opened the door to new passive/active system capabilities and applications. The combination of the dual-band and avalanche gain functionalities is another technological challenge that will enable many applications, such as dual band detection over large temperature range.

8.4.5 Lead Salt Photodiodes

In comparison with photoconductive detectors, lead salt photodiodes have been less exploited in the IR region and have not found numerous military and commercial applications.

A wide variety of techniques have been used to form p–n junctions in lead salts. They have included interdiffusion, diffusion of donors, ion implantation, proton bombardment, Schottky barrier, and creation of n-type layers on p-type material by VPE or LPE. A summary of works for high-quality photodiode fabrication is given in Table 8.4 of the monograph *Infrared Photon Detectors* [131].

Most of the implant work in IV–VI semiconductors has been devoted to the formation of n-on-p junction photodiodes, mainly covering the 3–5 μm spectral region. An alternative technology, adopted especially to prepare long-wavelength photodiodes, was the use of heterojunctions of n-type PbTe (PbSeTe) deposited onto p-type PbSnTe substrates by LPE, VPE, MBE, and hot-wall epitaxy (HWE) [17,131]. A considerably simpler technique for photodiode fabrication involves evaporating a thin metal layer onto the semiconductor surface. It is a planar technology with the potential for cheap fabrication of large arrays. An excellent review of Schottky-barrier IV–VI photodiodes with emphasis on thin-film devices has been given by Holloway [132]. Recent progress in the development of lead salt photodiodes has been reviewed by Zogg and Ishida [133].

Detector arrays made from lead salts tend to have very high uniformity in their cutoff wavelengths because of the relatively slow variation in bandgap with composition.

For metal/semiconductor sensors, Pb has proven to be a good blocking material. The p–n junction works well with tellurides. Selenides exhibit a too-high diffusion in order to obtain reliable devices. For both types of junctions, the performance is limited by the dislocation densities in the layers. The low-temperature R_0A products scale linearly with the inverse dislocation density. The optimal carrier concentration in the active region is in the low 10^{17} cm^{-3} range.

A first attempt to realize quasi-monolithic lead salt detector arrays was described by Barrett, Jhabvala, and Maldari, who elaborated on direct integration of PbS photoconductive detectors with MOS transitions [134,135]. In this process, the PbS films were chemically deposited on the overlaying SiO$_2$ and metallization. Detectivity at 2.0–2.5 μm of 10^{11} cm Hz$^{1/2}$/W was measured at 300 K on an integrated photoconductive PbS detector-Si MOSFET preamplifier. Elements of 25×25 μm^2 were easily fabricated.

The research group at the Swiss Federal Institute of Technology demonstrated the first realization of monolithic PbTe FPA (96×128) on a Si substrate containing the active addressing electronics [136,137]. The monolithic approach used here overcomes the large mismatch in the thermal coefficient of expansion between the group IV–VI materials and Si. Large lattice mismatches between the detector active region and the Si did not impede fabrication of the high-quality layers because the easy plastic deformation of the IV–VIs by dislocation glide on their main glide system without causing structural deterioration.

A schematic cross section of a PbTe pixel grown epitaxially by MBE on a Si readout structure is shown in Figure 8.29. A 2–3 nm thick CaF$_2$ buffer layer is employed for compatibility with the Si substrate. Active layers of 2–3 μm thickness suffice to obtain only a near reflection loss limited quantum efficiency. The spectral response curves of PbTe photodiodes are shown in Figure 8.30. Typical quantum efficiencies

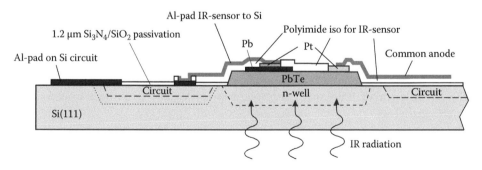

FIGURE 8.29 Schematic cross section of one pixel showing the PbTe island as a backside-illuminated photovoltaic IR detector, the electrical connections to the circuit (access transistor), and a common anode. (After Alchalabi, K. Zimin, D., Zogg, H., and Buttler, W., Monolithic heteroepitaxial PbTe-on-Si infrared focal plane array with 96×128 pixels, *IEEE Electron Dev. Lett.*, 22, 110–112, 2001. With permission. © 2001 IEEE.)

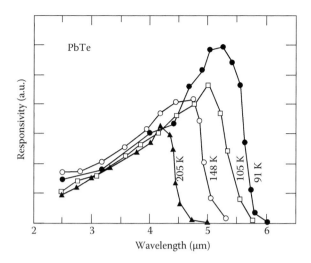

FIGURE 8.30 Spectral responses of a PbTe photodiode. (After Zogg, H., Blunier, S., Hoshino, T., Maissen, C., Masek, J., and Tiwari, A.N., Infrared sensor arrays with 3–12 μm cutoff wavelengths in heteroepitaxial narrow-gap semiconductors on silicon substrates, *IEEE Trans. Electron. Dev.*, 38, 1110–1117, 1991. With permission. © 1991 IEEE.)

are around 50% without an AR coating. Metal-semiconductor Pb/PbTe detectors are employed. Each Pb-cathode contact is fed to the drain of the access transistor while the anode (sputtered Pt) is common for all pixels. Figure 8.31 shows a complete array.

Despite typical dislocation densities in the 10^7 cm^{-2} range in epitaxial lattice and thermal-expansion-mismatched IV–VI on Si(111) layers, useful photodiodes can be fabricated with R_oA products to $200\,\Omega$ cm^2 at 95 K (with a 5.5 μm cutoff wavelength). This is due to the high permittivities of the IV–VI materials, which shield the electric field from charged defects over short distances.

Lead salt heterostructure PV detectors also have been fabricated. The heterojunctions were formed directly between the substrate and the IV–VI films. This solution has been employed to fabricate high-density monolithic PbS–Si heterojunction arrays [139]. However, the performance of heterostructure arrays is inferior to the Schottky barrier and p–n junction arrays.

FIGURE 8.31 Part of the completely processed monolithic 96×128 PbTe-on-Si IR FPA for the MWIR with the readout electronics in the Si substrate. Pixel pitch is 75 μm. (After Zogg, H., Alchalabi, K., Zimin, D., and Kellermann, K., Two-dimensional monolithic lead chalcogenide infrared sensor arrays on silicon read-out chips and noise mechanisms, *IEEE Trans. Electron Dev.*, 50, 209–214, 2003. With permission. © 2003 IEEE.)

8.4.6 InAs/GaInSb Superlattice Photodiodes

High performance InAs/GaInSb SL photovoltaic detectors are predicted by the theoretical promise of longer intrinsic lifetimes due to the suppression of Auger recombination mechanism. Significant progress has been made on the quality of the material during the past several years. The wafer surface roughness achieved 1–2 Å.

At present InAs/GaInSb SL photodiodes are typically based on p-i-n double heterostructures with an unintentionally doped, intrinsic region between the heavily doped contact portions of the device. Figure 8.32 shows a cross section scheme of a completely processed mesa detector. The main technological challenge for the fabrication of photodiodes is the growth of thick SLS structures without degrading the materials quality. High-quality SLS materials thick enough to achieve acceptable quantum efficiency is crucial to the success of the technology. Surface passivation is also serious problem. Considerable surface leakage is attributed to the discontinuity in the periodic crystal structure caused by mesa delineation. Several materials and processes have been explored for device passivation. Some of the more prominent thin films studied have been silicon nitride, silicon oxides, ammonium sulfide, aluminum gallium antimonide alloys, and polyimide [93].

The performance of LWIR photodiodes in the high temperature range is limited by diffusion process. Space charge recombination currents dominate reverse bias at 78 K and taking the dominant recombination centers to be located at the intrinsic Fermi level. At low temperature, the currents are diffusion limited near zero bias voltage. At larger biases, trap-assisted tunneling currents dominate. Figure 8.33 shows the experimental data and theoretical prediction of the R_0A product as a function of temperature for InAs/GaInSb photodiode with 10.5 μm cutoff wavelength at 78 K. The photodiodes are depletion region (generation–recombination) limited in temperature range below 100 K. The trap-assisted tunneling is dominant at $T \leq 40$ K.

Optimization of the SL photodiode architectures is still an open area. Since some of the device design parameters depend on material properties, like carrier lifetime and diffusion lengths, these properties are still being improved. Also additional design modifications dramatically improve the photodiode performance. For example, Aifer et al. [140] have reported W-structured type II superlattice (WSL) LWIR photodiodes with R_0A values comparable for state-of-the-art HgCdTe. In this design illustrated in Figure 8.34a, the AlSb barriers are replaced with shallower $Al_{0.40}Ga_{0.49}In_{0.11}Sb$ quaternary barrier layers. In such structure two InAs "electron-wells" are located on either side of an InGaSb "hole-well" and are bound on either side by AlGaInSb "barrier" layers. The barriers confine the electron wavefunctions symmetrically about the hole-well, increasing the electron–hole overlap while nearly localizing the wavefunctions. The resulting quasi-dimensional densities of states give the WSL its characteristically strong absorption near band edge.

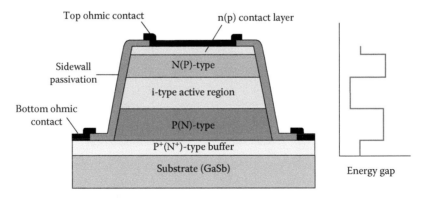

FIGURE 8.32 Schematic structure of p-i-n double heterojunction photodiode.

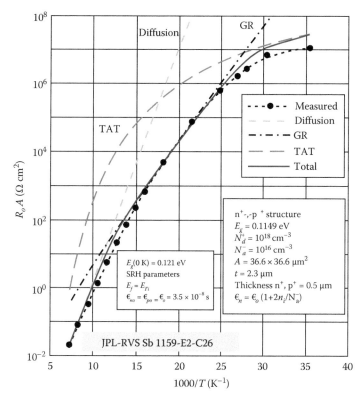

FIGURE 8.33. R_oA product as a function of temperature for InAs/GaInSb photodiode with $\lambda_c = 10.5\,\mu m$ at 78.5 K. (After Pellegrini, J. and DeWames, R., *Proc. SPIE*, 7298, 7298, 2009. With permission from International Society for Optical Engineering.)

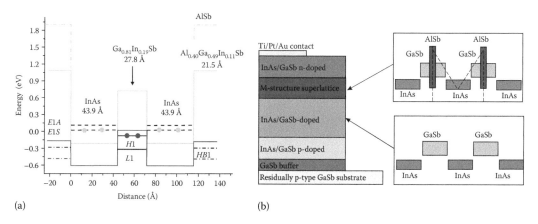

FIGURE 8.34 Schematic diagrams of modified type-II LWIR photodiodes: (a) band profiles at $k = 0$ of enhanced WSL (With permission from Aifer, E.H., Tischler, J.G., Warner, J.H., Vurgaftman, I., Bewley, W.W., Meyer, J.R., Canedy, C.L., and Jackson, E.M., W-structured type-II superlattice long-wave infrared photodiodes with high quantum efficiency, *Appl. Phys. Lett.*, 89, 053510, 2006. Copyright 2006, American Institute of Physics) and (b) p-π-M-n SL (band alignment of standard and M shape superlattices are shown). (After Nguyen, B.-M. et al., *Appl. Phys. Lett.*, 91, 163511, 2007. With permission.)

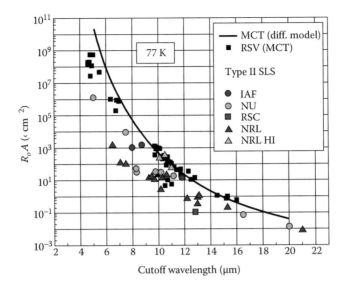

FIGURE 8.35 Dependence of the R_oA product of InAs/GaInSb SLS photodiodes on cutoff wavelength compared to theoretical and experimental trend lines for comparable HgCdTe photodiodes at 77 K. (With kind permission from Springer Science+Business Media: *J. Electron. Mater.*, Antimonide type-II W photodiodes with long-wave infrared R_oA comparable to HgCdTe, 36, 2007, 852–856, Candey, C.L., Aifer, H., Vurgaftman, I., Tischler, J.G., Meyer, J.R., Warner, J.H., and Jackson, E.M.)

Another type II superlattice photodiode design with the M-structure barrier is shown in Figure 8.34b. This structure significantly reduces the dark current, and on the other hand, does not shown a strong effect on the optical properties of the devices [141]. The AlSb layer in one period of the M structure, having a wider energy gap, blocks the interaction between electrons in two adjacent InAs wells, thus, reducing the tunneling probability and increasing the electron effective mass. The AlSb layer also acts as a barrier for holes in the valence band and converts the GaSb hole-quantum well into a double quantum well. As a result, the effective well width is reduced, and the hole's energy level becomes sensitive to the well dimension. Device with a cutoff wavelength of 10.5 μm exhibits a R_oA product of 200 Ω cm² when a 500 nm thick M structure was used. Recently, using double M-structure heterojunction at the single device level, the R_oA product up to 5300 Ω cm² has been obtained for a 9.3 μm cutoff at 77 K [142].

Figure 8.35 compares the R_oA values of InAs/GaInSb SL and HgCdTe photodiodes in the long wavelength spectral range [143]. The solid line denotes the theoretical diffusion limited performance of p-type HgCdTe material. As it can be seen in the figure, the most recent photodiode results for SL devices rival that of practical HgCdTe devices, indicating substantial improvement has been achieved in SL detector development.

Type-II based InAs/GaInSb based detectors have been rapid progress over the past few years. The presented results indicate that fundamental material issues of InAs/GaInSb SLs fulfill practical realization of high performance FPAs.

8.5 Focal Plane Arrays

The term "FPA" refers to an assemblage of individual detector picture elements ("pixels") located at the focal plane of an imaging system. Although the definition could include one-dimensional ("linear") arrays as well as 2D arrays, it is frequently applied to the latter. Usually, the optics part of an optoelectronic images device is limited only to focusing of the image onto the detectors array. These so-called "staring arrays" are scanned electronically usually using circuits integrated with the arrays. The architecture of detector-readout assemblies has assumed a number of forms. The types of ROICs include the function of pixel deselecting, antiblooming on each pixel, subframe imaging, output preamplifiers, and may include yet other functions. IR imaging systems, which use 2D arrays, belong to so-called "second-generation" systems.

In general, the architectures of FPAs may be classified as monolithic and hybrid. When the detector material is either silicon or a silicon derivative (such as, e.g., platinum silicide PtSi), the detector and ROIC can be built on a single wafer. There are a few obvious advantages to this structure, principally in the simplicity and lower cost associated with a directly integrated structure. Common examples of these FPAs in the visible and near IR ($0.7–1.0\,\mu m$) are found in camcorders and digital cameras. Two generic types of silicon technology provide the bulk of devices in these markets: charge-coupled devices (CCDs) and complementary metal-oxide-semiconductor (CMOS) imagers. CCD technology has achieved the highest pixel counts or largest formats with numbers approaching 10^{10} (see Figure 8.36). CMOS imagers are also rapidly moving to large formats and are expected to compete with CCDs for the large format applications within a few years. Because the CCD imager market is much smaller than that for CMOS devices in general, it may be difficult for CCD to remain competitive in the long-term. For the last 25 years array size has been increasing at an exponential rate, following a Moore's Law grow path (see insert of Figure 8.36), with the number of pixels doubling every 19 months.

Because the efforts to develop monolithic FPAs using narrow-gap semiconductors failed in the 1980s, at present are fabricated mostly hybrid FPAs using these material systems. Hybrid FPAs detectors and multiplexers are fabricated on different substrates and mated with each other by flip-chip bonding or loophole interconnection (see Figure 8.37). In this case, we can optimize the detector material and multiplexer independently. Other advantages of the hybrid FPAs are near 100% fill factor and increased signal-processing area on the multiplexer chip. Indium bump bonding of readout electronics, first demonstrated in the mid-1970s, provides for multiplexing the signals from millions of pixels onto a few output lines, greatly simplifying the interface between the sensor and the system electronics.

For IR FPAs, the relevant figure of merit is the noise equivalent temperature difference (*NEDT*). Noise equivalent difference temperature of a detector represents the temperature change, for incident radiation, that gives an output signal equal to the rms noise level. *NEDT* is defined

$$NEDT = \frac{V_n(\partial T/\partial Q)}{(\partial V_s/\partial Q)} = V_n \frac{\Delta T}{\Delta V_s} \tag{8.35}$$

where
V_n is the rms noise
ΔV_s is the signal measured for the temperature difference ΔT

It can be approximated that

$$NEDT = \left(C\eta\sqrt{N_w}\right)^{-1} \tag{8.36}$$

where
C is the thermal contrast
N_w is the number of photogenerated carriers integrated for one integration time, t_{int}

$$N_w = \eta A t_{int} \Phi_B \tag{8.37}$$

where Φ_B is the photon flux density incident on detector area A.

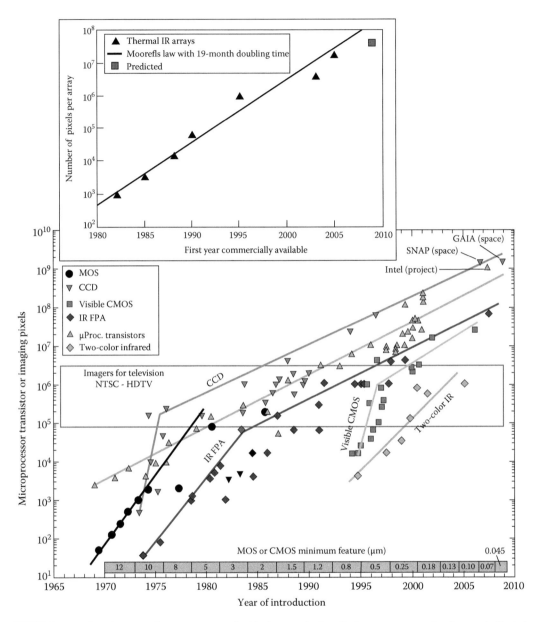

FIGURE 8.36 Imaging array formats compared with the complexity of microprocessor technology as indicated by transistor count. The timeline design rule of MOS/CMOS features is shown at the bottom. (After Norton, P., Detector focal plane array technology, in R. Driggers (ed.), *Encyclopedia of Optical Engineering*, Marcel Dekker Inc., New York, 2003, pp. 320–348. With permission.)

8.5.1 Narrow-Gap Semiconductor Focal Plane Arrays

Although efforts have been made to develop monolithic structures using a variety of IR detector materials (including narrow-gap semiconductors) over the past 30 years, only a few have matured to a level of practical use. These included PtSi, and more recently PbS, PbTe, and uncooled silicon microbolometers. Other IR material systems (InGaAs, InSb, HgCdTe, GaAs/AlGaAs quantum-well IR photodetector [QWIP], and extrinsic silicon) are used in hybrid configurations. The arrays of photodiodes are

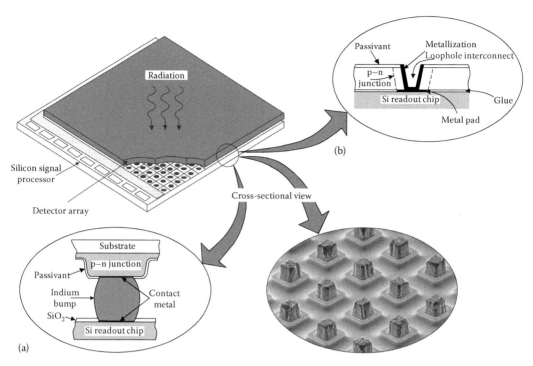

FIGURE 8.37 Hybrid FPA with independently optimized signal detection and readout: (a) indium bump techniques, (b) loophole technique.

hybridized to CMOS ROICs and then integrated into cameras for video-rate output to a monitor or for use with a computer for quantitative measurements and machine vision. Table 8.8 contains a description of representative narrow-gap semiconductor IR FPAs that are commercially available as standard products and/or catalogue items from the major manufacturers.

At present, InGaAs FPAs are fabricated by several manufacturers including Goodrich Corporation, Indigo Systems, Teledyne Judson Technologies, Aerius Photonics, and XenICs. The largest and finest pitched imager in $In_{0.53}Ga_{0.47}As$ material system has been demonstrated recently. The arrays with more than one million detector elements (1280×1024 and 1024×1024 formats) for low background applications have been developed by Raytheon Vision Systems with detectors provided by Goodrich [145]. The detector elements are as small as 15 μm pitch [146].

InSb arrays can be scaled to large formats with small pixel sizes because the InSb detectors are thinned after hybridization and are fabricated using a planar, ion-implanted process. Thinning the bulk InSb to ≤10 μm is necessary for operation with backside illumination (to obtain sufficient quantum efficiencies and minimize cross talk), but it also forces the InSb to match the thermal expansion of the Si ROIC, thereby providing a reliable hybrid structure with respect to repeated thermal cycling.

Large staring InSb focal plane evolution has been driven by astronomy applications. While this is somewhat surprising given the comparative budget sizes of the defense market and the astronomical community, astronomers have funded large focal plane development to dramatically improve telescope throughput. A challenge for large focal planes is maintaining optical focus, and in consequence maintaining the flatness of the detector surface over a large area. Many of the packaging concepts are shared with the three-side buttable $2k \times 2k$ FPA InSb modules developed by RVS for the James Webb Space Telescope (JWST) mission [147]. Four ORION arrays (see Figure 8.38) were deployed as a 4096×4096 focal plane in the NOAO near-IR camera, currently in operation at the Mayall 4 m telescope on Kit Peak [148].

TABLE 8.8   Representative Narrow-Gap Semiconductor IR FPAs Offered by Some Major Manufactures

Manufacturer/ Web Site	Size/ Architecture	Pixel Size (µm)	Detector Material	Spectral Range (µm)	Oper. Temp. (K)	$D^*(\lambda_p)$ (cm Hz$^{1/2}$/W)/ NETD (mK)
Goodrich	320 × 240/H	25 × 25	InGaAs	0.9–1.7	300	1×10^{13}
Corporation/www. sensorsinc.com	640 × 512/H	25 × 25	InGaAs	0.4–1.7	300	$>6 \times 10^{12}$
Raytheon/www. raytheon.com	1024 × 1024/H	30 × 30	InSb	0.6–5.0	50	
	2048 × 2048/H (Orion II)	25 × 25	HgCdTe	0.6–5.0	32	
	2048 × 2048/H (Virgo-2k)	20 × 20	HgCdTe	0.8–2.5	4–10	
Teledyne	2048 × 2048/H	18 × 18	HgCdTe	1.0–1.7	140	
	2048 × 2048/H	18 × 18	HgCdTe	1.0–2.5	77	
	2048 × 2048/H	18 × 18	HgCdTe	1.0–5.4	40	
Sofradir/infrared. sofradir.com	1280 × 1024/H (Jupiter)	15 × 15	HgCdTe	3.7–4.8	77–110	18
	384 × 288/H (Venus)	25 × 25	HgCdTe	7.7–9.5	77–80	17
Selex	1024 × 768/H (Merlin)	16 × 16	HgCdTe	3–5	up to 140	15
	640 × 512/H (Eagle)	24 × 24	HgCdTe	8–10	up to 90	24
SCD/www.scd.co.il	1280 1024/H	15 × 15	InSb	3–5	77	20
Northrop Grumman	320 × 240/M	30 × 30	PbS	1–4	77	8×10^{10} (ambient) 3×10^{11} (220 K)

H, hybrid; M, monolithic.

FIGURE 8.38   Demonstration of the two-sided buttable ORION modules to create a $4k \times 4k$ focal plane. One module contains an InSb SCA while the others have bare readouts. (After Hoffman, A.W. et al., *Proc. SPIE*, 5499, 59, 2004. With permission from International Society for Optical Engineering.)

The main mode of operation of HgCdTe detectors used in FPAs is photovoltaic effect. Spectral cutoff can be tailored by adjusting the HgCdTe alloy composition. Most applications are concentrated in the SWIR (1–3 µm), MWIR (3–5 µm), and LWIR (8–12 µm). Also development work on improving performance of very LWIR (VLWIR) photodiodes in the 13–18 µm region for important earth-monitoring applications are undertaken.

(a)   (b)

FIGURE 8.39 Large HgCdTe FPAs: (a) a mosaic of four Hawaii-2RGs as is being used for astronomy observations (From Beletic, J.W. et al., *Proc. SPIE*, 7021, 70210H, 2008. With permission from International Society for Optical Engineering) and (b) 16 2048×2048 HgCdTe arrays assembled for the VISTA telescope. (From Hoffman, A., *Laser Focus World*, pp. 81–84, February 2006. With permission.)

The best results have been obtained using hybrid architecture. Higher density detector configuration leads to higher image resolution as well as greater system sensitivity. HgCdTe IR FPAs have been made in linear (240, 288, 480, 960, and 1024), 2D scanning with TDI (with common formats of 256×4, 288×4, 480×6), and various 2D staring formats with size from 64×64 up to 4096×4096 pixels. Efforts are also underway to develop APD capabilities in the 1.6 µm and at longer wavelength region. Pixel sizes ranging from 15 µm square to over 1 mm have been demonstrated. While the size of individual arrays continues to grow, the very large FPAs required for many space missions by mosaicking a large number of individual arrays. An example of a large mosaic developed by Teledyne Imaging Sensors, is a 147 megapixel FPA that is comprised of 35 arrays, each with 2048×2048 pixels. This is currently the world's largest IR focal plane [149]. Figure 8.39 shows an example of large HgCdTe FPAs.

8.5.2 Third-Generation Infrared Detectors

In the IR spectral region, third-generation systems are now being developed. Third-generation IR systems considered to be those that provide enhanced capabilities like larger number of pixels, higher frame rates, better thermal resolution as well as multicolor functionality and other on-chip functions. Multicolor capabilities are highly desirable for advance IR systems. Systems that gather data in separate IR spectral bands can discriminate both absolute temperature and unique signatures of objects in the scene. By providing this new dimension of contrast, multiband detection also offers advanced color processing algorithms to further improve sensitivity compared to that of single-color devices.

In the IR regions of interest such as SWIR, MWIR, and LWIR, four detector technologies present multicolor capability: HgCdTe photodiodes, QWIPs, antimonide based type II SLS photodiodes, and quantum dot IR photodetectors (QDIPs).

The unit cell of integrated multicolor FPAs consists of several co-located detectors, each sensitive to a different spectral band (see Figure 8.40). Radiation is incident on the shorter band detector, with the longer wave radiation passing through to the next detector. Each layer absorbs radiation up to its cutoff, and hence transparent to the longer wavelengths, which are then collected in subsequent layers. In the case of HgCdTe, this device architecture is realized by placing a longer wavelength HgCdTe photodiode optically behind a shorter wavelength photodiode.

Back-to-back photodiode two-color detectors were first implemented using quaternary III–V alloy ($Ga_xIn_{1-x}As_yP_{1-y}$) absorbing layers in a lattice matched InP structure sensitive to two different SWIR bands [151]. A variation on the original back-to-back concept was implemented using HgCdTe at Rockwell and Santa Barbara Research Center [152]. Following the successful demonstration of multispectral detectors

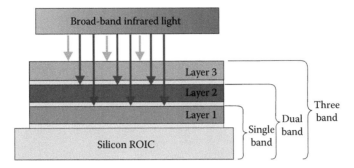

FIGURE 8.40 Structure of a three-color detector pixel. IR flux from the first band is absorbed in Layer 3, while longer wavelength flux is transmitted through the next layers. The thin barriers separate the absorbing bands.

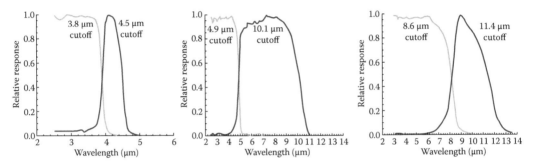

FIGURE 8.41 Spectral response curves for two-color HgCdTe detectors in various dual-band combinations of MWIR and LWIR spectral bands. (After Norton, P.R., *Proc. SPIE*, 3379, 102, 1998. With permission from International Society for Optical Engineering.)

in LPE-grown HgCdTe devices, the MBE and MOCVD techniques have been used for the growth of a variety of multispectral detectors at Raytheon, BAE Systems, Leti, Selex, QinetiQ, DRS, Teledyne, and NVESD. For more than a decade, steady progression has been made in a wide variety of pixel sizes (to as small as $20\,\mu m$), array formats (up to 1280×720) and spectral-band sensitivity (MWIR/MWIR, MWIR/LWIR, and LWIR/LWIR).

Figure 8.41 illustrates examples of the spectral response from different two-color devices. Note that there is minimal cross talk between the bands, since the short-wavelength detector absorbs nearly 100% of the shorter wavelengths. Test structures indicate that the separate photodiodes in a two-color detector perform exactly like single-color detectors in terms of achievable R_0A variation with wavelength at a given temperature.

Raytheon Vision Systems has developed two-color, large-format IR FPAs to support the US Army's third generation FLIR systems in both 640×480 and "high-definition" 1280×720 formats with $20 \times 20\,\mu m$ unit cells. The ROICs share a common chip architecture and incorporate identical unit cell circuit designs and layouts; both FPAs can operate in either dual-band or single-band modes. High-quality MWIR/LWIR 1280×720 FPAs with cutoffs ranging out to $11\,\mu m$ at 78K have demonstrated excellent sensitivity and pixel operabilities exceeding 99.9% in the MW band and greater than 98% in the LW band. Median 300 K *NEDT* values at f/3.5 of approximately $20\,mK$ for the MW and $25\,mK$ for the LW have been measured for dual-band TDMI operation at 60 Hz frame rate with integration times corresponding to roughly 40% (MW) and 60% (LW) of full well charge capacities. As shown in Figure 8.42 [154], excellent high resolution IR camera imaging with *f*/2.8 FOV broadband refractive optics at 60 Hz frame rate has been achieved.

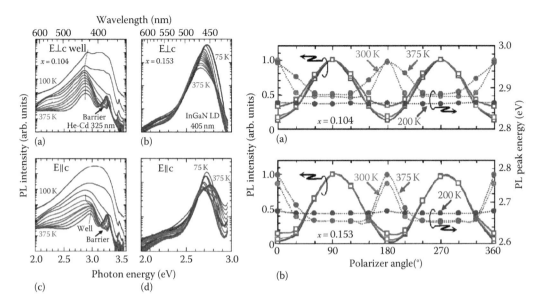

FIGURE 3.17 Polarized PL spectra of In$_{0.10}$Ga$_{0.90}$N QWs for (a) E⊥c and (b) E∥c, and those of In$_{0.15}$Ga$_{0.85}$N QWs for (c) E⊥c and (d) E∥c as functions of T (left). Normalized PL intensity and PL peak energy as functions of the polarizer angle at various temperatures for InN molar fractions x of (a) 0.104 and (b) 0.153 (right). (With permission from Kubota, M., Okamoto, K., Tanaka, T., and Ohta, H., *Appl. Phys. Lett.*, 92, 011920, 2008. Copyright 2008, American Institute of Physics.)

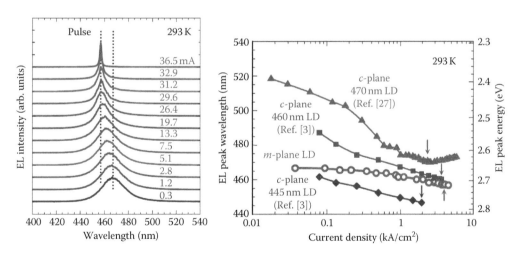

FIGURE 3.18 EL spectra (left) and EL peak wavelength of *m*-plane LD (open circles) as a function of current density (right). Data for *c*-plane LDs with lasing wavelengths of 445 nm (closed diamonds), 460 nm (closed squares), and 470 nm (closed triangles) are also shown for comparison. The vertical arrows indicate the threshold current density for each LD. The references in the right figure can be found in the original paper.) (From Kubota, M., Okamoto, K., Tanaka, T., and Ohta, H. *Appl. Phys. Express*, 1, 011102, 2008. With permission from Japan Society of Applied Physics.)

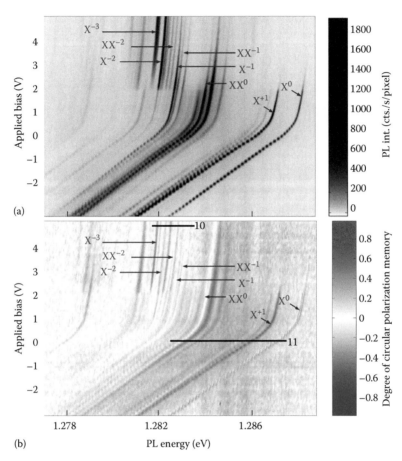

FIGURE 12.9 Bias-dependent PL spectra (a) and DCPM (b) from a single QD excited at 1.369 eV. The black horizontal lines marked 10 and 11 indicate the bias and spectral ranges from which Figures 12.10 and 12.11 were obtained. (From Poem, E. et al., *Solid State Commun.*, 149, 1493, 2009. With permission.)

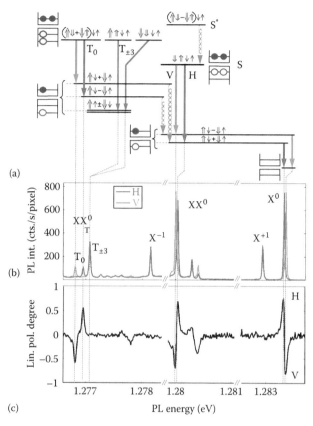

(a)

(b)

(c)

FIGURE 12.18 (a) Energy level diagram for excitons and biexcitons in a neutral QD. Single-carrier level occupations are given alongside each many-carrier level. The spin wavefunctions are depicted above each level. The symbol ↑ (⇓) represents spin up (down) electron (hole). Short (long) symbols represent charge carriers in the first (second) energy level. S (S*) indicates the ground (excited) biexciton hole-singlet state. T_0 ($T_{\pm3}$) indicates the metastable spin-triplet biexciton state with z-axis spin projection of 0 (±3). The solid (curly) vertical arrows indicate spin preserving (non-) radiative transitions. Dark- (light-) gray arrows represent photon emission in horizontal, H (vertical, V) polarization. (b) Polarized PL spectra. H (V) in dark (light) gray. Spectral lines that are relevant to this work are marked and linked to the transitions in (a) by dashed lines. (c) Linear polarization spectrum. The value 1 (−1) means full H (V) polarization.

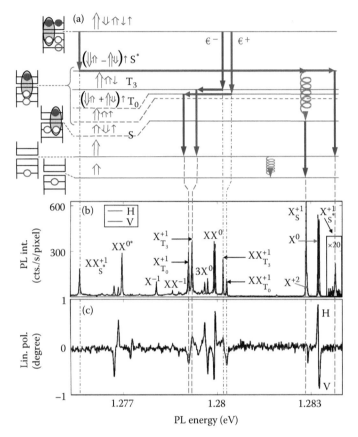

FIGURE 12.22 (a) Schematic description of the energy levels of a singly positively charged QD. Vertical (curly) arrows indicate radiative (non-radiative) transitions between these levels. State occupation and spin wavefunctions are described to the left of each level where ↑ (⇓) represents an electron (hole) with spin up (down). A short blue (long red) arrow represents a carrier in its first (second) level. S (T) stands for two holes' singlet (triplet) state and 0 (3) for $S_z = 0$ ($S_z = \pm 3$) total holes' pseudo-spin projection on the QD growth direction. The excited state singlet is indicated by S*. Only one out of two (Kramers) degenerate states is described. (b) Measured PL spectrum on which the actual transitions are identified. Transitions that are not discussed here are marked by gray letters. (c) Measured degree of linear polarization spectrum, along the in-plane symmetry axes of the QD. Positive (negative) value represents polarization along the QD's major (minor) axis. (Reprinted with permission from Poem, E., Kodriano, Y., Tradonsky, C., Gerardot, B.D., Petroff, P.M., and Gershoni, D., Radiative cascades from charged semiconductor quantum dots, *Phys. Rev. B*, 81, 085306, 2010. Copyright 2010 by the American Physical Society.)

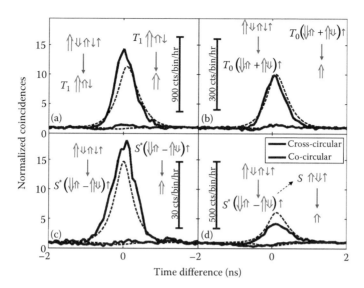

FIGURE 12.23 (See color insert.) Measured and calculated time-resolved, polarization-sensitive intensity correlation functions for the four radiative cascades described in Figure 12.22. The states involved in the first (second) photon emission are illustrated to the left (right) side of each panel. All symbols and labels are as in Figure 12.22. Solid blue (red) line stands for measured cross-(co-)circularly polarized photons. Dashed lines represent the corresponding calculated functions. The bar presents the acquisition rate in coincidences per time bin (80 ps) per hour. (Reprinted with permission from Poem, E., Kodriano, Y., Tradonsky, C., Gerardot, B.D., Petroff, P.M., and Gershoni, D., Radiative cascades from charged semiconductor quantum dots, *Phys. Rev. B*, 81, 085306, 2010. Copyright 2010 by the American Physical Society.)

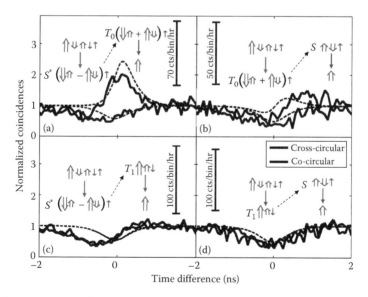

FIGURE 12.24 (See color insert.) Measured and calculated time-resolved, polarization-sensitive intensity correlation functions, across the radiative cascades. (a and c) Correlations between the singlet biexciton transition and the exciton transition from the T_0, (T_3) state. (b and d) Correlations between the T_0, (T_3) biexciton transition and the ground X^{+1} exciton transition. All symbols and labels are as in Figure 12.22. The meanings of all line types and colors are as in Figure 12.23. (Reprinted with permission from Poem, E., Kodriano, Y., Tradonsky, C., Gerardot, B.D., Petroff, P.M., and Gershoni, D., Radiative cascades from charged semiconductor quantum dots, *Phys. Rev. B*, 81, 085306, 2010. Copyright 2010 by the American Physical Society.)

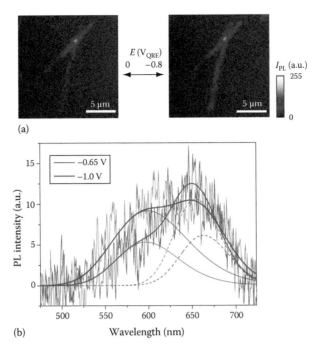

FIGURE 14.5 (a) PL images of single TiO$_2$ nanowires acquired without (left) and with (right) applied potential (*E*) of −0.8 V vs. Ag wire (= −0.37 V vs. NHE). Upon the applied potential of −0.8 V, a significant increase in the PL intensity was observed. (b) PL spectra observed at different times. The solid lines indicate the Gaussian distributions fitted to the spectra.

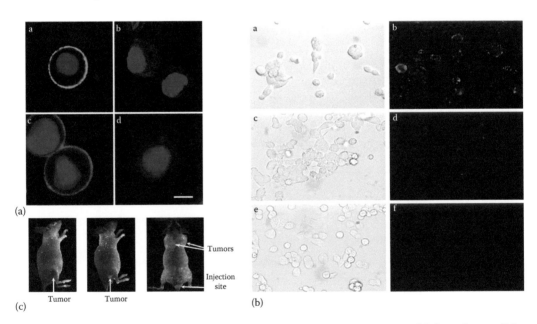

FIGURE 15.5 QDs in cellular and whole animal in vivo imaging. Immunofluorescent labeling of extracellular Her2 cancer marker with antibody-conjugated QDs is shown in (a). (Reprinted by permission from MacMillan Publishers Ltd. *Nat. Biotechnol.*, Wu, X., Liu, H. et al., Immunofluorescent labeling of cancer marker Her2 and other cellular targets with semiconductor quantum dots, 21(1), 41–46, Copyright 2003.) Visualization of hSERT in HEK-293T cells with serotonin-conjugated CdSe/ZnS core/shell nanocrystals is shown in (b). (Reprinted with permission from Rosenthal, S.J., Tomlinson, I. et al., Targeting cell surface receptors with ligand-conjugated nanocrystals, *J. Am. Chem. Soc.*, 124(17), 4586–4594. Copyright 2002 American Chemical Society.) The use of QDs to target cancer tumors in living mice is shown in (c). (Reprinted from by permission from MacMillan Publishers Ltd. *Nat. Biotechnol.*, Gao, X., Cui, Y. et al., In vivo cancer targeting and imaging with semiconductor quantum dots, *Nat Biotechnol.*, 22(8), 969–976, Copyright 2004.)

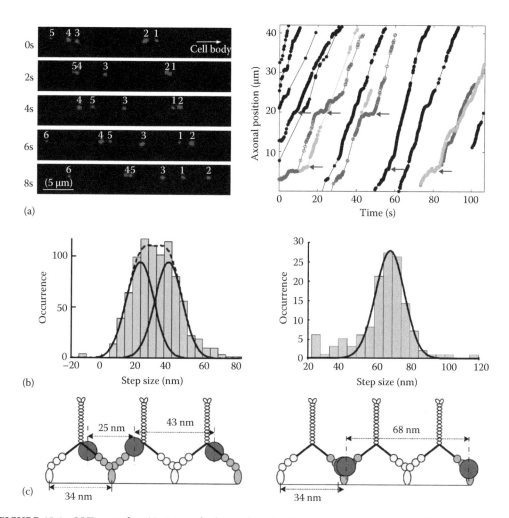

FIGURE 15.6 SQT examples. (a) Cui et al. observed unidirectional retrograde transport of endosomes, containing QDs conjugated to NGF dimer, along the neuronal axon. (Reprinted from Cui, B., Wu, C. et al., One at a time, live tracking of NGF axonal transport using quantum dots, *Proc. Natl. Acad. Sci.*, 104(34), 13666–13671, 2007. Copyright 2007 National Academy of Sciences, U.S.A.) (b) Sun et al. visualized the movement of biotinylated myosin X along actin filaments and bundles with QD–streptavidin conjugates. (Reprinted by permission from MacMillan Publishers Ltd. *Nat. Struct. Mol. Biol.*, Sun, Y., Sato, O. et al., Single-molecule stepping and structural dynamics of myosin X, *Nat. Struct. Mol. Biol.*, 17(4), 485–491, Copyright 2010.)

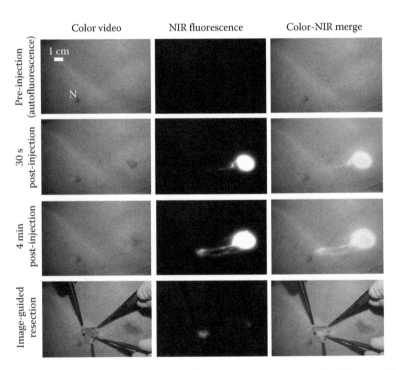

FIGURE 15.9 SLN mapping surgical procedure aided by NIR-emitting type II QDs. (Reprinted by permission from MacMillan Publishers Ltd. *Nat. Biotechnol.*, Kim, S., Lim, Y.T. et al., Near-infrared fluorescent type II quantum dots for sentinel lymph node mapping, 22(1), 93–97, Copyright 2004.)

(a)   (b)

FIGURE 8.42 Still camera image taken at 78 K with $f/2.8$ FOV and 60 Hz frame rate using two-color 20 μm unit-cell MWIR/LWIR HgCdTe/CdZnTe TLHJ 1280×720 FPA hybridized to a 1280×720 TDMI ROIC. (After King, D.F. et al., *Proc. SPIE*, 6206, 62060W, 2006. With permission from International Society for Optical Engineering.)

8.6 Conclusions

The future applications of IR detector systems require

- Higher pixel sensitivity
- Further increase in pixel density
- Cost reduction in IR-imaging array systems due to less cooling sensor technology combined with integration of detectors and signal processing functions (with much more on-chip signal processing)
- Photon counting in the wide spectral range (e.g., development of e-APD and h-APD HgCdTe photodiodes)
- Improvement in functionality of IR-imaging arrays through development of multispectral sensors

Array sizes will continue to increase but perhaps at a rate that falls below the Moore's Law curve. An increase in array size is already technically feasible. However, the market forces that have demanded larger arrays are not as strong now that the megapixel barrier has been broken.

There are many critical challenges for future civilian and military IR detector applications. For many systems, such as night-vision goggles, the IR image is viewed by the human eye, which can discern resolution improvements only up to about one megapixel, roughly the same resolution as high-definition television. Most high-volume applications can be completely satisfied with a format of 1280 × 1024. Although wide-area surveillance and astronomy applications could make use of larger formats, funding limits may prevent the exponential growth that was seen in past decades.

Despite serious competition from alternative technologies and slower progress than expected, narrow-gap semiconductors and especially HgCdTe ternary alloys, are unlikely to be seriously challenged for high-performance applications, applications requiring multispectral capability and fast response. It is predicted that HgCdTe technology will continue in the future to expand the envelope of its capabilities because of its excellent properties. Type II InAs/GaInSb superlattice structure is a relatively new alternative IR material system and has great potential for LWIR/VLWIR spectral ranges with performance comparable to HgCdTe with the same cutoff wavelength.

Third-generation IR imagers are beginning a challenging road to development. For multiband sensors, boosting the sensitivity in order maximize identification range is the primary objective. The goal for dual-band MW/LW IR FPAs are 1920 × 1080 pixels, which due to lower cost should be fabricated on silicon wafers. The challenges to attaining those specifications are material uniformity and defects, heterogeneous integration with silicon and ultra well capacity (on the order of a billion in the LWIR).

Abbreviations

APD	avalanche photodiode
BLIP	background-limited performance
CBD	chemical bath deposition
CCD	charge-coupled device
CMOS	complementary metal-oxide-semiconductor
EPD	etch-pit density
FOV	field of view
FPA	focal-plane array
HWE	hot-wall epitaxy
IC	integrated circuit
IR	infrared
LPE	liquid-phase epitaxy
LWIR	long wavelength infrared
MBE	molecular-beam epitaxy
MIS	metal-insulator-semiconductor
ML	monolayer
MOCVD	metalorganic chemical vapor deposition
MOSFET	metal-oxide-semiconductor field-effect transistor
MWIR	middle wavelength infrared
NEDT	noise equivalent difference temperature
NEP	noise equivalent power
PC	photoconductive
PEM	photoelectromagnetic
PV	photovoltaic
QDIP	quantum dot infrared photodetector
QWIP	quantum-well infrared photodetector
ROIC	readout integrated circuits
SL	superlattice
SLS	strained-layer superlattice
SPRITE	Signal PRocessing In The Element
SR	Shockley–Read
SWIR	short-wavelength infrared
TDI	time delay and integration
TEC	thermal expansion coefficient
VLWIR	very long wavelength infrared
VPE	vapor-phase epitaxy

References

1. W.D. Lawson, S. Nielson, E.H. Putley, and A.S. Young, Preparation and properties of HgTe and mixed crystals of HgTe-CdTe, *J. Phys. Chem. Solids* **9**, 325–329 (1959).
2. P.W. Kruse, M.D. Blue, J.H. Garfunkel, and W.D. Saur, Long wavelength photoeffects in mercury selenide, mercury telluride and mercury telluride-cadmium telluride, *Infrared Phys.* **2**, 53–60 (1962).
3. J. Melngailis and T.C. Harman, Single-crystal lead-tin chalcogenides, in *Semiconductors and Semimetals*, Vol. 5, pp. 111–174, R.K. Willardson and A.C. Beer (Eds.), Academic Press, New York (1970).

4. T.C. Harman and J. Melngailis, Narrow gap semiconductors, in *Applied Solid State Science*, Vol. 4, pp. 1–94, R. Wolfe (Ed.), Academic Press, New York, 1974.

5. A. Rogalski and J. Piotrowski, Intrinsic infrared detectors, *Prog. Quant. Electron.* **12**, 87–289 (1988).

6. J.N. Schulman and T.C. McGill, The CdTe/HgTe superlattice: Proposal for a new infrared material, *Appl. Phys. Lett.* **34**, 663–665 (1979).

7. G.C. Osbourn, InAsSb strained-layer superlattices for long wavelength detector applications, *J. Vac. Sci. Technol.* **B2**, 176–178 (1984).

8. D.L. Smith and C. Mailhiot, Proposal for strained type II superlattice infrared detectors, *J. Appl. Phys.* **62**, 2545–2548 (1987).

9. C. Mailhiot and D.L. Smith, Long-wavelength infrared detectors based on strained InAs–GaInSb type-II superlattices, *J. Vac. Sci. Technol.* **A7**, 445–449 (1989).

10. D. Long and J.L. Schmit, Mercury-cadmium telluride and closely related alloys, in *Semiconductors and Semimetals*, Vol. 5, pp. 175–255, R.K. Willardson and A.C. Beer (Eds.), Academic Press, New York, 1970.

11. P.W. Kruse, The emergence of $Hg_{1-x}Cd_xTe$ as a modern infrared sensitive material, in *Semiconductors and Semimetals*, Vol. 18, pp. 1–20, R.K. Willardson and A.C. Beer (Eds.), Academic Press, New York, 1981.

12. W.F.H. Micklethweite, The crystal growth of mercury cadmium telluride, in *Semiconductors and Semimetals*, Vol. 18, pp. 48–119, R.K. Willardson and A.C. Beer (Eds.), Academic Press, New York, 1981.

13. R. Dornhaus, G. Nimtz, and B. Schlicht, *Narrow-Gap Semiconductors*, Springer Verlag, Berlin, Germany, 1983.

14. P. Capper (Ed.), *Properties of Narrow Gap Cadmium-based Compounds*, Emis Datareviews Series No. 10, Inspec, IEE, London, U.K., 1994.

15. A. Rogalski (Ed.), *Infrared Photon Detectors*, SPIE Optical Engineering Press, Bellingham, WA, 1995.

16. P. Capper (Ed.), *Narrow-Gap II–VI Compounds for Optoelectronic and Electromagnetic Applications*, Chapman & Hall, London, U.K., 1997.

17. A. Rogalski, K. Adamiec, and J. Rutkowski, *Narrow-Gap Semiconductor Photodiodes*, SPIE Optical Engineering Press, Bellingham, WA, 2000.

18. J. Chu and A. Sher, *Device Physics of Narrow Gap Semiconductors*, Springer, New York, 2009.

19. K.F. Hulme and J.B. Mullin, Indium antimonide—A review of its preparation, properties and device applications, *Solid-State Electronics* **5**, 211–247 (1962).

20. K.F. Hulme, Indium antimonide, in *Materials Used in Semiconductor Devices*, pp. 115–162, C.A. Hogarth (Ed.), Wiley-Interscience, New York, 1965.

21. S. Liang, Preparation of indium antimonide, in *Compound Semiconductors*, pp. 227–237, R.K. Willardson and H.L. Georing (Eds.), Reinhold, New York, 1966.

22. J.B. Mullin, Melt-growth of III–V compounds by the liquid encapsulation and horizontal growth techniques, in *III–V Semiconductor Materials and Devices*, pp. 1–72, R.J. Malik (Ed.), North Holland, Amsterdam, the Netherlands, 1989.

23. W.F.M. Micklethwaite and A.J. Johnson, InSb: Materials and devices, in *Infrared Detectors and Emitters: Materials and Devices*, pp. 178–204, P. Capper and C.T. Elliott (Eds.), Kluwer Academic Publishers, Boston, MA, 2001.

24. S. Kasap and P. Capper (Eds.), *Springer Handbook of Electronic and Photonic Materials*, Springer, Heidelberg, Germany, 2006.

25. S. Adachi, *Physical Properties of III–V Semiconducting Compounds: InP, InAs, GaAs, GaP, InGaAs, and InGaAsP*, Wiley-Interscience, New York, 1992; *Properties of Group-IV, III–V and II–VI Semiconductors*, John Wiley & Sons, Ltd, Chichester, U.K., 2005.

26. I. Vurgaftman, J.R. Meyer, and L.R. Ram-Mohan, Band parameters for III–V compound semiconductors and their alloys, *J. Appl. Phys.* **89**, 5815–5875 (2001).

27. P. Bhattacharya (Ed.), *Properties of Lattice-Matched and Strained Indium Gallium Arsenide*, IEE, London, U.K., 1993.

28. G.H. Olsen and V.S. Ban, InGaAsP: The next generation in photonics materials, *Solid State Technol.* **30**, 99–105 (February 1987).

29. A. Rogalski, *New Ternary Alloy Systems for Infrared Detectors*, SPIE Optical Engineering Press, Bellingham, WA, 1994.

30. M. Razeghi, Overview of antimonide based III–V semiconductor epitaxial layers and their applications at the center for quantum devices, *Eur. Phys. J. AP* **23**, 149–205 (2003).

31. G.B. Stringfellow and P.R. Greene, Liquid phase epitaxial growth of $InAs_{1-x}Sb_x$, *J. Electrochem. Soc.* **118**, 805–810 (1971).

32. Y.P. Varshni, Temperature dependence of the energy gap in semiconductors, *Physica* **34**, 149 (1967).

33. P.W. Kruse, Indium antimonide photoconductive and photoelectromagnetic detectors, in *Semiconductors and Semimetals*, Vol. 5, pp. 15–83, R.K. Willardson and A.C. Beer (Eds.), Academic Press, New York, 1970.

34. A. Rogalski, InAsSb infrared detectors, *Prog. Quant. Electron.* **13**, 191–231 (1989).

35. A. Joullie, F. Jia Hua, F. Karouta, H. Mani, and C. Alibert, III–V alloys based on GaSb for optical communications at 2.0–4.5 μm, *Proc. SPIE* **587**, 46–57 (1985).

36. M.B. Reine, Mercury cadmium telluride, in *Encyclopedia of Modern Optics*, pp. 392–402, R.D. Guenther, D.G. Steel, and L. Bayvel (Eds.), Elsevier, Oxford, U.K., 2004.

37. P. Norton, HgCdTe infrared detectors, *Opto-Electron. Rev.* **10**, 159–174 (2002).

38. J.M. Peterson, J.A. Franklin, M. Readdy, S.M. Johnson, E. Smith, W.A. Radford, and I. Kasai, High-quality large-area MBE HgCdTe/Si, *J. Electron. Mater.* **36**, 1283–1286 (2006).

39. R. Bornfreund, J.P. Rosbeck, Y.N. Thai, E.P. Smith, D.D. Lofgreen, M.F. Vilela, A.A. Buell et al., High-performance LWIR MBE-grown HgCdTe/Si focal plane arrays, *J. Electron. Mater.* **37**, 1085–1091 (2007).

40. G.L. Hansen, J.L. Schmit, and T.N. Casselman, Energy gap versus alloy composition and temperature in $Hg_{1-x}Cd_xTe$, *J. Appl. Phys.* **53**, 7099–7101 (1982).

41. G.L. Hansen and J.L. Schmit, Calculation of intrinsic carrier concentration in $Hg_{1-x}Cd_xTe$, *J. Appl. Phys.* **54**, 1639–1640 (1983).

42. M.H. Weiler, Magnetooptical properties of $Hg_{1-x}Cd_xTe$ alloys, in *Semiconductors and Semimetals*, Vol. 16, pp. 119–191, R.K. Willardson and A.C. Beer (Eds.), Academic Press, New York, 1981.

43. J.P. Rosbeck, R.E. Star, S.L. Price, and K.J. Riley, Background and temperature dependent current–voltage characteristics of $Hg_{1-x}Cd_xTe$ photodiodes, *J. Appl. Phys.* **53**, 6430–6440 (1982).

44. P.N.J. Dennis, C.T. Elliott, and C.L. Jones, A method for routine characterization of the hole concentration in p-type cadmium mercury telluride, *Infrared Phys.* **22**, 167–169 (1982).

45. S. Barton, P. Capper, C.L. Jones, N. Metcalfe, and N.T. Gordon, Electron mobility in p-type grown $Cd_xHg_{1-x}Te$, *Semicond. Sci. Technol.* **10**, 56–60 (1995).

46. E.O. Kane, Band structure of InSb, *J. Phys. Chem. Solids* **1**, 249–261 (1957).

47. W.W. Anderson, Absorption constant of $Pb_{1-x}Sn_xTe$ and $Hg_{1-x}Cd_xTe$ alloys, *Infrared Phys.* **20**, 363–372 (1980).

48. R. Beattie and A.M. White, An analytic approximation with a wide range of applicability for band-to-band radiative transition rates in direct, narrow-gap semiconductors, *Semicond. Sci. Technol.* **12**, 359–368 (1997).

49. E. Finkman and S.E. Schacham, The exponential optical absorption band tail of $Hg_{1-x}Cd_xTe$, *J. Appl. Phys.* **56**, 2896–2900 (1984).

50. S.E. Schacham and E. Finkman, Recombination mechanisms in p-type HgCdTe: Freezeout and background flux effects, *J. Appl. Phys.* **57**, 2001–2009 (1985).

51. C.A. Hougen, Model for infrared absorption and transmission of liquid-phase epitaxy HgCdTe, *J. Appl. Phys.* **66**, 3763–3766 (1989).

52. V.C. Lopes, A.J. Syllaios, and M.C. Chen, Minority carrier lifetime in mercury cadmium telluride, *Semicond. Sci. Technol.* **8**, 824–841 (1993).

53. S.M. Johnson, D.R. Rhiger, J.P. Rosbeck, J.M. Peterson, S.M. Taylor, and M.E. Boyd, Effect of dislocations on the electrical and optical properties of long-wavelength infrared HgCdTe photovoltaic detectors, *J. Vac. Sci. Technol.* **B10**, 1499–1506 (1992).

54. K. Jóźwikowski and A. Rogalski, Effect of dislocations on performance of LWIR HgCdTe photodiodes, *J. Electron. Mater.* **29**, 736–741 (2000).

55. M.C. Chen, L. Colombo, J.A. Dodge, and J.H. Tregilgas, The minority carrier lifetime in doped and undoped p-type $Hg_{0.78}Cd_{0.22}Te$ liquid phase epitaxy films, *J. Electron. Mater.* **24**, 539–544 (1995).

56. M.A. Kinch, Fundamental physics of infrared detector materials, *J. Electron. Mater.* **29**, 809–817 (2000).

57. M.A. Kinch, F. Aqariden, D. Chandra, P.-K. Liao, H.F. Schaake, and H.D. Shih, Minority carrier lifetime in p-HgCdTe, *J. Electron. Mater.* **34**, 880–884 (2005).

58. R.G. Humpreys, Radiative lifetime in semiconductors for infrared detectors, *Infrared Phys.* **23**, 171–175 (1983).

59. R.G. Humpreys, Radiative lifetime in semiconductors for infrared detectors, *Infrared Phys.* **26**, 337–342 (1986).

60. T. Elliott, N.T. Gordon, and A.M. White, Towards background-limited, room-temperature, infrared photon detectors in the $3-13\,\mu m$ wavelength range, *Appl. Phys. Lett.* **74**, 2881–2883 (1999).

61. N.T. Gordon, C.D. Maxey, C.L. Jones, R. Catchpole, and L. Hipwood, Suppression of radiatively generated currents in infrared detectors, *J. Appl. Phys.* **91**, 565–568 (2002).

62. C.T. Elliott and C.L. Jones, Non-equilibrium devices in HgCdTe, in *Narrow-Gap II–VI Compounds for Optoelectronic and Electromagnetic Applications*, pp. 474–485, P. Capper (Ed.), Chapman & Hall, London, U.K., 1997.

63. A.R. Beattie and P.T. Landsberg, Auger effect in semiconductors, *Proc. Roy. Soc.* **A249**, 16–29 (1959).

64. J.S. Blakemore, *Semiconductor Statistics*, Pergamon Press, Oxford, U.K., 1962.

65. M.A. Kinch, M.J. Brau, and A. Simmons, Recombination mechanisms in $8-14\,\mu m$ HgCdTe, *J. Appl. Phys.* **44**, 1649–1663 (1973).

66. T.N. Casselman and P.E. Petersen, A comparison of the dominant Auger transitions in p-type (Hg,Cd)Te, *Solid State Commun.* **33**, 615–619 (1980).

67. T.N. Casselman, Calculation of the Auger lifetime in p-type $Hg_{1-x}Cd_xTe$, *J. Appl. Phys* **52**, 848–854 (1981).

68. P.E. Petersen, Auger recombination in mercury cadmium telluride, in *Semiconductors and Semimetals*, Vol. 18, 121–155, R.K. Willardson and A.C. Beer (Eds.), Academic Press, New York, 1981.

69. A.M. White, The characteristics o minority-carrier exclusion in narrow direct gap semiconductors, *Infrared Phys.* **25**, 729–741 (1985).

70. S. Krishnamurthy and T.N. Casselman, A detailed calculation of the Auger lifetime in p-type HgCdTe, *J. Electron. Mater.* **29**, 828–831 (2000).

71. S. Krishnamurthy, M.A. Berding, Z.G. Yu, C.H. Swartz, T.H. Myers, D.D. Edwall, and R. DeWames, Model of minority carrier lifetimes in doped HgCdTe, *J. Electron. Mater.* **34**, 873–879 (2005).

72. R. Truboulet, Alternative small gap materials for IR detection, *Semicond. Sci. Technol.* **5**, 1073–1079 (1990).

73. A. Sher, A.B. Chen, W.E. Spicer, and C.K. Shih, Effects influencing the structural integrity of semiconductors and their alloys, *J. Vac. Sci. Technol. A* **3**, 105–111 (1985).

74. A. Wall, C. Caprile, A. Franciosi, R. Reifenberger, and U. Debska, New ternary semiconductors for infrared applications: $Hg_{1-x}Mn_xTe$, *J. Vac. Sci. Technol. A* **4**, 818–822 (1986).

75. K. Guergouri, R. Troboulet, A. Tromson-Carli, and Y. Marfaing, Solution hardening and dislocation density reduction in CdTe crystals by Zn addition, *J. Crystal Growth* **86**, 61–65 (1988).

76. P. Maheswaranathan, R.J. Sladek, and U. Debska, Elastic constants and their pressure dependences in $Cd_{1-x}Mn_xTe$ with $0 \leq x \leq 0.52$ and in $Cd_{0.52}Zn_{0.48}Te$, *Phys. Rev. B* **31**, 5212–5216 (1985).

77. A. Rogalski, $Hg_{1-x}Zn_xTe$ as a potential infrared detector material, *Prog. Quant. Electron.* **13**, 299–253 (1989).
78. A. Rogalski, $Hg_{1-x}Mn_xTe$ as a new infrared detector material, *Infrared Phys.* **31**, 117–166 (1991).
79. A. Rogalski, Hg-based alternatives to MCT, in *Infrared Detectors and Emitters: Materials and Devices*, pp. 377–400, P. Capper and C.T. Elliott (Eds.), Kluwer Academic Publishers, Boston, MA, 2001.
80. J. Piotrowski and A. Rogalski, *High-Operating Temperature Infrared Photodetectors*, SPIE Press, Bellingham, WA, 2007.
81. R. Dalven, A review of the semiconductor properties of PbTe, PbSe, PbS and PbO, *Infrared Phys.* **9**, 141–184 (1969).
82. Yu.I. Ravich, B.A. Efimova, and I.A. Smirnov, *Semiconducting Lead Chalcogenides*, Plenum Press, New York, 1970.
83. H. Maier and J. Hesse, Growth, properties and applications of narrow-gap semiconductors, in *Crystal Growth, Properties and Applications*, pp. 145–219, H.C. Freyhardt (Ed.), Springer Verlag, Berlin, Germany, 1980.
84. D.E. Bode, Lead salt detectors, *Physics of Thin Films*, Vol. 3, pp. 275–301, G. Hass and R.E. Thun (Eds.), Academic Press, New York, 1966.
85. T.S. Moss, G.J. Burrel, and B. Ellis, *Semiconductor Optoelectronics*, Butterworth, London, U.K., 1973.
86. R.H. Harris, Lead-salt detectors, *Laser Focus/Electro-Optics*, pp. 87–96 (December 1983).
87. T.H. Johnson, Lead salt detectors and arrays: PbS and PbSe, *Proc. SPIE* **443**, 60–94 (1984).
88. S.G. Parker and R.E. Johnson, Preparation and properties of (Pb,Sn)Te, in *Preparation and Properties of Solid State Material*, Vol. 6, pp. 1–65, W.R. Wilcox (Ed.), Marcel Dekker, Inc., New York, 1981.
89. D.L. Partin, Molecular-beam epitaxy of IV–VI compound heterojunctions and superlattices, in *Semiconductors and Semimetals*, Vol. 33, pp. 311–336, R.K. Willardson and A.C. Beer (Eds.), Academic Press, Boston, MA, 1991.
90. D. Genzow, A.G. Mironow, and O. Ziep, On the interband absorption in lead chalcogenides, *Phys. Stat. Sol. (b)* **90**, 535–542 (1978).
91. O. Ziep and D. Genzow, Calculation of the interband absorption in lead chalcogenides using a multiband model, *Phys. Stat. Sol. (b)* **96**, 359–368 (1979).
92. L. Bürkle and F. Fuchs, InAs/(GaIn)Sb superlattices: a promising material system for infrared detection, in *Handbook of Infrared Detection and Technologies*, pp. 159–189, M. Henini and M. Razeghi (Eds.), Elsevier, Oxford, U.K., 2002.
93. G.J. Brown, F. Szmulowicz, K. Mahalingam, S. Houston, Y. Wei, A. Gon, and M. Razeghi, Recent advances in InAs/GaSb superlattices for very long wavelength infrared detection, *Proc. SPIE* **4999**, 457–466 (2003).
94. C. Mailhiot, Far-infrared materials based on InAs/GaInSb type II, strained-layer superlattices, in *Semiconductor Quantum Wells and Superlattices for Long-Wavelength Infrared Detectors*, pp. 109–138, M.O. Manasreh (Ed.), Artech House, Boston, MA, 1993.
95. J.P. Omaggio, J.R. Meyer, R.J. Wagner, C.A. Hoffman, M.J. Yang, D.H. Chow, and R.H. Miles, Determination of band gap and effective masses in $InAs/Ga_{1-x}In_xSb$ superlattices, *Appl. Phys. Lett.* **61**, 207–209 (1992).
96. C.H. Grein, P.M. Young, M.E. Flatté, and H. Ehrenreich, Long wavelength InAs/InGaSb infrared detectors: Optimization of carrier lifetimes, *J. Appl. Phys.* **78**, 7143–7152 (1995).
97. E.R. Youngdale, J.R. Meyer, C.A. Hoffman, F.J. Bartoli, C.H. Grein, P.M. Young, H. Ehrenreich, R.H. Miles, and D.H. Chow, Auger lifetime enhancement in $InAs-Ga_{1-x}In_xSb$ superlattices, *Appl. Phys. Lett.* **64**, 3160–3162 (1994).
98. O.K. Yang, C. Pfahler, J. Schmitz, W. Pletschen, and F. Fuchs, Trap centers and minority carrier lifetimes in InAs/GaInSb superlattice long wavelength photodetectors, *Proc. SPIE* **4999**, 448–456 (2003).
99. J. Pellegrini and R. DeWames, Minority carrier lifetime characteristics in type II InAs/GaSb LWIR superlattice n+πp+ photodiodes, *Proc. SPIE* **7298**, 7298-67 (2009).

100. M.Z. Tidrow, L. Zheng, and H. Barcikowski, Recent success on SLS FPAs and MDA's new direction for development, *Proc. SPIE* **7298**, 7298-61 (2009).

101. A. Rogalski, *Infrared Detectors*, 2nd edn., Taylor & Francis, Boca Raton, FL, 2010.

102. T.W. Case, Notes on the change of resistance of certain substrates in light, *Phys. Rev.* **9**, 305–310 (1917); The thalofide cell—A new photoelectric substance, *Phys. Rev.* **15**, 289 (1920).

103. E.W. Kutzscher, Letter to the editor, *Electro-Opt. Syst. Des.* **5**, 62 (June 1973).

104. R.J. Cashman, Film-type infrared photoconductors, *Proc. IRE* **47**, 1471–1475 (1959).

105. J.N. Humphrey, Optimization of lead sulfide infrared detectors under diverse operating conditions, *Appl. Opt.* **4**, 665–675 (1965).

106. S. Espevik, C. Wu, and R.H. Bube, Mechanism of photoconductivity in chemically deposited lead sulfide layers, *J. Appl. Phys.* **42**, 3513–3529 (1971).

107. C. Nascu, V. Vomir, I. Pop, V. Ionescu, and R. Grecu, The study of PbS films: Influence of oxidants on the chemically deposited PbS thin films, *Mater. Sci. Eng.* **B41**, 235 (1996).

108. I. Grozdanov, M. Najdoski, and S.K. Dey, A simple solution growth technique for PbSe thin films, *Mater. Lett.* **38**, 28 (1999).

109. S. Horn, D. Lohrmann, P. Norton, K. McCormack, and A. Hutchinson, Reaching for the sensitivity limits of uncooled and minimally-cooled thermal and photon infrared detectors, *Proc. SPIE* **5783**, 401–411 (2005).

110. E.L. Dereniak and G.D. Boreman, *Infrared Detectors and Systems*, Wiley, New York, 1996.

111. C.T. Elliott and N.T. Gordon, Infrared detectors, in *Handbook on Semiconductors*, Vol. 4, pp. 841–936, C. Hilsum (Ed.), North-Holland, Amsterdam, the Netherlands, 1993.

112. http://www.judsontechnologies.com/files/pdf/MCT_shortform_Dec2002.pdf

113. A. Blackburn, M.V. Blackman, D.E. Charlton, W.A.E. Dunn, M.D. Jenner, K.J. Oliver, and J.T.M. Wotherspoon, The practical realization and performance of SPRITE detectors, *Infrared Phys.* **22**, 57–64 (1982).

114. C.T. Elliott, Infrared detectors with integrated signal processing, in *Solid State Devices*, pp. 175–201, A. Goetzberger and M. Zerbst (Eds.), Verlag Chemie, Weinheim, Germany, 1983.

115. L.J. Kozlowski, K. Vural, J.M. Arias, W.E. Tennant, and R.E. DeWames, Performance of HgCdTe, InGaAs and quantum well GaAs/AlGaAs staring infrared focal plane arrays, *Proc. SPIE* **3182**, 2–13 (1997).

116. J.T. Wimmers and D.S. Smith, Characteristics of InSb photovoltaic detectors at 77 K and below, *Proc. SPIE* **364**, 123–131 (1983).

117. J.T. Wimmers, R.M. Davis, C.A. Niblack, and D.S. Smith, Indium antimonide detector technology at Cincinnati Electronics Corporation, *Proc. SPIE* **930**, 125–138 (1988).

118. M. Zandian, J.D. Garnett, R.E. DeWames, M. Carmody, J.G. Pasko, M. Farris, C.A. Cabelli, D.E. Cooper, G. Hildebrandt, J. Chow, J.M. Arias, K. Vural, and D.N.B. Hall, Mid-wavelength infrared p-on-on $Hg_{1-x}Cd_xTe$ heterostructure detectors: 30–120 Kelvin state-of-the-art performance, *J. Electron. Mater.* **32**, 803–809 (2003).

119. M.B. Reine, Photovoltaic detectors in MCT, in *Infrared Detectors and Emitters: Materials and Devices*, pp. 279–312, P. Capper and C.T. Elliott (Eds.), Kluwer Academic Publishers, Boston, MA, 2000.

120. O. Gravrand, L. Mollard, C. Largeron, N. Baier, E. Deborniol, and Ph. Chorier, Study of LWIR and VLWIR focal plane array developments: Comparison between p-on-n and different n-on-p technologies on LPE HgCdTe, *J. Electron. Mater.* **38**, 1733–1740 (2009).

121. W.E. Tennant, D. Lee, M. Zandian, E. Piquette, and M. Carmody, MBE HgCdTe technology: A very general solution to IR detection, described by "Rule 07", a very convenient heuristic, *J. Electron. Mater.* **37**, 1407–1410 (2008).

122. J.A. Stobie, S.P. Tobin, P. Norton, M. Hutchins, K.-K. Wong, R.J. Huppi, and R. Huppi, Update on the imaging sensor for GIFTS, *Proc. SPIE* **5543**, 293–303 (2004).

123. T. Chuh, Recent developments in infrared and visible imaging for astronomy, defense and homeland security, *Proc. SPIE* **5563**, 19–34 (2004).

124. A.S. Gilmore, J. Bangs, A. Gerrish, A. Stevens, and B. Starr, Advancements in HgCdTe VLWIR materials, *Proc. SPIE* **5783**, 223–230 (2005).

125. C.L. Jones, L.G. Hipwood, C.J. Shaw, J.P. Price, R.A. Catchpole, M. Ordish, C.D. Maxey et al., High performance MW and LW IRFPAs made from HgCdTe grown by MOVPE, *Proc. SPIE* **6206**, 620610 (2006).

126. C. Verie, F. Raymond, J. Besson and T. Nquyen Duy, Bandgap spin-orbit splitting resonance effects in $Hg_{1-x}Cd_xTe$ alloys, *J. Cryst. Growth* **59**, 342–346 (1982).

127. J. Vallegra, J. McPhate, L. Dawson, and M. Stapelbroeck, Mid-IR counting array using HgCdTe APDs and the Medipix 2 ROIC, *Proc. SPIE* **6660**, 66600O (2007).

128. D.N.B. Hall, B.S. Rauscher, J.L. Pipher, K.W. Hodapp, and G. Luppino, HgCdTe optical and infrared focal plane array development in the next decade, *Astro2010: The Astronomy and Astrophysics Decadal Survey*, Technology Development Papers, no. 28.

129. G. Leveque, M. Nasser, D. Bertho, B. Orsal, and R. Alabedra, Ionization energies in $Cd_xHg_{1-x}Te$ avalanche photodiodes, *Semicond. Sci. Technol.* **8**, 1317–1323 (1993).

130. M.A. Kinch, *Fundamentals of Infrared Detector Materials*, SPIE Press, Bellingham, WA, 2007.

131. A. Rogalski, IV–VI detectors, in *Infrared Photon Detectors*, pp. 513–559, A. Rogalski (Ed.), SPIE Press, Bellingham, WA, 1995.

132. H. Holloway, Thin-film IV–VI semiconductor photodiodes, in *Physics of Thin Films*, Vol. 11, pp. 105–203, G. Haas, M.H. Francombe, and P.W. Hoffman (Eds.), Academic Press, New York, 1980.

133. H. Zogg and A. Ishida, IV–VI (lead chalcogenide) infrared sensors and lasers, in *Infrared Detectors and Emitters: Materials and Devices*, pp. 43–75, P. Capper and C.T. Elliott (Eds.), Kluwer, Boston, MA, 2001.

134. M.D. Jhabvala and J.R. Barrett, A monolithic lead sulfide-silicon MOS integrated-circuit structure, *IEEE Trans. Electron. Dev.* **ED-29**, 1900–1905 (1982).

135. J.R. Barrett, M.D. Jhabvala, and F.S. Maldari, Monolithic lead salt-silicon focal plane development, *Proc. SPIE* **409**, 76–88 (1988).

136. K. Alchalabi, D. Zimin, H. Zogg, and W. Buttler, Monolithic heteroepitaxial PbTe-on-Si infrared focal plane array with 96×128 pixels, *IEEE Electron. Dev. Lett.* **22**, 110–112 (2001).

137. H. Zogg, K. Alchalabi, D. Zimin, and K. Kellermann, Two-dimensional monolithic lead chalcogenide infrared sensor arrays on silicon read-out chips and noise mechanisms, *IEEE Trans. Electron. Dev.* **50**, 209–214 (2003).

138. H. Zogg, S. Blunier, T. Hoshino, C. Maissen, J. Masek, and A.N. Tiwari, Infrared sensor arrays with $3–12\,\mu m$ cutoff wavelengths in heteroepitaxial narrow-gap semiconductors on silicon substrates, *IEEE Trans. Electron. Dev.* **38**, 1110–1117 (1991).

139. A.J. Steckl, H. Elabd, K.Y. Tam, S.P. Sheu, and M.E. Motamedi, The optical and detector properties of the PbS–Si heterojunction, *IEEE Trans. Electron. Dev.* **ED-27**, 126–133 (1980).

140. E.H. Aifer, J.G. Tischler, J.H. Warner, I. Vurgaftman, W.W. Bewley, J.R. Meyer, C.L. Canedy, and E.M. Jackson, W-Structured type-II superlattice long-wave infrared photodiodes with high quantum efficiency, *Appl. Phys. Lett.* **89**, 053510 (2006).

141. B.-M. Nguyen, D. Hoffman, P-Y. Delaunay, and M. Razeghi, Dark current suppression in type II InAs/GaSb superlattice long wavelength infrared photodiodes with M-structure barrier, *Appl. Phys. Lett.* **91**, 163511 (2007).

142. E.K. Huang, D. Hoffman, B.-M. Nguyen, P.-Y. Delaunay, and M. Razeghi, Surface leakage reduction in narrow band gap type-II antimonide-based superlattice photodiodes, *Appl. Phys. Lett.* **94**, 053506 (2009).

143. C.L. Canedy, H. Aifer, I. Vurgaftman, J.G. Tischler, J.R. Meyer, J.H. Warner, and E.M. Jackson, Antimonide type-II W photodiodes with long-wave infrared R_oA comparable to HgCdTe, *J. Electron. Mater.* **36**, 852–856 (2007).

144. P. Norton, Detector focal plane array technology, in *Encyclopedia of Optical Engineering*, pp. 320–348, R. Driggers (Ed.), Marcel Dekker Inc., New York, 2003.

145. A. Hoffman, T. Sessler, J. Rosbeck, D. Acton, and M. Ettenberg, Megapixel InGaAs arrays for low background applications, *Proc. SPIE* **5783**, 32–38 (2005).

146. M.D. Enriguez, M.A. Blessinger, J.V. Groppe, T.M. Sudol, J. Battaglia, J. Passe, M. Stern, and B.M. Onat, Performance of high resolution visible-InGaAs imager for day/night vision, *Proc. SPIE* **6940**, 69400O (2008).

147. A.W. Hoffman, E. Corrales, P.J. Love, and J. Rosbeck, M. Merrill, A. Fowler, and C. McMurtry, 2K × 2K InSb for astronomy, *Proc. SPIE* **5499**, 59–67 (2004).

148. E. Beuville, D. Acton, E. Corrales, J. Drab, A. Levy, M. Merrill, R. Peralta, and W. Ritchie, High performance large infrared and visible astronomy arrays for low background applications: Instruments performance data and future developments at Raytheon, *Proc. SPIE* **6660**, 66600B (2007).

149. J.W. Beletic, R. Blank, D. Gulbransen, D. Lee, M. Loose, E.C. Piquette, T. Sprafke, W.E. Tennant, M. Zandian, and J. Zino, Teledyne imaging sensors: Infrared imaging technologies for astronomy & civil space, *Proc. SPIE* **7021**, 70210H (2008).

150. A. Hoffman, Semiconductor processing technology improves resolution of infrared arrays, *Laser Focus World*, pp. 81–84 (February 2006).

151. J.C. Campbell, A.G. Dentai, T.P. Lee, and C.A. Burrus, Improved two-wavelength demultiplexing InGaAsP photodetector, *IEEE J. Quantum Electron.* **QE-16**, 601 (1980).

152. J.A. Wilson, E.A. Patten, G.R. Chapman, K. Kosai, B. Baumgratz, P. Goetz, S. Tighe, R. Risser, R. Herald, W.A. Radford, T. Tung, and W.A. Terre, Integrated two-color detection for advanced FPA applications, *Proc. SPIE* **2274**, 117–125 (1994).

153. P.R. Norton, Status of infrared detectors, *Proc. SPIE* **3379**, 102–114 (1998).

154. D.F. King, W.A. Radford, E.A. Patten, R.W. Graham, T.F. McEwan, J.G. Vodicka, R.F. Bornfreund, P.M. Goetz, G.M. Venzor, and S.M. Johnson, 3rd-Generation 1280 × 720 FPA development status at Raytheon Vision Systems, *Proc. SPIE* **6206**, 62060W (2006).

9

Solid-State Lighting

Lekhnath Bhusal
National Renewable Energy Laboratory
and
Philips Lumileds Lighting Company

Angelo Mascarenhas
National Renewable Energy Laboratory

9.1 Introduction

Since the development of the first electric lamp based on principles of a black body radiator by Thomas Edison (and independently by Joseph Swan) in 1879, the three traditional light sources, viz., incandescent lamps, compact fluorescent lamps (CFLs), and high-intensity discharge (HID) lamps, have evolved to the crest of their performance levels. Currently, any possible incremental improvement in the efficiency of these sources is insufficient to fulfill the growing energy demand and consumption; hence the development of new alternatives for lightening is imminent. In this regard, solid-state lighting technology, such as light-emitting diodes (LEDs) based on semiconductor material, has the potential to achieve a significant enhancement over today's most efficient lighting sources.

Historically, Nick Holonyak Jr. provided the first practical demonstration of a visible-spectrum LED in 1962 (with extremely dim emission ~0.1 lm/W).[1] After this breakthrough, the first commercial red LEDs based on gallium arsenide phosphide (GaAsP) were fabricated in high volume in the late 1960s, initially by Monsanto and Hewlett-Packard. In the 1970s, red-emitting aluminum gallium arsenide (AlGaAs) LEDs became popular for outdoor applications with performance comparable to the red-filtered incandescent lights, for instance, in red traffic lights and vehicle tail lights. By the end of 1970s, LEDs started penetrating the mainstream lighting industries by replacing incandescent lamps for indicator lamps and plasma discharge vacuum tubes for numeric displays. Due to various material and growth issues, after a long pause of almost two decades, Hewlett Packard[2] and Toshiba[3] used a quaternary aluminum indium gallium phosphide (AlInGaP) to develop high-brightness (HB) red and amber LEDs, followed by the first blue LEDs developed by Nakmura et al.[4] using gallium nitride (GaN). These discoveries stirred new research interest at the industrial level due to the possibility of combining red, green, and blue LEDs or coating the blue LED with a yellow phosphor leading to the creation of white LEDs for general lighting purposes. Presently, commercially available LED packages are comparable to the efficacies (a term, defined in Section 9.6.1, is used in exchange of efficiency in the lighting industry) of fluorescent and certain HID lamps (see for instance Ref. [5]). LED devices are vibration and shock resistant, exceptionally long-lived and allow for a wide variety of lighting, with control over color and intensity. This is an interesting time in the development of lighting technology as the predominant Edison-type lighting technology is slowly being replaced by the emerging solid-state lighting technology.

The chapter provides an overview of solid-state lighting technology, its physics, and the most common terminology used in this technology, along with a brief historical review of more than four decades of gradual development in the field. Extensive references provided at the end can be explored for further in-depth understanding of the various topics covered.

9.2 Low Brightness Applications

At the beginning of their development, LEDs were exclusively used for low brightness applications, where the efficiency and the power consumption were not an issue. Low-brightness LED systems can be categorized as LED systems emitting visible wavelengths that can be seen in the dark for indoor applications, such as on-off indicators in electronic devices and numeral digits in clocks. In general, an efficacy of a few lumens/watt is sufficient for these applications and can be achieved with very inefficient electric to light conversion (<1%) in semiconductor material systems. A lumen is a unit (SI unit system) of luminous flux, which represents the power of light source as perceived by the human eye and is defined in Section 9.6.1.

9.2.1 GaAsP and GaP Material System

The $GaAs_{1-x}P_x$ alloy was one of the first to be used for low-brightness visible spectrum LEDs following the first practical demonstration of visible light emission.[1] GaAs is a direct bandgap semiconductor with a bandgap in the near infrared spectral region (1.424 eV and 300 K), whereas GaP is an indirect bandgap semiconductor with a bandgap in the green portion of the spectrum (2.272 eV and 300 K). A large lattice-mismatch between the binaries GaAs and GaP (~3.6%) causes a lattice-mismatch between the GaAs substrate and the $GaAs_{1-x}P_x$ alloy. $GaAs_{1-x}P_x$ alloys have a direct bandgap for $x \sim 0.45$–0.50. For higher values of x, the alloy becomes an indirect bandgap semiconductor. As shown in Figure 9.1, for direct bandgap $GaAs_{1-x}P_x$, the Γ-valley provides the conduction band minima, whereas for the indirect bandgap, the X-valley is the lowest conduction band minimum.[6,7] As a result of the lattice mismatch and also due to the proximity to the indirect–direct crossover, the luminescence efficiency decreases substantially for higher x.[8] These restrictions limit the use of GaAsP alloys to a fundamental limit of wavelengths >650 nm for low brightness applications.

The first improvement in LED material, surpassing the performance of GaAsP-based LEDs, was the addition of an isoelectronic impurity N to GaP and later to GaAsP (Figure 9.1). Because N substitutes for P isoelectronically, GaP-based LEDs doped with the isoelectronic impurity N leads to new momentum selection rules due to the formation of an optically active level within the forbidden gap of the semiconductor.[9] The new optically active nitrogen level formed in the forbidden gap of the host material provides a recombination path with emission in the green spectral region. A similar effect can be achieved using the Zn-O isoelectronic pair, creating a deeper level in the forbidden gap of GaP, providing a recombination path for a longer wavelength emission in the red spectrum.[10] These LED devices were grown using the technique of liquid phase epitaxy (LPE). LPE is a simple method for growing high-purity semiconductor crystal layers from the melt on solid substrates.[11–13] This solid–liquid equilibrium-based growth process, however, is difficult to use to grow some alloys where the distribution coefficients of the constituent materials are significantly different, for instance, for AlP and GaP. It is also very difficult to control the layer thickness smaller than ~1 μm. Nonetheless, this technique continues to play an important role till today to grow very thick high-purity current-spreading layers in HB LEDs due to the high deposition rate (~2 μm/min) and also for the high-volume production

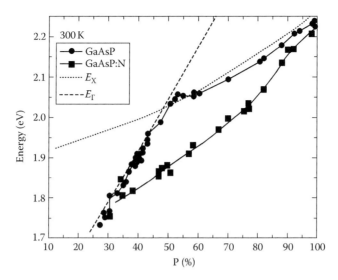

FIGURE 9.1 Peak emission energy (room-temperature) versus P composition for undoped and nitrogen-doped GaAsP LEDs (injected current density ~5 A/cm^2). Also shown is the energy gap of the direct-to-indirect (E_Γ–E_X) transition. (With permission from Craford, M.G. Shaw, R.W. Herzog, A.H., and Groves W.O., *J. Appl. Phys.*, 43, 4075, 1972. Copyright 1972, American Institute of Physics.)

of various low brightness LEDs based on GaP:N, GaP:Zn:O, and AlGaAs material systems. Another growth technique, called vapor phase epitaxy (VPE) was also developed in the 1960s and 1970s for high-volume production of various GaAsP-based LEDs. The advantage of VPE over LPE is the flexibility to grow various alloys since VPE is not limited by solid–liquid equilibrium conditions. In contrast to the use of melts in LPE, this technique involves the concept of gas-phase chemistry in an open tube configuration with the ability to grow alloys several hundred degrees below their melting point. Analogous to the GaP:N LEDs produced by LPE, Craford et al. in the early 1970s used the VPE growth technique to incorporate significant concentrations of N isoelectronic impurities into indirect bandgap GaAsP alloys, creating amber, yellow, and yellow-green GaAsP:N LEDs.[6,14] Indirect bandgap GaAsP:N and GaP:N or Zn-O yielded LEDs with efficiencies in the range 2–3 lm/W that dominated the indicator lamp markets for several years. The quantum efficiency of such LEDs is limited due to various fundamental reasons such as the lattice mismatch of GaAsP to GaAs, and the indirect bandgap of GaP and GaAsP (for $x > 0.5$), which could not be counteracted by improved materials technologies or device design. Consequently, the search for alternate approaches and other materials continued. A detailed discussion of efficiency is given in Section 9.6.1.

9.2.2 AlGaAs Material System

Due to the very similar atomic radii of Al and Ga, unlike the case for GaAsP, the AlGaAs material system can be grown nearly lattice-matched to GaAs (and Ge) substrates for all Al mole fractions. AlAs is an indirect bandgap semiconductor with a bandgap in the yellow-green portion of the spectrum ($E_g = 2.153$ eV and 300 K). As the Al mole fraction in AlGaAs is increased, the bandgap of the resulting alloy increases from that of the GaAs to that of AlAs. For Al fractions >45%, the AlGaAs alloy becomes indirect. As shown in Figure 9.2 (assuming the suggested values of bowing and bandgap corrected for 300 K in Ref. [15]), for Al mole fractions <45%, the Γ-valley provides the conduction band minimum, while for x > 45%, the X-valley is the lowest conduction band minimum. The direct–indirect crossover occurs at a wavelength of ~620 nm. It is important to note that to prevent the transfer of carriers from the Γ-valley to the X-valley, the bandgap of the active layer should be several kT smaller than

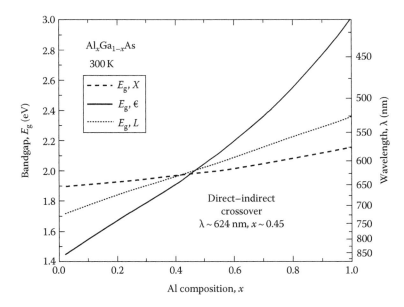

FIGURE 9.2 Bandgap energy and emission wavelength of AlGaAs at room temperature versus Al composition. E_Γ is the direct gap at the Γ point, and E_L and E_X are the indirect gap at the L and X point of the Brillouin zone, respectively.

the direct–indirect crossover bandgap value (k = Boltzmann constant and T = temperature). The first AlGaAs epitaxial growth using LPE was reported in 1967,[16] followed by the successful development of heterostructures (Section 9.5.2) lasers based on these alloys.[17,18] The AlGaAs alloy system was a perfect alloy for heterostructure development as it is lattice matched to the GaAs substrate for all Al concentrations. Hence, AlGaAs-based lattice-matched heterostructures were developed for the first time to produce relatively HB LEDs.[19,20] Though the AlGaAs alloy was useful for visible LEDs, this material was initially problematic because of the strong oxygen (O) affinity of Al, which caused high concentrations of deep traps associated with O deep donor states in the crystal. Only after well-developed high-volume LPE reactors were available in the late-1970s, did high-quality multilayer device structures based on AlGaAs-based LEDs become mainstream LEDs.[21] This development in growth reactors put AlGaAs LEDs in the mainstream market by the early 1980s as the dominant red-emitting LEDs. The AlGaAs LEDs were the first LEDs to exceed the luminous efficiency level of filtered incandescent light bulbs,[22,23] enabling them to begin replacing incandescent bulbs in exterior applications suitable for red light, such as red traffic lights, center-high-mount stop lights on motor vehicles, and outdoor message signs. These AlGaAs-based LEDs constituted the mainstream red LED until supplanted by the higher efficiency and brighter red InAlGaP LED device technology.

9.3 High Brightness Applications

High brightness (HB) LED systems are visible light emitters that can be used for bright outdoor applications and also possibly for indoor lighting. For these applications, an output emission of a few tens to a few hundreds of lumens/watt is required. In contrast to the low brightness LEDs, very efficient electric to light conversion is important in HB LED systems. The development of AlGaAs systems discussed before could be considered as the beginning of HB LEDs, as they exceeded the intensity of colored glass filtered incandescent lamps for applications in traffic lights and automobile tail lights. But true application of HB LEDs only appeared after the development of AlGaInP and InGaN-based LED systems.

9.3.1 AlGaInP Material System

Development of quaternary alloys enabled the efficient use of alloying to access various visible wavelength regimes and lattice matching on the available substrates and are increasingly used in the design of HB LEDs and advanced lasers. Concomitant with the start of the quaternary alloy GaAlAsP/GaAsP growth and the demonstration of the first $Ga_xIn_{1-x}P$-based laser,[24,25] the direct-bandgap $(Al_yGa_{1-y})_xIn_{1-x}P$ quaternary alloy became of interest for HB LEDs in the red, orange, and yellow spectral regions and is currently the leading material system for this spectral range. The $Ga_xIn_{1-x}P$ alloy is lattice matched to GaAs substrates for a Ga concentration of ~51%, with a bandgap of ~1.9 eV for the disordered alloy. The phenomenon of spontaneous long-range ordering in the alloy causes a reduction in the bandgap and will be discussed later in Section 9.4. For Ga concentrations >51%, for instance, the lattice constant of $Ga_xIn_{1-x}P$ becomes smaller than the lattice constant of the GaAs substrate, causing various material issues due to lattice mismatch. To resolve the issue of lattice mismatch, it was noted that AlP and GaP have virtually identical lattice constants, and hence Al and Ga can be substituted for one another without affecting the alloy lattice constant (Figure 9.3, assuming the suggested values of bowing and bandgap corrected for 300 K in Ref. [15]). Lattice matching requires the In-composition in AlGaInP to be fixed at 0.49. The addition of Al to the ternary alloy $Ga_{0.51}In_{0.49}P$ enables reaching shorter wavelengths, whilst keeping the quaternary lattice matched to GaAs. However, AlGaInP becomes an indirect semiconductor at an Al composition of ~53%. The interest in this material system is obvious because of the possibility of covering the red to yellow-green wavelength regions of the visible spectrum and the ability to maintain the lattice constant matched to the GaAs substrate. However, this material was difficult to grow using the then conventional methods of growth, such as LPE and VPE. AlP is much more thermodynamically stable compared to GaP, making composition control very difficult due to Al segregation. These technical difficulties delayed the development of AlGaInP materials for LEDs until the late-1980s, when good-quality AlGaInP material was reported using a kinetically controlled growth process such as organometallic vapor phase epitaxy (OMVPE) (also called metal-organic chemical vapor deposition, MOCVD). The OMVPE technique is a nonequilibrium growth technique introduced for GaAs growth in 1968[26] and later modified significantly. This technique uses vapor transport of source materials, which are reacted in the heated zone before getting deposited on the substrate. The MOCVD technique

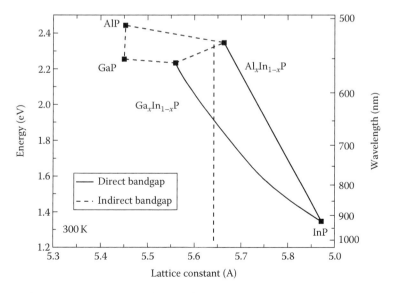

FIGURE 9.3 Bandgap energy and wavelength versus lattice constant of $(Al_xGa_{1-x})_yIn_{1-y}P$ (300 K). Vertical line is $(Al_xGa_{1-x})_{0.5}In_{0.5}P$ lattice matched to GaAs.

is highly versatile as complex heterostructures containing multiple layers with abrupt interfaces can be grown reproducibly. The method is also very suitable to precisely control composition and doping. The many advantages of the technique have made this the method of choice for almost all kinds of LED growth, although a few disadvantages exist, such as slow growth rate and the use of hazardous gases.

The combination of advances in kinetically controlled crystal growth process such as MOCVD and the capability of growing DH opened up a pathway for high-efficiency AlGaInP LED devices. Today, these high-power AlGaInP LEDs[2,3,27] are produced commercially in high volume on both absorbing substrates[28] and transparent substrates,[29] with device performance exceeding that of an unfiltered incandescent lamp. Unfortunately, due to various material issues, these advanced technologies were still limited for the production of high-efficiency red LEDs. One of the main challenges in the growth of high quality AlGaInP using MOVPE is the incorporation of residual oxygen during growth. As discussed in Section 9.2.2, oxygen is known to have very high affinity for Al, causing degradation of quality by introduction of deep level states that acts as nonradiative recombination centers. The problem of unintentional oxygen incorporation during growth is especially challenging for high-Al-containing devices (for instance, an Al concentration >30% needed to reach green wavelengths using AlGaInP). The oxygen contamination problem still exists to various extents for high Al-concentration. In the use of AlGaInP active layers for green and yellow-green emission, this problem constitutes a major setback for the realization of high efficiency devices when the emission wavelength is reduced from the red to the green. In Section 9.4, we will discuss our efforts on the development of Al-free GaInP active layers to reach the green emission wavelengths.

Until the early 1990s, the lack of development of HB LEDs providing critical colors of the visible spectrum other than the red diminished the potential of LEDs for becoming the ultimate white light source. Various attempts to produce efficient blue and green-emitting LEDs proved futile until a breakthrough by Nakamura and coworkers successfully produced bright blue LEDs based on the high bandgap InGaN material system.

9.3.2 InGaN System

Historically, the use of high bandgap material for blue emission can be traced down to the 1970s, when J. Pankove and coworkers at the Radio Corporation of America (RCA) laboratory studied GaN films grown on sapphire by VPE.[30–32] In 1971, Manasevit et al.[33] successfully used today's dominant MOCVD growth technology to grow single-crystal GaN films on sapphire, although the film had a large number of dislocations. During that period, a range of issues related to the material and its doping caused various groups to abandon what was then assumed futile research. Though the MOCVD technology was later revived by Dupius and coworkers in late-1977,[34,35] the interest in the material was not revived until 1986, when Akasaki and coworkers published their results on very high-quality GaN material and *p–n* junctions on sapphire using MOCVD.[36–38] The key to their success was successful *p*-type doping of GaN using Mg as a dopant. They used electron beam irradiation to activate the Mg *p*-type doping successfully for the first time.[39] In spite of their success, the slow and expensive method of electron beam irradiation to dope the device hindered further interest in GaN-based LED production until Shuji Nakamura and coworkers at Nichia Corporation, Japan, demonstrated a GaN blue LED using a thermally activated Mg dopant.[40–42] The thermal technique for doping GaN was fast and inexpensive in comparison to the use of the electron beam irradiation method. The solution of the *p*-type doping issues in high bandgap GaN-based material can be considered to be the stepping stone toward the solid-state white-lighting revolution, as within the next few years, many labs worldwide including Cree Research and Hewlett-Packard Optoelectronics (now Philips Lumileds Lighting Co) in the United States, Osram in Germany, and Toyoda Gosei in Japan joined the race for developing high-volume high-performance blue LEDs. The blue LEDs were soon used to create white LEDs by exciting a phosphor to generate a band of visible spectrum light emission. Currently, the performance of InAlGaN-based white LEDs is either on par or exceeds that of conventional lighting technologies.

Unlike the case of the III–V arsenide and phosphide material systems, the GaInN material system is found to surprisingly have a very high radiative recombination efficiency despite the presence of a very high concentration of threading dislocations (TDs) (~10^9–10^{10}/cm^2).[43,44] The cause of these dislocations is primarily the lattice mismatch between the commonly used sapphire and SiC substrates and the GaInN epitaxial layer. In contrast, the III–V arsenide and phosphide material systems with dislocation density greater than ~10^6/cm^2 would be optically inactive because of very low radiative efficiency. The key for high performance of InGaN alloys is carrier localization, which limits the carrier diffusion lengths, limiting carrier interactions with TDs, and suppressing nonradiative recombination. In this scenario, several localization mechanisms have been proposed. The first is based on the immiscibility of In in GaN that causes a fluctuation In content in the bulk GaInN system. This fluctuation causes carriers to be localized in potential minima generated by regions exhibiting InN-like properties, thus preventing carriers from reaching the dislocations.[45,46] Another emerging idea is that the monolayer size fluctuations in the InGaN quantum well width could serve as an effective mechanism to provide carrier localization.[47] Further studies are required to understand localization mechanisms that may explain the behavior of the GaInN alloy under high current densities (droop) and higher operating temperatures.

The direct bandgap of GaN is 3.44 eV. The bandgap of InN was accepted as 1.9 eV, until rediscovered by Davydov et al. to be 0.7 eV in 2001.[48,49] Theoretically, the GaInN alloy is suitable for covering the entire visible spectrum. However, the growth of GaInN becomes very difficult at higher In compositions because of the large difference in interatomic spacing between GaN and InN.[50] This limits the use of this alloy system for blue emission only, as its green and red emissions are very nonefficient. The current status and details of GaN-based devices are described in Section II of this book and can also be found in Refs. [51,52].

The performance of nitride-based LEDs is additionally marred by the phenomenon of "droop." In general, a high current is required to drive LEDs to emit enough light for HB applications needed for general lighting. Droop is the fall in the internal quantum efficiency of a device as the current increases, preventing the use of high currents in currently available InGaN LEDs. Understanding what causes droop in InGaN has become a topic of intense debate in recent years. Several research groups offer different explanations for this phenomenon. Researchers at Philips Lumileds believe that Auger recombination is the dominant cause for droop,[53,54] Schubert and coworkers from Rensselaer Polytechnic Institute suggest that carrier leakage in the presence of built in electric field across quantum wells is responsible,[55,56] and Monemar from Linköping University attributes this problem to defect related contributions to the reduction of the internal quantum efficiency under high forward bias.[57] Though the Auger recombination concept is emerging as the most plausible explanation, currently, there is a lack of consensus and understanding of the physics underlying droop, and this is considered a very serious problem in the development of HB LEDs as an efficient lighting solution.

9.4 Lattice-Mismatched Strain-Free GaInP Material System

$Ga_xIn_{1-x}P$ alloys are an important technological material for various applications including photovoltaics and LEDs and have been studied extensively for many years.[58] The bandgap of this alloy at room temperature when lattice matched to the GaAs substrate is ~1.91 eV ($x \sim 0.51$). For various applications and for significant improvements in the performance of multijunction solar cells, however, a higher bandgap in the range of ~2.0–2.2 eV is desired.[59,60] These higher bandgap alloys could also be used for the development of LEDs for solid-state white-light applications, by providing various color components for tri- or quad-color approaches.[60] To circumvent the limitation on the bandgap of GaInP, higher bandgaps could be achieved by introducing aluminum and forming the quaternary alloy AlGaInP. As discussed earlier in Section 9.3.1, oxidation of the high Al-containing material has been widely observed leading to short-term degradation.[61] Higher bandgaps can also be reached by increasing the Ga content ($x > 0.51$), causing the $Ga_xIn_{1-x}P$ epilayer to be lattice-mismatched to GaAs. Some experimental work on the growth of the lattice-mismatched alloy with and without the ordering phenomena had been performed previously.[62–64]

A very thin $Ga_xIn_{1-x}P$ epilayer lattice-mismatched to GaAs will tend to grow coherently with the substrate and accommodate the in-plane strain by a tetragonal distortion of the lattice. As the critical thickness of $Ga_xIn_{1-x}P$ is reached, the epilayer relaxes. This relaxation varies in degree and is mediated by the formation of TDs, which can thread through the entire structure damaging device performance. The performance of this lattice-mismatched epilayer can be enhanced significantly by growing a compositionally step-graded $GaAs_{1-z}P_z$ buffer, resulting in relatively low defect densities.[65–67] The technique of grades to change the lattice constant gradually has been employed at the National Renewable Energy Lab (NREL), USA, to obtain ~1.0 eV bandgap semiconductor material for use in triple and quadruple junction solar cells. The alloy InGaAs could provide the required 1.0 eV bandgap, but with the penalty of significant lattice-mismatch to the GaAs substrate. The detrimental consequences associated with metamorphic growth to achieve 1.0 eV bandgap InGaAs alloys were resolved at NREL by pioneering strain management in the InGaAs alloy through the use of GaInP step grades, leading to world's most efficient multijunction solar cells.[65,66] The advantage of this graded buffer approach is the ability to grow the material with little or no strain. The ability to change the lattice constant of the growth surface, in effect creating a pseudo-substrate on which to grow the active layers, is an important development that has wide applications for photovoltaics and LEDs. A well-designed graded layer minimizes the density of dislocations that thread through the active layers, primarily by confining them to the step-graded layers.[68] Using a similar technique, $GaAs_{1-z}P_z$ grades can be used to produce a pseudo-substrate to grow strain-free lattice-mismatched $Ga_xIn_{1-x}P$ alloys, opening a pathway to reach a higher bandgap (~1.9–2.2 eV) without a substantial degradation of the material.[67]

It is well established that $Ga_xIn_{1-x}P$ epilayers grown by MOVPE tend to exhibit a partially ordered CuPt crystal structure,[69,70] leading to a bandgap reduction of ~100 meV and a valence band splitting, providing an additional variable for tailoring the electronic structure via the order parameter, η.[71] The bandgap reduction has been understood to result predominantly from the repulsion between the folded L-point and the Γ-point of the conduction band, with a relatively small effect on the valence band.[72] The ordered $Ga_xIn_{1-x}P$ alloy is a $\langle 111 \rangle$ monolayer superlattice of $Ga_{x+\eta/2}In_{1-x-\eta/2}P/Ga_{x-\eta/2}In_{1-x+\eta/2}P$, where the order parameter η has a maximum value $\eta_{max}(x) = \min[2x, 2(1-x)]$. At $x = 0.5$, the fully ordered structure is a superlattice of alternating $\{1\bar{1}1\}$ or $\{\bar{1}11\}$ monolayers of GaP and InP. Trace amounts of the surfactant Sb can be used to disorder an epilayer that would otherwise have grown ordered, thus enabling control of the order parameter and the bandgap.[73,74]

The ability to grow highly lattice mismatched $Ga_xIn_{1-x}P$ epilayers as well as to control the spontaneous alloy ordering to tailor the high bandgaps was demonstrated recently (Ref. [67] and references therein). The samples were grown by atmospheric pressure MOVPE in a custom-built vertical reactor. Other growth details can be found in the reference. An Sb surfactant was used to disorder the $Ga_xIn_{1-x}P$ during growth. Figure 9.4 shows a low-magnification cross-sectional TEM image of $Ga_xIn_{1-x}P$ layers grown at 750°C with and without the grades. Prevalent generation of TDs in the $Ga_{0.72}In_{0.28}P$ epilayer without the GaAsP grades is apparent in top Figure 9.4. On the other hand, most of the dislocations are absorbed by the GaAsP grades before the final $Ga_{0.71}In_{0.29}P$ epilayer, as shown in the bottom of Figure 9.4.

Figure 9.5 shows a [110]-pole diffraction pattern for a selected area of a $Ga_{0.74}In_{0.26}P$ epilayer grown over step-graded layers. The large spots represent the fundamental reflections, and small spots are caused by the alloy ordering induced superlattice reflections. This confirms the presence of a single CuPt *B* variant ordering in the material. Images of the orthogonal [1–10]-pole do not show any superlattice reflections, indicating that there is no CuPt*A* variant. The bandgap of each sample was measured by room temperature photoluminescence (PL) using a 532 or 405 nm laser to excite the sample and a 0.27 m spectrometer followed by an air-cooled Si-CCD detector. Figure 9.6 shows the bandgaps extracted from room temperature PL measurements for $Ga_xIn_{1-x}P$ epilayers with (triangles) and without (circles) the step-grades and are compared to the theoretical expectations for various Ga concentrations. An unusual trend of the bandgap versus Ga concentration for the epilayers without the grades can be noticed, probably caused by large number of defects due to the increased strain for higher Ga

FIGURE 9.4 A low-magnification cross-sectional TEM image of (a) GaInP epilayer without grades and (b) with GaAsP grades.

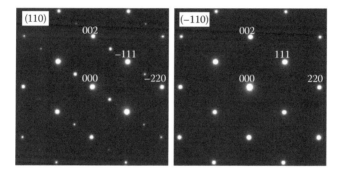

FIGURE 9.5 [110]-pole transmission electron diffraction pattern for a $Ga_{0.74}In_{0.26}P$ sample with $\eta = 0.32$. The larger points represent fundamental reflections as indicated. The smaller points represent the superlattice reflections for the [−111] CuPt B variant. No superlattice reflections are visible from the [11−1] variant.

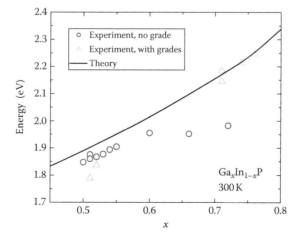

FIGURE 9.6 Bandgap as extracted form the room-temperature PL spectra for GaInP epilayers grown with (triangles) and without (circles) GaAsP grades compared to the theoretical expectation (solid line).

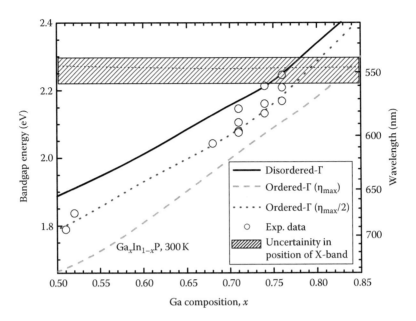

FIGURE 9.7 Bandgap energy of the $Ga_xIn_{1-x}P$ versus Ga concentration and order parameter at room tempera-ture. Symbols are the experimental data points. Solid lines are theoretical calculations for various ordering level. The semitransparent strip shows the uncertainty in the position of X-band. (With permission from Steiner, M.A., Bhusal, L., Geisz, J.F., Norman, A.G., Romero, M.J., Olavarria, W.J., Zhang, Y., and Mascarenhas, A., *J. Appl. Phys.*, 106, 063525, 2009. Copyright 2009, American Institute of Physics.)

concentrations. Figure 9.7 shows how the bandgap $E_g(x, \eta)$ of the GaInP alloy varies with composition and order parameter. The experimental points show the bandgaps extracted from room temperature PL and modulated photo/electro reflectance measurements. Also shown are the calculated curves[75,76] for the bandgap of the disordered ($\eta = 0$), half-ordered ($\eta = \eta_{max}/2$) and fully ordered ($\eta = \eta_{max}$) alloys. Overall, the experimental data are in good agreement with the calculated curves over the measurement range. We have shown that samples with $x > 0.76$ and fully disordered appear to have indirect band-gaps.[77] The figure also shows a band of uncertainty in the location of the X-band, as there is a spread of data available in the literature for the direct–indirect crossover point. Our measurements suggest values between $x \sim 0.74$ and 0.75. As the work is still in progress, further studies and more careful probing of the dynamics of the direct–indirect crossover using lifetime measurements is required to pin-point the location of the transition point.

The ability to grow lattice-mismatched GaInP as an active layer to achieve various high bandgaps has applications for the development of solid-state white lighting. As various colors with wavelength ~690–570 nm can be achieved with this alloy, a bi-, tri-, or quad-color mixing approach, in conjunction with GaN-based blue light LEDs could provide pathways to achieve a device with high color-rendering index and high efficacy.

9.5 Physics of Solid-State Light-Emitting Diodes

Conventional lighting sources are based on the phenomena of incandescence or discharge in gases and are basically associated with large energy losses either due to the high temperature or large Stokes shift. Solid-state light emitting diodes (LEDs), on the other hand, offer an alternative way of light generation, using a *p–n* junction made of differently doped semiconductor materials. The spontaneous emission of light occurs due to radiative recombination of excess electrons and holes across the junction. The excess electron and holes in the *p–n* junction are generated by injecting current from an external source. This phenomenon of luminescence caused by the radiative recombination of injected carriers is often called injection luminescence and is the basis of all solid-state lamps.

9.5.1 *p–n* Junction

The most basic component of a solid-state LED involves at least one layer of luminescent capable semiconductor material for radiative recombination and regions of *p*- and *n*-conductivity type semiconductors for carrier injection. The simplest design could be a *p–n* homojunction with at least one of the injection regions being used as the recombination region. This structure is illustrated in Figure 9.8 to elucidate the basic physics behind carrier injection and recombination. Near the unbiased *p–n* junction, electrons on the *n*-type semiconductor diffuse over to the *p*-type and recombine with majority carrier holes. Similarly, holes diffuse in the opposite direction from *p*-type to *n*-type and recombine with majority carrier electrons. The process creates a region depleted of free charge carriers across the junction and is called a *depletion region*. The absence of free charge carriers leaves behind ionized donors (*n*-side) and acceptors (*p*-side) forming a space charge region. This space charge region creates an internal electric field that blocks further diffusion of the carrier, leading to electrical equilibrium. The space charge region creates a potential barrier that electrons or holes must overcome to cross the junction. An external forward (reverse) bias can decrease (increase) this barrier. Hence, under forward bias, electrons and holes are injected into the space charge region or possibly into the region of opposite conductivity, causing an increase in current flow. The injected carriers eventually recombine radiatively, thereby emitting photons, or nonradiatively. On the other hand, a reverse bias causes the potential barrier to increase, suppressing the injection of carriers. Under a reverse bias, only a small reverse current (reverse in comparison to the current caused by the flow of injected minority carriers) called saturation current flows across the junction. The current–voltage (*I–V*) characteristic of a *p–n* junction is described by the Shockley equation, giving a current density (current/area)

$$J(V) = J_0 \left(e^{qV/kT} - 1 \right), \quad J_0 = \left(\sqrt{\frac{D_p}{\tau_p}} \frac{n_i^2}{N_D} + \sqrt{\frac{D_n}{\tau_n}} \frac{n_i^2}{N_A} \right) \tag{9.1}$$

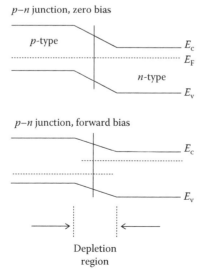

FIGURE 9.8 A *p–n* homojunction.

The Shockley equation gives the *I–V* characteristic of an ideal *p–n* junction. In practice, a diode has unavoidable parasitic resistances in series and parallel to the device. A series resistance R_s, in general, is caused by the contact resistance and the resistance of various layers across the structure. A parallel resistance R_p, on the other hand, is caused by current paths bypassing the *p–n* junction. After considering the inclusion of nonideal factors, the modified Shockley equation is given by

$$J = \frac{(V - JR_s)}{R_p} + J_0\left(e^{q(V-JR_s)/nkT} - 1\right)$$

(9.2)

Here, $1 \le n \le 2$ is the ideality factor of the diode. For an ideal diode, $n = 1$. A value of $n > 2$ is not very common, but it has been observed in some very high bandgap diodes.[78]

9.5.2 Heterostructures

Heterostructures are artificial structures composed of semiconductor material of different chemical compositions and hence of different bandgaps. In contrast to the homojunctions as described in an earlier section, the material composition in heterostructures is a function of distance. The change in material properties with distance provides control over the potential profile across the device, resulting in better control of carrier injection, recombination mechanisms, and absorption properties of the device. Currently, almost all HB LEDs use heterostructures of various complexity to enhance the radiative recombination and photon emission. A single heterostructure (SH) or a *p–n* heterojunction is the simplest of such heterostructures and is shown in Figure 9.9. The *p*-type region, for instance, can be made of a semiconductor with bandgap $E_{g1} < E_{g2}$, where E_{g2} is the bandgap of the *n*-type region. The band discontinuity can be designed such that the potential barrier for holes that diffuse from *p*- to *n*-type is increased, while the barrier for electrons is reduced. This nonuniformity increases the ratio of the injected electrons to the injected holes, thus increasing radiative recombination and hence the luminescence. Double heterostructures (DHs) as shown in Figure 9.10 offer better bandgap engineering useful for the HB LEDs. The DH is comprised of an undoped or lightly doped narrow gap active region sandwiched between an *n*- and *p*-doped wide bandgap conducting or confining region. In contrast to a SH, a DH enables bidirectional carrier injection into the active region. Also, the larger bandgap conducting regions provide a barrier to confine the injected carriers within the active region, thus increasing the injected carrier density in the active layer. The conducting regions are often called *cladding* or *confinement layers*. Higher carrier density in the active region eventually improves the radiative recombination rate. Often, it is not possible to design a DH due to constraints on the availability of a semiconductor with suitable properties. For a device to work optimally as a LED, the material is required to have a

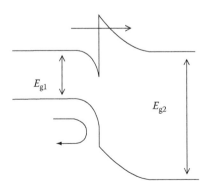

FIGURE 9.9   A SH or a *p–n* heterojunction.

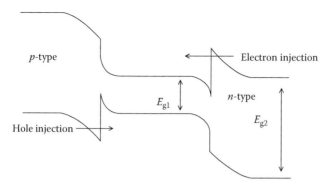

FIGURE 9.10 A DH enables bidirectional carrier injection into the active region.

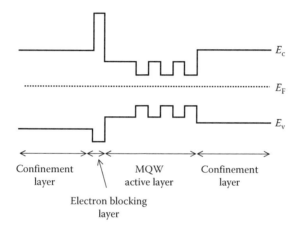

FIGURE 9.11 Concept of EBL and multiquantum well in heterostructure-based LEDs.

direct bandgap with lattice constant of the active and confinement layers matched to the substrate. Large lattice mismatch could cause a large defect density in the material giving rise to nonradiative recombination. Various engineering methods can be applied to achieve desirable properties from various semiconductors to be used for LEDs. The thickness of the active layer also plays a critical role in device performance.[79] If the thickness of the layer becomes comparable to the de Broglie wavelength of electrons, quantum mechanical effects start playing a critical role. In this situation, the DH is called a quantum well structure. Various HB LEDs now use single or multiple quantum wells for improving their performance (Figure 9.11). In a DH, it is important to keep the injected carriers confined within the active region or QW region to maximize the recombination probability. Unfortunately, because of very high electron mobility, electrons can leak into the *p*-type contact layer without recombination in the active layer. This effect is less pronounced for holes as their mobility is much smaller than for electrons. A specially designed layer called an electron blocking layer (EBL) made of a material with a high bandgap providing a higher band offset in the conduction band solves this problem, as shown in Figure 9.11.

9.6 Solid-State Lamps

During the early developments of low brightness LEDs, few may have envisioned that the technology could eventually lead to the replacement of conventional incandescent and fluorescent lamps. An LED emits monochromatic light in a narrow band of wavelength, but a broadband spectrum of visible wavelengths is required to produce white light for indoor and outdoor general purpose lighting. Producing efficient white light using monochromatic LEDs is a current technological challenge in the field. Only after the development of blue GaN-based emitters, did solid-state white-light LEDs become practical, and the ultimate goal of solid-state lighting technology get diverted to create the ultimate white lamp with properties superior to that of conventional lamps. Fundamentally, there are two ways to produce white light emitters based on solid-state LEDs. Both approaches have their unique technological challenges to produce an ultimate white lamp. The first approach is to mix red, blue, and green colors emitted by different LEDs. This approach suffers from the presence of the "green gap" problem. Also, because of the narrow bandwidth of individual color components, a careful choice of more than three colors is required to produce a better color-rendering index (defined later),[60] which adds to the complexity and cost with this approach. In the second approach, emission from a blue or UV LED is down converted (down conversion is the conversion of high energy photons to low energy photons) using phosphors to emit light in the visible spectrum. Currently, this approach suffers from the unavailability of appropriate phosphor materials for efficient down conversion and unavailability of phosphors emitting a red spectrum to produce a warmer white light.[60] Both the LED and the phosphor approaches, however, are nontoxic, unlike the situation with mercury used in CFLs. In contrast to other practical devices, such as solar cells and transistors, LEDs for white-light applications are complex to design as the quality of white light relates to color perception and human vision.

Understanding of photometry, colorimetry, and color rendering is important for solid-state lighting design. More details can be found, for instance, in Ref. [80]. In 1924,[81] the International Commission on Illumination (CIE) introduced the relative luminous efficiency function (or eye sensitivity function), $V(\lambda)$ for photopic vision (at day time or during the abundance of light when cone receptors of the eye mediate vision, in contrary to scotopic vision where the eye adapts to low light situations and rods receptors mediate vision). This function is referred to as the CIE 1931 $V(\lambda)$ function,[82] which was modified in 1978 by Judd and Vos to correct the previous underestimation of the human eye sensitivity in the blue spectral region by the CIE 1931 $V(\lambda)$ function. The modified function is called CIE 1978 $V(\lambda)$. The CIE $V(\lambda)$ functions are shown in Figure 9.12. The photopic eye sensitivity function has maximum sensitive at the green wavelength of 555 nm. The function $V(\lambda)$ is defined in the wavelength interval of 380–780 nm, as this interval of electromagnetic radiation is capable of creating sensation in the human eye and is defined as the *visual spectrum*.

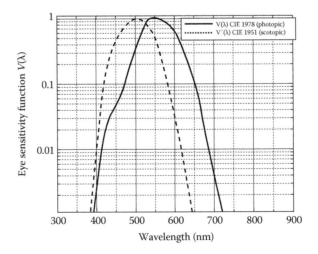

FIGURE 9.12 CIE eye sensitivity function for photopic and scotopic vision regimes.

9.6.1 Photometry

The visual spectrum ranging from 380 to 780 nm is a part of electromagnetic radiation. The physical properties of any radiation such as number of photons, photon energy, and power are characterized by *radiometric units*. As radiation outside the visual spectrum is unable to cause any luminous sensation in the human eye, it is important to characterize the visual spectrum with a separate system of units called *photometric units*. The following photometric quantities are often used in characterizing any light source.

The *luminous intensity* gives the light intensity of a light source as perceived by the human eye and is measured in *candela* (cd) in SI unit systems. The *luminous intensity* of one candela is defined as a monochromatic light source emitting an optical power of 1/683 W at 555 nm per steradian (sr) solid angle. The *luminous flux* represents the light power of a source perceived by the human eye and is measured in units of the lumen (lm) in the SI units system. A *luminous flux* of 1 lm is defined as a monochromatic light source emitting an optical power of 1/683 W at 555 nm. The *luminous flux*, Φ_{lum}, is determined using the radiometric light power as

$$\Phi_{lum} = 683\left(\frac{lm}{w}\right)\int_\lambda V(\lambda)P(\lambda)d\lambda \tag{9.3}$$

where
$P(\lambda)$ is the light power emitted per unit wavelength and is called the *power spectral density*
The factor 683 lm/W is a normalizing factor
$V(\lambda)$ is the CIE eye sensitivity function defined earlier

The *luminosity function* or the *luminous efficacy* of optical radiation measures the ability of radiation to produce a visual sensation in lm/W and is defined as

$$\text{Luminous efficacy} = \frac{\Phi_{lum}}{\text{Optical power}} = \frac{683\left(\frac{lm}{w}\right)\int_\lambda V(\lambda)P(\lambda)d\lambda}{\int_\lambda P(\lambda)d\lambda} \tag{9.4}$$

where, the total optical power, $P = \int P(\lambda)d\lambda$. Note that the *luminous efficiency* of a light source is also measured in lm/W, but it defines a conversion efficiency of electrical power to light power.

$$\text{Luminous efficiency} = \frac{\Phi_{lum}}{\text{Electric power}} = \frac{683\left(\frac{lm}{w}\right)\int V(\lambda)P(\lambda)d\lambda}{IV} \tag{9.5}$$

where
I is the current
$V(\lambda)$, in this case, is the voltage applied (V_{appl}) by an external electric power source

From Equations 9.4 and 9.5, it is important to note that the *luminous efficiency* is the product of *luminous efficacy* and the absolute electric-to-optical conversion efficiency,

$$\text{Luminous efficiency} = \text{Luminous efficacy} \times \frac{P}{IV} = \text{Luminous efficacy} \times \frac{\text{Optical power}}{\text{Electrical power}} \tag{9.6}$$

It is often confusing to distinguish between the *luminous efficiency* and *luminous efficacy* as both have the same units of lm/W. The *luminous efficacy* is the ability to produce a visual sensation in the human eye, while the *luminous efficiency* is the ability of the light source to convert the consumed electric power to sensation in the human eye. This distinction is clear from Equation 9.6. The *external quantum efficiency* of an LED is an important performance parameter for the device that is defined as the ratio of the number of photons emitted and the number of electrons passed through the LED and is a unit-less number. Here, again, it is important to distinguish clearly between efficacy defined earlier and the unit-less efficiency of an LED. According to the definitions, the efficiencies of UV or infrared emitting LEDs, for example, could be high, but for the solid-state lighting proposes, their efficacy will be zero as UV and infrared radiation do not produce any visual sensation in our eye. External quantum efficiency can be broken down as the product of the nonexplicit parameters, *internal quantum efficiency*, *injection efficiency*, and *light-extraction efficiency*, $\eta_{external} = \eta_{internal} \times \eta_{injection} \times \eta_{extraction}$. The *injection efficiency*, $\eta_{injection}$, is the fraction of the electrons passed through the device that are injected into the active region, where radiative or nonradiative recombination takes place. The *internal quantum efficiency* is the ratio of the number of electron-hole pairs that recombine radiatively to the total number of pairs that recombine in the active region. For an LED as a commercial product, the consumer-oriented definition of efficiency is termed *wall-plug efficiency*. The *wall-plug efficiency* is the efficiency of converting electrical to optical power and is defined as the ratio of the radiant flux (the total optical output power of the device, in watts) and the electrical input power. When an LED is connected to an external power source, each electron-hole pair acquires a certain amount of energy from the source. The term, *feeding efficiency* is defined as the ratio of the mean energy of the photons emitted and the total energy that an electron-hole pair acquires from the power source. The *wall-plug efficiency* can be determined using the *external quantum efficiency* and the *feeding efficiency* as $\eta_{wall-plug} = \eta_{external} \times \eta_{feeding}$. In the case of a white lamp using down conversion in a phosphor to generate white light, this should also be multiplied by the phosphor conversion efficiency. The internal quantum efficiency is one of the key parameters for judging the quality suitable for HB LEDs. Good material quality and efficient carrier confinement (Section 9.5.2) are required to increase the radiative recombination rates in the active region. For details of recombination phenomena in semiconductor materials, a topic that is beyond the scope of this chapter, one can refer to text books, e.g., by Sze.[83]

9.6.2 Colorimetry

Colorimetry is the science of measurement of color. Though the sensation of color is very subjective, a numerical description of colors is well formulated with definitions of *tristimulus value*, *chromaticity coordinates*, *color temperature* (CT), and *color-rendering index* of a light source. These definitions are crucial to describe any light source used for various applications. Color-matching functions *x*, *y*, and *z* were introduced by CIE in 1931. The concept is based on the fact that any monochromatic color can be produced from just three different colors. This property of human color vision is called *trichromacy*. The three different colors corresponding to a given monochromatic color can be determined optically by comparing a human visual sensation created by two light sources. One source is the mixture of three colors, say, red, green, and blue, and the second source is the monochromatic color. A person with normal color vision can adjust the relative intensities of the red, green, and blue lights to make the mixed light and the monochromatic light appear identical. A series of such matches can then mathematically be transformed to a set of three dimensionless color-matching functions (CMFs) $\bar{x}(\lambda)$, $\bar{y}(\lambda)$, and $\bar{z}(\lambda)$ (Figure 9.13). The green CMF $\bar{y}(\lambda)$ is matched to the eye sensitivity function $V(\lambda)$. Note that due to human vision subjectivity, the CMFs are not unique, and there are different versions of the CMFs (see Refs. [84,85] by Judd Voss). For a given power spectral density $P(\lambda)$, the tristimulus values *X*, *Y*, and *Z* are obtained by integrating the spectrum with the CMFs $\bar{x}(\lambda)$, $\bar{y}(\lambda)$, and $\bar{z}(\lambda)$.

$$X = \int \bar{x}(\lambda)P(\lambda)d\lambda, \quad Y = \int \bar{y}(\lambda)P(\lambda)d\lambda, \quad Z = \int \bar{z}(\lambda)P(\lambda)d\lambda, \tag{9.7}$$

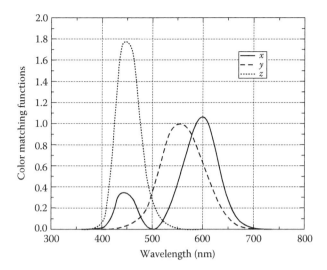

FIGURE 9.13   CIE 1931 color-matching functions.

X, Y, and Z provide the power of each of three primary colors (red, blue, and green) required to match the color of a given monochromatic light. The tristimulus values for a given power spectral density $P(\lambda)$ are normalized to calculate chromaticity coordinates x, y, and z as

$$x = \frac{X}{X+Y+Z}, \quad y = \frac{Y}{X+Y+Z}, \quad z = \frac{Z}{X+Y+Z} \tag{9.8}$$

The value of the chromaticity index z can be determined from x and y ($z = 1 - x - y$), hence it is redundant with no additional information. Thus, a two-dimensional description of colors can be made using chromaticity coordinates (x, y). The (x, y) chromaticity diagram is shown in Figure 9.14. Monochromatic colors are located on the perimeter of the chromaticity diagram, and white light is found in the center. All the colors can be located with the correct chromaticity coordinates (x, y) within the diagram. In the chromaticity diagram defined by (x, y), the color difference between two points is not proportional to the geometrical difference between those two points. Defining uniform chromaticity coordinates (u', v'), CIE (1976) solves this problem, as the uniform chromaticity diagram now provides the color difference between two points in the diagram as directly proportional to the geometrical distance between these two points (Figure 9.15). The uniform chromaticity coordinates are determined using the nonlinear transformation of chromaticity coordinates as

$$u' = \frac{4x}{-2y+12y+3}, \quad v' = \frac{9y}{-2x+12y+3} \tag{9.9}$$

9.6.3 Color-Rendering Index

A given chromaticity coordinate (x, y) describes quantitatively the color of the light source. The same chromaticity coordinates can be assigned to sources of different spectrum such as a Planck radiator (e.g., hot filament) or for a combination of different monochromatic sources. After the light from those different sources reflects from an object, the reflected spectrum alters depending on the properties of the reflecting surface. Hence a shift in the chromaticity coordinate occurs after the reflection, and the

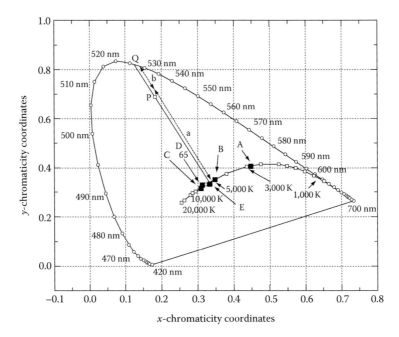

FIGURE 9.14 Chromaticity diagram showing coordinates of various points of interest, A,B,C,D,[65] and E. Monochromatic colors are located on the perimeter and white light is located at the center point E. Also shown is the process of determining dominant wavelength (Q) and color purity of a light source with coordinates given by the point P.

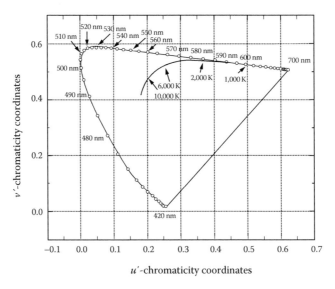

FIGURE 9.15 CIE 1976 $u'v'$ uniform coordinates.

same object can create a different color sensation to the eye under different sources (even though those sources have the same chromaticity coordinate). The ability to render the true colors of an object is assessed by the *color-rendering index* (CRI). By convention, the Planckian black body is assumed to have a perfect CRI of 100. The special CRIs are defined for the set of eight test-colors samples, where the special CRI for the ith sample is defined as $CRI_i = 100 - 4.6\Delta C_i$. Here, ΔC represents the quantitative color change of the sample when illuminated first with the reference source and then with the test source. The *general CRI* is then calculated as the average of eight special CRIs as

$$\text{CRI} = \frac{1}{8} \sum_{i=1}^{8} \text{CRI}_i \qquad (9.10)$$

If there is no difference in the color appearance of the test samples under illumination by the reference sample and the test sample, then $\Delta C = 0$ and the CRI of the test sample is 100. The selection of a reference source and the calculation of the quantitative color change ΔC are complex process, and the details are beyond the scope of this chapter; readers are referred to the Refs. [86,87]. Recently, various concerns were raised about the validity of CRI numbers for defining the properties of LEDs, as this definition is several decades old and was defined without any anticipation of solid-state lighting (Ref. [88] and references therein). In this regard, NIST is taking the initiative to standardize the color quality scale as detailed in Ref. [88].

9.6.4 Color Purity

The monochromatic colors ($\Delta\lambda \approx 0$) are located on the perimeter of the chromaticity diagram. As the spectral width, $\Delta\lambda$, of the source increases, the chromaticity coordinate of the source moves inward to the center of the diagram. For $\Delta\lambda \to \infty$, the light source becomes white, and the coordinates are located at the center of the diagram. The *dominant wavelength* in the light source (if $\Delta\lambda \neq 0$) is defined as the wavelength of the point on the perimeter of the diagram, which is closest to the color coordinates of the source. A line connecting the center of the diagram ($x = 1/3$ and $y = 1/3$) and the chromaticity coordinates of the light source can be extended to intersect the perimeter of the diagram. The coordinate of such an intersecting point gives the *dominant wavelength* of the source. The *color purity* of a light source is determined by the closeness of the color coordinates of the source to the dominant wavelength. Numerically, it is determined by dividing the distance, a, between color coordinates of the source and the color coordinate of an equal energy point ($x = 1/3$ and $y = 1/3$) by the distance, b, between the coordinates of the source and the coordinates of the *dominant wavelength*. Hence the color purity is 100% for a monochromatic light located on the perimeter of the diagram (see Figure 9.14).

The Planckian black body radiation spectrum is considered as a standard source in many applications since the spectrum is characterized only by temperature of the body. The maximum intensity of black body radiation emanating at temperature T occurs at a specific wavelength given by *Wien's law*

$$\lambda_{\max} \text{ (nm)} = \frac{2.8977 \times 10^6}{T(K)} \qquad (9.11)$$

The *chromaticity coordinates* of the black body radiation at different temperatures form a Planckian locus as shown in Figures 9.14 and 9.15. Some standard illuminants such as A (incandescent source, $T = 2856\,\text{K}$, $x = 0.4476$, and $y = 0.4074$), B (direct sunlight, $T = 4870$, $x = 0.3484$, and $y = 0.3516$), C (overcast sky, $T = 6770$, $x = 0.3101$, and $y = 0.3162$), D$_{65}$ (daylight, $T = 6500\,\text{K}$, $x = 0.3128$, and $y = 0.3292$), and E (equal energy point, $x = 1/3$, and $y = 1/3$) as defined by CIE fall on this locus (Figure 9.14). The CT of a white light source is the temperature of a Planckian black body radiator with the same *chromaticity coordinate* as the white light source. If the *chromaticity coordinates* of the white light source do not fall on the Planckian locus, the CT is determined as the temperature of the Planckian black body radiator whose coordinates are closest to the color of white light. The CT in this case is termed as the *correlated color temperature* (CCT). It is important to note that the CT or CCT is not an indicator of LED heat in the conventional sense where yellowish or reddish colors, as in very hot body, are considered warm. In contrast, light sources with higher CT or CCT (~3600–5500 K) are considered to produce *cool light* and is preferred for visual tasks because it produces higher contrast in comparison to lower CCT light sources. On the other hand, *warm light* corresponds to lower CCT (~2700–3000 K). In the $u'–v'$ uniform chromaticity diagram, the point on the Planckian locus that is at the shortest geometrical distance from the coordinates of the white light source gives the CCT of the white light source. However, on the x, y chromaticity diagram, this is not true due to the nonuniformity of the diagram as discussed in Section 9.6.2.

9.7 Conclusions and Future Outlook

Lighting is one of the biggest sources for producing greenhouse gas emission, a fact that was in denial for many decades, probably because of the lack of other lighting alternatives. Lighting all over the world not only consumes a significant amount of resources, but also does it in a very inefficient manner. With the recent unprecedented advances in III–V semiconductor technology, the future of solid-state lighting has become quite exciting and gaining prominence for its potential to replace the predominant power-consuming Edison-type lighting technology. The new LED lighting technology is drawing attention at various levels as improving energy efficiency in the lighting sector will eventually lead to reduced energy use and reduced carbon emission. Recently, the European Commission adopted new regulations that will effectively phase out incandescent light bulbs in Europe by 2012. It is expected that soon various other states will follow this phase out. The upfront cost of the solid-state lighting device could be a deterrent for high-scale market penetration. Extensive R&D efforts by various private entities and support from various government agencies should bring the cost down, along with improving the various features of lighting. Technological challenges, such as heat management, color rendering, lifetime, and low efficiencies are expected to be overcome in a time frame of 5–10 years so as to increase the efficacy from the current values of ~100 lm/W to more than 200 lm/W,[89] thus making the solid-state LED lamp an ultimate light source for the future.

This chapter is designed to serve as an overview and reference for students as well as researchers involved in the growing field of solid-state lighting with extensive supplementary bibliography to facilitate further understanding as desired.

Acknowledgments

We would like to acknowledge Myles Steiner of NREL for providing lattice mismatched GaInP alloys samples and data and Andrew Norman of NREL for the TEM images. This work was supported in part by SC/BES/DMS, LDRD, and EERE Solid-State Lighting program funding at NREL.

References

1. N. Hollonyak Jr. and S. F. Bevacqua, (1962), *Appl. Phys. Lett.* **1**, 82.
2. C. P. Kuo, R. M. Fletcher, T. D. Osentowski, M. C. Lardizabal, M. G. Craford, and V. M. Robbins, *Appl. Phys. Lett.* **57**, 2937 (1990).
3. H. Sugawara, M. Ishikawa, and G. Hatakoshi, *Appl. Phys. Lett.* **58**, 1010 (1991).
4. S. Nakamura, M. Senoh, N. Iwasa, S. Nagahama, T. Yamaka, and T. Mukai, *Jpn. J. Appl. Phys.* **34**, L1332 (1995).
5. R. D. Dupuis and M. R. Krames, *J. Lightwave Technol.* **26**, 1154 (2008).
6. M. G. Craford, R. W. Shaw, A. H. Herzog, and W. O. Groves, *J. Appl. Phys.* **43**, 4075 (1972).
7. M. Bugajski, A. M. Kontkiewicz, and H. Mariette, *Phys. Rev. B* **28**, 7105 (1983).
8. J. C. Campbell, N. Holonyal, M. G. Craford, and D. L. Keune, *J. Appl. Phys.* **45**, 4543 (1974).

9. R. A. Logan, H. G. White, and W. Wiegmann, *Appl. Phys. Lett.* **13**, 139 (1968).

10. W. Gershenzon and R. M. Mikulyak, *Appl. Phys. Lett.* **8**, 245 (1966).

11. J. M. Woodall, H. Rupprecht, and W. Reuter, *J. Electrochem. Soc.* **116**, 899 (1969).

12. J. Nishizawa, S. Shinozaki, and K. J. Ishida, *J. Appl. Phys.* **44**, 1638 (1973).

13. G. B. Stringfellow, In *Crystal Growth: A Tutorial Approach*, IPE of III/V semiconductors, eds. W. Brandsley, D. T. J. Hurle, and I. B. Mullin, p. 217 (North Holland Pub, Amsterdam, the Netherlands, 1979).

14. M. G. Craford, *IEEE Trans. Electron Dev.* **ED-24**, 935 (1977).

15. I. Virgaftman, J. R. Meyer, and L. R. Ram-Mohan, *J. Appl. Phys.* **89**, 5815 (2001).

16. H. Rupprecht, J. M. Woodall, and G. D. Petit, *Appl. Phys. Lett.* **11**, 81 (1967).

17. Z. I. Alferov, V. M. Andreev, V. I. Borodulin, D. Z. Garbuzov, E. P. Morozov, T. G. Pak, A. I. Petrov, E. L. Portnoi, N. P. Chernousov, V. I. Shveikin, and I. V. Yashumov, *Fiz. Tekh. Poluprov.* **5**, 972 (1971).

18. Z. I. Alferov, V. M. Andreev, D. Z. Garbuzov, Y. V. Zhilyaev, E. P. Morozov, E. L. Portnoi, and V. G. Trofim, *Fiz. Tekh. Poluprov.*, **4**, 1826 (1970).

19. Z. I. Alferov, V. M. Andreev, D. Z. Garbuzov, and V. D. Rumyantsev, *Sov. Phys. Semicond. (Engtransl)* **9**, 305 (1975).

20. J. Nishizawa, S. Shinozaki, and K. J. Ishida, *J. Appl. Phys.* **44**, 1638 (1973).

21. J. Nishizawa and K. Suto, *J. Appl. Phys.* **48**, 3484 (1977).

22. S. Ishiguro, K. Sawa, S. Nagao, H. Yamanaka and S. Koike, *Appl. Phys. Lett.* **43**, 1034 (1983).

23. F. M. Steranka, High brightness light emitting diodes, In *Semiconductor and Semimetals*, eds. G. B. Stringfellow and M. G. Craford, Vol. 48 (Academic Press, San Diego, CA, 1997).

24. R. D. Burnham, N. Holonyak, Jr., and D. R. Scifres, *Appl. Phys. Lett.* **17**, 455 (1970).

25. D. R. Scifres, N. Holonyak, Jr., H. M. Macksey, and R. D. Dupuis, *Appl. Phys. Lett.*, **20**, 184 (1972).

26. H. M. Manasevit, *Appl. Phys. Lett.,* **12**, 156 (1968).

27. H. Sugawara, K. Itaya, M. Ishikawa, G. Hatakoshi, *Jpn. J. Appl. Phys.* **31**, 2446 (1992).

28. R. M. Fletcher, C. P. Kuo, T. D. Osentowski, K. H. Huang, M. G. Craford, and V. M. Robbins, *J. Electron. Mater.* **20**, 1125 (1991).

29. F. A. Kish, F. M. Steranka, D. C. DeFevere, D. A. Vanderwater, K. G. Park, C. P. Kuo, T. D. Osentowski et al., *Appl. Phys. Lett.* **64**, 2839 (1994).

30. J. I. Pankove, H. P. Maruska, and J. E. Berkeyheiser, Optical properties of GaN, In *Proc. 10th Int. Conf. Physics of Semiconductors*, Cambridge, MA, p. 593 (1970).

31. H. P. Maruska, D. Stevenson, and J. I. Pankove, *Appl. Phys. Lett.* **22**, 303 (1973).

32. J. I. Pankove, *IEEE Trans. Electron Dev.* **ED-22**, 721 (1975).

33. H. M. Manasevit, F. M. Erdman, and W. I. Simpson, *J. Electrochem. Soc.* **118**, 1864 (1971).

34. R. D. Dupius and P. D. Dapkus, *IEEE Trans. Electron Dev.* **ED-24**, 1195 (1977).

35. N. Holonyak, R. M. Kolbas, W. D. Laidig, B. A. Vojak, R. D. Dupius and P. D. Dapkus, *Appl. Phys. Lett.* **33**, 737 (1978).

36. H. Amano, N. Sawaki, I. Akasaki, and Y. Toyoda, *Appl. Phys. Lett.* **48**, 353 (1986).

37. I. Akasaki, H. Amano, K. Hiramatsu, and N. Sawaki, High efficiency blue LED utilizing GaN film with AlN buffer layer grown by MOVPE, In *Gallium Arsenide and Related Compounds 1987, Proc. 14th Int. Symp.*, Bristol, p. 633 (1988).

38. H. Amano, T. Asahi, and I. Akasaki, *Jpn. J. Appl. Phys.* **29**, L205 (1990).

39. H. Amano, M. Kito, K. Hiramatsu, and I. Akasaki, *Jpn. J. Appl. Phys.* **28**, L2112 (1989).

40. S. Nakamura, M. Senoh, and T. Mukai, *Jpn. J. Appl. Phys.* **30**, L1708 (1991).

41. S. Nakamura, T. Mukai, and M. Senoh, *Jpn. J. Appl. Phys.* **30**, L1998 (1991).

42. S. Nakamura, T. Mukai, M. Senoh, and N. Iwasa, *Jpn. J. Appl. Phys.* **31**, L139 (1992).

43. S. D. Lester, F. A. Ponce, M. G. Craford, and D. A. Steigerwald, *Appl. Phys. Lett.* **66**, 1249 (1995).

44. F. A. Ponce and D. P. Bour, *Nature* **386**, 351 (1997).

45. S. Chichibu, T. Azuhata, T. Sota, and S. Nakamura, *Appl. Phys. Lett.* **69**, 4188 (1996).

46. S. F. Chichibu, A. Uedono, T. Onuma, B. A. Haskell, A. Chakraborty, T. Koyama, P. T. Fini et al., *Nature Mater.* **5**, 810 (2006).
47. C. J. Humphreys, *Philos. Mag.* 87, 1971 (2007).
48. V. Yu Davydov et al., *Phys. Stat. Solidi (b),* **229**, R1 (2002).
49. V. Yu Davydov et al., *Phys. Stat. Solidi (b),* **230**, R4 (2002).
50. I. Ho, G. B. Stringfellow, *Appl. Phys. Lett.* **69**, 2701 (1996).
51. S. Nakamura, *MRS Bull.,* **34**, 101 (2009).
52. S. Nakamura, S. Pearton, and G. Fasol, *The Blue Laser Diode: The Complete Story* (Springer, Heidelberg, Germany, 2000).
53. N. F. Gardner, G. O. Muller, Y. C. Shen, G. Chen, S. Watanabe, W. Gotz, and M. R. Krames, *Appl. Phys. Lett.* **91**, 243506 (2007).
54. Y. C. Shen, G. O. Mueller, S. Watanabe, N. F. Gardner, A. Munkholm, and M. R. Krames, *Appl. Phys. Lett.* **91**, 141101 (2007).
55. M. F. Schubert, J. Xu, J. K. Kim, E. F., Schubert, M. H. Kim, S. Yoon, S. M. Lee, C. Sone, T. Sakong, and Y. Park, *Appl. Phys. Lett.* **93**, 041102 (2008).
56. Y.-L. Li, Y.-R. Huang, Y.-H. Lai, *Appl. Phys. Lett.* **91**, 181113 (2007).
57. B. Monemar and B. E. Sernelius, *Appl. Phys. Lett.* **91**, 181103 (2007).
58. M. Bugajski, A. M. Kontkiewicz, and H. Mariette, *Phys. Rev. B* **28**, 7105 (1983).
59. I. Tobias and A. Luque, *Prog. Photovoltaics* **10**, 323 (2002).
60. J. M. Phillips, M. E. Coltrin, M. H. Crawford, A. J. Fischer, M. R. Krames, R. Mueller-Mach, G. O. Mueller, Y. Ohno, L. E. S. Rohwer, J. A. Simmons and J. Y. Tsao, *Laser Photonics Rev.* **1**, 307 (2007).
61. C. H. Chen, S. A. Stockman, M. J. Peanasky, and C. P. Kuo, In *Semiconductors and Semimetals*, Vol. 48 (Academic Press, San Diego, CA, 1997, Chapter 4).
62. M. Kondow, H. Kakibayashi, T. Tanaka and S. Minagawa, *Phys. Rev. Lett.* **63**, 884 (1989).
63. J.-F. Lin, M.-C. Wu, M.-J. Jou, C.-M. Chang, C.-Y. Chen and B.-J. Lee, *J. Appl. Phys.* **74**, 1781 (1993).
64. J. S. Yuan, M. T. Tsai, C. H. Chen, R. M. Cohen and G. B. Stringfellow, *J. Appl. Phys.* **60**, 1346 (1986).
65. J. F. Geisz, D. J. Friedman, J. S. Ward, A. Duda, W. J. Olavarria, T. E. Moriarty, J. T. Kiehl, M. J. Romero, A. G. Norman, and K. M. Jones, *Appl. Phys. Lett.* **93**, 123505 (2008).
66. J. F. Geisz, et al., *Appl. Phys. Lett.* **91**, 023502 (2007).
67. M. A. Steiner, L. Bhusal, J. F. Geisz, A. G. Norman, M. J. Romero, W. J. Olavarria, Y. Zhang, and A. Mascarenhas, *J. Appl. Phys.* **106**, 063525 (2009).
68. S. P. Ahrenkiel et al., *J. Electronic Mater.* **33**, 185 (2004).
69. A. Gomyo, T. Suzuki and S. Iijima, *Phys. Rev. Lett.* **60**, 2645 (1988).
70. A. Mascarenhas, S. R. Kurtz, A. Kibbler and J. M. Olson, *Phys. Rev. Lett.* **63**, 2108 (1989).
71. A. Mascarenhas and Y. Zhang, In *Spontaneous Ordering in Semiconductor Alloys*, ed. A. Mascarenhas, p. 283 (Kluwer Academic, Dordrecht, the Netherlands/Plenum, New York, 2002).
72. Y. Zhang, A. Mascarenhas and L.-W. Wang, *Appl. Phys. Lett.* **80**, 3111 (2002).
73. J. K. Shurtleff, R. T. Lee, C. M. Fetzer and G. B. Stringfellow, *Appl. Phys. Lett.* **75**, 1914 (1999).
74. J. M. Olson, W. E. McMahon and S. Kurtz, *Conference Record of the 2006 IEEE 4th World Conference on Photovoltaic Energy Conversion*, Waikoloa, HI, p. 787 (2006).
75. Y. Zhang, A. Mascarenhas, and L.-W. Wang, *Phys. Rev. B* **90**, 235202 (2008).
76. Y. Zhang, C.-S. Jiang, D. J. Friedman, J. F. Geisz, and A. Mascarenhas, *Appl. Phys. Lett.* **94**, 091113 (2009).
77. L. Bhusal, B. Fluegel, M. A. Steiner, and A. Mascarenhas, *J. Appl. Phys.* **106**, 114909 (2009).
78. J. M. Shah, Y.-L. Li, T. Gessmann, and E. F. Schubert, *J. Appl. Phys.,* **94**, 2627 (2003).
79. H. Sugawara, M. Ishikawa, Y. Kokubun and S. Naritsuka, U.S. Patent no. 5048035 (1991).
80. E. F. Schubert, *Light Emitting Diodes* (Cambridge University Press, Cambridge, U.K., 2006).
81. CIE. *Commission Internationale de l'Éclairage Proceedings, 1924* (Cambridge University Press, Cambridge, MA, 1926).

82. CIE. *Commission Internationale de l'Éclairage Proceedings, 1931* (Cambridge University Press, Cambridge, MA, 1932).

83. S. M. Sze, *Physics of Semiconductor Devices* (Wiley-Interscience, New York, 2006).

84. Judd, D. B. Report of U.S. Secretariat Committee on Colorimetry and artificial daylight, *Proceedings of the Twelfth Session of the CIE, Stockholm* (p. 11). (Bureau Central de la CIE, Paris, 1951).

85. Vos, J. J., Colorimetric and photometric properties of a 2-deg fundamental observer. *Color Res. Appl.*, **3**, 125–128(1978).

86. CIE. Method of measuring and specifying colour rendering properties of light source. Publication no: 13.3 (1995).

87. G. Wyszecki and W. S. Stiles, *Color Science: Concept and Methods, Quantitative Data and Formulae* (Wiley, New York, 2000).

88. Ohno, Y. and W. Davis, Rational of color quality index, NIST, June 2010.

89. Multi-year program plan, Solid State Research and Development, EERE, U.S. DOE, March 2010. http://apps1.eere.energy.gov/buildings/publications/pdfs/ssl/ssl_mypp2010_web.pdf

10

Fundamentals of the Quantum Confinement Effect

Patanjali
Kambhampati
McGill University

10.1 Introduction to Quantum Dot

10.1.1 What Is a Quantum Dot?

A quantum dot (QD) is a semiconductor crystal with a size on the nanometer length scale. The nanometer length scale is key as this is the regime in which quantum confinement effects arise (the topic of this chapter). The simplest picture is that the radius of the crystallite should be approximately that of the Bohr radius of the electron (more precisely the exciton). The Bohr radius of an exciton (a_0) in a semiconductor will depend on material parameters which distinguish it from the Bohr radius of an electron in a hydrogen atom (Alivisatos, 1996a). These parameters (effective mass and dielectric constant) yield Bohr radii that range from ~1–20 nm, with $a_0 = 4\pi\varepsilon_0\varepsilon\hbar^2 m^{-1}q^{-2}$, with the permittivity of vacuum being ε_0, dielectric constant ε, reduced Planck constant $\hbar$, mass m, and charge q. When the crystallite is on the order of the excitonic Bohr radius, quantum confinement effects will arise (Efros and Rosen, 2000) (Figure 10.1).

The second qualitative point about QDs is their shape. The semiconductor QD is often called a semiconductor nanocrystal. The terms "quantum dot" and "nanocrystal" are used somewhat interchangeably. Historically, "quantum dot" is largely used to describe these semiconductor crystallites which are grown by epitaxial methods in ultra-high vacuum chambers. These are the epitaxial or self-assembled quantum dots (SAQD). Alternatively, one may grow these crystallites by solution phase chemistry. This form of crystallite is largely called a "nanocrystal" or a colloidal quantum dot (CQD). While there are

FIGURE 10.1 Image of CdSe colloidal quantum dots. (Reprinted with permission from Alivisatos, A.P., Perspectives on the physical chemistry of semiconductor nanocrystals, *J. Phys. Chem.*, 100, 13226–13239, 1996a. Copyright 1996 American Chemical Society.)

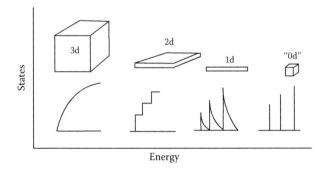

FIGURE 10.2 Illustration of the effect of spatial confinement of electrons on the electronic density of states. (Reprinted with permission from Alivisatos, A.P., Perspectives on the physical chemistry of semiconductor nanocrystals, *J. Phys. Chem.*, 100, 13226–13239, 1996a. Copyright 1996 American Chemical Society.)

important differences between SAQDs and CQDs such as shape, passivation, and size, there are also many similarities. It is the similarities that fall under the ideas of quantum confinement effect; hence both should be viewed as related lines within the family of quantum-confined nanostructures.

Based on the general idea of quantum-confined nanostructure, we view the term "quantum dot" as the preferred term since it captures the main idea of quantum confinement effects and also notes the geometry of the structure. For example, one may have a nanocrystal which does not exhibit quantum confinement effects due to Bohr radius arguments. Hence such a nanocrystal would not be a QD. Similarly, one has other quantum-confined geometries such as nanorods, shells, and tetrapods (Talapin et al., 2007). These nanostructures similarly may or may not show quantum confinement effects. In this chapter, we summarize the main ideas of quantum confinement effects in nanostructures by focusing on the CQD to illustrate the concepts (Figure 10.2).

10.1.2 Basic Science

The semiconductor QD offers much interest in basic science studies. Much of the interest arises from the fact that the QD is physically intermediate between the size limits of molecules and bulk solids

(Alivisatos, 1996a; Klimov, 2007). Decades of physical chemistry have provided a detailed understanding of electronic structure and dynamics in molecules. Similarly, condensed matter physics has led to understanding of bulk crystals. Not surprisingly, there are many differences between these limits, such as resolvable eigenstates in molecules and many-body interactions in solids. To what extent does the QD support molecule-like and bulk-like features? This question suggests much of the interest in semiconductor nanostructures in general, and QDs as a specific test bed with which to explore these ideas.

10.1.2.1 Electronic Structure

The main feature in QDs is the presence of quantum confinement effects, the topic reviewed in this chapter. The most striking and intuitive aspect of QDs is the ability to tune the absorption and emission spectra across the visible range by simply changing the particle size (Alivisatos, 1996b; Kippeny et al., 2002). This phenomenon can be qualitatively understood in terms of the simple particle in a sphere approach where the energy levels increase their spacing as the particle becomes small. In this manner, the QD may be considered an "artificial atom," a term which saw use in the early literature. Of course this artificial atom concept is merely qualitative and is not an exact analogy. The realistic electronic structure of the QD is quite rich, as will be discussed later. Much of the richness arises from the interpolation between the quantized limit of atoms and molecules, and the continuum limit of bulk solids which supports many-body interactions (Figure 10.3).

Another qualitative point of electronic structure is the manner in which one treats the system. While quantum mechanics is general, its implementation falls into two categories. Physical chemists may expect wavefunctions and potential energy surfaces based on the position space representation of quantum mechanics appropriate for aperiodic systems. Alternatively, periodic systems like bulk crystals are treated in the momentum space representation, thereby yielding band structure. One might naturally ask how to best treat the electronic structure of the dot, which may have on the order of 10 lattice periods. The localization in physical space renders a delocalization in momentum space via Fourier transformation. Consider the analogy of light waves. Normally one considers a laser to be monochromatic at some wavelength. Now suppose one has a pulse of light centered at 800 nm with 10 periods (like a QD of 3 nm diameter with a 0.3 nm period). This pulse of 10 periods Fourier transforms to a spectrum of 35 nm bandwidth. Hence, size (i.e., the periodicity) of the system will have implications on the manner in which one treats the electronic structure.

The first step in the discussion of the electronic structure of QDs is to note the nature of the excited states. By virtue of the many-electron nature of bulk solids and nanocrystals, it is convenient to introduce

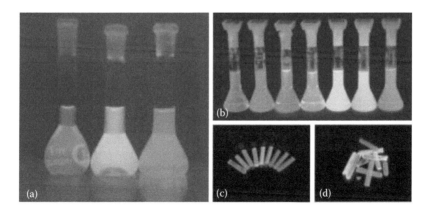

FIGURE 10.3 Images of photoluminescence (PL) from colloidal CdSe quantum dots in dispersion. The PL spectra redshift as the dot becomes larger. (Rogach et al. 2002, Organization of matter on different size scales: Monodisperse nanocrystals and their superstructures, *Adv. Func. Mater.*, 12(10), 653, 2002. Copyright Wiley VCH Verlag GmbH & Co. KGaA. Reprinted with permission.)

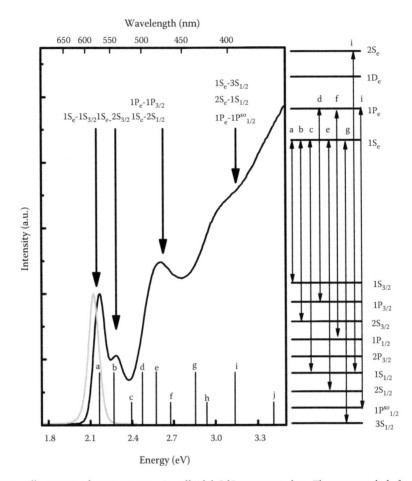

FIGURE 10.4 Illustration of excitonic states in colloidal CdSe quantum dots. The term symbols for the electron (upper manifold) and hole (lower manifold) are from the multiband effective mass approximation level of theory. Linear absorption spectrum of CdSe dots revealing excitonic transitions which are assigned as noted earlier.

excitations which appear like single particles. Promotion of an electron from its valence band (VB) to the conduction band (CB) is treated as creation of a quasiparticle called an electron, residing in the CB, and a hole residing in the VB. The uppermost states in the VB correspond to a molecular highest occupied molecular orbital (HOMO) and the lowest energy states in the CB correspond to the lowest unoccupied molecular orbital (LUMO). Hence, the energy gap between the CB and the VB edges corresponds to the HOMO–LUMO transition in a molecular picture. Since the electron is negatively charged, it will attract the positively charged hole that is left behind in the VB. This Coulombic attraction lowers the total energy of the electron-hole pair forming a bound state called an exciton (Scholes and Rumbles, 2006). These (Wannier) excitons are the relevant excitation in QDs. They account for the absorption features at low energy, the emission spectrum, and a large fraction of the linear and nonlinear optical properties of the dot. The main challenge in the electronic structure and spectroscopy of QDs is to obtain a realistic picture of excitons (Figure 10.4).

10.1.2.2 Dynamics

The real electronic structure of excitons in QDs results in a wide variety of dynamical processes which are similarly of interest for basic science as well as applications. For example, absorption of a photon can produce a high-energy exciton. This excited exciton will dissipate its energy as it relaxes to its lowest

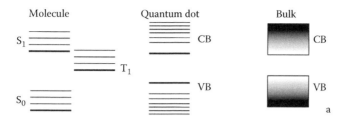

FIGURE 10.5 Illustration of the evolution from the continuous density of states (DOS) in a bulk semiconductor to the quantized DOS in a quantum-confined structure. As the structure gets small and undergoes quantum confinement effects, the energy gap increases as does the level spacing. A molecule has discrete electronic states such as excited singlet and triplet states, dressed by vibronic progressions. The quantum dot has a manifold of such states, with large spacings at low energy and smaller spacings with increasing energy. Like molecules, these excitonic states are similarly dressed with phonon progressions and singlet-triplet splittings (fine structure). In the bulk, there is a continuous density of electronic states.

energy state: the band edge exciton. The manner in which this process proceeds is similar to molecular concepts of radiationless transitions (Avouris et al., 1977). The paths by which the exciton can relax include phonon-based channels, electron–hole interactions, and coupling to surface ligands (Cooney et al., 2007a,b). Hence, the coupling between excitons and phonons is of importance. The phonons are the normal modes of the lattice vibrations. The excitonic wavefunctions enable one to compute exciton-phonon coupling strengths which gives rise to Frank–Condon progressions and may control optical nonlinearities and carrier relaxation dynamics (Sagar et al., 2008a,b). Exciton–exciton interactions are similarly important. For example, two excitons can recombine nonradiatively through pathways mediated by quantum confinement effects. This process is called multi-carrier recombination (MCR) (Klimov, 2000, 2006, 2007; Pandey and Guyot-Sionnest, 2007). The time reversal of MCR is multiple exciton generation (MEG), a point which has received considerable interest due to its relevance to QD-based solar cells (Klimov, 2006, 2007; Nozik, 2008). Finally, the very presence of multiple excitations per dot offers interesting science and informs applications. For example, two bound excitons form a biexciton (Sewall et al., 2006, 2008, 2009b). These biexcitons are key features in optical amplification by QDs (Cooney et al., 2009a,b) (Figure 10.5).

10.1.3 Applications

Certainly a large part of the interest in QDs is from their promise in a wide variety of applications. One of the earliest applications for CQDs is in biological imaging (Michalet et al., 2005). These materials have several appealing features for imaging such as: high absorption cross sections, narrow emission spectra, continuous absorption spectra, and photostability. Hence, CQDs have been replacing traditional dyes in biological imaging applications. These same characteristics suggest their importance in lighting applications such as light-emitting diodes (LEDs) and lasers. In the case of lasers, the QD work is a natural evolution from the early work on semiconductor quantum wells. In fact, the quantum well literature suggested the need to create a QD earlier than the first developments of the dot. In this application, the appeal lies in the large oscillator strengths and wide level spacings which should be helpful for developing novel optical gain media.

10.2 Hierarchies of Theory Overview

The objective of this chapter is to introduce the main theoretical approaches to understanding the electronic structure of QDs. We begin with the lowest level of theory and work toward the most sophisticated atomistic treatments of real QDs. Both as a student and as an experimentalist, it is useful to have a sense of the hierarchies of the theories. The idea of these hierarchies alerts the reader as to when each level of theory becomes necessary to understand some phenomenon. Since there has been extensive discussion of each of the methods elsewhere, our objective is not to provide an exhaustive survey of the methods and their predictions. Instead, we will refer the interested reader to the primary literature as well as some excellent reviews that focus on each specific method. With a basic overview of the methods, we are poised to understand the increasing levels of complexity of the electronic structure of QDs. Each level of complexity has some cost of abstraction and complexity. Our aim is to recognize the value provided by these increases in complexity by providing an explanation for the function of realistic QDs.

The simplest picture of the QD is the particle-in-a-sphere (PIS) (Brus, 1983, 1984). This picture enables one to qualitatively understand the size tunable absorption spectrum, and the existence of discrete features in the absorption spectrum. However, this simple picture misses much of the tremendous richness and complexity of these systems. Hence, it is important to note the hierarchies of theory necessary to understand aspects of the excitonics of QDs. In many cases, a higher level of theory does not merely produce a more precise number, but can qualitatively change the basic understanding of some observable.

The main levels of theory are: PIS, multiband effective mass approximation (EMA) approach, and atomistic approaches such as empirical pseudopotential method (EPM) and ab initio. The PIS approach provides no insight into many of the simplest aspects of the dot, such as the fluorescence Stokes shift or the nature of the excitonic transitions in the absorption spectrum. Hence, the minimal level of theory required to design and interpret these experiments is the EMA approach. EMA has been applied to these colloidal CdSe QDs by Efros, Bawendi, and Norris (Ekimov et al., 1993; Efros et al., 1996; Norris and Bawendi, 1996; Norris et al., 1996; Efros and Rosen, 2000; Norris, 2004). Those studies have yielded tremendous insight into the electronic structure and linear optical properties of these materials. These EMA results have furthermore informed the design and much of the interpretation of our experiments (Sewall et al., 2006, 2008, 2009a,b; Cooney et al., 2007a,b, 2009a,b; Sagar et al., 2008a,b), some of which are summarized here. We note, however, that the atomistic (whether EPM or ab initio) approaches often yield qualitatively different results (e.g., ordering of states, bright/dark states, piezoelectricity, exciton cooling). Hence, the level of theory used to understand the observables is quite important (Figure 10.6).

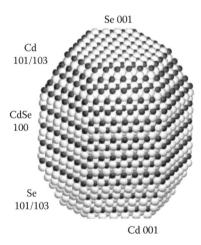

FIGURE 10.6 Illustration of lattice structure of colloidal CdSe quantum dots. The PIS approach considers only a sphere of some radius with a single electron band and a single hole band. The EMA approach includes additional bands as well as lattice asymmetries as perturbations. The atomistic approaches (e.g., EPM) include the realistic physical structure as a starting point. (Reprinted with permission from Kippeny, T., Swafford, L.A., and Rosenthal, S.J., Semiconductor nanocrystals: A powerful visual aid for introducing the particle in a box, *J. Chem. Educ.*, 79, 1094, 2002. Copyright 2002 American Chemical Society.)

10.3 Particle in a Sphere

10.3.1 Summary of the Approach

The PIS approach was first described by Brus (Brus, 1983, 1984; Kippeny et al., 2002) to describe the lowest absorption feature (band edge exciton) of small semiconductor crystallites. The PIS as well as the EMA methods are continuum models which do not retain the atomistic details of the dot. The PIS model can qualitatively describe the size dependence of the absorption edge, but it has many limitations, as will be discussed later.

The starting point of the PIS model is to assume (1) a sphere of radius R; (2) a homogeneous medium of some dielectric constant, with no point charges and no potential; and (3) an infinite potential outside the sphere. These conditions give the following Hamiltonian

$$\widehat{H} = -\frac{h^2}{8\pi m_c^2}\nabla_c^2 + V \tag{10.1}$$

with

$$V = \begin{cases} 0 & r < R \\ \infty, & r \geq R \end{cases}. \tag{10.2}$$

Here,

    R corresponds to the radius of the dot

    r is the radial coordinate

    m_c is the effective mass of the carrier (electron or hole)

We note that m_c is more generically noted as m^*, and will be discussed in more detail in the section on the EMA method. Briefly, m^* is called the effective mass as it describes the interaction of the electron and hole quasiparticles with the crystal lattice. The solutions to this Hamiltonian are

$$\Psi_n(r) = \frac{1}{r\sqrt{2\pi R}}\sin\left(\frac{n\pi r}{R}\right) \tag{10.3}$$

$$E_n = \frac{h^2 n^2}{8m_c R^2}, \quad n = 1, 2, 3\ldots \tag{10.4}$$

These equations correspond to an isolated electron in a sphere, a situation which might represent a real electron injected into a dot. In the more common case of optical excitation, an exciton is produced which corresponds to an interacting electron–hole pair. In the case of the exciton, the Hamiltonian is now

$$\widehat{H} = -\frac{h^2}{8\pi m_e^2}\nabla_e^2 - \frac{h^2}{8\pi m_h^2}\nabla_h^2 + V(\vec{S_e}, \vec{S_h}). \tag{10.5}$$

$\vec{S_e}$ and $\vec{S_h}$ correspond to the positions of the electron/hole point charges

$m_e = 0.13\,m_0$

$m_h = 0.45\,m_0$ for CdSe

m_0 is the real electron mass

The potential energy term has two contributions. The first is from direct Coulomb interaction between the electron and hole. The second is from polarization based on dielectric solvation. Essentially, each point charge polarizes the dielectric medium (ε) which produces a back reaction on the counter charge. Now, the excitonic PIS Hamiltonian reads

$$\widehat{H} = -\frac{h^2}{8\pi m_e^2}\nabla_e^2 - \frac{h^2}{8\pi m_h^2}\nabla_h^2 - \frac{e^2}{4\pi\varepsilon_{CdSe}\varepsilon_0\left|\vec{s_e}-\vec{s_h}\right|} + \frac{e^2}{2}\sum_{k=1}^{\infty}\alpha_k\frac{S_e^{2k}+S_h^{2k}}{R^{2k+1}}, \qquad (10.6)$$

with

$$\alpha_k = \frac{(\varepsilon-1)(k+1)}{4\pi\varepsilon_{CdSe}\varepsilon_0\,(\varepsilon k+k+1)}. \qquad (10.7)$$

The electron charge is e and $\varepsilon=\varepsilon_{CdSe}/\varepsilon_{medium}$. As R becomes large, the polarization term (fourth term) becomes small and one recovers a hydrogenic Hamiltonian for the exciton. Using a simple product wavefunction (no correlation)

$$\Phi_X\left(\vec{S_e},\vec{S_h}\right) = \psi\left(\vec{S_e}\right)\psi\left(\vec{S_h}\right) \qquad (10.8)$$

Brus performed a variational calculation which yielded

$$E_X = \frac{h^2}{8R^2}\left(\frac{1}{m_e}+\frac{1}{m_h}\right) - \frac{1.8e^2}{4\pi\varepsilon_{CdSe}\varepsilon_0 R} + \frac{e^2}{R}\sum_{k=1}^{\infty}\alpha_k\left(\frac{S}{R}\right)^{2k}. \qquad (10.9)$$

The mean separation between the charges is given by

$$\left|\vec{S_e}-\vec{S_h}\right| = \frac{R}{1.8}. \qquad (10.10)$$

So the total energy is given by $E_x = E_{kinetic} + E_{Coulombic} + E_{polarization}$. The polarization term is the smallest and is convenient to omit, given the qualitative nature of the calculation. Hence, the first term is the kinetic energy of two particles of unequal mass in a sphere, and the second term is the attraction between these two charges. Now, the total energy of the system is the energy required to excite the electron across the band gap (E_g) plus the energy of the exciton (E_x):

$$E_{total} = E_g + E_x. \qquad (10.11)$$

This simple model developed by Brus is essentially a single band particle in an infinite sphere model with Coulomb interaction between the electron and hole treated as a mean field perturbation.

10.3.2 Predictions and Limitations of the Approach

With this simple model, one can obtain qualitative insight into the electronic structure of spherical QDs. The main prediction of this model is the strong size dependence of the energy of the band edge exciton (the onset of absorption for the dot). For large R, there will be a $1/R$ dependence of the total energy, and for small R there will be a $1/R^2$ dependence. The comparison to experiment is reasonably

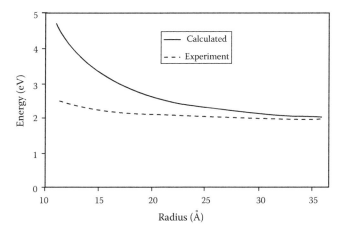

FIGURE 10.7 Comparison between experiment and theory for the PIS approach. The PIS approach yields a qualitative understanding of the size dependence to the intraband energy gap. (Reprinted with permission from Kippeny, T., Swafford, L.A., and Rosenthal, S.J., Semiconductor nanocrystals: A powerful visual aid for introducing the particle in a box, *J. Chem. Educ.*, 79, 1094, 2002. Copyright 2002 American Chemical Society.)

good for large R ($R > 2.5$ nm) and fails for smaller R ($R < 2$ nm). This model is not appropriate for treating the higher excitonic states which are observed in the linear absorption spectrum. And it does not explain the origin of the photoluminescence (PL) Stokes shift (energy gap between absorbing and emitting states). However, the value of this model is that it gives a qualitative estimate of the most obvious aspect of the quantum confinement effect: the ability to tune the spectra of colloidal CdSe QDs across the visible spectrum simply by changing their size. In order to obtain an understanding of the linear absorption spectrum and the emission spectrum, one must turn to the next level of theory: the multiband effective mass approach (Figure 10.7).

10.4 Multiband Effective Mass Approximation Approach

The PIS approach discussed earlier is contingent upon two main assumptions which are connected to the idea of energy bands in periodic systems. In addition to any assumptions noted in the summary of Brus' derivation, it is assumed that there are single bands for the electron and hole, and that these bands are parabolic. The idea is that within the dot there is no potential so the carriers behave in a free electron-like manner: the energy goes as $p^2/2m_0$, or $\hbar^2 k^2/2m_0$. For the electron or hole in the semiconductor, the *real* electron mass (m_0) is replaced by the carrier effective mass (m^* or m_c)—the effective mass (tensor) is given by the reciprocal of the second derivative of energy with respect to wavevector. Hence, the use of an effective mass with a single band is called a nearly free electron which results from the many-body nature of the electronic structure of the semiconductor.

However, it is known that the VB may comprise several closely lying bands in the case of II–IV semiconductors. Hence, the coupling between these bands may be of some importance. Furthermore, it is known that bands are generally not parabolic. This non-parabolicity of the CB and VB manifolds will be important as well. Should one aim for a realistic picture of the states and transitions in a real QD; one must move beyond the PIS approach and use the multiband EMA approach. The EMA approach was applied to spherical "nanocrystal" QDs by Efros and coworkers. This work was subsequently applied to high-quality CQDs by Norris, Bawendi, and Efros (Ekimov et al., 1993; Efros et al., 1996; Norris and Bawendi, 1996; Norris et al., 1996; Efros and Rosen, 2000; Norris, 2004). We will follow the excellent discussion by Norris (Norris, 2004) to which the reader is referred for a more thorough treatment (Figure 10.8).

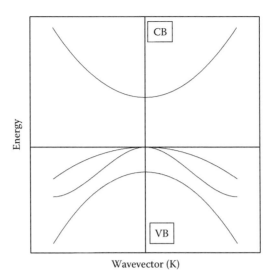

Wavevector (K)

FIGURE 10.8 Illustration of energy bands in periodic solids. This is a schematic of the band structure of bulk CdSe which has a single low-energy conduction band (CB) and three low-energy valence bands (VB). The bands can be degenerate, non-parabolic, and anisotropic. These bands form the basis for creating the quantized excitonic states in quantum dots.

The concept of energy bands was implicit in the PIS discussion. Here, these ideas will be explicitly discussed. Bloch's theorem states that the electronic wavefunction in a periodic system can be written as

$$\Psi_{nk}(\vec{r}) = u_{nk}(\vec{r})\exp(i\vec{k}\cdot\vec{r}), \tag{10.12}$$

where u is a function with the periodicity of the lattice and the functions are labeled by the band index n and the wavevector k. Now, the energy of an electron in the CB and a hole in the VB is treated as nearly free electron, with

$$E_{CB}(k) = \frac{\hbar^2 k^2}{2m_g} + E_g \tag{10.13}$$

and

$$E_{VB}(k) = -\frac{\hbar^2 k^2}{2m_h}. \tag{10.14}$$

So the effective masses are given by the curvature in *CB* and *VB*, which are not necessarily parabolic in a more realistic case. The idea of the effective mass is that it accounts for the interaction between the carrier and the periodic potential of the crystal. Now, one assumes that the single particle (electron or hole) wavefunctions can be written in terms of these Bloch functions. This *envelope approximation* yields

$$\psi(\vec{r}) = \sum_k C_{nk} U_{nk}(\vec{r})\exp\left(i\vec{k}\cdot\vec{r}\right). \tag{10.15}$$

Assuming the u_{nk} functions have a weak k dependence (non-dispersive), one obtains

$$\psi(\vec{r}) = U_{no}(\vec{r})\sum_k C_{nk}\exp\left(i\vec{k}\cdot\vec{r}\right) = U_{no}(\vec{r})f(\vec{r}). \tag{10.16}$$

The functions f are the single particle envelope functions (1S, 1P, etc.). And the functions u can be represented as a sum of atomic wavefunctions, φ, using the linear combination of atomic orbitals (LCAO) approximation:

$$u_{nk} = \sum_i C_{nl}\varphi_n\left(\vec{r} - \vec{r_i}\right).$$ (10.17)

At this point, one needs to obtain the envelope functions, f, which were previously obtained solving the PIS problem.

At this point, we introduced the basic ideas of dispersive states which yield energy bands in periodic systems. In the case of one filled band in the VB and one empty band in the CB, one obtains the particle in a sphere solution to the single band model. Of course these bands are furthermore isotropic and parabolic. These three approximations can be lifted to give further insight into the real electronic structure of the QD. In the case of II–VI semiconductors (e.g., CdSe), the CB arises from Cd 5S atomic orbitals and is twofold (spin) degenerate at $k=0$. The VB arises from Se 4p atomic orbitals and is sixfold degenerate at $k=0$. The degeneracy within the VB is the first point to address in moving beyond the PIS model.

Prior to the introduction of quantum confinement effects, the VB has an additional structure. There is strong spin–orbit interaction which creates a split off band by separating the $p_{3/2}$ and $p_{1/2}$ bands. The subscript refers to the total spin–orbit angular momentum $J = L + S$. In this case, $L = 1$ and $S = 1/2$. Secondly, the $p_{3/2}$ bands are further split at non-zero wavevector ($k \neq 0$) into bands with $J_z = \pm 3/2$ and $\pm 1/2$. These two are called the heavy hole (hh) and light hole (lh) bands, respectively. The notation of heavy/light comes from the effective mass of a hole in each band which has dissimilar curvature. Finally, the hh and lh bands are further split at $k=0$ due to crystal field splitting in the wurtzite CdSe lattice. These effects of multiple bands with differing curvatures can be incorporated into the envelope picture by using $k \cdot p$-perturbation theory (Figure 10.9).

The idea of $k \cdot p$ (k dot p)-perturbation theory is that one can expand the bulk bands about $k=0$ and obtain analytic solutions using a perturbative approach. With the single particle Hamiltonian

$$H_0 = \frac{p^2}{2m_0} + V(x),$$ (10.18)

one can show that

$$\left[H_0 + \frac{1}{m_0}\left(k \cdot p\right)\right]u_{nk} = \lambda_{nk}u_{nk},$$ (10.19)

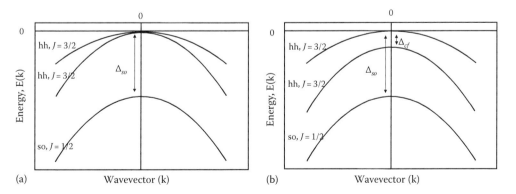

FIGURE 10.9 Illustration of the electronic structure of the VB. These three VBs mix to form the hole states in the quantum dot. The VBs do not have the same curvature (effective mass). The bands with less curvature are denoted as heavy-hole (hh). The degeneracy can be lifted by the crystal field (Δ_{cf}) and spin–orbit (Δ_{so}) interactions.

with

$$\lambda_{nk} = E_{nk} - \frac{k^2}{2m_0}. \tag{10.20}$$

Knowing u_{n0} and E_{n0}, one can treat (10.19) using second-order perturbation theory to obtain energies

$$E_{nk} = E_{n0} + \frac{k^2}{2m_0} + \frac{1}{m_0^2} \sum_{m \neq n} \frac{\left| \vec{k} \cdot \overrightarrow{p_{nm}} \right|^2}{E_{n0} - E_{m0}} \tag{10.21}$$

and wavefunctions

$$u_{nk} = u_{n0} + \frac{1}{m_0} \sum_{m \neq n} u_{m0} \frac{\vec{k} \cdot \overrightarrow{p_{nm}}}{E_{n0} - E_{m0}} \tag{10.22}$$

with

$$\overrightarrow{p_{nm}} = \left\langle u_{n0} \left| \vec{p} \right| u_{m0} \right\rangle. \tag{10.23}$$

Equations 10.18 through 10.23 outline the general approach to a perturbative treatment of band effects such as non-parabolicity and mixing.

This general approach was applied to specific features in the realistic electronic structure of II-IV nanocrystal quantum dots within the multiband EMA method by Efros, Rosen, Bawendi, and Norris. We provide a brief summary of this work here. The reader is referred to the chapter by Norris for an introduction, the review by Califano and Mulvaney for further details, and the reviews by Efros for the full picture within EMA. In order to obtain the total Hamiltonian, one takes the zeroth-order Hamiltonian (Equation 10.8) and adds, E_g to include the bandgap, and a term D, which treats the perturbations in the VB:

$$H = H_0 + E_g + D. \tag{10.24}$$

Since the VB arises from p atomic states, these states are denoted as $\{|X\rangle, |Y\rangle, |Z\rangle\}$ and $\{|\alpha\rangle, |\beta\rangle\}$ for spin. The aim is to introduce additional perturbations such as spin–orbit interactions, crystal field splittings, electron–hole exchange, and shape asymmetries.

Within this basis, one can treat the spin–orbit interaction as a perturbation and see the manner in which excitonic fine structure arises from the angular momentum interactions in the VB. The spin–orbit term in the Hamiltonian is

$$H_{SO} = \varepsilon_{SO} L \cdot S. \tag{10.25}$$

The spin–orbit coupling constant is ε_{SO} and the L and S are the orbit and spin operators. The eigenstates of (10.25) are the eigenstates of the total angular momentum, $J = L + S$. With $L = 1$ and $S = 1/2$, J becomes 3/2 and 1/2. Now, the eigenstates $\{|J, J_z\rangle\}$ are

$$\left| \frac{3}{2}, \frac{3}{2} \right\rangle = \frac{1}{\sqrt{2}} (X + iY)\alpha \tag{10.25a}$$

$$\left|\frac{3}{2},\frac{1}{2}\right\rangle = \frac{1}{\sqrt{6}}\left[(X+iY)\beta - 2Z\alpha\right] \tag{10.25b}$$

$$\left|\frac{3}{2},-\frac{1}{2}\right\rangle = \frac{1}{\sqrt{6}}\left[(X-iY)\alpha + 2Z\beta\right] \tag{10.25c}$$

$$\left|\frac{3}{2},-\frac{3}{2}\right\rangle = \frac{1}{\sqrt{2}}(X-iY)\beta \tag{10.25d}$$

$$\left|\frac{1}{2},\frac{1}{2}\right\rangle = \frac{1}{\sqrt{3}}\left[(X-iY)\beta + Z\alpha\right] \tag{10.25e}$$

$$\left|\frac{1}{2},-\frac{1}{2}\right\rangle = \frac{1}{\sqrt{3}}\left[-(X-iY)\alpha + Z\beta\right]. \tag{10.25f}$$

The $J = 1/2$ states are lower in energy in the VB and are called the "split off" band. One can now write an expression for the VB perturbations by recasting D in this basis. Doing so yields

$$D = \begin{pmatrix} P+Q & L & M & O & iL/\sqrt{2} & -i\sqrt{2}M \\ L^{\star} & P-Q & O & M & i\sqrt{2}Q & i\sqrt{3/2}L \\ M^{\star} & O & P-Q & -L & -i\sqrt{3/2}L^{\star} & -i\sqrt{2} \\ O & M^{\star} & -L^{\star} & P+Q & -i\sqrt{2}M & -iL^{\star}/\sqrt{2} \\ -iL^{\star}/\sqrt{2} & -i\sqrt{2}Q & i\sqrt{3/2}L & i\sqrt{2}M & P-\Delta & O \\ i\sqrt{2}M^{\star} & -i\sqrt{3/2}L^{\star} & -i\sqrt{2}Q & iL\sqrt{2} & O & P-\Delta \end{pmatrix}. \tag{10.26}$$

The specific elements of this Luttinger–Kohn Hamiltonian for the case of crystals with inversion symmetry are detailed in the reviews by Efros and Califano and Mulvaney (Gomez et al., 2006). The problem is now one of the motion of a spin 3/2 particle in a degenerate VB. Following some tensor algebra, one can reduce the term D to a simplified form which is proportional to the dot product, $\boldsymbol{P} \cdot \boldsymbol{J}$. As this dot product resembles spin–orbit interaction, one notes the *total angular momentum* $\boldsymbol{F} = \boldsymbol{L} + \boldsymbol{J}$, with the states labeled nL_F.

Implementing the approach summarized above enabled Ekimov and Xia to reproduce many of the coarse features in the linear absorption spectra. However, additional effects are needed to correctly describe the emissive state. In addition to the above mixing of VB states, one must include the crystal field, shape anisotropy, and finally electron–hole exchange in order to fully describe the band edge exciton. The interested reader is referred to the excellent review by Califano and Mulvaney (Gomez et al., 2006) for further detail and the original works by Efros (Ekimov et al., 1993; Efros and Rosen, 2000). Since these details have been presented extensively elsewhere, we will simply summarize the results (Figure 10.10).

The above discussion of fine structure within the VB has as its main point the creation of fine structure within the band edge exciton. Consider the band edge exciton being created from a 1S electron in the CB and a 1S hole created in the VB. This bound electron–hole pair creates the lowest energy exciton $(X_1) = 1S_e - 1S_h$, which is eightfold degenerate. The VB is fourfold degenerate, followed by twofold spin

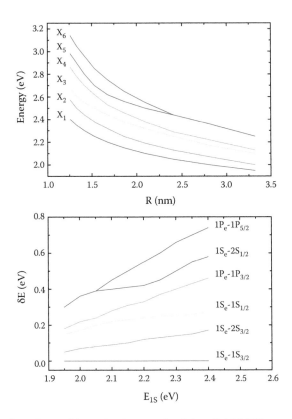

FIGURE 10.10 The size dependence of the excitonic energy levels in colloidal CdSe quantum dots (top panel). The energy of the excited excitonic states as a function of the first excitonic state, X_1. The generic excitonic states (X_i) are related to their assignments in the EMA model.

degeneracy. Within EMA, the band edge exciton is now labeled $1S_e$–$1S_{3/2}$, where the subscript on the hole state denotes F. The above perturbations split the band edge exciton into five levels. It is this degeneracy breaking which gives rise to the Stokes shift between the absorbing and emitting states, as well as the lifetime of these states.

In detail, the problem of the electronic structure of the band edge (1S) exciton becomes one of the perturbations which create coupling between momenta. Both the VB and CB are constructed from an atomic orbital basis of some (l, s). Through spin–orbit interaction, these momenta are coupled to produce J for each band. In both cases, there is an envelope function which has an angular momentum, L. Due to the degeneracy of the VB, the VB bands are mixed and a total angular momentum of the hole is now the constant of motion, F. Finally, electron–hole Coulomb and exchange interactions yield a total exciton momentum, N (Figure 10.11).

At one limit there is the unperturbed, degenerate band edge exciton ($1S_e$–$1S_{3/2}$ in EMA) which is eightfold degenerate. Including only electron–hole exchange yields the exchange-correlated exciton momentum, N. The $N = 2$ exciton is dark and lower in energy, relative to $N = 1$. This exchange splitting is analogous to singlet-triplet splitting in molecules. It is also seen in all forms of dots, including the larger SAQD. The other effects such as crystal field and shape-based splittings are case specific. This discussion is centered about colloidal CdSe QDs as the test case to illustrate the premises. These additional perturbations split the doublet into five levels based on the projection of N onto the unique c-axis of wurtzite CdSe. Now, one has from lowest to highest energy $|N_z| = (2, 1, 0, 1, 0)$. These sub-states are (dark, bright, dark, bright, dark), respectively (Figure 10.12).

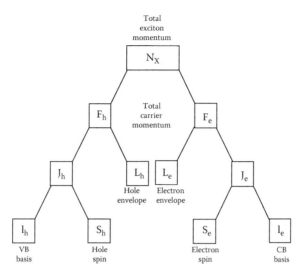

FIGURE 10.11 The creation of "good quantum numbers" to denote excitonic states. The quantum numbers are used to denote electron and hole states as term symbols, e.g., $1S_e-1S_{3/2}$. Alternatively, one can describe the excitonic state by combining the electron and hole states.

10.5 Atomistic Approaches

The EMA approach is the most widely used method to calculate the electronic structure of QDs. Unlike the PIS method, it can explain the fluorescence Stokes shift, the temperature dependence of the lifetimes, and a reasonable assignment of the nature of the excitonic transitions in the linear absorption spectra. The EMA offers a convenient notation for designing and interpreting experiments. In particular, the EMA approach, as applied to colloidal CdSe QDs by Efros and Bawendi (Ekimov et al., 1993; Efros and Rosen, 2000), has guided many experiments. This approach has also been applied to dots of other compositions such as InAs (Banin et al., 1998; Efros and Rosen, 2000) and PbSe (Kang and Wise, 1997).

Despite its widespread use, the EMA does have shortcomings both on theoretical grounds and also in the ability to explain aspects of recent experiments. The EMA assumes infinite boundaries, a continuum-like crystal structure, VB warping, CB non-parabolicity (Efros and Rosen, 2000; Norris, 2004; Gomez et al., 2006). The EMA completely neglects the surface of the dots; it cannot treat effects like piezoelectricity and does not give the same ordering of excitonic states, as suggested by atomistic theories and new experiments (Franceschetti and Zunger, 1997; Wang and Zunger, 1998; Fu et al., 1998; Franceschetti et al., 1999; Zunger, 2001, 2002; Gomez et al., 2006; Bester, 2009; Kilina et al., 2009; Prezhdo, 2009). It furthermore does not give a qualitatively correct picture of the electronic structure in PbSe dots. In addition to the continuum methods of EMA, atomistic methods have been implemented to understand the electronic structure of realistic QDs and to better explain experiments. It is important to note that these higher levels of theory often give more than simply a more precise number for some observable. In some cases, these atomistic theories can give qualitatively different pictures of the electronic structure and resultant dynamics. These issues will be summarized in Section 10.6 (Figure 10.13).

The atomistic methods have been reviewed in detail by Califano (Gomez et al., 2006), Zunger (Zunger, 2001, 2002), Bester (Bester, 2009), and Prezhdo (Prezhdo, 2009) and will only be briefly discussed. The objective is to make note of the differences in how physical effects are treated and their implication on the electronic structure and dynamics of excitons in QDs.

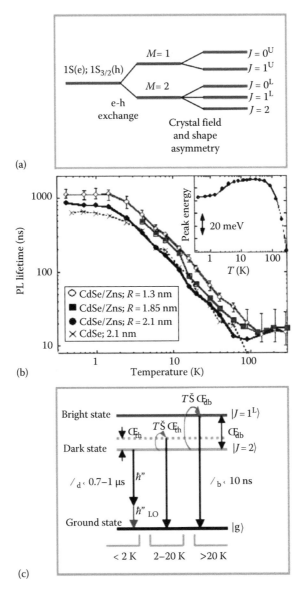

(a)

(b)

(c)

FIGURE 10.12 The creation of fine structure in the band edge exciton: $1S_e$–$1S_{3/2}$ in the EMA picture. The various perturbations mix states and break degeneracies, thereby yielding a ladder of states within the band edge exciton (a). At ~300 K, emission is from the optically bright upper states and at ~10 K, emission is from the optically forbidden (in EMA) dark states (b). This ladder of states accounts for the Stokes shift between absorption and emission, (c). (Reprinted with permission from Klimov, V.I., Mechanisms for photogeneration and recombination of multi-excitons in semiconductor nanocrystals: Implications for lasing and solar energy conversion, *J. Phys. Chem. B*, 110, 16827–16845, 2006. Copyright 2006 American Chemical Society.)

The EPM was developed to describe the band structure of bulk semiconductors. The starting point is the positions of the constituent atoms based on experimentally determined crystal structures. Based on geometry, the single particle Schrodinger equation is solved as follows:

$$\left[-\frac{\hbar^2}{2m_o}\nabla^2 + V_{PS}(r) + V_{nl} \right]\psi_i(r,\sigma) = \varepsilon_i\psi_i(r,\sigma). \qquad (10.27)$$

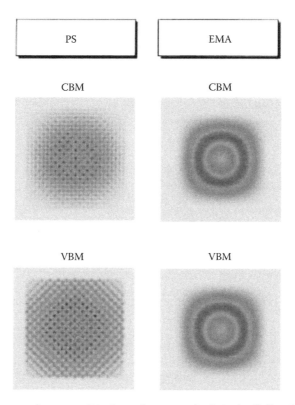

FIGURE 10.13 Comparison of atomistic (EPM) wavefunctions of a GaAs dot (left) and EMA calculation of the same (right). (Reprinted with permission from Franceschetti, A. and Zunger, A., Direct pseudopotential calculation of exciton Coulomb and exchange energies in semiconductor quantum dots, *Phys. Rev. Lett.*, 78, 915–918, 1997. Copyright 1997 by the American Physical Society.)

Here, V_{ps} is the screened microscopic pseudopotential obtained from a superposition of screened atomic potentials as follows:

$$V_{ps}(r) = \sum_{i\alpha} v_\alpha (r - R_{i,\alpha}). \tag{10.28}$$

The v's are the screened atomic potentials of atom type α at the position $\mathbf{R}_{i,\alpha}$, and V_{nl} is a short-range operator that treats the nonlocal part of the potential including spin–orbit interaction. The atomic pseudopotentials are derived from bulk density functional theory (DFT) pseudopotentials and are fitted to reproduce bulk features such as transition energies, deformation potentials, effective masses, and bulk wavefunctions. Since the approach is derived from the treatment of periodic solids, it can be extended to treat quantum-confined solids of arbitrary dimensionality and shape while retaining atomistic details. The excitonic wavefunctions are, $\Psi^{(\rho)}$, where ρ denotes the set of excitonic quantum numbers. These wavefunctions are obtained from single substitution Slater determinants following

$$\psi^{(\rho)} = \sum_{v=1}^{N_v} \sum_{c=1}^{N_c} C_{v,c}^{(\rho)}, \tag{10.29}$$

with the excited states

$$\Phi_{V,C}(r_1,\sigma_1,\ldots,r_N,\sigma_N) = \left[\psi_1(r_1,\sigma_1),\ldots \psi_c(r_V,\sigma_V),\ldots \psi_N(r_N,\sigma_N) \right], \tag{10.30}$$

and the ground state

$$\Phi_0(r_1,\sigma_1,\ldots,r_N,\sigma_N) = [\psi_1(r_1,\sigma_1),\ldots\psi_V(r_V,\sigma_V),\ldots\psi_N(r_N,\sigma_N)]. \tag{10.31}$$

Here
  N is the total number of electrons
  a subscript denotes the numbers in the VB and CB
  σ is the spin state

The excited states are obtained by promoting an electron from the valence state (ψ_v) to the conduction state (ψ_c). The excitonic states are then obtained by solving the secular equation in this basis:

$$\sum_{v'=1}^{N_v}\sum_{c'=1}^{N_c} H_{vc,v'c'}C_{v',c'}^{(\rho)} = E^{(\rho)}C_{v,c}^{(\rho)}. \tag{10.32}$$

The matrix elements in this basis are

$$H_{vc,v'c'} \equiv \langle\Phi_{V,C}|H|\Phi_{V'C'}\rangle = (\varepsilon_c - \varepsilon_v)\delta_{v,v'}\delta_{c,c'} - J_{vc,v'c'} + K_{vc,v'c'}. \tag{10.33}$$

where the J's and K's are the direct and exchange Coulomb integrals

The details of the approach will not be further discussed as the objective is to highlight the different predictions and explanations for select phenomena.

In addition to the EPM method, one can perform fully ab initio atomistic calculations using for example, DFT (Prezhdo, 2008, 2009; Kilina et al., 2009). The ab initio methods introduced increased precision, decreased reliance on external parameters, at the cost of a smaller model system. For example, it may be tractable to use DFT methods on dots with 10^2–10^3 atoms, whereas EPM methods can be used for dots with 10^3–10^6 atoms. Hence, the ab initio methods are now approaching the capacity to treat model dots which are nearly as large as real dots. At present, these DFT methods are still applied toward high-level calculations of dots smaller than what are experimentally studied. Nonetheless, one expects these methods will soon become possible for accurately treating electronic structure in realistically sized dots. A particular appeal of this approach is the ability to treat dynamics on an appropriate quantum mechanical level using time-dependent density functional theory (TDDFT) and nonadiabatic molecular dynamics. These points will be discussed later (Figure 10.14).

10.6 Predictions and Limitations of the Approaches

The main result of the EMA treatment is an understanding of the linear optical properties of the dots, in particular the emitting states. These results are summarized here, in light of the original experimental work by Bawendi and coworkers (Norris et al., 1994, 1996; Nirmal et al., 1995; Norris and Bawendi, 1995, 1996; Efros et al., 1996). It is worth noting that recent experiments are once again causing this problem to be revisited. Hence, the following section will note deviations from the EMA picture.

While the earlier discussion focused on the electronic structure of the band edge exciton, there are also implications on the ordering of excitonic states within the "coarse structure." The coarse structure leads to the features in the linear absorption spectrum. Some of the predictions are quite qualitative—Is some feature S-like or P-like in symmetry? These issues will have impact on nonlinear experiments such as THz probing of hole cooling (Hendry et al., 2006), optical gain (Cooney et al., 2009a,b), and multiexciton processes (Klimov, 2006, 2007; McGuire et al., 2008).

Aspects of the spectroscopy of QDs will be discussed in greater length in a different chapter in this book. In this section, we will briefly note several experimental observations which can be rationalized in

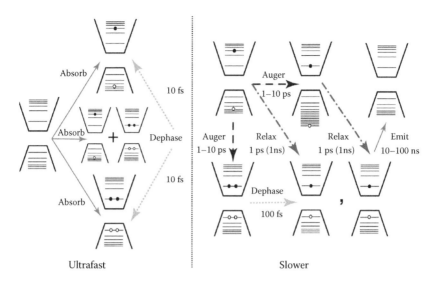

FIGURE 10.14 Dynamical processes in quantum dots which may be computed by high-level ab initio methods. (Reprinted with permission from Prezhdo, O.V., Photoinduced dynamics in semiconductor quantum dots: Insights from time-domain ab initio studies, *Acc. Chem. Res.*, 42, 2005–2016, 2009. Copyright 2009 American Chemical Society.)

different ways based on the theoretical approach used. As earlier, the objective here is not to exhaustively detail the experimental or the theoretical methods, but to note how the hierarchies in the theories relate to the experiments.

The simple size dependence of the first absorption feature (band edge exciton) is qualitatively described using the PIS approach. Despite its simplicity, this approach is reasonably accurate in providing an estimation of the excitation energies and the functional form of its size dependence. As expected, higher levels of theory yield better matches with experiment. This is an unsurprising result and does not suggest the point at which higher levels of theory are required.

The first instance, which requires invoking a higher level of theory, is the nature of the emitting state. One can absorb a photon into the 1S band edge exciton, or alternatively absorb into higher excitonic states followed by intraband relaxation to the band edge exciton. The emitted photon is redshifted from the absorption band of the band edge exciton by 10–80 meV. This is the fluorescence Stokes shift as is familiar from molecules. It was originally proposed that this Stokes shift arose from emission from defect or surface states within the band gap (Nirmal and Brus, 1999). But subsequent experiments and theory (within EMA) suggested an alternative explanation. As summarized earlier, the eightfold degenerate band edge exciton undergoes degeneracy breaking to produce five sub-levels. It is the presence of these sub-levels which gives rise to the observed Stokes shift. We note that the Stokes shift is described as a global Stokes shift (or nonresonant Stokes shift) and a resonant Stokes shift (Klimov, 2007). The former describes the mean energy shift between the absorption band and the emissive band. And the latter describes the energy shift between the lowest absorbing band and the emissive band. Both types of shifts can be rationalized by the presence of this excitonic fine structure within the band edge exciton.

One of the main observations in the emissive spectroscopy of the single exciton is that the EMA predicts the lowest level to be optically dark with a total angular momentum of $N = 2$. This state is equivalent to a triplet state and this singlet-triplet splitting between the lowest bright state and lowest dark state arises from electron–hole exchange. Hence, the lowest state is called the "dark exciton" and its existence was predicted by EMA and rationalized by the PL experiments by Bawendi et al. (Nirmal et al., 1995; Efros et al., 1996; Norris et al., 1996). These PL experiments showed that the lifetime became much longer at low temperature, as expected from the forbidden nature of emission from the dark exciton.

Emission from this state should be accompanied by creation of a phonon in order to conserve momentum. Again, the early experiments confirmed this prediction by showing large longitudinal optical (LO) phonon progressions in fluorescence line narrowing experiments. In this case, it is clear that the PIS approach could not describe the phenomena and the EMA approach was both essential and seemed to successfully describe all the main phenomena.

However, the EPM calculations suggested that the lowest energy dark states were not truly dark (Califano et al., 2005; Gomez et al., 2006). These states are mixed with the higher lying bright states which increase the oscillator strength of the states which were formerly dark in the EMA picture. In this case, the atomistic calculations do not merely give a different number to some observable (e.g., energy splittings or lifetimes) but it gives a qualitatively different physical picture. In EMA, the lowest state is dark, and in EPM it is simply less bright. The most recent single dot PL experiments on well-passivated colloidal CdSe QDs now show that there is very weak coupling to the LO phonons (Chilla et al., 2008). Hence, emission from the putative dark state is not enabled by emission from LO phonons. It is simply an allowed process. This experimental data is reconciled by the EPM calculations but not the EMA ones.

The EMA and EPM pictures also yield a different ordering of the excitonic states which are observable in the linear absorption spectra. The linear absorption spectra clearly show resolvable features which correspond to specific excitonic transition. The EMA picture has been able to assign the observable peaks in colloidal CdSe QDs. Much like the nature of the dark exciton, the EPM approaches yield a qualitative difference in terms of the ordering of the states. In particular, Wang and coworkers have computed the higher lying excitonic transitions using both EPM and EMA approaches (Wang and Zunger, 1998). They find certain transitions to be assigned to excitons of S-type symmetry within EMA and that the same features are assigned to transitions of P-type symmetry in the EPM approach. This is clearly a qualitative difference between approaches which should be experimentally verifiable. Briefly, one can use a femtosecond pump pulse to fill in the CB states (Klimov, 2000, 2007). At increasing fluence, the S-type states in the CB will fill, followed by the P-type states. We monitored the state filling signals which reflect shell filling and we found that our femtosecond pump/probe experiments are completely consistent with the ordering of states predicted by EPM (Sewall et al., 2009a). There is a similar problem in PbSe CQDs which has been tested by pump/probe experiments (Trinh et al., 2008) (Figure 10.15).

PbSe has proven to be a particularly interesting system in which to compare the need to invoke a higher level of theory. The main idea comes from the band structure in the CB and VB. In the case of CdSe, based on effective masses and band degeneracy, the VB has a higher density of states than the CB. This situation has profound impact on the relaxation dynamics of the excitons, a point which will be briefly discussed in Section 10.6. When the EMA approach is applied to PbSe, one finds a symmetric band structure which results in an equally sparse VB. This mirror symmetry in PbSe which is predicted to be absent in CdSe makes certain implications on the timescales and pathways by which hot electrons relax to their band edge state (Harbold et al., 2005; Schaller et al., 2005). In contrast, the atomistic approaches predict a sparse CB and a dense VB for PbSe (An et al., 2006, 2008; Kilina et al., 2009; Prezhdo, 2009). Once again, there is a qualitative difference between the approaches. This difference will be discussed later (Figure 10.16).

A final point is in the treatment of the surface and the symmetry of the lattice. For example, the real dot may be piezoelectric. But the symmetry imposed by an EMA envelope will not yield a piezoelectric dot. Hence, this effect must be included after the fact. In contrast, atomistic calculations can predict the existence of anisotropic effects like piezoelectricity (Zunger, 2001, 2002). Similarly, the atomistic calculations can reveal the electronic structure of the surface states.

10.7 Dynamical and Higher-Order Processes

Thus far we have discussed the hierarchies of theories in terms of the electronic structure of the single exciton. The single exciton (X) has a coarse electronic structure which manifests itself in the linear

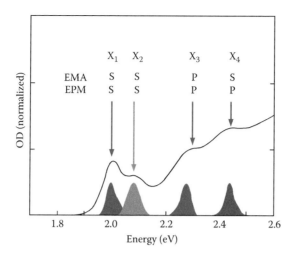

FIGURE 10.15 Representative linear absorption spectrum of colloidal CdSe quantum dots. The peaks correspond to specific excitonic transitions which can be described by various levels of theory. The character of the higher excitonic transitions (e.g., X_4) is qualitatively different for the levels of theory. EMA predicts an s-type electron and EPM predicts a p-type electron. (Reprinted with permission from Sewall, S.L., Cooney, R.R., and Kambhampati, P., Experimental tests of effective mass and atomistic approaches to quantum dot electronic structure: Ordering of electronic states, *Appl. Phys. Lett.*, 94, 243116-3, 2009a; Sewall, S.L., Franceschetti, A., Cooney, R.R., Zunger, A., and Kambhampati, P., Direct observation of the structure of band-edge biexcitons in colloidal semiconductor CdSe quantum dots, *Phys. Rev. B Cond. Matter Mater. Phys.*, 80, 081310(R), 2009b. Copyright 2009, American Institute of Physics.)

absorption spectrum, the PL excitation spectrum, and the photoconductance spectrum. It also has a fine electronic structure which manifests itself in its PL spectrum, particularly the resonant and non-resonant Stokes shifts. These are all static properties of single excitons. Given the static level structure of excitonic eigenstates, there will also be transitions between these states. This topic is often called hot electron relaxation dynamics (Klimov, 2000; Nozik, 2001), but is more generally hot exciton relaxation as the electron or the hole may be undergoing transitions. This problem is also akin to molecular radiationless transitions. Based on the electronic structure and physical structure of the dot, one aims to understand the timescales and pathways by which a hot exciton relaxes to its band edge state (Figure 10.17).

In addition to relaxation dynamics, QDs can support multiple excitations per dot at relatively modest fluences (Klimov, 2007). One may create multiple excitons by sequential or simultaneous absorption of multiple photons. Doing so will create multiple excitons which will sequentially fill up the excitonic shells. Two excitons create a biexciton (XX) and so on. The ground state of the biexciton comprises two S excitons and the ground state of the triexciton comprises two S and one P exciton. Much like the excitation spectrum of X, there are also excited states of multiexcitons (Sewall et al., 2006, 2008, 2009b). These multiexcitons can have unique electronic structure, they can relax to their lowest energy configuration, and they will ultimately annihilate each other through some form of MCR. The time reversal of MCR is MEG, a topic which has received considerable interest due to its importance in QD solar cells and its controversial magnitude and experimental signatures. All of these are of great importance to creating an understanding of a real QD. One might naively expect that the electronic structure of X is the majority of the problem. However, these dynamical and higher-order processes are the driving forces behind key applications such as optical amplification, solar cells, and entangled photons.

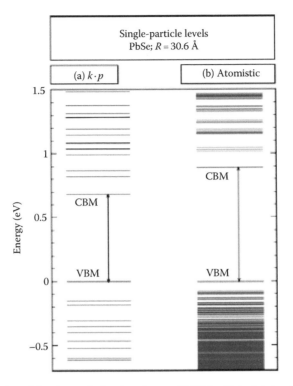

FIGURE 10.16 Illustration of the electronic density of states (DOS) for PbSe dots treated under the EPM (left) and EPM (right) levels of theory. In EMA, the DOS are symmetric about the band gap. In contrast, the EPM approach yields a denser VB manifold. (Reprinted with permission from An, J.M., Franceschetti, A., Dudiy, S.V., and Zunger, A., The peculiar electronic structure of PbSe quantum dots, *Nano Lett.*, 6, 2728–2735, 2006. Copyright 2006 American Chemical Society.)

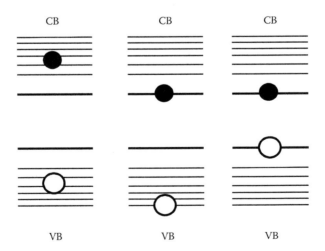

FIGURE 10.17 Illustration of hot exciton relaxation dynamics. The initially prepared exciton is electronically hot (left). The electron can transfer its energy to the hole (center). This Auger relaxation process, which dominates exciton cooling in quantum dots, enables the electron to relax by up-pumping the hole. The final state is a cold exciton in which both the electron and the hole are in the band edge exciton state.

10.7.1 Hot Exciton Relaxation Dynamics

An early topic in the electronic structure and dynamics of excitons in QDs was the question of the timescale of hot electron relaxation (Klimov, 2000; Nozik, 2001). In a bulk semiconductor, electrons undergo intraband cooling via emission of LO phonons. This process is fast due to the continuum of states in a bulk system. In a QD, the S-P energy gap can be 10 LO phonon quantum, thereby rendering the phonon-based relaxation channel as inefficient. Hence, it was anticipated that there would be a phonon-based bottleneck for carrier cooling in quantum-confined nanostructures. Much of the literature was then focused on the search for this phonon bottleneck. It was subsequently proposed that there may be additional relaxation channels that are efficient in QDs but not the bulk. The electron–hole Auger scattering mechanism was one such proposal put forth by Efros (Efros et al., 1995). In this scheme, the electron unidirectionally transfers energy to the hole. The directionality arises from the denser VB manifold and the efficiency arises from forced carrier overlap and loosening of momentum restrictions by confinement in position space. This theory was confirmed in experiments by Klimov and coworkers in colloidal CdSe QDs (Klimov and McBranch, 1998; Klimov et al., 1999; Klimov, 2000).

While the Auger channel within the EMA approach was successful in explaining the femtosecond relaxation times for a P electron in CdSe, it failed to explain the similarly fast relaxation in PbSe CQDs (Harbold et al., 2005; Schaller et al., 2005). Recalling that PbSe has a mirror symmetry to the CB and VB manifolds, it was expected that the Auger channel would not be relevant. Hence, alternative explanations were put forth to rationalize the experiments. Subsequently, the atomistic calculations (EPM and TDDFT) have shown that PbSe maintains a dense VB and can still support the ultrafast Auger relaxation channel (An et al., 2006, 2008; Kilina et al., 2009; Prezhdo, 2009). In addition, the TDDFT reveal additional relaxation pathways. In short, the carrier relaxation dynamics experiments also suggest the need for higher levels of theory even to qualitatively describe the observed phenomena (Figure 10.18).

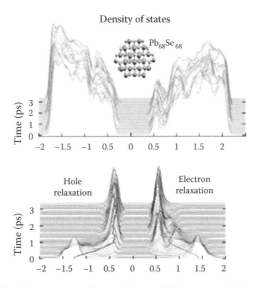

FIGURE 10.18 Higher levels of theory can predict qualitative differences in higher-order processes such as exciton cooling. For example, atomistic theory predicts a dense VB in PbSe. This situation, along with other dynamical processes, can yield fast exciton cooling in situations that are unanticipated from lower levels of theory. (Reprinted with permission from Kilina, S.V., Kilin, D.S., and Prezhdo, O.V., Breaking the phonon bottleneck in PbSe and CdSe quantum dots: Time-domain density functional theory of charge carrier relaxation, *ACS Nano*, 3, 93–99, 2009. Copyright 2009 American Chemical Society.)

10.7.2 Multiple Exciton Structure, Recombination, and Generation

Since QDs can support multiple excitons per dot, one asks what is the electronic structure of the lowest of these excitons—the biexciton (XX). This is a topic which extends the discussion of the electronic structure of X. One of the main difficulties has been in the experimental observation of the multiexcitonic states (Sewall et al., 2006; Klimov, 2007). This difficulty arises since the excited states of XX are constructed from excited states of X, which only survive on the sub-picosecond timescale (Sewall et al., 2006). Hence, the excited states of XX have lifetimes of 0.1–1 ps. Now, there is a ground state to XX, much like the ground state of X (the band edge exciton). This ground state can undergo non-radiative MCR. Similar to the Auger-based relaxation of hot electrons, it was proposed that there would be an Auger-based recombination where two ground state excitons (the ground state biexciton) would recombine to produce an excited exciton which would then cool. Experiments by Klimov have confirmed this picture. The time reversal of this MCR process is multiple MEG. In the MEG process, one high-energy photon creates a single high-energy exciton. This high-energy exciton creates two or more low-energy excitons. There have been several proposed mechanisms for MEG. But the simplest idea is a time reversal of MCR, with an inversion of the relevant initial and final densities of states in a Golden Rule calculation. Hence, it is immediately obvious that the electronic structure of multiexcitons (the initial state in MCR and the final state in MEG) becomes important. These structural and dynamical processes are topics unto themselves and are beyond the scope of this chapter. Our aim here is to note the richness of the structural and dynamical processes beyond the simple electronic structure of X. And much like the structure of X, each of these more complex processes places requirements on the level of theory needed to explain the observations.

10.8 Summary and Concluding Remarks

The semiconductor QD (nanocrystal) has been available for experimental and theoretical probing since the mid-1980s. Our understanding of the dot based on experiment and theory has advanced quite a bit in the last two decades since high-quality colloidal CdSe became available. One can find several excellent sources for descriptions of the electronic structure of the exciton in a QD. Our objective here is to provide a critical overview of the electronic structure of QDs, as opposed to a comprehensive review.

There are several approaches to understanding the electronic structure of the QD, ranging from simple PIS calculations to sophisticated TDDFT calculations. Each of these methods is commonly discussed in isolation. We take the view that each of these methods offers valuable perspective on the inner workings of the QD. The simplest approaches offer a convenient way of qualitatively thinking about the dots. And the most advanced methods can explain the state-of-the-art experiments and in some cases yield predictions that have yet to be confirmed by experiment. Conversely, the simplest methods should not be used beyond their point of relevance—they exist to provide a coarse roadmap to the salient physics. Similarly, the most advanced approaches are often not necessary to understand some phenomena, and can obfuscate some of the issues for the non-specialist.

Hence, effort is made throughout this chapter to identify the specific situations which cause one to invoke a specific level of theory. In describing three of the main approaches to understanding the electronic structure of QDs, we show that the simple PIS approach offers a convenient explanation for some of the immediately obvious phenomena. We then discuss the multiband EMA approach. This approach has yielded tremendous insight into experiments on real dots. Being a continuum approach, this EMA treatment eventually encounters limitations. At this point, the atomistic methods importantly yield qualitative insights in addition to the expected improvements in precision. Our discussion of the available approaches in far from exhaustive. And our discussion of each approach is incomplete. Hence, apologies are made to those whose important contributions were omitted. We hope that this chapter will serve to place the important theoretical developments in an experimental context so as to assist the student and the experimentalist to continue to explore the rich physics of semiconductor QDs.

References

Alivisatos, A. P. (1996a) Perspectives on the physical chemistry of semiconductor nanocrystals. *Journal of Physical Chemistry*, 100, 13226–13239.

Alivisatos, A. P. (1996b) Semiconductor clusters, nanocrystals, and quantum dots. *Science*, 271, 933–937.

An, J. M., Califano, M., Franceschetti, A., and Zunger, A. (2008) Excited-state relaxation in PbSe quantum dots. *The Journal of Chemical Physics*, 128, 164720–164727.

An, J. M., Franceschetti, A., Dudiy, S. V., and Zunger, A. (2006) The peculiar electronic structure of PbSe quantum dots. *Nano Letters*, 6, 2728–2735.

Avouris, P., Gelbart, W. M., and El-Sayed, M. A. (1977) Nonradiative electronic relaxation under collision-free conditions. *Chemical Reviews*, 77, 793–833.

Banin, U., Lee, C. J., Guzelian, A. A., Kadavanich, A. V., Alivisatos, A. P., Jaskolski, W., Bryant, G. W., Efros, A. L., and Rosen, M. (1998) Size-dependent electronic level structure of InAs nanocrystal quantum dots: Test of multi-band effective mass theory. *Journal of Chemical Physics*, 109, 2306–2309.

Bester, G. (2009) Electronic excitations in nanostructures: An empirical pseudopotential based approach. *Journal of Physics: Condensed Matter*, 21, 023202.

Brus, L. E. (1983) A simple model for the ionization potential, electron affinity, and aqueous redox potentials of small semiconductor crystallites. *Journal of Chemical Physics*, 79, 5566–5571.

Brus, L. E. (1984) Electron-electron and electron-hole interactions in small semiconductor crystallites: The size dependence of the lowest excited electronic state. *Journal of Chemical Physics*, 80, 4403–4409.

Califano, M., Franceschetti, A., and Zunger, A. (2005) Temperature dependence of excitonic radiative decay in CdSe quantum dots: The role of surface hole traps. *Nano Letters*, 5, 2360–2364.

Chilla, G., Kipp, T., Menke, T., Heitmann, D., Nikolic, M., Fromsdorf, A., Kornowski, A., Forster, S., and Weller, H. (2008) Direct observation of confined acoustic phonons in the photoluminescence spectra of a single CdSe-CdS-ZnS core-shell-shell nanocrystal. *Physical Review Letters*, 100, 057403–057404.

Cooney, R. R., Sewall, S. L., Anderson, K. E. H., Dias, E. A., and Kambhampati, P. (2007a) Breaking the phonon bottleneck for holes in semiconductor quantum dots. *Physical Review Letters*, 98, 177403–177404.

Cooney, R. R., Sewall, S. L., Dias, E. A., Sagar, D. M., Anderson, K. E. H., and Kambhampati, P. (2007b) Unified picture of electron and hole relaxation pathways in semiconductor quantum dots. *Physical Review B (Condensed Matter and Materials Physics)*, 75, 245311–245314.

Cooney, R. R., Sewall, S. L., Sagar, D. M., and Kambhampati, P. (2009a) Gain control in semiconductor quantum dots via state-resolved optical pumping. *Physical Review Letters*, 102, 127404.

Cooney, R. R., Sewall, S. L., Sagar, D. M., and Kambhampati, P. (2009b) State-resolved manipulations of optical gain in semiconductor quantum dots: Size universality, gain tailoring, and surface effects. *Journal of Chemical Physics*, 131, 164706.

Efros, A. L., Kharchenko, V. A., and Rosen, M. (1995) Breaking the phonon bottleneck in nanometer quantum dots: Role of Auger-like processes. *Solid State Communications*, 93, 281–284.

Efros, A. L. and Rosen, M. (2000) The electronic structure of semiconductor nanocrystals. *Annual Review of Materials Science*, 30, 475–521.

Efros, A. L., Rosen, M., Kuno, M., Nirmal, M., Norris, D. J., and Bawendi, M. (1996) Band-edge exciton in quantum dots of semiconductors with a degenerate valence band: Dark and bright exciton states. *Physical Review B: Condensed Matter*, 54, 4843–4856.

Ekimov, A. I., Hache, F., Schanne-Klein, M. C., Ricard, D., Flytzanis, C., Kudryavtsev, I. A., Yazeva, T. V., Rodina, A. V., and Efros, A. L. (1993) Absorption and intensity-dependent photoluminescence measurements on cadmium selenide quantum dots: Assignment of the first electronic transitions. *Journal of the Optical Society of America B: Optical Physics*, 10, 100–107.

Franceschetti, A., Fu, H., Wang, L. W., and Zunger, A. (1999) Many-body pseudopotential theory of excitons in InP and CdSe quantum dots. *Physical Review B: Condensed Matter and Materials Physics*, 60, 1819–1829.

Franceschetti, A. and Zunger, A. (1997) Direct pseudopotential calculation of exciton Coulomb and exchange energies in semiconductor quantum dots. *Physical Review Letters*, 78, 915–918.

Fu, H., Wang, L.-W., and Zunger, A. (1998) Applicability of the k.p method to the electronic structure of quantum dots. *Physical Review B: Condensed Matter and Materials Physics*, 57, 9971–9987.

Gomez, D. E., Califano, M., and Mulvaney, P. (2006) Optical properties of single semiconductor nanocrystals. *Physical Chemistry Chemical Physics*, 8, 4989–5011.

Harbold, J. M., Du, H., Krauss, T. D., Cho, K.-S., Murray, C. B., and Wise, F. W. (2005) Time-resolved intraband relaxation of strongly confined electrons and holes in colloidal PbSe nanocrystals. *Physical Review B: Condensed Matter and Materials Physics*, 72, 195312/1–195312/6.

Hendry, E., Koeberg, M., Wang, F., Zhang, H., De Mello Donega, C., Vanmaekelbergh, D., and Bonn, M. (2006) Direct observation of electron-to-hole energy transfer in CdSe quantum dots. *Physical Review Letters*, 96, 057408/1–057408/4.

Kang, I. and Wise, F. W. (1997) Electronic structure and optical properties of PbS and PbSe quantum dots. *Journal of the Optical Society of America B: Optical Physics*, 14, 1632–1646.

Kilina, S. V., Kilin, D. S., and Prezhdo, O. V. (2009) Breaking the phonon bottleneck in PbSe and CdSe quantum dots: Time-domain density functional theory of charge carrier relaxation. *ACS Nano*, 3, 93–99.

Kippeny, T., Swafford, L. A., and Rosenthal, S. J. (2002) Semiconductor nanocrystals: A powerful visual aid for introducing the particle in a box. *Journal of Chemical Education*, 79, 1094.

Klimov, V. I. (2000) Optical nonlinearities and ultrafast carrier dynamics in semiconductor nanocrystals. *Journal of Physical Chemistry B*, 104, 6112–6123.

Klimov, V. I. (2006) Mechanisms for photogeneration and recombination of multiexcitons in semiconductor nanocrystals: Implications for lasing and solar energy conversion. *Journal of Physical Chemistry B*, 110, 16827–16845.

Klimov, V. I. (2007) Spectral and dynamical properties of multiexcitons in semiconductor nanocrystals. *Annual Review of Physical Chemistry*, 58, 635–673.

Klimov, V. I. and Mcbranch, D. W. (1998) Femtosecond 1P-to-1S electron relaxation in strongly confined semiconductor nanocrystals. *Physical Review Letters*, 80, 4028–4031.

Klimov, V. I., Mcbranch, D. W., Leatherdale, C. A., and Bawendi, M. G. (1999) Electron and hole relaxation pathways in semiconductor quantum dots. *Physical Review B: Condensed Matter and Materials Physics*, 60, 13740–13749.

Mcguire, J. A., Joo, J., Pietryga, J. M., Schaller, R. D., and Klimov, V. I. (2008) New aspects of carrier multiplication in semiconductor nanocrystals. *Accounts of Chemical Research*, 41, 1810–1819.

Michalet, X., Pinaud, F. F., Bentolila, L. A., Tsay, J. M., Doose, S., Li, J. J., Sundaresan, G., Wu, A. M., Gambhir, S. S., and Weiss, S. (2005) Quantum dots for live cells, in vivo imaging, and diagnostics. *Science*, 307, 538–544.

Nirmal, M. and Brus, L. (1999) Luminescence photophysics in semiconductor nanocrystals. *Accounts of Chemical Research*, 32, 407–414.

Nirmal, M., Norris, D. J., Kuno, M., Bawendi, M. G., Efros, A. L., and Rosen, M. (1995) Observation of the "dark exciton" in CdSe quantum dots. *Physical Review Letters*, 75, 3728–3731.

Norris, D. J. (2004) In *Semiconductor and Metal Nanocrystals: Synthesis and Electronic and Optical Properties*, Electronic structure in semiconductor nanocrystals: Optical experiment, ed. V. I. Klimov, New York: Marcel Decker.

Norris, D. J. and Bawendi, M. G. (1995) Structure in the lowest absorption feature of CdSe quantum dots. *Journal of Chemical Physics*, 103, 5260–5268.

Norris, D. J. and Bawendi, M. G. (1996) Measurement and assignment of the size-dependent optical spectrum in CdSe quantum dots. *Physical Review B: Condensed Matter*, 53, 16338–16346.

Norris, D. J., Efros, A. L., Rosen, M., and Bawendi, M. G. (1996) Size dependence of exciton fine structure in CdSe quantum dots. *Physical Review B: Condensed Matter*, 53, 16347–16354.

Norris, D. J., Sacra, A., Murray, C. B., and Bawendi, M. G. (1994) Measurement of the size dependent hole spectrum in CdSe quantum dots. *Physical Review Letters*, 72, 2612–2615.

Nozik, A. J. (2001) Spectroscopy and hot electron relaxation dynamics in semiconductor quantum wells and quantum dots. *Annual Review of Physical Chemistry*, 52, 193–231.

Nozik, A. J. (2008) Multiple exciton generation in semiconductor quantum dots. *Chemical Physics Letters*, 457, 3–11.

Pandey, A. and Guyot-Sionnest, P. (2007) Multicarrier recombination in colloidal quantum dots. *Journal of Chemical Physics*, 127, 111104/1–111104/4.

Prezhdo, O. V. (2008) Multiple excitons and the electron-phonon bottleneck in semiconductor quantum dots: An ab initio perspective. *Chemical Physics Letters*, 460, 1–9.

Prezhdo, O. V. (2009) Photoinduced dynamics in semiconductor quantum dots: Insights from time-domain ab initio studies. *Accounts of Chemical Research*, 42, 2005–2016.

Rogach, A. L., Talapin, D. V., Shevchenko, E. V., Kornowski, A., Haase, M., and Weller, H. (2002) Organization of matter on different size scales: Monodisperse nanocrystals and their superstructures. *Advanced Functional Materials*, 12(10), 653–664.

Sagar, D. M., Cooney, R. R., Sewall, S. L., Dias, E. A., Barsan, M. M., Butler, I. S., and Kambhampati, P. (2008a) Size dependent, state-resolved studies of exciton-phonon couplings in strongly confined semiconductor quantum dots. *Physical Review B (Condensed Matter and Materials Physics)*, 77, 235321–235314.

Sagar, D. M., Cooney, R. R., Sewall, S. L., and Kambhampati, P. (2008b) State-resolved exciton-phonon couplings in CdSe semiconductor quantum dots. *Journal of Physical Chemistry C*, 112, 9124–9127.

Schaller, R. D., Pietryga, J. M., Goupalov, S. V., Petruska, M. A., Ivanov, S. A., and Klimov, V. I. (2005) Breaking the phonon bottleneck in semiconductor nanocrystals via multiphonon emission induced by intrinsic nonadiabatic interactions. *Physical Review Letters*, 95, 196401/1–196401/4.

Scholes, G. D. and Rumbles, G. (2006) Excitons in nanoscale systems. *Nature Materials*, 5, 683–696.

Sewall, S. L., Cooney, R. R., Anderson, K. E. H., Dias, E. A., and Kambhampati, P. (2006) State-to-state exciton dynamics in semiconductor quantum dots. *Physical Review B: Condensed Matter and Materials Physics*, 74, 235328.

Sewall, S. L., Cooney, R. R., Anderson, K. E. H., Dias, E. A., Sagar, D. M., and Kambhampati, P. (2008) State-resolved studies of biexcitons and surface trapping dynamics in semiconductor quantum dots. *Journal of Chemical Physics*, 129, 084701.

Sewall, S. L., Cooney, R. R., and Kambhampati, P. (2009a) Experimental tests of effective mass and atomistic approaches to quantum dot electronic structure: Ordering of electronic states. *Applied Physics Letters*, 94, 243116–243113.

Sewall, S. L., Franceschetti, A., Cooney, R. R., Zunger, A., and Kambhampati, P. (2009b) Direct observation of the structure of band-edge biexcitons in colloidal semiconductor CdSe quantum dots. *Physical Review B (Condensed Matter and Materials Physics)*, 80, 081310(R).

Talapin, D. V., Nelson, J. H., Shevchenko, E. V., Aloni, S., Sadtler, B., and Alivisatos, A. P. (2007) Seeded growth of highly luminescent CdSe/CdS nanoheterostructures with rod and tetrapod morphologies. *Nano letters*, 7, 2951–2959.

Trinh, M. T., Houtepen, A. J., Schins, J. M., Piris, J., and Siebbeles, L. D. A. (2008) Nature of the second optical transition in PbSe nanocrystals. *Nano Letters*, 8, 2112–2117.

Wang, L.-W. and Zunger, A. (1998) High-energy excitonic transitions in CdSe quantum dots. *Journal of Physical Chemistry B*, 102, 6449–6454.

Zunger, A. (2001) Pseudopotential theory of semiconductor quantum dots. *Physica Status Solidi B: Basic Research*, 224, 727–734.

Zunger, A. (2002) On the farsightedness (hyperopia) of the standard $k \cdot p$ model. *Physica Status Solidi (a)*, 190, 467–475.

11

Selenide and Sulfide Quantum Dots and Nanocrystals: Optical Properties

Andrea M. Munro
Pacific Lutheran University

11.1 Introduction

The optical and electronic properties of colloidal nanocrystals have been extensively studied over the past three decades.[1-4] Many of the early studies of CdS and CdSe colloidal nanocrystals focused on understanding the intrinsic properties of the nanocrystals themselves and on improving the reproducibility and size-distribution of nanocrystal syntheses. Colloidal nanocrystals have been made by reverse micelle syntheses,[5] hot injection methods,[6,7] and by microwave syntheses.[8,9] Many of the theoretical studies of CdSe nanocrystals were performed using data from experiments with nanocrystals synthesized by variations of the hot injection method. These theoretical studies focused on how the energy levels of CdSe nanocrystals change as a function of nanocrystal size and shape.[1,10,11]

More recently, people have been investigating the use of colloidal nanocrystals in applications ranging from biological markers[12] to their use as chromophores in light-emitting diodes,[13-16] dye-sensitized solar cells,[17,18] and photovoltaics,[19,20] and as charge transport materials in field-effect transistors.[3] In order to utilize colloidal nanocrystals in these applications, it has become increasingly important to determine how the local environment[21,22] and the surface chemistry of the nanocrystals can affect the optical and electronic properties of individual nanocrystals and nanocrystal films. The surface of a CdSe nanocrystal is capped with either an inorganic shell of a wider band gap material (i.e., CdS, ZnS) in order to confine excitons to the core of the nanocrystal or the surface is capped with organic ligands (Figure 11.1). Inorganic shells and organic ligands that are used to cap a nanocrystal can alter the photoluminescence of the nanocrystal by (1) creating or passivating charge trap states on the nanocrystal surface, (2) forming an insulating, dielectric layer around the nanocrystal, (3) acting as electron or hole acceptors, or (4) allowing energy transfer between the ligands and nanocrystal (increasing or decreasing

307

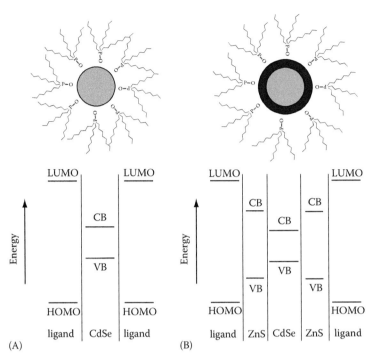

FIGURE 11.1 Schematics and energy band diagrams of (A) a CdSe nanocrystal and (B) a CdSe/ZnS core/shell nanocrystal. The ZnS shell is grown to confine the electron and hole in the CdSe core, away from surface states.

the nanocrystal photoluminescence). This chapter will describe the current understanding of colloidal CdSe and CdS nanocrystal photoluminescence and how nanocrystal photoluminescence can be affected by the capping ligands and the local environment of the nanocrystals.

One of the most well-known characteristics of colloidal nanocrystals is the size-tunability of their optical and electronic properties. The relationship between nanocrystal diameter and optical band gap energy was first modeled using the effective mass approximation.[1,11,23] Since that time, there have been a number of theoretical and experimental studies that demonstrate that the size and shape of a nanocrystal not only affect the nanocrystal band gap,[24] but also the dielectric constant[24] and molar absorptivity[25] of the nanocrystals.

Semiconductor nanocrystals absorb photons with energy equal to or greater than the nanocrystal band gap. The absorption spectra of CdS nanocrystals have a shoulder that indicates the onset of absorption, while CdSe nanocrystals have four distinct absorption transitions, shown in Figure 11.2.[23] The transitions are labeled to indicate that the electron and hole wave functions can be considered separately. The transitions occur between the electron states and light and heavy hole states as described by Ekimov et al.[23] The photoluminescence lifetime of CdSe nanocrystals is commonly reported as being tens of nanoseconds at room temperature and as long as 50 ns at 10 K; this lifetime is longer than fluorescence lifetimes of small molecules and shorter than phosphorescence lifetimes.[26,27] Due to the long lifetime of the nanocrystal photoluminescence, there has been controversy about whether it is best described as fluorescence or phosphorescence. Although, CdSe nanocrystals have large spin-orbit coupling and are not expected to have singlet and triplet states. However, in 1996 Efros et al. calculated the excited state energy of CdSe and CdTe nanocrystals as a function of nanocrystal diameter and shape, taking into account the differences between spheres, rods, and ellipsoids.[26] They predicted that spherical CdSe nanocrystals would have long lifetimes, because the states lying lowest in energy were so called "dark" states (i.e., states with an angular momentum of 2). In addition, they predicted that

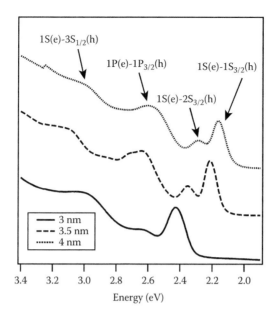

FIGURE 11.2 Absorption spectra of 2.4 nm (solid line), 3.2 nm (dashed line), and 3.4 nm (dotted line) diameter CdSe nanocrystals. The absorption transitions are indicated on the 3.4 nm diameter CdSe spectrum.

for nanocrystal rods, the lowest lying excited states would be so called "light" states and that the lifetimes of nanorods would be much shorter than the lifetimes of spherical nanocrystals. Although there have been reports of photoluminescence lifetime data for CdSe nanorods,[28] the measured lifetimes are not single exponential, making it difficult to conclude if the short lifetimes are due to the presence of defects in the nanorods or due to an intrinsically faster radiative decay rate for nanorods compared to spherical nanocrystals.

11.2 Colloidal Nanocrystal Photoluminescence

One of the earliest questions that had to be addressed concerning colloidal nanocrystal photoluminescence was the nature of the emitting state. The nanocrystal batches produced by early synthetic routes had large size and shape distributions and exhibited both a narrow band emission and a broad (>100 nm), redshifted emission band (Figure 11.3).[29,30] Due to the large intensity of this broad emission compared to the band edge emission in early nanocrystal samples, there was speculation as to whether the observed photoluminescence was due to recombination of carriers localized on the nanocrystal surface or due to recombination through internal core states.[27,30] In 1997, Kuno et al. chemically modified the surface of CdSe nanocrystals ranging in diameter from 2.4 to 8.6 nm with trioctylphosphine oxide (TOPO), ZnS shells, 4-methylpyridine, 4-(trifluoromethyl)thiophenol, and tri(2-ethylhexyl)phosphate to determine how chemical modification of the nanocrystal surface would affect nanocrystal photoluminescence.[30] The authors found that the band edge emission energy did not change with nanocrystal surface chemistry. However, they noted that the broad "deep-trap" emission was surface sensitive. Both 4-methylpyridine and 4-(trifluoromethyl)thiophenol decreased the nanocrystal photoluminescence quantum yield, while growing an inorganic ZnS shell on the nanocrystals, increased the nanocrystal band edge quantum yield and decreased the broad emission feature. The authors concluded that the band edge emission was due to intrinsic core levels in the nanocrystals, and assigned the broad, redshifted emission to surface recombination at defect sites.[5,30]

The relative intensities of band gap emission and the "deep-trap" emission feature were shown to change as the nanocrystal surface ligands were exchanged and as the samples were allowed to age by

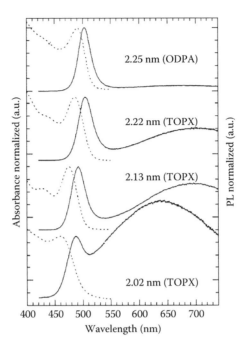

FIGURE 11.3 UV-vis absorbance (dashed lines) and photoluminescence (solid line) spectra of CdSe nanocrystals of various sizes. The figure shows both narrow band edge emission and a broad, redshifted emission feature. The capping ligands are octadecylphosphonic acid and a mixture of TOPO and TOP-Se (TOPX). (Reprinted with permission from Kalyuzhny, G. and Murray, R.W. *J. Phys. Chem. B*, 109, 7012–7021, 2005. Copyright 2005 American Chemical Society.)

Kalyuzhny and Murray.[29] They found that the relative intensities of the "deep-trap" to band edge emission features could be increased by adding TOP-Se to solutions of CdSe nanocrystals in chloroform, as shown in Figure 11.3. The authors attributed the "deep-trap" luminescence to selenium surface atoms bound with TOP (trioctylphosphine). With the improvement of nanocrystal synthetic routines, this feature is not often reported for room temperature photoluminescence measurements, although surface trap emission is still observed for ultrasmall CdSe nanocrystals.[31,32]

Since 2005, Rosenthal et al. have published a series of articles describing white light emission from ultrasmall CdSe nanocrystals.[31–33] The white light observed arises from blue band edge nanocrystal emission and a broad (>100 nm) band of luminescent trap emission that spans the visible spectrum. The authors define ultrasmall nanocrystals as those with diameters <1.7 nm. Although it is not surprising that such small nanocrystals are likely to have increased surface emission (as the nanocrystal diameter decreases, the fraction of the total nanocrystal atoms that are surface atoms increases); they also report ligand-dependent emission pinning. The authors reported that both the nanocrystal absorbance and photoluminescence spectra blueshift with decreasing nanocrystal diameter, until the nanocrystal absorbance reaches ~410 nm. After this point, as the nanocrystal diameter decreases, the absorbance blueshifts, but the photoluminescence energy is pinned. The energy at which the nanocrystal photoluminescence is pinned was found to vary with the nanocrystal surface ligands. Schreuder et al. compared the photoluminescence of ultrasmall nanocrystals with phosphonic acid ligands with varying alkane chain length.[32] They found that as the ligand became more electronegative (as the ligand length decreased), the photoluminescence pinning wavelength shifted to higher energy. These observations led the authors to conclude that in the case of ultrasmall nanocrystals the material begins to approach the molecular limit and it might not be possible to consider the nanocrystal and ligand as separate materials. It should be noted that in this study the phosphonic acids were used during growth of the nanocrystals and the ligands were not exchanged after the synthesis. Thus, it is possible that the differences in the pinned emission could result from differences in the syntheses and not just from the use of the various phosphonic acids as capping ligands.

11.3 Quenching by Surface States

One of the challenges when reading through the ligand exchange literature for CdS and CdSe nanocrystals, is that the nanocrystals typically have photoluminescence quantum yields of >30% for newer synthetic routes and ~10% for nanocrystals synthesized by older hot injection routes. The nanocrystal photoluminescence quantum yield can vary from batch to batch and is affected by ligand purification and dilution of the nanocrystal solution. In 2001, Talapin et al.[34] reported an improvement in the hot injection method by the addition of amines to the CdSe nanocrystal synthesis. The hot injection method originally was performed by heating TOPO to 370°C under nitrogen and quickly injecting a solution of dimethylcadmium and Se-TOP. Talapin et al. found that adding hexadecylamine to the TOPO decreased the size distribution of the nanocrystals and increased their final quantum yield. In that article, the authors also reported that the addition of hexadecylamine to a batch of nanocrystals synthesized only in TOPO resulted in increased photoluminescence and a slight blueshift in the emission band. The nanocrystals were then precipitated from solution and redispersed in a TOPO/TOP mixture, at which point the emission spectrum redshifted and the photoluminescence intensity decreased. The article illustrated the dual functions of the organic molecules used during nanocrystal synthesis to control nanocrystal nucleation and growth as well as to bind to the nanocrystal surface after synthesis.

A number of studies have been conducted to determine whether different ligands can be used to increase or decrease nanocrystal photoluminescence. The effects of adding a given ligand to a nanocrystal solution is influenced by the ligands that capped the nanocrystal surface after synthesis. If the initial ligands create surface states on the nanocrystals, it is likely that the addition of a new ligand will increase the nanocrystal photoluminescence intensity. This can be observed in reports of increases in CdSe nanocrystal photoluminescence when alkanethiols are added to nanocrystals synthesized by reverse micelle methods,[5] compared to nanocrystal photoluminescence quenching when thiols are added to CdSe nanocrystal solutions made by hot injection methods.[29,35,36] The synthetic routes to produce CdS and CdSe nanocrystals have developed over the last few decades, initially nanocrystals were synthesized in reverse micelles and then by hot injection methods. Since the early hot injection method syntheses involving the injection of dimethylcadmium and selenium-TOP precursors into 90% TOPO, the synthesis has evolved, changing the cadmium precursor and the organic ligands. The most widely used synthetic routes often utilize a combination of organic ligands including phosphonic acids, alkylamines, phosphines, and carboxylic acids.

After Kuno et al.[30] reported that exchanging the surface ligands on CdSe nanocrystals did not alter the band edge energy, but did alter the photoluminescence intensity of the nanocrystals, a number of groups began studying the effects of surface ligands on the band edge photoluminescence of CdSe nanocrystals. Once it was established that the narrow emission feature was the size-tunable, band edge emission, researchers turned their focus to analyzing only the band edge photoluminescence; the remainder of this chapter will discuss nanocrystal band edge photoluminescence, unless otherwise specified. Landes et al.[37] systematically studied the effects of adding butylamine to solutions of CdSe nanocrystals in toluene. They reported the systematic quenching of CdSe photoluminescence as butylamine was added to the solution; the butylamine concentrations ranged from 10^{-4} to 4×10^{-2} M. They analyzed their data noting that butylamine is a hole acceptor and modeled the photoluminescence quenching with a Stern-Volmer formalism.

In 2003, Sharma et al.[38] measured the effects of the addition of butylamine and *p*-phenylenediamine on CdSe nanocrystal photoluminescence. In contrast to the report by Landes et al. they found that the addition of butylamine (at concentrations ranging from 5×10^{-5} to 10^{-3} M) increased the photoluminescence intensity of CdSe nanocrystals. Kalyuzhny and Murray reported that the addition of alkanethiols to solutions of CdSe nanocrystals (synthesized by the hot injection method and capped with TOPO) quenched CdSe nanocrystal photoluminescence and that the addition of alkylamines increased CdSe nanocrystal photoluminescence;[29] this second result was consistent with the observations by Talapin et al. discussed above.[34]

11.4 Ligand–Nanocrystal Binding Constants

After these initial reports of ligand effects on solution nanocrystal photoluminescence, other researchers set out to calculate the binding constants of organic ligands commonly used to cap CdSe nanocrystals.[35,36,39–43] The initial attempts to determine the binding constants of alkylamines and alkanethiols used the solution photoluminescence intensity to determine the fraction of bound ligands.[35,39] Both papers reported that alkanethiols quench CdSe photoluminescence and that alkylamines increase CdSe photoluminescence at low concentration. Additionally, Munro et al. found that although adding butylamine increased CdSe photoluminescence at low concentrations (<10 mM); the addition of high concentration of butylamine quenches and blueshifts CdSe photoluminescence (consistent with reports from both Sharma et al.[38] and Landes et al.[37]).

The papers from Bullen and Mulvaney[35] and Munro et al.[39] both investigated changes in CdSe nanocrystal photoluminescence as a function of concentration of added alkylamines and alkanethiols. Both also assumed a linear relationship between the fraction of bound surface sites on the nanocrystals and the nanocrystal photoluminescence. By assuming a linear relationship between photoluminescence intensity and added ligand concentration Bullen and Mulvaney calculated binding constants for alkanethiols and alkylamines. They also reported that the binding constants varied with nanocrystal diameter, potentially due to differences in surface energy. Munro et al. used a similar method to calculate a binding constant for octadecanethiol; however, they found that the midpoint of the quenching isotherms changed with nanocrystal concentration and for nanocrystal synthesized with different cadmium precursors. They explained these observations by claiming that alkanethiols strongly bind to CdSe nanocrystals, such that the concentration of total added thiol could not be approximated as the concentration of free thiol in solution. They also speculated that nanocrystals synthesized by different routes may have different numbers of cadmium surface atoms.

To date there is no established way to determine the fraction of bound and unbound ligands or the fraction of bound and unbound surface sites on each nanocrystal. Groups have made estimates, but this is currently one of the greatest challenges when studying ligand effects on nanocrystal photoluminescence.[39–41] Most often it is assumed that a linear relationship exists between the photoluminescence intensity of a nanocrystal and the fraction of bound surface sites on the nanocrystal; however, this assumption was questioned by Munro and Ginger from their report of single-molecule quenching studies of adding alkanethiols to CdSe nanocrystals.[36]

Subsequently, to avoid the challenge inherent in assuming the number of binding sites per nanocrystal Koole et al.[42] and Ji et al.[41] examined changes in nanocrystal photoluminescence as ligands were added to a nanocrystal solution and allowed to equilibrate. Both authors used the change in nanocrystal intensity with time to estimate adsorption and desorption rates for alkanethiols and alkylamines, respectively. They could then calculate a binding constant for each ligand from the rates of adsorption and desorption. The results of these different studies were summarized in a theoretical paper by Schapotschnikow et al.[43] Ji et al.[41] noted that the ligand binding constants determined from each of these studies depend on competition between the ligand of interest and the "native" ligands from synthesis that compete during ligand exchange. In order to mitigate competition between the original ligands covering the nanocrystals and the added ligands, Ji et al. used pyridine-capped CdSe nanocrystals to determine the binding constant and adsorption rate for octylamine, since pyridine is a weak binder to CdSe nanocrystals.[41]

In these studies of ligand effects on CdSe nanocrystal photoluminescence, insulating ligands with long alkane chains were primarily used and the authors attributed all changes in nanocrystal photoluminescence to the ligands passivating or creating traps on the nanocrystal surface. Organic ligands can also alter nanocrystal photoluminescence by acting as electron or energy acceptors or donors. For example, the addition of *p*-substituted phenylamines to solutions of CdSe nanocrystals has been shown to quench CdSe nanocrystal photoluminescence by Sharma et al. in 2003 and Knowles et al. in 2009. These articles illustrate the challenge of determining how ligands alter nanocrystal photoluminescence; whether ligand binding creates a surface trap or whether the organic ligand acts as an electron donor or acceptor, quenching nanocrystal photoluminescence.

Knowles et al. took one of the most sophisticated approaches to understanding the influence of ligands on CdSe nanocrystal photoluminescence to date in their study of CdSe nanocrystal photoluminescence quenching by a variety of *p*-substituted phenylanilines. Their new approach was to consider the process of photoluminescence quenching as a function of ligand concentration, as having two separate components (1) the binding constant of each ligand with the nanocrystal surface and (2) the rate of non-radiative decay associated with binding of each ligand. The authors found that in studies of photoluminescence quenching as a function of ligand concentration for *N*,*N*-dimethyl-*p*-phenylenediamine, *p*-methoxyaniline, aniline, *p*-bromoaniline, and trifluoromethoxyaniline, the midway point of their quenching curves varied over four orders of magnitude (Figure 11.4). However, they estimated that the binding constants for each ligand varied over only one order of magnitude. The authors estimated the binding constants for CdSe nanocrystals using the calculated binding constant of each ligand with Cd^{2+} ions; as the *p*-substituted group became more electron withdrawing, the binding constant to Cd^{2+} decreased. Using a Freudlich isotherm to model ligand binding and quenching, Knowles et al. concluded that the *p*-substitution on each phenylaniline could quench CdSe photoluminescence in two different ways, either by (1) creating an electron trapping surface state upon ligand exchange or (2) by the ligands functioning as hole scavengers with the photoexcited nanocrystals. Their conclusions that dimethyl-*p*-phenylenediamine and *p*-methoxyaniline act as hole scavengers is consistent with the conclusions of Sharma et al.[38] for *p*-phenylenediamine quenching of CdSe.

Sharma et al. monitored the nanocrystal photoluminescence intensity as a function of *p*-phenylenediamine concentration and also monitored the nanocrystal photoluminescence lifetime and the transient

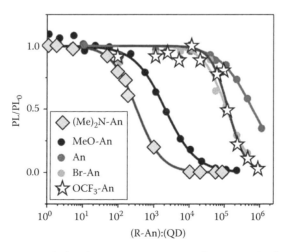

FIGURE 11.4 Plot of the integrated photoluminescence intensity of nanocrystal solutions with different para-substituted anilines (R-An) as a function of the concentration ratio of aniline and the nanocrystals [R-An]:[QD], the solution intensities are normalized with respect to the initial nanocrystal photoluminescence before addition of aniline. (Adapted with permission from Knowles, K.E., Tice, D.B., McArthur, E.A., Solomon, G.C., Weiss, E.A., *J. Am. Chem. Soc.*, 132, 1041–1050, 2010. Copyright 2010 American Chemical Society.)

absorption spectrum of their solutions. They observed a long-lived nanocrystal absorbance bleach (on the order of nanoseconds) in the presence of *p*-phenylenediamine and a long-lived absorption feature that they identified as the oxidized form of *p*-phenylenediamine. The authors concluded that *p*-phenylenediamine quenches CdSe nanocrystal photoluminescence by hole transfer from the photoexcited nanocrystal to the ligand.

11.5 Nanocrystal Photoluminescence Quenching by Charge and Energy Transfer

Although nanocrystal photoluminescence is strongly influenced by the presence of nanocrystal surface states, nanocrystal photoluminescence can also be quenched by charge transfer or energy transfer to another material. Organic capping ligands can be used as linkers between a charge or energy acceptor and CdS or CdSe nanocrystals. Due to the large extinction coefficient, size-tunable absorption, and solution processability of colloidal nanocrystals, they have been studied as materials for photovoltaics and dye-sensitized solar cells.[2,3,17] The process of electron transfer from CdSe nanocrystals to metal oxides has been heavily investigated by Kamat et al.[44,45] The authors have demonstrated that CdSe nanocrystals tethered to TiO_2 films behave as sensitizers for the metal oxide. They also found that the photocurrent efficiency of their CdSe-TiO_2 devices varies with nanocrystal size. Smaller nanocrystals absorb fewer photons due to their wider band gap, although the energy level offset between the conduction band of the nanocrystals and metal oxide is predicted to increase with decreasing nanocrystal diameter. As the nanocrystal diameter increases the total light absorbed increases, while the rate of electron transfer to TiO_2 decreases.

Dibell and Watson[46] measured the rate of charge transfer from CdS nanocrystals to TiO_2 nanocrystals as a function of distance. The authors tethered the CdS and TiO_2 nanocrystals together in solution with mercaptoalkanoic acids of varying length. Their work utilized steady-state photoluminescence, photoluminescence lifetime, and transient absorption measurements to determine an upper bound for the rate of electron transfer from photoexcited CdS nanocrystals to TiO_2. Their paper also illustrated the challenges inherent in making these measurements. Surface traps on both the CdS and TiO_2 nanocrystals compete with charge transfer between the nanocrystals and can allow relaxation of the nanocrystals from excited and reduced or oxidized states. For this reason Dibell and Watson could only calculate an upper bound for the rate of electron transfer from CdS to TiO_2.

Nanocrystal photoluminescence can also be quenched by energy transfer from the nanocrystals to an energy acceptor. Studies of energy transfer to and from nanocrystals are primarily conducted by researchers in order to develop sensors (i.e., for immunoassays, detection of protein folding events, detection of specific toxins or analytes). These systems are often comprised of nanocrystals with small molecule dyes that are either free or bound to the nanocrystal. Förster resonant energy transfer (FRET) is often used to explain the phenomenon of energy transfer from nanocrystals, but there have been reports that Dexter[47] or "non-Förster"[48] energy transfer may occur. Most researchers have discussed the difficulty of determining an acceptor-donor distance in such systems, because the nanocrystal diameter is on the order of a typical FRET distance (in the FRET model, the acceptor and donor are point charges).[13] There have been reports of CdSe nanocrystal photoluminescence quenching due to energy transfer to small molecules and between nanocrystals. Energy transfer between nanocrystals causes a redshift in the observed nanocrystal photoluminescence when nanocrystals are processed from solution into a thin film.[49,50]

The determination of FRET efficiencies from nanocrystals to small molecules is complicated by (1) the challenge of determining the number of energy acceptors attached to each nanocrystal, (2) the difficulty in determining how the attachment chemistry can alter the nanocrystal photoluminescence (i.e., creating or passivating surface states), and (3) determining the average distance between the acceptor and donor. Each of these are well illustrated in a 2004 article by Clapp et al.,[51] which described FRET between CdSe/ZnS core/shell nanocrystals and Cy3 dye molecules bound to the nanocrystal with maltose binding protein. The authors discuss in detail how binding maltose binding protein without Cy3 increases the photoluminescence efficiency of the nanocrystals in solution. The authors kept the nanocrystal: maltose binding protein ratio constant, while varying the nanocrystal: dye ratio from 1:0 to 1:10 using varying amounts of maltose binding protein with and without Cy3. The authors monitored the steady-state photoluminescence intensity and photoluminescence lifetime of both the nanocrystal and the dye. This last step is essential in order to determine whether photoluminescence quenching of the nanocrystal is due to energy transfer to the dye or another quenching process, such as charge transfer. They found that energy transfer occurred between the nanocrystal donor and the Cy3 dye acceptor and characterized the energy transfer as FRET. Additionally, they reported increasing FRET efficiency as the number of dye molecules per nanocrystal increased. They concluded that the sensitivity of detection methods that employ FRET could be increased using nanocrystals with multiple attached acceptors.

Interestingly, in a subsequent report, Clapp et al.[52] were unable to observe energy transfer *from* a small molecule dye (AlexaFluor 488 or Cy3) to CdSe/ZnS core/shell nanocrystals and attributed this to the fast fluorescence decay rate of the donor fluorophores. The authors speculated that in their system the dye fluorescence decay rate was faster than the rate of FRET. Despite the absence of energy transfer from small dye molecules to CdSe/ZnS core/shell nanocrystals, Anikeeva et al.[53] and Achermann et al.[54] demonstrated energy transfer from a phosphorescent dye and from a GaAs quantum well to CdSe/ZnS nanocrystals, respectively. Both groups investigated energy transfer from a thin energy donor layer to a nanocrystal monolayer.

11.6 Nanocrystal Intraband Charge Relaxation

The size-tunability of the CdS and CdSe band gap arises from the fact that for very small nanocrystals, the nanocrystal diameter is less than the Bohr radius of the material. In the strong confinement regime, the energy bands of the nanocrystals become more narrow or "atom-like."[1,23,55] As nanocrystal diameter decreases the spacing between energy levels increases, such that the spacing between electron energy levels is greater than the energy of a phonon in the crystal. This led to the expectation of a "phonon bottleneck" for CdSe nanocrystals that would prevent fast relaxation to the nanocrystal band edge.[56]

However, the "phonon bottleneck" was not experimentally observed until 2008 by Pandey and Guyot-Sionnest.[57] Researchers have primarily relied on transient absorption measurements to monitor the rate of intraband electron relaxation for CdSe nanocrystals. El-Sayed et al. monitored intraband relaxation of CdSe nanocrystals from high-energy excited states to the nanocrystal band edge. The authors also monitored the effects of charge transfer from the photoexcited nanocrystal to an electron acceptor. By monitoring the fluorescence lifetime and the transient absorption spectra of nanocrystals and nanocrystals with methyl viologen, 1,4-benzoquinone, and 1,2-naphthoquinone, they found that after excitation, the exciton relaxes to the nanocrystal band edge and that the electron can be trapped in emissive deep traps. The addition of electron acceptors quenched CdSe photoluminescence by charge transfer of an electron to the small molecule electron acceptor. The authors noted that nanocrystals excitons relaxed to the band edge within picoseconds and did not report the presence of a "phonon bottleneck."

In 1999, Klimov et al.[56] attempted to observe the "phonon bottleneck" for CdSe nanocrystals by transient absorption and found that the electrons in the nanocrystals relaxed quickly (within picoseconds) from the 1P(e) to the 1S(e) state. The authors reported that the relaxation rate was independent of nanocrystal surface chemistry, comparing TOPO and ZnS shell capped CdSe nanocrystals. Auger processes were invoked to explain fast electron relaxation in these systems. The absence of the phonon bottleneck

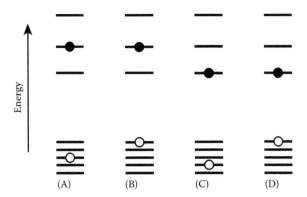

FIGURE 11.5 Intraband relaxation of an electron to the band edge in the presence of a hole. (A) A photon is absorbed, creating an exciton on the nanocrystal. (B) The hole rapidly relaxes to the band edge by phonon emission. (C) The electron relaxes to the band edge by transferring energy to the hole. (D) The hole rapidly relaxes to the band edge.

was attributed to relaxation of the electron by Auger processes, as had been predicted by Efros et al.[58] The electron can rapidly relax by coupling with the hole; when the electron relaxes to the band edge, its energy is transferred to the hole (Figure 11.5). The spacing of the hole energy levels is much smaller than for the electron, so the hole can rapidly relax to the band edge by phonon emission. To test this theory, Klimov et al. repeated the experiments with PbSe nanocrystals.[59] They hypothesized that the phonon bottleneck would be observed for PbSe nanocrystals, since the energy level spacing for electron and hole energy levels are symmetric in PbSe. However, they found that the intraband relaxation was fast for both CdSe and PbSe nanocrystals and that the rate of intraband relaxation increased with decreasing nanocrystal diameter.

Guyot-Sionnest et al.[60] observed surface chemistry dependent intraband relaxation for electrons in CdSe nanocrystals by comparing nanocrystals capped with dodecanethiol, oleylamine, oleic acid, TOPO, and tetradecylphosphonic acid. The authors used dodecanethiol to create hole traps on the nanocrystal surface; in order to monitor the dynamics of intraband electron relaxation in the absence of a hole. The authors reported a difference in the electron intraband relaxation rate as a function of surface ligand coverage, but did not observe the phonon bottleneck. They hypothesized that the electron could still relax in the absence of the hole, by coupling with ligand phonons. In 2009, Kilina et al. reported density functional theory simulations of intraband electron and hole relaxation for CdSe and PbSe nanocrystals.[61] They showed that electron–phonon interactions were stronger for quantum confined nanocrystals. They also demonstrated that the electronic state degeneracies were lifted due to surface reconstruction and thermal fluctuations in the nanocrystals, which produced a distribution of energy levels.

It was not until 2008 that Pandey and Guyot-Sionnest[57] observed the "phonon bottleneck" for CdSe nanocrystals capped with a thick inorganic shell. By engineering the electronic structure of the core/shell nanocrystal the authors were able to slow intraband electron relaxation lifetime to over 1 ns. Pandey and Guyot-Sionnest grew thick multilayer shells on the CdSe nanocrystals, the most successful nanocrystals had a shell comprised of a one monolayer ZnS inner layer, nine monolayers of ZnSe, and an outer monolayer of CdSe. The authors found that if the ZnSe layer was sufficiently thick, it would act as a hole acceptor, separating the electron and hole wavefunctions and allowing the authors to observe the "phonon bottleneck," while decreasing coupling between the excited electron and the ligand phonons. The authors noted that the ZnS layer was needed to maintain the shape of the nanocrystal absorption spectrum and that the CdSe outer layer was needed to prevent electron surface traps on the core/shell nanocrystals.

11.7 Nanocrystal Photoluminescence Intermittency

Most nanocrystal photoluminescence experiments are conducted with large ensembles of nanocrystals, but there are a number of reports of single-nanocrystal experiments, such as single-nanocrystal imaging and spectroscopy.[62-64] The purpose of single-nanocrystal imaging or single-nanocrystal spectroscopy is to analyze the properties (e.g., nanocrystal photoluminescence bandwidth, emission energy) of a series of individual nanocrystals in time. These experiments can indicate the heterogeneity within a larger ensemble. For example, the bandwidth of a batch of nanocrystals is broader than the bandwidth of an individual nanocrystal within that batch,[62,63] due to the size distribution of the nanocrystals synthesized in a single batch. Single-nanocrystal experiments are generally performed with nanocrystals on a solid surface, they can be tethered to the surface or embedded in a polymer matrix. Single-nanocrystal experiments monitor the emission intensity or both the emission intensity and emission energy of individual nanocrystals over time. Two notable phenomena that are observed during single-nanocrystal experiments are photoluminescence intermittency and spectral diffusion.[62,63] Photoluminescence intermittency describes variations in emission intensity of a single nanocrystal with time, while spectral diffusion occurs when the emission energy of a single nanocrystal varies in time.

In this chapter, we examine the process of photoluminescence intermittency (i.e., blinking) of CdSe nanocrystals.[62-64] The initial reports of nanocrystal photoluminescence intermittency were performed with CdSe and CdSe/ZnS nanocrystals embedded in an insulating polymer matrix. These reports focused on the subset of nanocrystals that exhibited digital "on"/"off" blinking (Figure 11.6). The behavior was thought to be analogous to photoluminescence intermittency of small molecules; in which the fluorescence of a small molecule is "on" when it is in a singlet state and the molecule is "off" when it is in a triplet state.[65,66] For small molecules, the rate of intersystem crossing can be determined by fitting the probability that the molecule will spend time in the "on" or "off" state.

When the "on" and "off" time statistics of CdSe nanocrystals were analyzed in a similar manner, the on and off distributions were better fit with power-laws and not single exponentials. A number of theories were developed to account for nanocrystal blinking.[55,64,67-69] CdSe nanocrystals have large spin-orbit coupling and are therefore not expected to have singlet and triplet states. Instead a theory of long-lived trap states was developed to account for the off times ranging from milliseconds to seconds. It was hypothesized that when a CdSe nanocrystal was photoexcited, one of the charge carriers (e.g., the electron or the hole) could tunnel off the nanocrystal; leaving the nanocrystal charged (Figure 11.7). As long as the nanocrystal remained charged it would stay "off" due to Auger processes. If an exciton was formed on the charged nanocrystal, its energy would be transferred to the remaining charge carrier when the exciton recombined. The nanocrystal would only return to the "on" state when the nanocrystal was neutralized. The power-law behavior was explained as resulting from variations in the height and

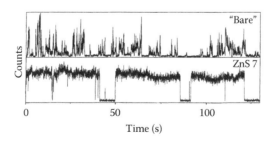

FIGURE 11.6 Photoluminescence intensity traces of two single nanocrystals with over time. The upper trace is of a "bare" nanocrystal, the lower trace is of a nanocrystal with ~7 monolayers of a ZnS shell. (Reprinted by permission from Macmillan Publishers Ltd. *Nature*, Nirmal, M., Dabbousi, B.O., Bawendi, M.G., Macklin, J.J., Trautman, J.K., Harris, T.D., and Brus, L.E., 383, 802–804, 1996. Copyright 1996.)

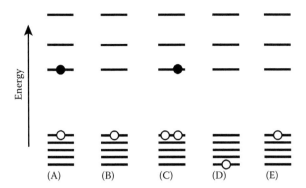

FIGURE 11.7 The charged "off" state during nanocrystal blinking according to the long-lived trap state model. (A) The nanocrystal is photoexcited. (B) One of the charge carriers tunnels away from the nanocrystal, leaving the nanocrystal charged. (C) When another photon is absorbed, the exciton can recombine and transfer its energy to the third carrier (D). (E) The charge carrier can relax to the band edge.

width of the tunneling barrier for the electron or hole over time.[64] Newer theories have been developed that explain the distributions for the blinking data, using a distribution of quenching pathways or trap states and not a single "on" and "off" state.[69] In fact, there have been papers theorizing that the trap states and defects on a given nanocrystal fluctuate with time during single-nano-crystal experiments.[61,69]

It is worth noting that while early studies of nanocrystal blinking focused on nanocrystals with digital "on"/"off" trajectories, it has become increasingly acknowledged that single CdSe nanocrystals blink with a distribution of intensities.[36,69] Although two state blinking between an "on" and an "off" state has been observed, digital blinking generally occurs for CdSe/ZnS core/shell nanocrystals.[67] Additionally, there have been reports of non-blinking CdSe nanocrystals, but all of these reports are for nanocrystals with an inorganic shell. Either the nanocrystals have an inorganic shell of ZnS and are surrounded by large organic ligands[70] or are "giant" nanocrystals with a very thick (~30–40 monolayers) inorganic shell that insulates the nanocrystal from its local environment.[71]

Although it is clear that much is understood about colloidal CdS and CdSe nanocrystal photoluminescence one of the current challenges will continue to be determining whether the nanocrystal photoluminescence spectra, efficiency, and relaxation rates are intrinsic to the nanocrystals themselves or if they can be tuned or manipulated by synthetic methods or processing of the nanocrystals. Some applications call for nanocrystals with high photoluminescence quantum yields, while others require charge transfer from the nanocrystals. Determining how to control the rate of charge transfer and energy transfer to and from a nanocrystal will continue to be intense areas of research. The work of Knowles et al.[40] and other researchers attempting to quantify the effects of ligands on nanocrystal photoluminescence and the binding dynamics and binding constants illustrates an important frontier for basic science research.

References

1. Efros, A. L.; Rosen, M. *Annual Review of Material Science* 2000, *30*, 475–521.
2. Nirmal, M.; Brus, L. *Accounts of Chemical Research* 1999, *32*, 407–414.
3. Talapin, D. V.; Lee, J.-S.; Kovalenko, M. V.; Shevchenko, E. V. *Chemical Reviews* 2010, *110*, 389–458.
4. Yin, Y.; Alivisatos, A. P. *Nature* 2005, *437*, 664–670.
5. Majetich, S. A.; Carter, A. C. *Journal of Physical Chemistry* 1993, *97*, 8727–8731.
6. Murray, C. B.; Norris, D. J.; Bawendi, M. G. *Journal of the American Chemical Society* 1993, *115*, 8706–8715.
7. Peng, Z. A.; Peng, X. G. *Journal of the American Chemical Society* 2002, *124*, 3343–3353.

8. Gerbec, J. A.; Magana, D.; Washington, A.; Strouse, G. F. *Journal of the American Chemical Society* 2005, *127*, 15791–15800.

9. Murugan, A. V.; Sonawane, R. S.; Kale, B. B.; Apte, S. K.; Kulkarni, A. V. *Materials Chemistry and Physics* 2001, *71*, 98–102.

10. Brus, L. E. *Journal of Chemical Physics* 1983, *79*, 5566–5571.

11. Efros, A. L.; Efros, A. L. *Soviet Physics of Semiconductors* 1982, *16*, 772–775.

12. Somers, R. C.; Bawendi, M. G.; Nocera, D. G. *Chemical Society Reviews* 2007, *36*, 579–591.

13. Colvin, V. L.; Schlamp, M. C.; Alivisatos, A. P. *Nature* 1994, *370*, 354–357.

14. Dabbousi, B. O.; Bawendi, M. G.; Onitsuka, O.; Rubner, M. F. *Applied Physics Letters* 1995, *66*, 1316–1318.

15. Coe-Sullivan, S. *Laser Focus World* 2007, *43*, 65–68.

16. Coe-Sullivan, S.; Steckel, J. S.; Woo, W. K.; Bawendi, M. G.; Bulovic, V. *Advanced Functional Materials* 2005, *15*, 1117–1124.

17. Kamat, P. V. *Journal of Physical Chemistry C* 2007, *111*, 2834–2860.

18. Buhbut, S.; Itzhakov, S.; Tauber, E.; Shalom, M.; Hod, I.; Geiger, T.; Garini, Y.; Oron, D.; Zaban, A. *ACS Nano* 2010, *4*, 1293–1298.

19. Greenham, N. C.; Peng, X. G.; Alivisatos, A. P. *Physical Review B* 1996, *54*, 17628–17637.

20. Johnston, K. W.; Pattantyus-Abraham, A. G.; Clifford, J. P.; Myrskog, S. H.; MacNeil, D. D.; Levina, L.; Sargent, E. H. *Applied Physics Letters* 2008, *92*, 151115.

21. Cui, S.-C.; Tachikawa, T.; Fujitsuka, M.; Majima, T. *Journal of Physical Chemistry C* 2010, *114*, 1217–1225.

22. Rabani, E.; Hetenyi, B.; Berne, B. J.; Brus, L. E. *Journal of Chemical Physics* 1999, *110*, 5355–5369.

23. Ekimov, A. I.; Hache, F.; Schanne-Klein, M. C.; Ricard, D.; Flytzanis, C.; Kudryavtsev, I. A.; Yazeva, T. V.; Rodina, A. V.; Efros, A. L. *Journal of the Optical Society of America B* 1993, *10*, 100–107.

24. Wang, L. W.; Zunger, A. *Physical Review B* 1996, *53*, 9579–9582.

25. Yu, W. W.; Qu, L. H.; Guo, W. Z.; Peng, X. G. *Chemistry of Materials* 2003, *15*, 2854–2860.

26. Efros, A. L.; Rosen, M.; Kuno, M.; Nirmal, M.; Norris, D. J.; Bawendi, M. *Physical Reviews B* 1996, *54*, 4843–4856.

27. Bawendi, M. G.; Wilson, W. L.; Rothberg, L.; Carroll, P. J.; Jedju, T. M.; Stegerwald, M. L.; Brus, L. E. *Physical Review Letters* 1990, *65*, 1623.

28. Jiang, Z.-J.; Leppert, V., Kelley, D. F. *Journal of Physical Chemistry C* 2009, 113, 19161–19171.

29. Kalyuzhny, G.; Murray, R. W. *Journal of Physical Chemistry B* 2005, *109*, 7012–7021.

30. Kuno, M.; Lee, J. K.; Dabbousi, B. O.; Mikulec, F. V.; Bawendi, M. G. *Journal of Chemical Physics* 1997, *106*, 9869–9882.

31. Dukes III, A. D.; Schreuder, M. A.; Sammons, J. A.; McBride, J. R.; Smith, N. J.; Rosenthal, S. J. *The Journal of Chemical Physics* 2008, *129*, 121102.

32. Schreuder, M. A.; McBride, J. R.; Dukes III, A. D.; Sammons, J. A.; Rosenthal, S. J. *Journal of Physical Chemistry C* 2009, *113*, 8169–8176.

33. Bowers II, M. J.; McBride, J. R.; Rosenthal, S. J. *Journal of the American Chemical Society* 2005, *127*, 15378–15379.

34. Talapin, D. V.; Rogach, A. L.; Kornowski, A.; Haase, M.; Weller, H. *Nano Letters* 2001, *1*, 207–211.

35. Bullen, C.; Mulvaney, P. *Langmuir* 2006, *22*, 3007–3013.

36. Munro, A. M.; Ginger, D. S. *Nano Letters* 2008, *8*, 2585–2590.

37. Landes, C.; Burda, C.; Braun, M.; El-Sayed, M. A. *Journal of Physical Chemistry B* 2001, *105*, 2981–2986.

38. Sharma, S. N.; Pillai, Z. S.; Kamat, P. V. *Journal of Physical Chemistry B* 2003, *107*, 10088–10093.

39. Munro, A. M.; Plante, I. J. L.; Ng, M. S.; Ginger, D. S. *Journal of Physical Chemistry C* 2007, *111*, 6220–6227.

40. Knowles, K. E.; Tice, D. B.; McArthur, E. A.; Solomon, G. C.; Weiss, E. A. *Journal of the American Chemical Society* 2010, *132*, 1041–1050.

41. Ji, X.; Copenhaver, D.; Sichmeller, C.; Peng, X. G. *Journal of the American Chemical Society* 2008, *130*, 5726–5735.
42. Koole, R.; Schapotschnikow, P.; de Mello Donega, C.; Vlugt, T. J. H.; Meijerink, A. *ACS Nano* 2008, *2*, 1703–1714.
43. Schapotschnikow, P.; Hommersom, B.; Vlugt, T. J. H. *Journal of Physical Chemistry C* 2009, *113*, 12690–12698.
44. Chakrapani, V.; Tvrdy, K.; Kamat, P. V. *Journal of the American Chemical Society* 2010, *132*, 1228–1229.
45. Kongkanand, A.; Tvrdy, K.; Takechi, K.; Kuno, M.; Kamat, P. V. *Journal of the American Chemical Society* 2008, *130*, 4007–4015.
46. Dibbell, R. S.; Watson, D. F. *Journal of Physical Chemistry C* 2009, *113*, 3139–3149.
47. Gotesman, G.; Naaman, R. *The Journal of Physical Chemistry Letters* 2010, *1*, 594–598.
48. Dayal, S.; Lou, Y.; Samia, A. C. S.; Berlin, J. C.; Kenney, M. E.; Burda, C. *Journal of the American Chemical Society* 2006, *128*, 13974–13975.
49. Kagan, C. R.; Murray, C. B.; Bawendi, M. G. *Physical Review B* 1996, *54*, 8633–8643.
50. Kagan, C. R.; Murray, C. B.; Nirmal, M.; Bawendi, M. G. *Physical Review Letters* 1996, *76*, 1517–1520.
51. Clapp, A. R.; Medintz, I. L.; Mauro, J. M.; Fisher, B. R.; Bawendi, M. G.; Mattoussi, H. *Journal of the American Chemical Society* 2004, *126*, 301–310.
52. Clapp, A. R.; Medintz, I. L.; Fisher, B. R.; Anderson, G. P.; Mattoussi, H. *Journal of the American Chemical Society* 2005, *127*, 1242–1250.
53. Anikeeva, P. O.; Madigan, C. F.; Coe-Sullivan, S. A.; Steckel, J. S.; Bawendi, M. G.; Bulovic, V. *Chemical Physics Letters* 2006, *424*, 120–125.
54. Achermann, M.; Petruska, M. A.; Koleske, D. D.; Crawford, M. H.; Klimov, V. I. *Nano Letters* 2006, *6*, 1396–1400.
55. Efros, A. L. *Nature Materials* 2008, *7*, 612–613.
56. Klimov, V. I.; McBranch, D. W.; Leatherdale, C. A.; Bawendi, M. G. *Physical Review B* 1999, *60*, 13740–13749.
57. Pandey, A.; Guyot-Sionnest, P. *Science* 2008, *322*, 929–932.
58. Efros, A. L.; Kharchenko, V. A.; Rosen, M. *Solid State Communications* 1995, *93*, 281–284.
59. Schaller, R. D.; Pietryga, J. M.; Goupalov, S. V.; Petruska, M. A.; Ivanov, S. A.; Klimov, V. I. *Physical Review Letters* 2005, *95*, 196401.
60. Guyot-Sionnest, P.; Wehrenberg, B.; Yu, D. *The Journal of Chemical Physics* 2005, *123*, 074709.
61. Kilina, S. V.; Kilin, D. S.; Prezhdo, O. V. *ACS Nano* 2009, *3*, 93–99.
62. Empedocles, S. A.; Norris, D. J.; Bawendi, M. G. *Physical Review Letters* 1996, *77*, 3873–3876.
63. Neuhauser, R. G.; Shimizu, K. T.; Woo, W. K.; Empedocles, S. A.; Bawendi, M. G. *Physical Review Letters* 2000, *85*, 3301–3304.
64. Kuno, M.; Fromm, D. P.; Hamann, H. F.; Gallagher, A.; Nesbitt, D. J. *Journal of Chemical Physics* 2001, *115*, 1028–1040.
65. Basche, T. *Journal of Luminescence* 1998, *76–7*, 263–269.
66. Basche, T.; Kummer, S.; Brauchle, C. *Nature* 1995, *373*, 132–134.
67. Nirmal, M.; Dabbousi, B. O.; Bawendi, M. G.; Macklin, J. J.; Trautman, J. K.; Harris, T. D.; Brus, L. E. *Nature* 1996, *383*, 802–804.
68. Frantsuzov, P. A.; Marcus, R. A. *Physical Review B* 2005, *72*, 155321.
69. Frantsuzov, P. A.; Volkan-Kacso, S.; Janko, B. *Physical Review Letters* 2009, *103*, 207402.
70. Hohng, S.; Ha, T. *Journal of the American Chemical Society* 2004, *126*, 1324–1325.
71. Chen, Y.; Vela, J.; Htoon, H.; Casson, J. L.; Werder, D. J.; Bussian, D. A.; Klimov, V. I.; Hollingsworth, J. A. *Journal of the American Chemical Society* 2008, *130*, 5026–5027.

12

Radiative Cascades in Semiconductor Quantum Dots

Eilon Poem
*The Technion—Israel
Institute of Technology*

David Gershoni
*The Technion—Israel
Institute of Technology*

12.1 Introduction

Semiconductor Quantum Dots (QDs) are nanometric scale regions of a narrow-bandgap semiconductor embedded within a wider-bandgap semiconductor. They confine charge carriers (electrons and/or holes) in all three dimensions. The QD's energy level spectrum is therefore discrete, much like that of the fundamental building blocks of nature—atoms and molecules. QDs are therefore often referred to as "artificial atoms" [1–3]. They have been extensively investigated recently as potential, technology-compatible quantum light emitters [4–6], providing single photons or "flying qubits" on demand. More recently, it has been shown that QDs can emit pairs of entangled photons [7,8]. Such capabilities are important for possible future applications such as quantum information processing [9] and cryptography [10]. Though similar effects were previously observed in the fluorescence of single atoms and molecules [11], semiconductor QDs offer many advantages. In particular, they exhibit large electrostatic capacitance, which enables a wide range of charge states [12]. This feature, among others, forms a sound base for the QDs' potential applications. At the same time it makes them an excellent stage for studying inter-charge-carrier interactions in confined spaces.

One of the experimental methods frequently used to investigate single QDs is micro-PL (μPL). In a μPL experiment, the QD is optically excited by laser light, focused on the QD trough a microscope objective. The light emitted by the QD is collected through the same microscope objective and

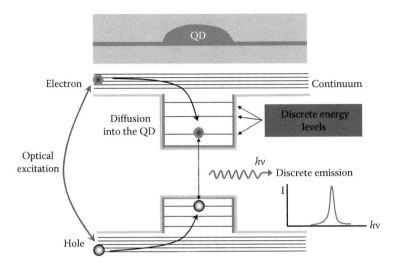

FIGURE 12.1 Schematic illustration of PL emission from a single semiconductor QD under nonresonant optical excitation.

spectrally analyzed by a spectrometer. Spatial filtering using a confocal aperture could be used in order to enhance the spatial resolution. The exciting light promotes an electron from the full valence band, leaving a hole there, to the empty conduction band, thus creating an electron–hole (e–h) pair or an "exciton." If the excitation is off resonant, the pair is generated in the vicinity of the QD, and it moves in the semiconductor until it either recombines first or is captured by the QD and only then recombines. Since the energy levels of a QD are discrete, the recombination of an e–h pair in a QD gives rise to a discrete spectral line (see Figure 12.1). In analogy to atomic physics, we denote PL lines resulting from the recombination of an e–h pair occupying the first (second, third) single-carrier levels by the letters $s(p, d)$.

A QD may be populated by more than one e–h pair or populated with an uneven number of electrons and holes. In such cases, the recombination energy of an e–h pair will be affected by the additional Coulombic interactions. Therefore, typically more than one spectral line is seen around the single e–h pair ground-state recombination line, as illustrated in Figure 12.2.

We name the various transitions as follows: The letter X (for eXciton) is written as many times as there are e–h pairs in the initial state. The number and sign of unpaired charge carriers is written as a superscript after the last X. For example, transitions starting from states that contain three holes and two electrons will be denoted XX^{+1}. Additional subscripts may be used to distinguish between different spin configurations of the initial and/or the final states, as will be explained later. The identification of a spectral line is thus the ability to identify an e–h pair recombination process, uniquely defined by its many-body initial and final states.

Each of the properties of the line's initial and final configurations can be obtained experimentally: The number of e–h pairs can be deduced from the dependence of the line's intensity upon the excitation power. The number and sign of excess charge carriers can be deduced from experiments in which the charge of the QD is controlled (by means of electrical and/or optical excitations). The spin configurations can be deduced by polarization-sensitive spectroscopy and polarization memory experiments. In addition, polarization-sensitive intensity correlation measurements can be used for verification of the line identification and for obtaining additional insight into the dynamical processes by which the excited QD relaxes, including the identification of radiative cascades.

The discussion of these processes is organized as follows: In the first section, we begin with theoretical description of the many-body states and energy levels of QD-confined charge carriers and

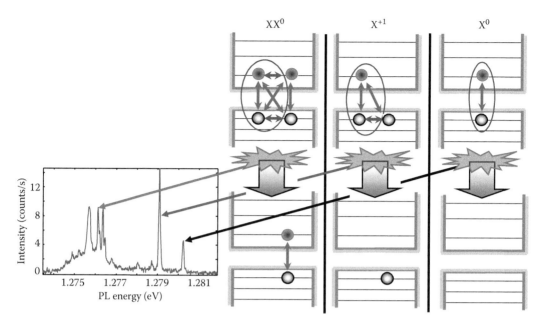

FIGURE 12.2 Schematic description of various radiative pair recombination processes in the presence of spectator charge carriers. The Coulombic interactions affect the many-body state energies. Therefore, several different spectral lines are expected for the recombination of an s-shell electron–hole pair.

radiative transitions between them. In the second section, we present measured polarization-sensitive PL spectra of a single QD and compare it to the theoretically predicted transitions. In the third section, we discuss charge carrier dynamics. We present results obtained by polarization memory experiments and present a theoretical model to explain them. We discuss intensity cross-correlation measurements as an experimental tool for probing carrier dynamics and present the theory behind it. In the last section, we discuss the notion of radiative cascades and present experimentally measured intensity cross-correlation functions, which reveal such cascades in both neutral and charged QDs. We show that important information about the dynamical processes can be extracted from these measurements.

12.2 Theoretical Model for Calculating Few Confined Carrier States in Semiconductor Quantum Dots

The goal of this section is to provide theoretical tools for analyzing the optical spectrum of semiconductor QDs and its polarization selection rules. We provide such a tool by a model that is based on a full-configuration interaction (FCI) method [13–15]. Traditionally, this method provides a straightforward tool for calculating many indistinguishable interacting carrier levels from their known single carrier states. Our approach is unique, however, in the sense that it includes also the exchange interaction between the seemingly distinguishable electrons and holes [16–18]. This interaction is orders of magnitude smaller than the exchange interactions between carriers of same charge [17,18], yet, as we show, it determines the polarization selection rules of the PL spectrum of semiconductor QDs.

The FCI method requires as an input the energies and wavefunctions of single carrier in the QD potential. We supply this information both for the electron and for the hole. There are few methods for solving the single carrier problem [19–27]. However, since the actual composition, strain, dimensions, and shape of a particular QD are not accurately known, we prefer the simplest approach that provides a consistent set of single carrier states, of which a finite set of states is chosen. The interaction energies between pairs of carriers in any one of the chosen states are then calculated. For the Coulomb interaction, these elements are given by

$$C^{p_1 p_2 p_3 p_4}_{n_1,n_2,n_3,n_4} = \iint d^3 r_1 d^3 r_2 \psi^{p_1^*}_{n_1}(\vec{r}_1) \psi^{p_2^*}_{n_2}(\vec{r}_2) \frac{e^2}{\varepsilon|\vec{r}_1 - \vec{r}_2|} \psi^{p_3}_{n_3}(\vec{r}_2) \psi^{p_4}_{n_4}(\vec{r}_1) \qquad (12.1)$$

where
  e is the electron charge
  ε is the dielectric constant
  $p_{1...4}$ can be either "*e*" (for electron) or "*h*" (for hole)
  the indices $n_{1...4}$ run over the appropriate single-particle states

At this stage, in order to insure convergence, it should be verified that the interaction elements that connect low energy states to higher energy states are smaller than the energy differences between these states. For a QD of length L, the Coulomb interaction energy is roughly proportional to $1/L$, while the single-particle confinement energy is roughly proportional to $1/L^2$. Therefore, the smaller the QD is, the smaller is the ratio between the two and the faster the calculations converge (less single-particle states should be considered).

The FCI many-carrier Hamiltonian is written in second-quantization formalism [14] as follows:

$$H = H_0 + H_{ee} + H_{hh} + H_{eh} \qquad (12.2)$$

where

$$H_0 = \sum_i \varepsilon^e_i a^\dagger_i a_i + \sum_j \varepsilon^h_j b^\dagger_j b_j$$

$$H_{ee} = \frac{1}{2} \sum_{i_1,i_2,i_3,i_4} C^{eeee}_{i_1,i_2,i_3,i_4} a^\dagger_{i_1} a^\dagger_{i_2} a_{i_3} a_{i_4}$$

$$H_{hh} = \frac{1}{2} \sum_{j_1,j_2,j_3,j_4} C^{hhhh}_{j_1,j_2,j_3,j_4} b^\dagger_{j_1} b^\dagger_{j_2} b_{j_3} b_{j_4} \qquad (12.3)$$

$$H_{eh} = \sum_{i_1,j_2,j_3,i_4} \left(-C^{ehhe}_{i_1,j_2,j_3,i_4} + C^{hehe}_{j_2,i_1,j_3,i_4} \right) a^\dagger_{i_1} b^\dagger_{j_2} b_{j_3} a_{i_4}$$

H_0 is the single-particle Hamiltonian, presumed to be already diagonalized, where ε^e_i (ε^h_j) is the energy of the single-electron (hole) state number i (j). H_{ee} (H_{hh}, H_{eh}) is the electron–electron (hole–hole, electron–hole) interaction Hamiltonian. The operator $a^\dagger_i$ (b_j) creates (annihilates) an electron (a hole) in the single-electron (hole) state number i (j). The electron and hole creation and annihilation operators have the usual Fermionic *anti*-commutation relations. This ensures the antisymmetry of the created states in the exchange of any pair of particles [14]. The second term in H_{eh} is the *e–h exchange interaction* (EHEI). We show below that this term, which we add to our many-carrier model for the first time [28] is the main source for linearly polarized spectral lines in the PL spectrum of single QDs.

Since it commutes with the electron number operator, $N_e = \sum_i a^\dagger_i a_i$ and with the hole number operator, $N_h = \sum_j b^\dagger_j b_j$ [14,15], the Hamiltonian in Equation 12.3 can be written as a separate matrix for each number of electrons and holes. We do so by using the population basis. Each wavefunction in this basis represents a Slater determinant (a complete particle-exchange antisymmetric state) composed of products of the chosen single-carrier states. The Hamiltonian matrix is then diagonalized, and the many-body states and eigenenergies are obtained. Finally, the energies, the oscillator strengths, and polarization selection rules for optical transitions between the obtained many-body states are calculated using the dipole approximation.

In the following sections, we discuss each step in the FCI model calculations.

12.2.1 Single-Carrier States

There are various methods for calculating single-carrier energies and states in semiconductor hetero-structures in general, and in QDs in particular [19–27]. These can be divided into two groups. The first group includes atomistic approaches, such as the pseudo-potential method [19] and the tight-binding method [20,21]. The second group includes envelope function and effective mass approximations, such as single and multiband $k \cdot P$ approximations [22–27] for the solution of the Schrödinger equation for a single carrier in the QD potential.

The methods in the first group consider the full atomic potential directly and thus obtain the full wavefunctions in atomic resolution. They require the information about the type and position of every atom in the structure and demand rather large amounts of computational resources.

In methods of the second group, it is assumed that the wavefunctions can be written as a sum of products of an intra-unit-cell Bloch function, $u_b(\vec{r})$, periodic over all unit cells and an inter-unit-cell envelope function, $\phi_n^b(\vec{r})$:

$$\psi_n = \sum_b \phi_n^b(\vec{r}) u_b(\vec{r}) \tag{12.4}$$

The summation is over all bands b. The problem is then reduced to finding the envelope functions, given the bulk material parameters (e.g., band-edge energies, carrier dispersion functions, bulk elastic moduli), as a function of position. Without magnetic fields, Kramers theorem applies, and the bands are arranged in degenerate pairs [29]. Generally, every pair of degenerate states can be mapped onto a pair of spin 1/2 states, even if the real spin of the original states is not 1/2, or even not defined at all [14]. The new spin component associated with a state is called the *pseudo-spin* component of the state. If only one (twice Kramers degenerate) band is considered, the equation for the envelope function reduces to a simple Schrödinger equation with an effective mass and an effective potential. For a zero magnetic field, this equation does not depend on the pseudospin of the state and yields the same envelope wavefunction for each pseudo-spin component. Taking into account only one band is justified when the band in question is nondegenerate (except for the Kramers degeneracy) and energetically isolated from other bands. This is usually the case for an electron in the conduction band of wide band III–V semiconductors. The case of a hole in the valence band is more complicated. In these bulk materials, the highest valence band states (the "heavy" and "light" hole states) are degenerate (at zero crystal momentum $k = 0$). In quantum structures, however, the confinement and the strain lift this degeneracy [23–25]. When this degeneracy removal is substantial, the single band description of the heavy hole band is still a good approximation.

Considering only a single band greatly reduces the amount of computational resources needed. It also offers direct intuitive insights into various model parameters and their influence on the calculated measurable properties of the QDs. It misses, though, some of the QD properties, notably properties that depend on orientation of the QD relative to the crystallographic axes of the semiconductor [28,30]. These and other properties may require more complicated methods. However, due to the lack of structural information needed for these methods we prefer this approach and we shall use it throughout this chapter.

For the effective potential acting on the single band carrier, we use a very simple model: a low constant potential inside a region in the shape of a rectangular or an elliptical slab and a higher constant potential outside that region. These shapes belong to the D_{2h} symmetry group. The D_{2h} is the highest symmetry group without a threefold or higher symmetry axis. Higher symmetries would not contain

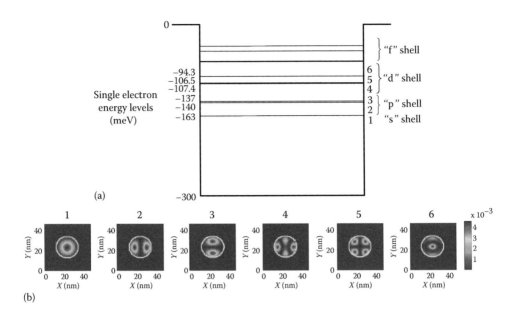

(a)

(b)

FIGURE 12.3 (a) Single-electron eigenenergies, numerically calculated for an elliptic slab model with height = 3 nm, length = 23.2 nm, width = 21.2 nm, and an electron effective mass of $0.085 m_0$. (b) In-plane cross sections of the calculated probability distributions (wavefunction squared) at half the slab's height.

in-plane anisotropy, which is crucial for the appearance of linear polarizations in the PL emission. Lower symmetries would require a more complicated model for describing the QD potential, possibly with more free structural parameters. Our model contains only three different geometrical parameters: the height, length, and width of the lower potential volume. Due to the different band offsets and different strain effects, the potential for the hole has different parameters than that for the electron [25]. In an attempt to reduce the number of free parameters, we express this freedom by only one additional parameter. Namely, the ratio between the length of the slab for the hole to that for the electron.

For epitaxially grown, self-assembled QDs, the height (defined to be along the growth direction) is usually much smaller than the length and width [31]. The single-particle wavefunctions are therefore quasi two dimensional and have no nodes along the growth axis (z). Each wavefunction can therefore be uniquely defined by two quantum numbers. These numbers, n_x and n_y respectively, reflect the number of the nodes in the carrier wavefunction along two axes perpendicular to the growth axis ($n_z = 0$). If the ratio between the width and length of the QD (the aspect ratio) is not far from 1 (a "nearly symmetric" QD), all wavefunctions having the same total number of nodes, $n = n_x + n_y$, will have close energies, well separated from the energies of wavefunctions with different total number of nodes. The single-particle spectrum is thus composed of "shells." Each shell contains $2(n + 1)$ states, where n is the total number of nodes ($n = 0, 1, 2, ...$). The extra factor of 2 is due to the pseudo-spin (Kramers) degeneracy. As an example, Figure 12.3 presents numerically calculated single-electron energies and wavefunctions for an elliptic slab. The model parameters where chosen to fit experimental measurements.

12.2.2 Few-Carrier States: The Role of the Electron–Hole Exchange Interaction

The effects of the Coulomb interaction between the carriers on the few-carrier energies and states can be quite generally divided into three types: classical, exchange, and correlation. In the classical picture, the few-particle states gain or lose energy due to the direct electrostatic interaction between their charge densities. The interaction elements corresponding to this effect, the "direct" Coulomb interactions, $C^{eeee}_{i_1,i_2,i_2,i_1}, C^{hhhh}_{j_1,j_2,j_2,j_1}, C^{ehhe}_{i_1,j_2,j_2,i_1}$, lie on the main diagonal of the Hamiltonian (1.2). Thus, they do not "mix" states of different symmetries and therefore do not have a direct effect on the form of the wavefunctions. The exchange Coulomb interaction elements result from the fact that in quantum mechanics, the carriers are indistinguishable. They have the forms $C^{eeee}_{i_2,i_1,i_2,i_1}, C^{hhhh}_{j_2,j_1,j_2,j_1}, C^{hehe}_{j_2,i_1,j_2,i_1}$. Elements of this type can be off diagonal, but then they connect only wavefunctions that have the *same* diagonal element (including the classical Coulomb terms) [14]. Thus, they can be viewed as first-order perturbations within degenerate subspaces. The form of the eigen wavefunctions within each degenerate subspace is determined by the diagonalization of the exchange interaction in that subspace. These "first-order" few-particle wavefunctions, which still contain a well-defined spatial wavefunction for each particle, are called "configurations" (due to subspace diagonalization, the pseudospin is often ill defined for each single carrier separately). Correlation Coulomb interaction elements are all the elements that are neither classical nor exchange. They are all off diagonal and connect different degenerate subspaces. Thereby, they mix states of different configurations. The inclusion of such elements in the calculation is the source of the name "configuration interaction." The influence of each type of interaction elements is illustrated in Figure 12.4.

In order to calculate the Coulomb interaction elements, Equation 12.1, in the envelope function approximation (Equation 12.4), it is convenient to separate the integration over the entire crystal into integration within a unit cell (of volume Ω, centered at the lattice vector $\vec{R}$) and summation over all unit cells:

$$\int d^3r = \sum_{\vec{R}} \int_{\Omega_{\vec{R}}} d^3r \qquad (12.5)$$

When the potential varies slowly over the length of one unit cell, it follows that the envelope function of the confined carrier varies slowly on that length scale as well. It could therefore be considered constant over the volume of one unit cell. This approximation is the slowly varying envelope function approximation (SVEFA). If only one (Kramers degenerate) band is included for each type of particle (band mixing is neglected), and the SVEFA is used, the integral in Equation 12.1 becomes

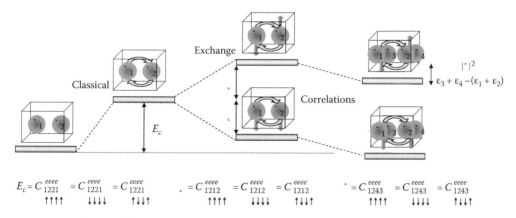

FIGURE 12.4 Schematic description of the three types of coulomb interactions between two confined charge carriers and their effect on the mutual two carriers' wavefunction and energy. Circles represent spatial single carrier wavefunctions. Arrows represent single carrier pseudospins. The wavefunction indices were separated into spatial (1, 2, …) and pseudo-spin (↑,↓) components.

$$C^{p_1,p_2,p_3,p_4}_{\substack{n_1,n_2,n_3,n_4 \\ s_1,s_2,s_3,s_4}} = \sum_{R_1,R_2} \phi^{p_1}_{n_1}{}^*(\vec{R}_1)\phi^{p_2}_{n_2}{}^*(\vec{R}_2)\phi^{p_3}_{n_3}(\vec{R}_2)\phi^{p_4}_{n_4}(\vec{R}_1)$$

$$\times \iint_\Omega \frac{e^2}{\varepsilon|\vec{R}_1 - \vec{R}_2 + \vec{l}_1 - \vec{l}_2|} u^*_{p_1,s_1}(\vec{l}_1)u^*_{p_2,s_2}(\vec{l}_2)u_{p_3,s_3}(\vec{l}_2)u_{p_4,s_4}(\vec{l}_1)d^3l_1d^3l_2 \qquad (12.6)$$

where we have defined $\vec{r} = \vec{R} + \vec{l}$, $\vec{R}$ being the lattice vector closest to $\vec{r}$, and the Bloch amplitude periodicity was used. We would now like to obtain the $\vec{R}$ dependence of the intra-unit-cell integral of Equation 12.6. For this purpose, the Coulomb operator is decomposed into a sum of products of a function of $\vec{R}_1 - \vec{R}_2$ and a function of $\vec{l}_1 - \vec{l}_2$. This is done by dividing the integration space into two subspaces: the *long-range* subspace, where $\vec{R}_1 - \vec{R}_2 > \vec{l}_1 - \vec{l}_2$, and the *short-range* subspace, where $\vec{R}_1 - \vec{R}_2 < \vec{l}_1 - \vec{l}_2$.

For the short-range subspace, since $\vec{l}_1$ and $\vec{l}_2$ are intra-unit-cell vectors, in most cases $\vec{R}_1 = \vec{R}_2$, and one can make the approximation

$$\frac{1}{|\vec{R}_1 - \vec{R}_2 + \vec{l}_1 - \vec{l}_2|} \cong \frac{1}{|\vec{l}_1 - \vec{l}_2|}v_0\delta(\vec{R}_1 - \vec{R}_2)$$

where v_0 is the volume of a unit cell.

For the long-range subspace, the Coulomb interaction operator can be expanded into a Taylor ("multipole") series:

$$\frac{1}{|\vec{R}_1 - \vec{R}_2 + \vec{l}_1 - \vec{l}_2|} \cong \frac{1}{|\vec{R}_1 - \vec{R}_2|} - \frac{(\vec{R}_1 - \vec{R}_2)^T}{|\vec{R}_1 - \vec{R}_2|^3}\cdot(\vec{l}_1 - \vec{l}_2)$$

$$+ \frac{1}{2}(\vec{l}_1 - \vec{l}_2)^T\cdot\frac{3(\vec{R}_1 - \vec{R}_2)(\vec{R}_1 - \vec{R}_2)^T - I}{|\vec{R}_1 - \vec{R}_2|^5}\cdot(\vec{l}_1 - \vec{l}_2) + \cdots \qquad (12.7)$$

where I is the 3×3 identity matrix.

The intra-unit-cell integration can now be performed independently for each element (of both the short- and long-range subspaces), yielding functions of known dependence of $\vec{R}_1$ and $\vec{R}_2$. Under the SVEFA, the summation over $\vec{R}_1$ and $\vec{R}_2$ can be replaced by an integral over the entire space, and it can be assumed that $\vec{R} \cong \vec{r}$. The Coulomb integrals thus have the general form

$$C^{p_1,p_2,p_3,p_4}_{\substack{n_1,n_2,n_3,n_4 \\ s_1,s_2,s_3,s_4}} = \sum_m \iint d^3r_1 d^3r_2 \phi^{p_1}_{n_1}{}^*(\vec{r}_1)\phi^{p_2}_{n_2}{}^*(\vec{r}_2)\hat{C}^{(m)}_{\substack{s_1,s_2,s_3,s_4 \\ p_1,p_2,p_3,p_4}}(\vec{r}_1 - \vec{r}_2)\phi^{p_3}_{n_3}(\vec{r}_2)\phi^{p_4}_{n_4}(\vec{r}_1) \qquad (12.8)$$

where

    m counts the orders of the long-range expansion and also the short-range term

    $\hat{C}^{(m)}_{\substack{s_1,s_2,s_3,s_4 \\ p_1,p_2,p_3,p_4}}(\vec{r}_1 - \vec{r}_2)$ represents the operator in inter-unit-cell space remaining after the intra-unit-cell

    integral has been performed

This operator depends on $s_{1...4}$, the pseudo-spin components, and on $p_{1...4}$, the particle identifiers, which together uniquely identify the Bloch amplitudes of the participating single carrier wavefunctions (see Equation 12.6). Other than the operator, the integral contains only single-carrier envelope functions, numbered by their own separate indices, $n_{1...4}$. Except for the case of the EHEI, the zeroth order of the multipole series, Equation 12.7 (the monopole–monopole term), when integrated, is much larger than all the other orders, including the short-range term [15]. In such cases, to first approximation, one can keep only this order. The operator $\hat{C}^{(0)}_{\substack{s_1,s_2,s_3,s_4 \\ p_1,p_2,p_3,p_4}}(\vec{r}_1 - \vec{r}_2)$ for the monopole–monopole long-range interaction is given by

$$\hat{C}^{(0)}_{\substack{s_1,s_2,s_3,s_4 \\ p_1,p_2,p_3,p_4}}(\vec{r}_1 - \vec{r}_2) = \delta_{p_1,p_4}\delta_{p_2,p_3}\delta_{s_1,s_4}\delta_{s_2,s_3}\frac{e^2}{\varepsilon|\vec{r}_1 - \vec{r}_2|} \tag{12.9}$$

This term gives rise to the well-known same-carrier isotropic (Hisenberg-type) exchange interaction, which arranges the states according to their total pseudospin (separately for each type of carrier). For example, it divides the fourfold degenerate subspace of two identical carriers occupying two different spatial orbitals into a pseudo-spin-zero singlet and a pseudo-spin-one triplet (as schematically shown in Figure 12.4). In this approximation, this interaction is not sensitive to the anisotropy in the potential shape. This is, however, not the case for the EHEI, since for it the operator in Equation 12.9 strictly vanishes. Thus, for the EHEI, the leading term includes a short-range term and the long-range dipole–dipole term (the monopole–dipole terms contribute only to correlation integrals and only when band mixing is not neglected). The short-range EHEI operator reads [17]

$$\hat{C}^{(short)}_{\substack{s_1,s_2,s_3,s_4 \\ h,e,h,e}}(\vec{r}_1 - \vec{r}_2) = v_0 E_{SR}\delta_{s_1,s_4}\delta_{s_2,s_3}\delta_{s_1,s_3}\delta(\vec{r}_1 - \vec{r}_2) \tag{12.10}$$

where

v_0 is the volume of one unit cell
E_{SR} is the following intra-unit-cell integral

$$E_{SR} = \int_\Omega d^3l_1 d^3l_2 u^*_{h,s}(\vec{l}_1)u^*_{e,s}(\vec{l}_2)\frac{e^2}{\varepsilon|\vec{l}_1 - \vec{l}_2|}u_{h,s}(\vec{l}_2)u_{e,s}(\vec{l}_1) \tag{12.11}$$

The operator in Equation 12.10 is an isotropic contact interaction between an electron and a hole. The term differentiates in energy between a state in which the electron and the hole have parallel pseudo-spins and that in which their pseudospins are antiparallel.

The EHEI long-range dipole–dipole operator reads [17]

$$\hat{C}^{(2)}_{\substack{s_1,s_2,s_3,s_4 \\ h,e,h,e}}(\vec{r}_1 - \vec{r}_2) = \vec{\mu}^\dagger_{s_1,s_2}\frac{e^2(I - 3\vec{n}\vec{n}^\dagger)}{\varepsilon|\vec{r}_1 - \vec{r}_2|^3}\vec{\mu}_{s_3,s_4} \tag{12.12}$$

where

I is the 3×3 identity matrix
$\vec{n} = (\vec{r}_1 - \vec{r}_2)/|\vec{r}_1 - \vec{r}_2|$
$\vec{\mu}_{s,s'} = \int_\Omega d^3l u^*_{h,s}(\vec{l})\vec{l}u_{e,s'}(\vec{l})$

From symmetry considerations, it can be shown that [17]

$$\vec{\mu}^\dagger_{\Uparrow,\downarrow} = \mu(\hat{x} - i\hat{y}); \quad \vec{\mu}^\dagger_{\Downarrow,\uparrow} = \mu(\hat{x} + i\hat{y}); \quad \vec{\mu}^\dagger_{\Uparrow,\uparrow} = \vec{\mu}^\dagger_{\Downarrow,\downarrow} = 0 \tag{12.13}$$

Here $\uparrow$ ($\Downarrow$) represents an electron (hole) pseudospin $+1/2$ [32]. Using Bloch amplitudes at $k = 0$ and the connection between the position and momentum expectation values [33], the value μ^2 can be estimated as follows [28]:

$$\mu^2 \cong \frac{\hbar^2 E_P}{4m_0 E_G^2} \qquad (12.14)$$

where

E_P is the conduction-valence band interaction energy [25]
E_G is the bandgap
m_0 is the mass of a free electron

When the operator in Equation 12.12 acts on the envelope functions, it resembles a quadrupole interaction. As such it vanishes completely if the QD has in-plane rotational symmetry, which is larger than two folds. As shown below, this interaction is the main mechanism that leads to the appearance of linearly polarized emission lines in the PL spectrum of self-assembled QDs.

The leading terms in the EHEI can thus be cast into an interaction Hamiltonian of the following form:

$$H_{EHEI} = \sum_{\substack{n_1,n_2,n_3,n_4 \\ s_1,s_2,s_3,s_4}} C^{hehe}_{\substack{n_1,n_2,n_3,n_4 \\ s_1,s_2,s_3,s_4}} a^\dagger_{n_1,s_1} b^\dagger_{n_2,s_2} b_{n_3,s_3} a_{n_4,s_4} \qquad (12.15)$$

For each combination $n_1 \ldots n_4$ of envelope functions, using Equations 12.10, 12.12, and 12.13, the elements $C^{hehe}_{\substack{n_1,n_2,n_3,n_4 \\ s_1,s_2,s_3,s_4}}$ can be arranged into the following 4×4 matrix in the e–h pseudo-spin basis, $s_2 s_1, s_4 s_3 \in \{\downarrow\Uparrow, \uparrow\Downarrow, \uparrow\Uparrow, \downarrow\Downarrow\}$ [17]:

$$C^{hehe}_{\substack{n_1,n_2,n_3,n_4 \\ s_1,s_2,s_3,s_4}} = \begin{pmatrix} \Delta_0^{n_1,n_2,n_3,n_4} & \frac{1}{2}\Delta_1^{n_1,n_2,n_3,n_4} & 0 & 0 \\ \frac{1}{2}\Delta_1^{n_1,n_2,n_3,n_4^*} & \Delta_0^{n_1,n_2,n_3,n_4} & 0 & 0 \\ 0 & 0 & 0 & \frac{1}{2}\Delta_2^{n_1,n_2,n_3,n_4} \\ 0 & 0 & \frac{1}{2}\Delta_2^{n_1,n_2,n_3,n_4^*} & 0 \end{pmatrix} \qquad (12.16)$$

This matrix represents also the e–h exchange Hamiltonian for the more familiar case of one electron and one hole [18], which is obtained from Equation 12.16 by the subtraction of $\frac{1}{2}\Delta_0^{n_1,n_2,n_3,n_4}$ from the main diagonal.

The terms $\Delta_0^{n_1,n_2,n_3,n_4}$ are often called the "isotropic e–h exchange." They are not sensitive to in-plane anisotropy and separate only states with parallel pseudospins from those with antiparallel pseudospins. They contain the short-range interaction and the "out-of-plane" component of the long-range interaction:

$$\Delta_0^{n_1,n_2,n_3,n_4} = v_0 E_{SR} \int d^3 r \phi_{n_1}^{h^*}(\vec{r}) \phi_{n_2}^{e^*}(\vec{r}) \phi_{n_3}^{h}(\vec{r}) \phi_{n_4}^{e}(\vec{r})$$

$$+ \frac{e^2 \mu^2}{\varepsilon} \iint d^3 r_1 d^3 r_2 \phi_{n_1}^{h^*}(\vec{r}_1) \phi_{n_2}^{e^*}(\vec{r}_2) \frac{2n_z^2 - n_x^2 - n_y^2}{|\vec{r}_1 - \vec{r}_2|^3} \phi_{n_3}^{h}(\vec{r}_2) \phi_{n_4}^{e}(\vec{r}_1) \qquad (12.17)$$

where

v_0 and E_{SR} are defined in Equation 12.11
$\vec{n}$ is defined in Equation 12.12
μ is defined in Equation 12.13

The terms $\Delta_2^{n_1,n_2,n_3,n_4}$ split the states $\uparrow\Uparrow$, $\downarrow\Downarrow$. In the single-band approximation, this term vanishes [17]. However, from symmetry considerations it appears that for zinc blende crystals of symmetry, which is lower than that of a sphere [16], it should not vanish. These terms result from the isotropic EHEI when band mixing is considered. These terms do not have an impact on the polarization selection rules and are generally quite small in respect to the other terms in Equation 12.16 [34]. We set them to zero (unless stated otherwise) in the discussion that follows.

The terms $\Delta_1^{l_1,l_2,l_3,l_4}$ are called the "anisotropic EHEI" [16–18].

In the single-band approximation, they are pure long range and are given by

$$\frac{1}{2}\Delta_1^{l_1,l_2,l_3,l_4} = \frac{3e^2\mu^2}{\varepsilon}\iint d^3r_1 d^3r_2 \phi_{l_1}^{h*}(\vec{r}_1)\phi_{l_2}^{e*}(\vec{r}_2)\frac{n_y^2 - n_x^2 + 2in_xn_y}{|\vec{r}_1 - \vec{r}_2|^3}\phi_{l_3}^{h}(\vec{r}_2)\phi_{l_4}^{e}(\vec{r}_1) \tag{12.18}$$

For electron and hole in their lowest single-particle state, the anisotropic EHEI removes the degeneracy between the states $\downarrow\Uparrow$, $\uparrow\Downarrow$ by $|\Delta_1^{1,1,1,1}|$. The resulting eigenstates are $(\downarrow\Uparrow \pm e^{i2\phi}\uparrow\Downarrow)/\sqrt{2}$ ("+" for the lower energy). Here ϕ is defined as the angle between the major axis of the QD and the x-axis of the frame of reference. Thus, $\Delta_1^{1,1,1,1} \equiv -|\Delta_1^{1,1,1,1}|e^{i2\phi}$. Obviously, by a proper choice of the coordinates ϕ can be set to 0. As we show below, optical transitions to and from these states are cross-linearly polarized.

The integrals in Equation 12.18 can be numerically calculated if the single-carrier envelope wavefunctions are accurately known [17]. In order to get insight and to save computational resources, we obtained approximate analytical model for these calculations. First, since lattice mismatch strain-induced QDs are typically quite flat and their dimension along the growth direction is an order of magnitude smaller than their lateral extent, we use a 2D model for describing the QD potential. The integrals' dimensionality is thus reduced to four. We then approximate the QD potential by a parabolic expression and obtain analytical wavefunctions for the single charge carriers [26–28]. With these wavefunctions, three of the four integrations can be analytically obtained. The last integral can then be easily calculated numerically. For a nearly symmetric QD, even this integral can be obtained analytically by expanding it into a power series in the aspect ratio of the model QD and then keeping terms up to first order only. The result of this derivation for $\Delta_1^{1,1,1,1}$ is

$$\Delta_1^{1,1,1,1} \cong \frac{3\sqrt{\pi}e^2\mu^2(\xi-1)}{2\varepsilon\beta\sqrt{1+\beta^2}(l_x^e)^3\xi^2} \tag{12.19}$$

where

μ^2 is given by Equation 12.14
l_x^e is the characteristic length (semiaxis) of the electron Gaussian wave function in the x-direction
β is the ratio between the characteristic length of the hole wave function to that of the electron
ξ is the length ratio between the minor and major axes of the QD—namely its aspect ratio

We chose the x-axis to be along the long axis of the QD, thereby setting ϕ to zero. For $\xi = 0.96$, $\beta = 0.72$, and $l_x^e = 72$ Å [28], which result in fairly good agreement between the calculated PL spectra and the measured one, Equation 12.19 yields $\Delta_1^{1,1,1,1} = 15\,\mu\text{eV}$, not too far from the measured value of $30\,\mu\text{eV}$ [7]. All the other $\Delta_1^{l_1,l_2,l_3,l_4}$ terms can be similarly obtained. Some examples for these terms are given in Table 12.1.

12.2.3 Optical Transitions between the States and Polarization Selection Rules

Once the many carrier states are known, optical transitions between these states can be quite accurately and straight forwardly calculated using the dipole approximation. The dipole operator can be expressed as [13], follows:

TABLE 12.1 Ratios of a Few Anisotropic e–h Exchange Terms to the Term of a Ground-State Exciton, for a Nearly Symmetric QD ($|\xi - 1| \ll 1$)

e–h Exchange Term	Ratio to $\Delta_1^{1,1,1,1}$	Value of the Ratio to $\Delta_1^{1,1,1,1}$
$\Delta_1^{1,2,1,2}$	$\dfrac{\beta^2}{1+\beta^2}\dfrac{2\xi-1}{\xi-1}$	-7.84
$\Delta_1^{2,1,2,1}$	$\dfrac{1}{1+\beta^2}\dfrac{2\xi-1}{\xi-1}$	-15.16
$\Delta_1^{1,3,1,3}$	$\dfrac{\beta^2}{1+\beta^2}\dfrac{\xi-2}{\xi-1}$	$+8.88$
$\Delta_1^{1,2,1,3}$	$i\dfrac{\beta^2}{1+\beta^2}\dfrac{\xi+1}{2(\xi-1)}$	$-8.36i$
$\Delta_1^{1,4,1,4}$	$\dfrac{\beta^4}{(1+\beta^2)^2}\dfrac{41-6\xi}{16}$	$+0.256$
$\Delta_1^{2,2,2,2}$	$\dfrac{\beta^2}{(1+\beta^2)^2}\dfrac{61\xi-45}{16(\xi-1)}$	-4.76

Source: Reprinted with permission from Poem, E., Shemesh, J., Marderfeld, I., Galushko, D., Akopian, N., Gershoni, D., Gerardot, B.D., Badolato, A., and Petroff, P.M., Polarization sensitive spectroscopy of charged quantum dots, *Phys. Rev. B*, 76, 235304, 2007. Copyright 2007 by the American Physical Society.
    The values given in the third column are for the case where $\xi = 0.96$ and $\beta = 0.72$.

$$\hat{\vec{P}} = \sum_{ij} \vec{p}_{ij} a_i b_j \tag{12.20}$$

In the single-band SVEFA, the inter-band transition momentum vector $\vec{p}_{ij}$ is given by

$$\vec{p}_{ij} \equiv \vec{p}_{l,l',s,s'} = \vec{M}_{s,s'} \int d^3 r \phi_l^{h*}(\vec{r}) \phi_l^e(\vec{r}) \tag{12.21}$$

where the electron and heavy hole state indices (*i* and *j*, respectively) are explicitly expressed in terms of their spatial (l,l') and pseudo-spin (s,s') components. For Bloch amplitudes at $k=0$, the momentum matrix elements $\vec{M}_{s,s'}$ are given by

$$\vec{M}_{\Uparrow,\downarrow} = \frac{i}{2}\sqrt{m_0 E_P}(\hat{x}-i\hat{y}); \quad \vec{M}_{\Downarrow,\uparrow} = \frac{i}{2}\sqrt{m_0 E_P}(\hat{x}+i\hat{y}); \quad \vec{M}_{\Uparrow,\uparrow} = \vec{M}_{\Downarrow,\downarrow} = 0 \tag{12.22}$$

The recombination of an e–h pair with antiparallel spins with hole spin $\Uparrow$ ($\Downarrow$) relative to the growth direction (*z*) therefore yields a right- (left-) hand circularly polarized photon (emitted in the *z*-direction). Thus, the recombination of an e–h pair (exciton) in a coherent state expressed as $\alpha\Uparrow\downarrow + \beta\Downarrow\uparrow$ yields a photon of polarization $\alpha\sigma^+ + \beta\sigma^-$. In particular, if $|\alpha| = |\beta|$, the emitted photon will be linearly polarized in a direction that is rotated by an angle $\phi = \frac{1}{2}\arg(\beta/\alpha)$ with respect to the *x*-axis. For $|\alpha| \neq |\beta|$, the polarization is elliptical. Exciton states with parallel electron and hole spins cannot recombine radiatively. Such states are called "dark excitons."

The rate of optical transition at energy ε and polarization $\vec{e}$ is calculated using Fermi golden rule and the dipole approximation [35]:

$$\Gamma_{\vec{e}}(\varepsilon) = \frac{4\alpha n \varepsilon}{3\hbar m_0^2 c^2} \sum_{i,f} \left| \left\langle f \left| \vec{e} \cdot \hat{\vec{P}} \right| i \right\rangle \right|^2 \delta_{\varepsilon, \varepsilon_i - \varepsilon_f} F_i \tag{12.23}$$

where

$\alpha = e^2/\hbar c \cong 1/137$ is the fine structure constant

n is the refraction index of the QD material

The indices i and f stand for initial and final state, respectively and ε_i and ε_f are the energies of these states. F_i is the probability to find the system in the initial state i.

We now proceed to discuss some specific examples.

1. *The neutral exciton*: As a simple example let us first examine the recombination of the ground bright neutral exciton (X^0) states. These states are mainly composed of the configuration of one electron and one hole in their respective lowest single carrier states. The EHEI results in four nondegenerate states. In the lowest energy pair of states, the electron and hole spins are parallel and the states are "dark." The highest energy pair of states (the "bright exciton") are $X_{H(V)}^0 = 1/\sqrt{2}(\Uparrow\downarrow \pm e^{i2\phi} \Downarrow\uparrow)$. They are separated in energy by $\left|\Delta_1^{1,1,1,1}\right|$. These states recombine radiatively by emitting cross-linearly polarized photons and give rise to a doublet in the PL spectrum (see Figure 12.5a). The lowest energy component of the doublet is polarized parallel to major axis of the QD and it is called "horizontal" (H). The highest one is polarized along the minor axis of the QD and is called "vertical" (V). Usually, the laboratory frame is chosen such that $\phi = 0$. In such a frame, we calculate [28]

$$\left| \left\langle 0 \left| \hat{x} \cdot \hat{\vec{P}} \right| X_H^0 \right\rangle \right|^2 = \left| \left\langle 0 \left| \hat{y} \cdot \hat{\vec{P}} \right| X_V^0 \right\rangle \right|^2 = 1.44 \times \frac{m_0 E_P}{2} \tag{12.24}$$

The factor 1.44 is due to the contributions of other spatial configurations to the ground-state wavefunction. Assuming equal population probabilities for the two radiative recombination X^0 states and the two dark X^0 states, we obtain a total bright X^0 rate of $(0.8\,\text{ns})^{-1}$, in excellent agreement with the measured lifetime [7].

The calculated rates of all other optical transitions presented below are given in units of this rate. We now proceed to examine a few other examples.

2. *Neutral biexciton recombination*: A neutral biexciton (XX^0) consists of two electrons and two holes. Its ground state is mostly composed of the configuration where all carriers are in their lowest single-carrier level. Since there are two carriers in each spatial state, due to Pauli's exclusion principle, the state must be antisymmetric under spin exchange—a spin-singlet state. Due mostly to the Coulomb correlation [36], the energy difference between the XX^0 and the X^0 states is different than that between the X^0 states and the 0 (empty QD—"vacuum") state. This leads to a difference in the recombination energies between the X^0 and the XX^0. This difference, mostly referred to as the "biexciton binding energy," is usually on the order of a few meV, and it can be either positive or negative [36]. Like in the X^0 transition, the XX^0 transition includes a cross-linearly polarized doublet. The doublet originates from the exciton states as well. Therefore, the XX^0 splittings is the same as that of the X^0. But, since in the biexciton transition the exciton is the final state, the energy order of the doublet's components is opposite to that of the exciton (Figure 12.5a).

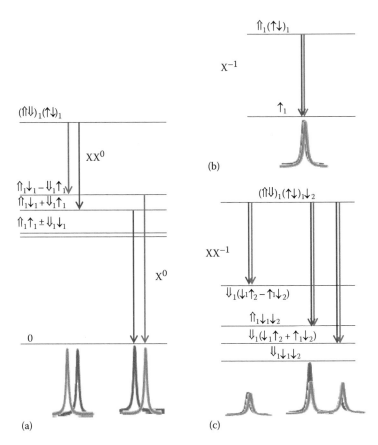

FIGURE 12.5 Schematic description of energy levels and inter-level optical transitions. (a) Neutral exciton (X^0) and biexciton (XX^0), (b) negatively charged exciton (X^{-1}), and (c) negatively charged biexciton (XX^{-1}). The most significant electronic configuration is described above each level. In (b) and (c) only one of the two Kramers degenerate states is described. Thin (thick) arrows represent electron (heavy hole) pseudospin (see text). The subscript numbers represent the orbital part of the single carrier wavefunction (see Figure 12.3). Note that the presented wavefunctions are not normalized. Below each transition diagram, a schematic polarization-sensitive PL spectrum is presented. Light-(dark-)gray line shows the spectrum in vertical (horizontal) polarizations.

3. *Singly charged excitons*: If the QD contains an additional electron (hole) alongside the exciton, the combined state is called a singly negatively (positively) charged exciton or a negative (positive) trion. We mark the relevant state as X^{-1} (X^{+1}). In its ground state, the three carriers are mostly in their respective lowest energy state. Due to Pauli's principle, the two carriers of same sign are in a spin-singlet state. The unpaired, minority carrier, can be in one of two possible spin states. Due to Kramers theorem [29], in the absence of magnetic field these states are degenerate.

   Upon recombination, one majority carrier is left in the QD. It, too, has two Kramers degenerate spin states. The selection rules for optical recombination, Equation 12.22, permit only two optical transitions between the trion and the single-charge state. These two transitions are in opposite polarizations. Thus, in the PL spectrum, only one, unpolarized PL line is expected for each charged exciton (Figure 12.5b). Applying external magnetic field, which lifts the degeneracy, results (in Faraday configuration) in a cross-circularly, polarized doublet [18].

4. *Singly charged biexcitons*: The main configuration consists of two electrons and two heavy holes in their lowest single-carrier level, and an additional charge in its second single-carrier level. We mark these charged biexcitons as XX^{+1} or XX^{-1} for positive or negative extra charge, respectively.

Like the case of the charged exciton, Kramers degeneracy holds here as well. Due to the small spatial overlap between the wavefunctions of charge carriers of different orbitals, radiative recombination of such e–h pairs is weak (see Equation 12.21). Therefore, we consider only recombinations between e–h pairs in their lowest levels. When such recombination occurs, the QD is left in an excited charged exciton state, with one of the majority carriers in its second orbital level (see Figure 12.5c). Since there are three unpaired carriers now, there are eight possible spin configurations. The strongest exchange interaction within this subspace is the same-particle exchange. It divides the eight states into two groups by the total spin of the two same-charge carriers. The high energy group consists of the states where the same-charge carriers form a spin-zero singlet (antisymmetric under same carrier exchange) state and in the low-energy group they form triplet (symmetric under same carrier exchange) spin states. We name the high-energy group S and the low-energy group T.

The spin of the minority carrier in each of these groups can be oriented either up or down.

The EHEI between this carrier and the other two defines the energy of the configuration. Among the S states, in both cases, the spin of the minority carrier is parallel to one of the majority carriers and antiparallel to the other one. Thus, the S group remains doubly (Kramers) degenerate. Among the T states there are three possible configurations, resulting in three different energies. The lowest energy is the one in which all three spins are oriented along the same direction. The highest energy is the one in which the minority carrier is antiparallel to both majority carrier spins. The one in between is the case in which it is parallel to one and antiparallel to the other (see Figure 12.5c).

It is easy to see that the lowest level, in which all the spins are parallel, cannot be reached by radiative annihilation of an e–h pair from the charged biexciton state. Therefore, the recombination of the charged biexciton results in three spectral lines: a single, low-energy line, which leads to the high-energy S level, and two higher energy lines, which lead to the two accessible T levels (Figure 12.5c). We mark these optical transitions by S, T_0, and T_1 (T_3), to associate the subscript with the total spin projection of the two electrons (heavy holes) majority carriers. We note that the S and T_0 lines have half the oscillator strength of the T_1 (or T_3) line, since only half of their wavefunctions is reached by annihilating e–h pair from the charged biexciton state (see Figure 12.5c).

In the absence of additional interactions, these lines should be unpolarized. This is because each level is doubly (Kramers) degenerate and two orthogonally polarized transitions between the levels are equally probable.

The anisotropic EHEI becomes important when the symmetry of the QD reduces. Due to its form in Equation 12.16, it mixes only states where there are antiparallel e–h spin components (assuming Δ_2 is negligible [34]). This mixing results in photon emission, which reflects the reduced symmetry of the system, that is, with *linear* polarization components testifying to the preferred directionality of the system. The degree of linear polarization reflects the magnitude of the level mixing term (the anisotropic EHEI), relative to the levels separation. Therefore, while the mixing between the T_0 and the $T_{1(3)}$ results in partial orthogonal linear polarizations, the S level remains essentially unpolarized. A more detailed quantitative description of this dependence can be found in Refs. [37,38].

5. *Doubly negatively charged exciton*: The main ground-state configuration of a doubly negatively charged exciton consists of two paired electrons in the lowest single-electron state, a hole in the lowest hole state, and an additional electron in the second electron state. The two paired electrons have spin zero and are exchange inert. The remaining is an e–h pair, where the hole is in its lowest state and the electron is in its second state. There are four spin configurations, similar to those of the neutral exciton. The main difference is in the actual value of the anisotropic EHEI, which, as can be seen in Table 12.1, is larger in its absolute value than that of ground-state neutral exciton and opposite in sign to it. After recombination of a ground level e–h pair, the QD is left

with two electrons, in different orbital states. The final states are therefore divided into a high-energy electronic singlet and low-energy electronic triplet. These states are energetically separated by the same-charge (elect.–elect.) exchange interaction. While the triplet states can be reached from all four initial states, the singlet state can be reached only from the two states where the unpaired electron and hole spins are antiparallel. Therefore, there are two groups of spectral lines: a low-energy cross-linearly polarized doublet and a high-energy triplet, made of a higher energy, cross-linearly polarized doublet and a lower energy, unpolarized single line (or more accurately a degenerate doublet). We note that within each doublet the two cross-linearly polarized components appear in opposite energy order. This is a consequence of the difference in the symmetry under spin exchange between the final states in each case (Figure 12.6a).

6. *Triply negatively charged exciton*: This configuration contains one hole and four electrons confined in the QD. There are two possibilities for the configuration of the lowest energy state. If the energy difference between the second and third single-electron levels is larger than the electron–electron exchange interaction between these levels, $C_{2,3,2,3}^{eeee}$, than the electronic spin configuration $\uparrow,\uparrow,\uparrow,\uparrow$ would be of two closed shells. In this case, the total electronic spin is zero (Figure 12.6b). In the opposite case, the two higher energy electrons would prefer to be in open shells, forming spin triplet configuration, thereby reducing the level energy by the electron–electron exchange energy

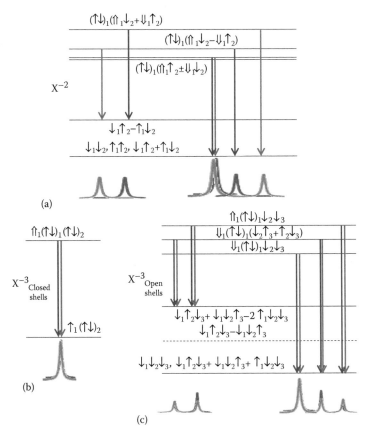

FIGURE 12.6 Schematic description of energy levels and inter-level optical transitions: (a) doubly negatively charged exciton (X^{-2}), (b) triply negatively charged exciton (X^{-3}) in closed-shell configuration, and (c) X^{-3} in open-shell configuration. The most significant electronic configuration is given above each level. The notations are explained in Figure 12.5. In (b) and (c) only one of the two Kramers degenerate states is given for each level. Below each transition diagram, a schematic polarization-sensitive PL spectrum is presented.

(see Figure 12.6c). As shown in the figure, for the closed shell case, both initial and final states are doubly degenerate, giving rise to a single, unpolarized spectral line (like in the case of the ground-state, singly charged exciton). For the open-shell case, the situation is markedly different, yielding a much richer spectrum. The initial states resemble the final states of the singly charged biexciton. They are arranged in three levels with a Kramers degenerate pair of states in each level. All the states are optically active. The final states are those of three electrons, each in its own spatial state. As shown in Figure 12.6c, the eight spin configurations are divided by the electron–electron exchange interaction according to their total spin to a lower energy quadruplet (total spin 3/2) and two higher energy doublets (total spin 1/2). The quadruplet can be reached optically from all three levels of the initial configuration. The higher doublet can be reached only from the two higher energy levels of the initial state. The intermediate level cannot be reached from any one of the three initial levels. This gives rise to two groups of spectral lines, a higher energy group consisting of three lines and a lower energy group consisting of two lines (Figure 12.6c). Considering the different spin configurations involved in each of the transitions, it can be shown [39] that the intensity ratio between the lines in the group of three is 3:2:1 (from lowest to highest energy) and 2:1 in the group of two lines.

Similar to the radiative decay of the singly charged biexciton, the two higher energy initial states of the open-shell, triply charged exciton are mixed by the anisotropic EHEI. This leads to partial linear polarizations in the spectral lines resulting from their recombinations. As we show below, our experimentally measured polarization-sensitive PL spectra are compatible with the second case: The triply negatively charged exciton is best described by the open-shell configuration.

12.3 Polarization-Sensitive Photoluminescence Spectrum

In this section, we present measured polarization-sensitive PL spectra and compare the details of the spectral lines observed with the model calculated transitions. This comparison, together with polarization memory measurements [40] and external-field-dependent PL measurements [28], is used for the identification of the spectral lines.

The sample used for the experiments presented here contains a single layer of In(Ga)As QDs embedded within a GaAs matrix. The QD layer was positioned at the middle of a one-wavelength cavity, made of two unequal distributed Bragg reflecting mirrors (DBR) [7,41]. In order to apply electric fields on the QDs, the top (bottom) DBR was n-(p-)type doped. For charging the QD in a controllable way, a 10 nm thick AlAs barrier was placed inside the cavity, between the QDs and the top, p-type doped mirror [28]. This barrier prolongs the tunneling time of heavy holes into the QDs at forward bias and out of the QDs at reverse bias, while marginally affecting the tunneling time of electrons. As a result, negative charging is facilitated upon forward bias and positive charging upon reverse bias.

Figure 12.7a through d shows the measured PL and linear polarization spectra of a single QD from this sample, under two external bias voltages. At zero bias (Figure 12.7a and b) the charge state is between −1 and +1, while at forward bias (Figure 12.7c and d) it is between −3 and −1. The various lines are marked by their initial state and where needed, their final state is also indicated. Figure 12.7e through h shows the corresponding calculated spectra.

In Figure 12.8, the measured polarized fine structure of various lines is compared with the calculated one. While most of the measured intensity ratios, relative PL energies, and polarization degrees and directions are reproduced by the theory discussed above, some measured spectral lines show unexpected polarization directionality. We find spectral lines with linear polarization axes rotated clockwise by 67.5° relative to the H and V axes. These rotated polarization directions, $V + \bar{D}$ and $H + D$, respectively, which roughly coincide with the [1, 2, 0] and [2, −1, 0] crystalline directions, appear only in spectral lines associated with configurations containing one unpaired p_x-shell charge carrier: X^{-2}, XX^{-1}, and XX^{+1}. Other configurations, which contain only s-shell charge carriers or either closed shells or two

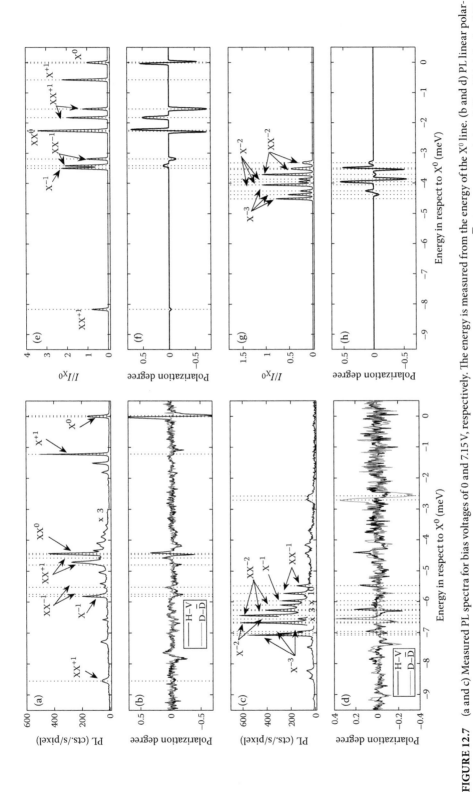

FIGURE 12.7 (a and c) Measured PL spectra for bias voltages of 0 and 7.15 V, respectively. The energy is measured from the energy of the X^0 line. (b and d) PL linear polarization spectra for bias voltages of 0 and 7.15 V, respectively. The black (orange) line presents projection on the H–V (D–$\bar{\text{D}}$) direction. (e and g) and (f and h) Calculated PL and PL linear polarization (projected on the H–V direction) spectra for various single QD excitonic transitions. Vertical dotted lines at various spectral lines are drawn to guide the eye. (Reprinted with permission from Poem, E., Shemesh, J., Marderfeld, I., Galushko, D., Akopian, N., Gershoni, D., Gerardot, B.D., Badolato, A., and Petroff, P.M., Polarization sensitive spectroscopy of charged quantum dots, *Phys. Rev. B*, 76, 235304, 2007. Copyright 2007 by the American Physical Society.)

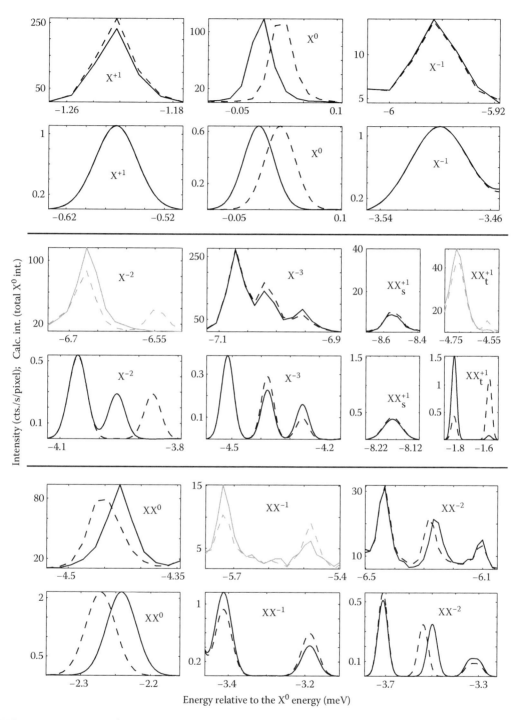

FIGURE 12.8 Measured (top panel in each pair) and calculated (bottom panel in each pair) high-resolution polarization-sensitive PL spectra for various spectral lines. The solid (dashed) black line represents H (V) polarized spectrum, while the solid (dashed) gray line represents V + $\bar{\text{D}}$(H + D) polarized spectrum. (Reprinted with permission from Poem, E., Shemesh, J., Marderfeld, I., Galushko, D., Akopian, N., Gershoni, D., Gerardot, B.D., Badolato, A., and Petroff, P.M., Polarization sensitive spectroscopy of charged quantum dots, *Phys. Rev. B*, 76, 235304, 2007. Copyright 2007 by the American Physical Society.)

unpaired p-shell carriers (p_x and p_y), the linear polarizations are along the H and V axes. The simple single band, single carrier, model that we presented above, is insensitive to the crystalline directions and cannot explain these observations.

12.4 Carrier Dynamics in Quantum Dots

In this section, we discuss two experimental methods for probing the dynamics of confined charge carriers in QD: polarization memory measurements and polarization-sensitive intensity correlation measurements. The theory that we use in order to gain intuition and to understand the dynamics of photoexcited carriers in these nanostructures from the analysis of these experimental results is thoroughly discussed.

12.4.1 Polarization Memory in Quasi-Resonant Excitation

In a polarization memory measurement, the system is excited by a polarized light, and the polarization of the PL emission is measured. If the polarization of a certain spectral line is the same as (opposite to) that of the exciting light, the line is said to have positive (negative) polarization memory. In general, the degree of polarization memory depends on the specific spectral line, its polarization, and the energy to which the exciting light is tuned. The degree of circular polarization memory (DCPM) and the degree of linear polarization memory (DLPM) along the major axes of the QD are defined as follows:

$$P_{circ} = \frac{I_+^+ - I_-^+}{I_+^+ + I_-^+} \quad P_{lin} = \frac{I_H^H - I_V^H}{I_H^H + I_V^H} \tag{12.25}$$

where
  I stands for the PL intensity
  the superscript (subscript) + (−) or H (V) stands for right-(left-)hand circular polarization or horizontal-(vertical-)linear polarization of the exciting (emitted) light

The horizontal (vertical) direction is determined by the polarization direction of the lower (higher) energy fine-structure component of the neutral exciton (X^0) line (see Figure 12.5a).

Usually, when a QD is excited well above the bandgap of the host material (nonresonant excitation), no polarization memory is observed. This is due to fast spin scattering of both types of charge carriers during their relaxation from the continuum to the QD states [42,43]. When the excitation energy is below the continuum of the QD or its wetting layer bandgap, we refer to it as to a quasi-resonant excitation. In this case, the situation is different and some polarization memory is observed. Resonant excitation into well-defined low-energy optical resonances of the QDs will not be reviewed here.

For quasi-resonant excitations, we find that different lines show different DCPM. However, none of the observed lines show any DLPM. The data and its discussion are presented below.

In Figure 12.9a, we present bias-dependent PL spectra from the same QD discussed above, optically excited at 1.369 eV. At this energy, a few meV below the bandgap of the InAs wetting layer, the QDs are quasi-resonantly excited [44]. At reverse bias, the spectral lines are redshifted due to the applied electric field, and lines due to optical transitions in the presence of positive charges are enhanced. At forward bias, flat-band conditions are reached and spectral lines due to the presence of negative charges appear, while lines in the presence of positive charges disappear. The various spectral lines are identified by their bias dependence and their order of appearance as the bias increases. As described above, their identifications also rely on their polarized fine structure [28].

In Figure 12.9b, we present the DCPM spectra as a function of the bias applied on the QD. Clearly, the DCPM of each and every spectral line is almost bias independent. In general, it is obvious that while all positive lines have positive DCPM, various negative lines have negative DCPM signs.

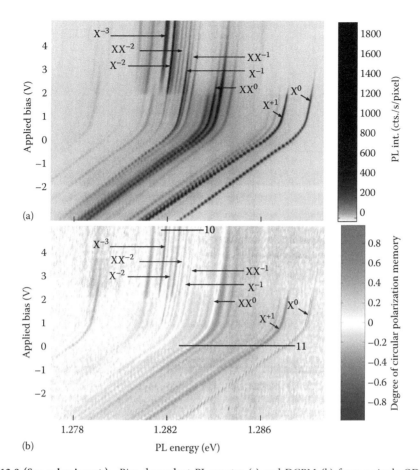

FIGURE 12.9 (See color insert.) Bias-dependent PL spectra (a) and DCPM (b) from a single QD excited at 1.369 eV. The black horizontal lines marked 10 and 11 indicate the bias and spectral ranges from which Figures 12.10 and 12.11 were obtained. (From Poem, E. et al., *Solid State Commun.*, 149, 1493, 2009. With permission.)

In Figure 12.10a, we present spectra of various spectral lines as obtained at a forward bias of 4.9 V. At this voltage, the QD is negatively charged with either one, two, or three electrons. The solid (dashed) line represents the spectrum obtained when the excitation and collection are co-(cross-)circularly polarized. In Figure 12.10b, we present the corresponding DCPM. In Figure 12.10, one clearly sees that the DCPM sign depends on the specific optical transition. Some spectral lines show positive memory, like all the lines associated with positive charges do. Some show no polarization memory and some show negative polarization memory. In Figure 12.11a, we present the spectrum obtained at 0 V. In Figure 12.11b, we present the measured DCPM and DLPM. The X^0 line shows no DCPM and in total no DLPM either, since its H and V polarized fine-structure components are equally visible upon H linearly polarized excitation. We note that the X^{+1} (positively charged exciton) shows strong positive DCPM but no DLPM.

12.4.2 Spin Scattering Rates

In order to explain these observations and to gain further insight into the phenomenon of polarization memory in quasi-resonantly optically excited single semiconductor QDs, we use the single band, full configuration interaction model described in the previous sections. We use the model to calculate the QD's confined many-carrier states and the selection rules for optical transitions between these states.

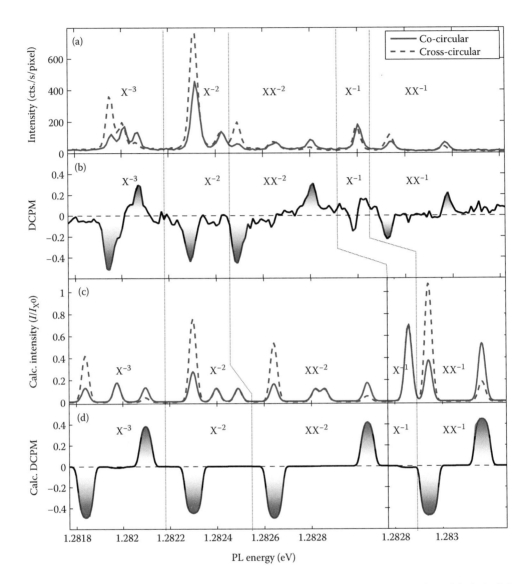

FIGURE 12.10 (a) Measured and (c) calculated polarization-sensitive spectra at 4.9 V. The solid (dashed) line represents spectrum obtained with co-(cross-)circularly polarized excitation and detection: $I_{co} = I_+^+$ ($I_{cross} = I_-^+$). (b) Measured and (d) calculated DCPM. The dotted vertical lines are guides to the eye. (From *Solid State Commun.*, 149, Poem, E., Khatsevich, S., Benny, Y., Marderfeld, I., Badolato, A., Petroff, P.M., and Gershoni, D., Polarization memory in single quantum dots, 1493, Copyright 2009, with permission from Elsevier.)

Prior to the excitation, the states within 1 meV from the ground state of a given number of N_h holes and N_e electrons are considered to be populated with equal probability. This assumption is compatible with thermal distribution at the ambient temperature in which the experiment was held (~20 K). We consider the polarized quasi-resonant excitation at a given polarization by adding an additional e–h pair to these states. The spin state of the additional carriers are defined by their initial spin polarization, S_{exc}, dictated by the polarization of the exciting light and by their spin-dephasing rate during thermalization. Quite generally, we describe the spin orientation loss by four probabilities, which apply to each carrier independently. The probabilities $p_j^{e,h}$ are for either spin orientation preservation, $j = 0$, or for spin rotations

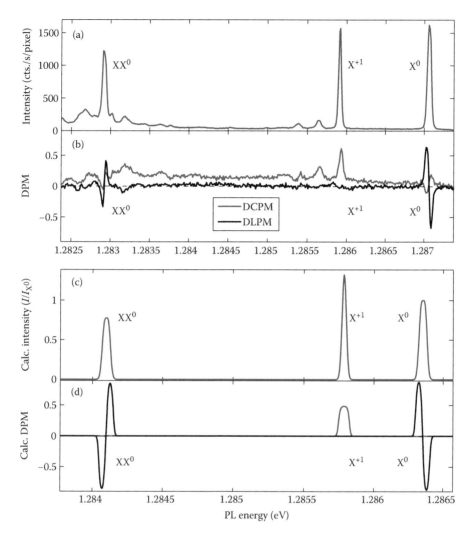

FIGURE 12.11 (a) Measured and (c) calculated polarized PL spectra at 0 V. (b) Measured and (d) calculated degrees of circular (gray line) and linear (black line) polarization memory. (From *Solid State Commun.*, 149, Poem, E., Khatsevich, S., Benny, Y., Marderfeld, I., Badolato, A., Petroff, P.M., and Gershoni, D., Polarization memory in single quantum dots, 1493, Copyright 2009, with permission from Elsevier.)

by π radians about the spatial directions x, y, and z for $j = 1$, 2, and 3, respectively. The spin states of the thermalized pair can now be represented by a 4×4 density matrix in the Hilbert space of the pair's spin states, $\{\downarrow\Uparrow, \uparrow\Downarrow, \uparrow\Uparrow, \downarrow\Downarrow\}$:

$$\rho^{th} = \sum_{j,j'=0}^{3} p_j^e p_{j'}^h \cdot \sigma_j^e \otimes \sigma_{j'}^h \big| S_{exc} \big\rangle \big\langle S_{exc} \big| \sigma_j^{e\dagger} \otimes \sigma_{j'}^{h\dagger} \tag{12.26}$$

where

$\sigma_j^{e(h)}$ are the Pauli matrices acting on the subspace of electron ($\uparrow$) (hole ($\Uparrow$)) spin states

σ_0 is the unit matrix

The operation $\otimes$ is the Kronecker product. In Equation 12.26, we used the identity: $\exp(i\pi \cdot 1/2\sigma_j) = i\sigma_j$. If one further assumes that the spin orientation loss (or dephasing) for both carrier types is isotropic, then the number of independent probabilities to be considered is reduced to two, p^e and p^h, where

$$p_1^{e(h)} = p_2^{e(h)} = p_3^{e(h)} \equiv p^{e(h)}; \quad p_0^{e(h)} = 1 - 3p^{e(h)} \tag{12.27}$$

We note here that the probabilities p^e and p^h can be written in terms of the spin-dephasing time T_2^* [45] and the thermal relaxation time τ as follows:

$$p^{e(h)} = \frac{\tau^{e(h)}}{(4\tau^{e(h)} + T_2^{*e(h)})} \tag{12.28}$$

These relations arise from the master equations (see below) describing a single charge carrier undergoing simultaneous spin dephasing and thermal relaxation from its orbital state.

The photogenerated pair increases the number of charge carriers to $N_h + 1$ holes and $N_e + 1$ electrons. The new many-carrier states are restricted to these many-carrier states, which accommodate the photogenerated carriers with their spin orientation. For an initial state $|A\rangle$ of N_e electrons and N_h holes, the resulting density matrix that defines the states with the additional thermalized pair is given by

$$\rho_A = \sum_{\alpha,\beta} \rho_{\alpha\beta}^{th} \hat{x}_\alpha^\dagger |A\rangle\langle A| \hat{x}_\beta \tag{12.29}$$

where $\hat{x}_\alpha^\dagger$ is the creation operator of an e–h pair with spin α in any combination of single-electron and single-hole spatial states

$$\hat{x}_\alpha^\dagger = \sum_{m,n} \hat{a}_{m,\alpha_e}^\dagger \hat{b}_{n,\alpha_h}^\dagger \tag{12.30}$$

Here, $\hat{a}_{m,\alpha_e}^\dagger$ ($\hat{b}_{n,\alpha_h}^\dagger$) is the creation operator of an electron (a hole) in the single electron (hole) spatial state m (n), with spin state α_e (α_h). With this notation the spin state of the e–h pair is given by $\alpha \equiv \{\alpha_e, \alpha_h\}$. With this description of the $N_e + 1$, $N_h + 1$ many-carrier state, we proceed by projecting it on all the possible "ground" states $|G\rangle$ within 1 meV of the lowest energy level of $N_e + 1$, $N_h + 1$ charge carriers. Recombination of one e–h pair form these states, actually gives rise to the PL signal.

We conclude by calculating the energies ε and the relative intensities of polarized optical transitions $I_{S_{em}}^G(\varepsilon)$ with polarization S_{em} from the ground states $|G\rangle$ to the many-body states of N_h holes and N_e electrons [29,46]. The S_{em} polarized spectrum resulting from S_{exc} polarized quasi-resonant excitation is finally obtained by summing over all the thermally populated initial states $|A\rangle$ and over all optically excited $|G\rangle$ states contributing to the photoluminescence:

$$I_{S_{em}}^{S_{exc}}(\varepsilon) = \sum_{G,A} \text{Tr}(\rho_A |G\rangle\langle G|) \cdot I_{S_{em}}^G(\varepsilon) \tag{12.31}$$

We note that ρ_A is obtained from the polarization of the exciting light $|S_{exc}\rangle$ via Equations 12.26 through 12.30. The two probabilities in Equation 12.27 p^e and p^h can now be extracted from the measured DCPM and DLPM by fitting them to the calculated ones. The values $p^e = 1/8$ and $p^h = 1/4$ describe very well the observations for this particular quasi-resonant excitation. In accord with Equation 12.28, these values mean that while the hole's dephasing time is much shorter than it's relaxation time, and thus it totally loses its spin polarization during the thermalization, the electron's degree of polarization is reduced only by one half, meaning that its dephasing and relaxation times are comparable. Kalevich et al. [47] used previously a similar assumption to successfully explain their observation of negative circular polarization memory in an ensemble of doubly negatively charged QDs.

The calculated spectra for co- and cross-circularly polarized emission from a negatively charged QD with 1 up to 3 charges were added together to form the calculated polarization-sensitive spectra in Figure 12.10c. Both single exciton and biexciton emissions were included. Gaussian broadening of $35\,\mu eV$ was assigned for each optically allowed transition. The calculated DCPM spectrum is presented in Figure 12.10d. By comparing the measured and calculated polarization-sensitive spectra and DCPM, one clearly notes that all the features of the measured DCPM are given by this simple model.

In Figure 12.11c, we present the calculated spectrum for the neutral exciton (X^0), the neutral biexciton (XX^0), and the singly positively charged exciton (X^{+1}). In Figure 12.11d, we present the corresponding calculated DCPM (gray) and DLPM (black). The H (V) directions are along the long (short) semiaxes of the model QD [28]. The positive DCPM of the X^{+1} spectral line and the lack of DCPM from the neutral excitonic transitions are clearly reproduced by our model. In addition, the model clearly reproduces the experimentally measured lack of DLPM from all the observed spectral lines at this quasi-resonant excitation energy. We note here, however, that DLPM is observed in some cases of resonant excitations [48–51]. In these cases, (to be presented elsewhere), both the electron and the hole retain at least part of their initial spin polarization during their thermalization, before they recombine.

Intuitively, one can easily comprehend the observed DCPM phenomena as a consequence of the iso-tropic-EHEI-induced energetic separation between states where the electron and hole spins are parallel to those where they are antiparallel. Since circularly polarized excitation and emission always involve e–h pairs with antiparallel spins, states with (anti-) parallel spins can be reached only in cases where one (none) of the carriers flips its spin. This simple reasoning leads to negative (positive) circular polarization memory. As an illustration for the processes, which lead to polarization memory, we schematically describe in Figure 12.12 the case of quasi-resonant excitation of the X^{-3} spectral line.

Before we conclude this discussion, we note that the appearance of negative DCPM in the lowest energy doublet of the doubly negatively charged exciton (X^{-2}), indicates that the energy splitting between the two components of this doublet is smaller than the radiative width of the lines [47,51]. Consequently, we set this particular EHEI energy to zero in our model [28].

12.4.3 Polarized Intensity Correlation Spectroscopy

The intensity (second-order) correlation function between two spectral lines, A and B, is defined as follows:

$$g_{A,B}^{(2)}(\tau) = \frac{\langle I_A(t)I_B(t+\tau)\rangle_t}{\langle I_A(t)\rangle_t \langle I_B(t)\rangle_t} \tag{12.32}$$

where
  $I_{A(B)}$ is the intensity of line A (B)
  $\langle\,\rangle_t$ means averaging over the time t

If B = A, the function is known as the intensity autocorrelation function and is commonly used to dem-onstrate quantum light or single photon sources [4–6].

Emission of a photon from line A at time t sets the system to the final state of the relevant optical transition. Therefore, $\langle I_A(t)I_B(t+\tau)\rangle_t = \langle I_A(t)\rangle_t \langle I_{B/A}(t+\tau)\rangle_t$, where $I_{B/A}(t+\tau)$ is the intensity of transition B at time $t+\tau$ *conditional* on the system being in the final state of transition A at time t. Furthermore,

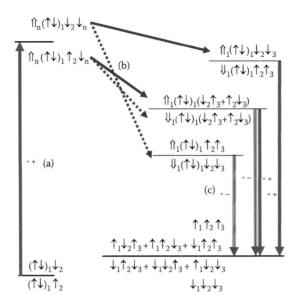

FIGURE 12.12 Schematic description of the processes that lead to the observed DCPM among the X^{-3} spectral lines. The symbol $\Uparrow$($\downarrow$) represents a spin up (down) hole (electron). The subscript near each symbol denotes the particle's energy level (see Figure 12.3). Gray color represents states that do not participate in the described process. (a) An e–h pair is photogenerated in high-lying energy levels (denoted by the subscript "n") by a quasi-resonant σ^+ polarized excitation and added to three QD electrons residing in their ground states. (b) During the thermalization, the hole spin projection along the growth direction is either preserved (solid dark-gray arrows) or flipped (dotted dark-gray arrows). The lowest (highest) energy level of the ground $N_e = 4$, $N_h = 1$ states, is reached only if the hole flips (preserves) its spin orientation. The intermediate level is reached in both cases. (c) All three levels return via radiative recombination of an s-shell e–h pair to the same fourfold degenerate $N_e = 3$, $N_h = 0$ level, giving rise to spectral lines with positive, negative, and no DCPM, respectively. (From *Solid State Commun.*, 149, Poem, E., Khatsevich, S., Benny, Y., Marderfeld, I., Badolato, A., Petroff, P.M., and Gershoni, D., Polarization memory in single quantum dots, 1493, Copyright 2009, with permission from Elsevier.)

the intensity of an optical transition is proportional to the population probability of its initial state (Equation 12.23). Therefore, for $\tau > 0$,

$$g_{A,B}^{(2)}(\tau > 0) = \frac{\langle I_A(t) \rangle_t \langle I_{B/A}(t + \tau) \rangle_t}{\langle I_A(t) \rangle_t \langle I_B(t) \rangle_t} = \frac{\langle I_{B/A}(t + \tau) \rangle_t}{\langle I_B(t) \rangle_t} = \frac{P_{B/A}(\tau)}{P_B^{SS}} \tag{12.33}$$

where $P_{B/A}(\tau)$ is the average population at time τ of the initial state of transition B with the condition that the state of the system at time 0 was the final state of transition A. The steady-state probability to find the system in the initial state of transition B, $P_{B/A}(\tau \to \infty)$, is denoted by P_B^{SS}.

At negative time differences ($\tau < 0$), the role of the two transitions is simply exchanged.

$$g_{A,B}^{(2)}(\tau < 0) = g_{B,A}^{(2)}(|\tau|) = \frac{P_{A/B}(|\tau|)}{P_A^{SS}} \tag{12.34}$$

This simple analysis shows the usefulness of intensity correlation measurements as a tool for studying QD-carrier population dynamics.

Experimentally, intensity correlation measurements are usually performed using a Hanbury-Brown and Twiss apparatus (HBTA) [52]. The apparatus that we constructed provides the necessary

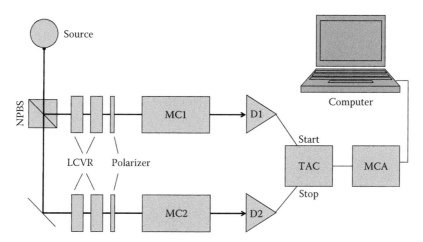

FIGURE 12.13 Schematic illustration of a Hanbury-Brown and Twiss apparatus. NPBS, nonpolarizing beam splitter; LCVR, liquid crystal variable retarder; MC, monochromator; D, detector; TAC, time to analog converter; MCA, multichannel analyzer.

means to temporally correlate between the emission intensities of two spectral lines, each one of them projected onto any desired polarization state. The apparatus is schematically described in Figure 12.13.

The emitted light from the sample is split into two beams by a nonpolarizing beam splitter. By two monochromators and two sets of polarizers and variable retarders, we select the desired spectral line and the desired polarization in each arm of the apparatus. A silicon avalanche photodetector in each arm detects a single photon and converts it to a current pulse. The time difference between the pulses from the two detectors is repeatedly measured using a time to analog converter. A multichannel analyzer then builds a histogram of time differences between the detection times of the two photons. In continuous wave (cw) measurements, the histogram can be straight forwardly normalized by its value at long time differences, to yield the intensity correlation function.

12.4.4 Rate Equations and Master Equations

In order to calculate the correlation functions, we calculate first the population probabilities.

For energetically distinct levels, coherence effects can be neglected, and the populations can be calculated using a set of coupled classical rate equations [53]. Where by "energetically distinct" we mean that the energetic distance between the levels is much wider than their total width.

The set of classical rate equations can be generally written in the following form [40,53]:

$$\dot{\vec{n}}(t) = \overset{\leftrightarrow}{R}(t)\vec{n}(t) \tag{12.35}$$

where
$\vec{n}(t)$ is a vector of level populations
$\overset{\leftrightarrow}{R}(t)$ is a matrix containing the transition rates between all the levels involved

In cw excitation, the detected first photon sets the starting conditions, and then, Equation 12.35 can be solved quite straightforwardly if the matrix $\overset{\leftrightarrow}{R}(t)$ is known [53].

When coherence matters, classical rate equations are not sufficient. A more general approach is formulated by a set of coupled master equations acting on the density matrix instead of on the populations only. Lindblad formulated the problem in the most general form [54]:

$$\dot{\rho}(t) = -\frac{i}{\hbar}[H,\rho(t)] + \sum_k \left(L_k \rho(t) L_k^\dagger - \frac{1}{2}\rho(t)L_k^\dagger L_k - \frac{1}{2}L_k^\dagger L_k \rho(t) \right) \tag{12.36}$$

where
  ρ is the density matrix
  H is the Hamiltonian
  $\{L_k\}$ describe non-unitary processes

For example, a transition between the state i and the state f with a rate Γ_{fi} is described by $L_{fi} = \sqrt{\Gamma_{fi}}|f\rangle\langle i|$; pure dephasing of the state i with a rate Γ_i^d is described by $L_i^d = \sqrt{\Gamma_i^d}|i\rangle\langle i|$. Similarly to Equation 12.35, Equation 12.36 can be written in a vector form

$$\dot{\vec{\rho}}(t) = \overleftrightarrow{\mathsf{L}}(t)\vec{\rho}(t) \tag{12.37}$$

where
  $\vec{\rho}(t)$ is a column vector formed from the columns of the density matrix
  $\overleftrightarrow{\mathsf{L}}(t)$ is the following matrix:

$$\overleftrightarrow{\mathsf{L}}(t) = -\frac{i}{\hbar}(I \otimes H - H^T \otimes I) + \sum_k \left((L_k^\dagger)^T \otimes L_k - \frac{1}{2}(L_k^\dagger L_k)^T \otimes I - \frac{1}{2}I \otimes L_k^\dagger L_k \right) \tag{12.38}$$

where
  I is the unit matrix
  $\otimes$ means Kronecker product
  superscript T means transposition

As is the case of the classical rate equations, Equation 12.37 can be solved quite straightforwardly.

Once the populations and the coherences are found, Equation 12.33 is used to find the intensity correlation function. Specifically, the intensity correlation function between a photon leading to the state ρ_1 (or $\vec{n}_1$) and the photon emitted from the state ρ_2 (or $\vec{n}_2$) is given by

$$g_{\rho_1,\rho_2}^{(2)}(\tau) = \frac{\mathrm{Tr}(\rho_2 \cdot \rho(\tau)|_{\rho(0)=\rho_1})}{\mathrm{Tr}(\rho_2 \cdot \rho_{SS})} \quad \text{or} \quad g_{\vec{n}_1,\vec{n}_2}^{(2)}(\tau) = \frac{\vec{n}_2 \cdot \vec{n}(\tau)|_{\vec{n}(0)=\vec{n}_1}}{\vec{n}_2 \cdot \vec{n}_{SS}} \tag{12.39}$$

where ρ_{SS} ($\vec{n}_{SS}$) is the density matrix (populations) at long time differences ($\tau \to \infty$)—the steady state solution.

12.5 Radiative Cascades

By radiative cascade we mean correlated emission of two or more photons. Radiative cascade usually occurs when a second photon is emitted within its radiative lifetime after the emission of a first photon. We distinguish here between two types of radiative cascades: direct ones and indirect ones. In a direct

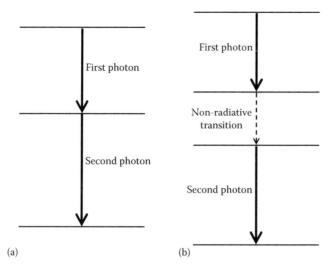

FIGURE 12.14 Schematic description of two types of radiative cascades: (a) a direct cascade, (b) an indirect cascade.

radiative cascade, the state of the system after the emission of the first photon is the state from which the second photon is emitted (Figure 12.14a). In an indirect cascade, a mediating process occurs between the two emissions, and the second photon is not emitted from the final state of the first emission (Figure 12.14b). Clearly, the mediating process rate should be faster or comparable to the radiative rate in order for the two emissions to form a radiative cascade.

Experimentally, a radiative cascade reveals itself in the measured intensity correlation function between two or more relevant spectral lines. It always gives rise to "bunching" at $\tau \geq 0$ and "antibunching" at $\tau \leq 0$. Here by "bunching" ("antibunching") we mean that the intensity correlation function obtains values that are greater (smaller) than 1. These reflect the fact that there is a definite order between the two emissions and that right after the emission of the first photon, the probability for the emission of the second photon is greater than its average value. In other words, emission of the second photon cannot precede the emission of the first one and that ordered emission of photon pairs is preferred over uncorrelated emissions.

In the following, we discuss several examples of direct and indirect radiative cascades in single semiconductor QDs.

12.5.1 Neutral Biexciton Cascade

The most studied [7,8,53,55–57] radiative cascade in semiconductor QDs is the neutral biexciton cascade. The initial state is a ground-state biexciton: two electrons and two holes occupying their respective lowest energy levels. Due to Pauli's exclusion principle, both carrier pairs form spin singlets with total spin 0. Optical recombination of one e–h pair results in the emission of a single photon, and a single e–h pair (bright exciton) remains in the QD. This pair then recombines within its radiative lifetime, resulting in emission of a second photon. At the end of the cascade, the QD is left empty of carriers. This cascade is a direct one, as the single bright exciton is both the state at which the first photon emission ends and the state from which the second photon emission occurs. As described in Figure 12.15, this intermediate level has two spin states. Therefore, since the transition rate from one spin state to the other is very slow (relative to the radiative lifetime) the cascade has two distinctive paths along which it can proceed.

Due to the different selection rules for the two bright exciton spin states, the two paths differ by the polarization of the emitted photons. If the intermediate state is degenerate, then this is the only difference

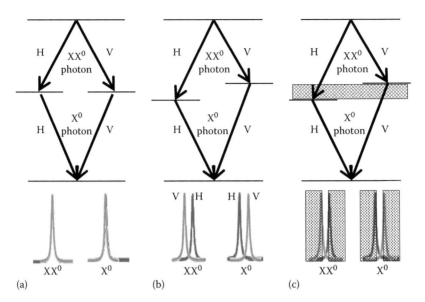

FIGURE 12.15 Schematic descriptions of the neutral biexciton radiative cascades (top) and PL spectra (bottom). (a) Degenerate exciton levels. (b) Split exciton levels. (c) Split exciton levels and spectral filters for entanglement distillation.

(Figure 12.15a) and the two photons are polarization entangled [55]. However, as discussed above, any anisotropy in the QD shape, strain distribution, etc., lifts the degeneracy between the two bright exciton levels through the anisotropic EHEI (Figure 12.15b). When the levels are split, the two paths are also distinct by their photon energies. This distinction reveals the information about "which path" the cascade occurs. Therefore, the degree of entanglement in the two photons polarization state becomes very low, and their polarization state is said to be classically correlated only: In each path the two photons are co-linearly polarized. The degree of entanglement depends on the ratio between the radiative width of the spectral lines and the splitting energy. There have been attempts to reduce the splitting energy by use of several techniques. Those include application of external fields (static [58] and alternating [59] electric fields, magnetic fields [60], strain [61]) or modification of the growth process (thermal annealing [62], growth along a different crystallographic direction [63]). Some of these techniques were successful, and recently there have been several demonstrations of the emission of entangled photon pairs from QDs using such techniques [8,59,63]. Alternatively, even if the splitting energy is larger than the radiative lifetime, entangled photons can still be produced by spectral filtering [7] (Figure 12.15c). Spectral filters postselect only photons emitted in the energy regions between the two fine-structure components of both the biexciton and exciton emissions. In these regions, the energies of the photons in both paths are similar such that they do not reveal the "which path" information, and the entanglement is partially restored. Indeed, the very first demonstration of entangled photons from radiative cascades in QDs [55] was achieved this way [7].

Figure 12.16a and b shows polarization-sensitive measured correlation functions between the biexciton (XX^0) and the exciton (X^0) lines, without any spectral filtering (photons from both fine-structure components arrive at the detectors), for co- and cross-linear polarizations (a) and for circular–circular and circular–diagonal polarization combinations (b). It is seen that for co-linear polarizations the exciton photons "bunch" immediately after the biexciton is detected. The probability of two photons to be emitted together (at zero time delay) is *at least* seven times larger than their average emission probability. The actual bunching is probably larger as the correlation function is smeared by the finite temporal resolution (~300 ps) of the detectors.

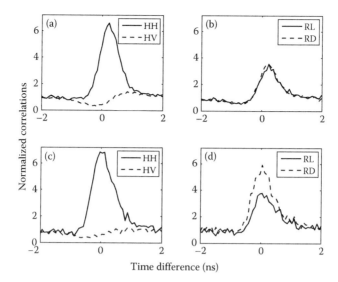

FIGURE 12.16 Measured polarization-sensitive intensity correlation functions between the XX^0 and the X^0 lines for various polarization combinations. (a and b) no spectral filtering. (c and d) with spectral filtering. The first (second) letter in each legend item specifies the polarization of the XX^0 (X^0) photon: H, horizontal; V, vertical; D, diagonal; R, right-hand circular; L, left-hand circular. The coincidence rate at long time differences for (a) and (b) (c and d) is 5 (0.5) per minute per time bin (80 ps).

For cross-linear polarizations the opposite happens: At small time differences the emission probability of a cross-linearly polarized photon pair almost vanishes. This "antibunching" proves experimentally that the rate of cross transitions between the exciton states is negligible relative to the radiative rate. The pairs of photons are always emitted with the same linear polarization. Other polarization-sensitive correlation measurements are all similar. For example, there is no difference between circular–circular and circular–diagonal polarization combinations. All show the same bunching signal, which amounts to one half of the signal in co-linear polarizations. This shows that preferred correlations exist only in the rectilinear basis, and thus the photons are only classically correlated, but not entangled [57].

The situation is different when spectral filtering is applied. In Figure 12.16c and d, we present measured correlation functions with spectral filtering. A pronounced difference is observed between correlations in circular–circular and circular–diagonal polarization combinations.

In order to unambiguously prove entanglement in radiative cascades, however, one should construct a set of measurements and demonstrate violation of the Bell inequality [64–67]. Another way is to construct by polarization tomography the entire density matrix of the two photon state [68]. This is done by measuring the XX^0–X^0 correlation function in 16 different polarization combinations. For each measurement, the number of events coming from the same cascade is extracted by integrating the background-reduced correlation function. The background in these measurements is readily available from the measurements in cross-linear polarizations [7]. In this case, all the measured events do not result from the same radiative cascade. Rather they result from cases in which the detected two photons were emitted in two different cascades. Therefore, they truly represent the background due to the cw nature of the experiment. Since this background is insensitive to the polarizations of the two photons, it can be safely subtracted from all the measured, normalized, intensity correlation functions.

The density matrix fully described the system, and there are a few tests that can be applied to it in order to quantify the degree of entanglement between the polarization states of the emitted pair of photons. We use the Peres criterion, which states that a density matrix is entangled if its partial transpose has negative eigenvalues [69].

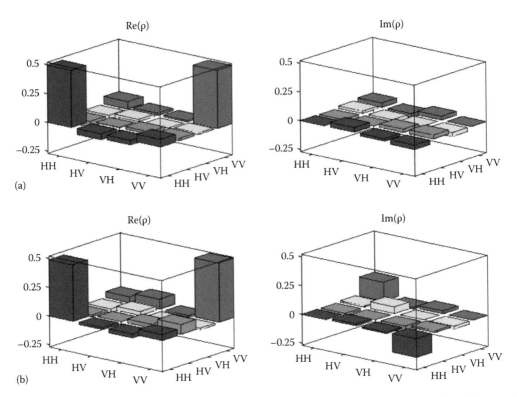

FIGURE 12.17 Measured two-photon polarization density matrices. (a) No spectral filtering. (b) With spectral filtering. Left (right)—real (imaginary) part. The Peres-criterion negativity for the matrix shown in (a) (b) is 0.03 ± 0.04 (0.18 ± 0.05). (Reprinted with permission from Akopian, N., Lindner, N.H., Poem, E., Berlatzky, Y., Avron, J., Gershoni, D., Gerardot, B.D., and Petroff, P.M., Entangled photon pairs from semiconductor quantum dots, *Phys. Rev. Lett.*, 96, 130501, 2006. Copyright 2006 by the American Physical Society.)

Figure 12.17 shows the measured density matrices without (a) and with (b) spectral filtering. While the matrix in Figure 12.17a fails to satisfy the Peres criterion within the measurement uncertainty, the matrix in Figure 12.17b satisfies it with a confidence level that is better than three standard deviations [7].

12.5.2 Spin-Blockaded Radiative Cascade

We recently discovered a novel radiative cascade that initiates from a metastable spin-blockaded biexcitonic state [70] rather than from the ground biexciton state. This cascade is an indirect one. The metastable biexciton state from which the cascade starts is composed of two electrons in their ground singlet state and two holes, one in its ground state and the other in an excited state. The two holes can form either a spin singlet or a spin triplet. The singlet state can relax to the ground state without changing its spin. The triplet states cannot, and they are thus metastable, blockaded from thermalization into their ground singlet state by Pauli's exclusion principle. These metastable biexciton states radiatively decay to form a single exciton that contains a hole in an excited state. This hole is now free to relax to its ground state, by the emission of a phonon, resulting in a ground-state exciton. The ground-state exciton then radiatively decays to the vacuum state. Consequently, this cascade involves the emission of three particles: two photons and one phonon. The intermediate non-radiative decay is fast (~30 ps [71]), and as shown below, preserves the exciton's spin. For a spin-preserving non-radiative decay, one expects all the "which-path" information carried by the intermediate phonon to reside in its energy. Therefore, one may expect this cascade to be another example where spectral filtering would be effective in restoring entanglement. We found that this is not the case. We applied the same filtering scheme to the two types of radiative cascades, in the same QD, and found that while for the ground-state biexciton cascade entanglement was restored, this was not the case for the spin-blockaded biexciton cascade. As discussed below, this is attributed to the "wrong" sign in the splitting of the two intermediate levels, combined with the fluctuating electrostatic environment ("spectral diffusion") [7].

The relevant level diagram of a neutral QD is presented in Figure 12.18a. The ground-state biexciton is marked by S (for Singlet), and the excited singlet biexciton is marked by S*. The metastable biexciton states are marked by T_0 and $T_{\pm 3}$. T stands for Triplet and the subscripts stand for the total two holes' spin projection on the QD's growth direction. The triplet biexciton states are not split by the EHEI since the total electronic spin is zero. High orders of the hole–hole exchange interaction can lower the $T_{\pm 3}$ states in respect to the T_0 state [72]. In our case however, the separation is smaller than the isotropic e–h exchange, which splits the dark and bright exciton states. The order of the T_0 and $T_{\pm 3}$ emission lines is thus the same as in the case of no hole–hole anisotropic exchange interaction (see Figure 12.18a).

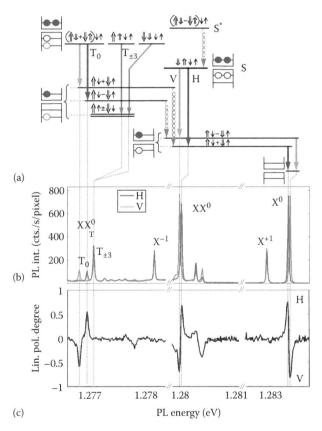

FIGURE 12.18 (See color insert.) (a) Energy level diagram for excitons and biexcitons in a neutral QD. Single-carrier level occupations are given alongside each many-carrier level. The spin wavefunctions are depicted above each level. The symbol ↑ (⇓) represents spin up (down) electron (hole). Short (long) symbols represent charge carriers in the first (second) energy level. S (S*) indicates the ground (excited) biexciton hole-singlet state. T_0 ($T_{\pm 3}$) indicates the metastable spin-triplet biexciton state with z-axis spin projection of 0 (±3). The solid (curly) vertical arrows indicate spin preserving (non-) radiative transitions. Dark- (light-) gray arrows represent photon emission in horizontal, H (vertical, V) polarization. (b) Polarized PL spectra. H (V) in dark (light) gray. Spectral lines that are relevant to this work are marked and linked to the transitions in (a) by dashed lines. (c) Linear polarization spectrum. The value 1 (−1) means full H (V) polarization.

We therefore neglect the contribution of these high orders to the hole–hole exchange in the following discussion. Recombinations from the $T_{\pm 3}$ states lead to the optically inactive ("dark") exciton intermediate states (see the spectral line in Figure 12.18b). Radiative decay does not proceed from these states and therefore they are not discussed here further [34]. The T_0 biexciton state recombines to form one of two *excited* exciton states. As in the case of the ground-state exciton discussed above, these states are the symmetric and antisymmetric combinations of spin projection eigenstates. Here however, the energetic ordering between the symmetric and antisymmetric combinations is the opposite of that between the ground-state exciton states. This is due to the opposite sign of the relevant anisotropic e–h exchange term, as seen in Table 12.1. It can be understood by noting that the quadrupole moment of the p_x-type spatial wavefunction of the excited hole is opposite in respect to that the s-type ground-state hole wavefunction.

Another difference between the T_0 and S biexciton cascades arises in the polarization selection rules for optical recombination of the biexciton. Due to the reversed spin-exchange symmetry of the T_0 biexciton in respect to that of the S biexciton, the polarization selection rules are reversed: While in the S biexciton case horizontal polarization leads to the symmetric combination, in the T_0 case it leads to the antisymmetric combination. This is similar to the case of the doubly charged exciton discussed above, where the selection rules depend on whether the final state belongs to a spin singlet or a spin triplet. Nevertheless, the order of polarizations of the emission line components will be the same for the two biexcitons. This is because the difference in the selection rules is compensated by the difference in the ordering of the levels. As seen in the polarized PL spectra presented in Figure 12.18b, and in the corresponding spectrum of linear polarization degree shown in Figure 12.18c, this is indeed the case.

The hole of the excited exciton states is not spin blockaded, and it quickly decays (nonradiatively) to its ground state, *while preserving its spin*. Therefore, horizontally (H) polarized T_0 biexciton recombination will be followed by vertically (V) polarized exciton recombination (see Figure 12.18a). This leads to correlated *cross*-linearly polarized photons, unlike the correlated co-linearly polarized photons emitted in the S biexciton cascade.

In Figure 12.19a (Figure 12.19c), we present measured time resolved intensity correlation functions between the T_0 (S) biexciton line and the exciton line, in both co- (dark gray) and cross- (light gray) linear polarizations. It is clearly seen that while for the cascade starting from the S biexciton, the emission of the exciton is "bunched" for co-linear polarizations and "anti-bunched" for cross-linear polarizations, the opposite happens for the cascade starting from the T_0 biexciton. This confirms that there is no change in the spin configuration during the decay of the hole.

In addition, we performed full polarization tomography for both cascades, both with and without spectral filtering. The resulting two-photon polarization density matrices for the case of no filtering are presented in Figure 12.19b (Figure 12.19d) for the T_0 (S) biexciton cascade. Since the imaginary parts of the matrices were zero to within the measurement uncertainty, only the real parts are displayed. In Figure 12.20, we present the density matrices obtained for the case of spectral filtering. While entanglement could be restored for photon pairs emitted from the S biexciton cascade (Figure 12.20a), no such restoration could be achieved for the T_0 cascade (Figure 12.20b). The differences between the two cases are discussed below.

Figure 12.21a shows the cascade initiated by the ground state of a biexciton in an ideal, symmetric QD. In this case, the two exciton energy levels are degenerate, and emitted photon pairs will be entangled [55]. Figure 12.21b shows the case of the ground-state biexciton cascade in an asymmetric QD, in which the exciton levels are split by an energy δ. Here, spectral filtering (cross-hatched rectangle) is necessary for the emitted photons to be entangled. Figure 12.21c shows a schematic plot of the two-photon probability distribution. The x-(y-)axis represents the energy of the exciton (biexciton) photon. The dark-gray spots show the regions of high emission probability. Their size and shape are determined by the radiative width of the exciton (γ_X) and biexciton (γ_{XX}) lines. The emission in these regions is dominated by un-entangled photon pairs. The energies of the two photons are related by total energy conservation: If the first photon has high energy, the second one will have low energy, and vice versa.

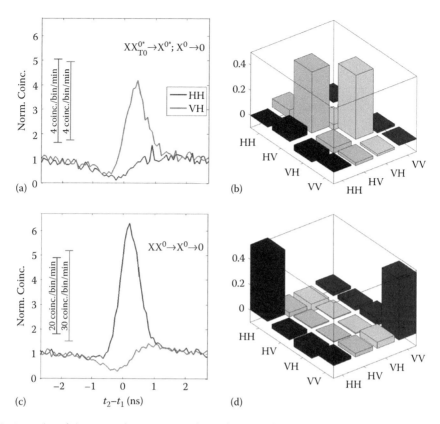

FIGURE 12.19 (a and c) Measured intensity correlation functions for the spin-blockaded and ground-state biexciton cascades, respectively. Dark- (light-) gray line represents the correlation in co-(cross-) linear polarizations. The coincidence rates are indicated by the scale bars in units of coincidences per time bin (80 ps) per minute. (b and d) Real parts of the two-photon polarization density matrices measured for the spin-blockaded and ground-state biexciton cascades, respectively.

This puts the two dark-gray spots on a line parallel to the (1,–1) direction. The cross-hatched rectangle represents an optimal spectral filter for entanglement restoration. It is δ-γ_X by δ-γ_{XX} in size. It rejects most of the un-entangled photon pairs while it keeps a measurable fraction of the entangled pairs, which lie mostly between the two dark-gray spots, on the connecting line. The degree of entanglement within the transmitted photon pairs is thus increased. A smaller filter would yield higher degree of entanglement, but will transmit considerably less photons.

Due to random fluctuations in the electrostatic environment of the QD, the energies of the spectral lines fluctuate with time. This "spectral diffusion" happens on timescales much longer than the radiative recombination time of the exciton. The random electric field is thus quasi static. Since all spectral lines experience almost the same shift in a given static electric field [44,73], see Figure 12.9, the energies of the exciton and biexciton lines will fluctuate in a correlated manner. The dark-gray spots of Figure 12.21c will thus randomly move along the dashed lines parallel to the (1,1) direction, as shown in the figure by the light-gray areas. As these areas are outside the filter, spectral diffusion does not interfere with the entanglement restoration. Indeed, as was discussed above (Figure 12.20a), entanglement can be restored by spectral filtering for the ground-state biexciton cascade.

The situation is different for an indirect radiative cascade. Figure 12.21d through f presents the case of the spin-blockaded biexciton radiative cascade. Here there is a fast, non-radiative (phononic) decay of the hole from its first excited state to its ground state between the biexciton and exciton radiative

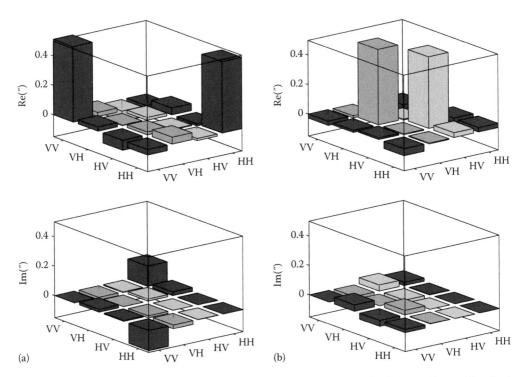

FIGURE 12.20   (a and b) Measured two-photon polarization density matrices for the ground-state (S) and spin-blockaded (T_0) biexciton cascades, respectively, as obtained with spectral filtering. Real (imaginary) parts are shown in the top (bottom) panels. The Peres-criterion negativity for the matrix in (a) (b) is 0.15 ± 0.03 (0.05 ± 0.1).

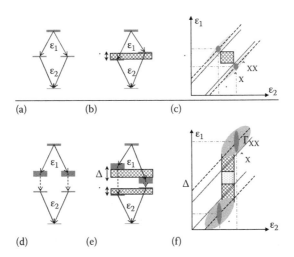

FIGURE 12.21   (a) Ideal direct cascade. The widths of the levels represent their decay rate. (b) Ground-state biexciton cascade in an anisotropic QD. (c) Schematic two-photon probability distribution. Dark gray—high-probability areas; cross hatched—spectral filter; light gray—inhomogeneous broadening due to spectral diffusion. (d through f) Same as (a through c) (respectively), for an indirect cascade. In (e) and (f) only the case of splittings in opposite directions is shown. The dotted rectangle in (f) is an example for a filter not penetrated by the high-probability areas for any amount of spectral diffusion.

recombinations. Figure 12.21d shows the ideal case where the excited and ground exciton states are each twofold degenerate. The short lifetime of the excited exciton states is represented by the large width of the energy levels. Since the spin of the exciton is conserved during the intermediate stage, and since the spatial parts of the exciton wavefunctions are identical for both decay paths, the emitted phonon does not carry any "which-path" information beyond its energy. In this case, one expects that appropriate filtering of the photon energies will restore the which-path ambiguity, resulting in entanglement of the polarization state of the photons. Figure 12.21e shows the case of an asymmetric QD. The degeneracy is lifted for both ground and excited exciton sates, in opposite manners. Due to the opposite splittings, the energies of the two photons are positively rather than negatively correlated. This is shown in Figure 12.21f, where the dark-gray spots again represent the regions of high probability. Their elongated shape is due to the larger width of the biexciton photon, which comes from the fast decay of its final state. Spectral diffusion will still shift these regions along the (1,1) direction, as shown by light gray. The analog of the optimal filter for this case is shown by the cross-hatched rectangle. It is $\Delta - \Gamma_{XX}$ by $\delta - \gamma_X$, where Δ is the excited exciton splitting (in absolute value) and Γ_{XX} is the width of the spin-blockaded biexciton transition. As in the previous case, such a filter excludes the dark-gray spots. However, it is not immune to spectral diffusion, as shown by the overlap of the cross-hatched rectangle and the light-gray areas. Indeed, as shown in Figure 12.20b, no entanglement could be detected even when this filter was applied. Further decreasing the filter width may solve the spectral diffusion problem, but then the photon pair collection rate would drastically decrease. We note that if the ground and excited exciton states would have split to the same direction, the situation would have been similar to that described in Figure 12.21c, and entanglement restoration by spectral filtering would not have been affected by spectral diffusion. A quantitative condition for spectral filtering to effectively erase the which-path information can be formulated by inspecting Figure 12.21c and f. Note first that spectral diffusion leads to motion of the dark gray spots along the (1,1) direction. Spectral filtering works if during this motion the spots stay strictly outside the filter area. The width of the filter that satisfies this condition is determined by looking at the projection of the filter on the orthogonal direction to the motion of the spots: the (1,–1) direction. Let $f > 0$ ($F > 0$) be the filter width for the exciton (biexciton) photon spectral line. The filter's projection on the (1,–1) direction is given by $(f + F)/\sqrt{2}$. The projection of the line connecting the centers of the two dark-gray spots on the (1,–1) direction is given by $|\Delta \pm \delta|/\sqrt{2}$, where the plus sign is for Figure 12.21c and the minus sign is for Figure 12.21f. For avoiding overlap one thus must have $f + F \leq |\Delta \pm \delta|$. The case of a minus sign forces narrow filter widths, which makes spectral filtering ineffective. Taking into account the widths of the lines (the sizes of the dark-gray spots) leads to an even stronger constraint

$$f + F \leq |\Delta \pm \delta| - (\Gamma_{XX} + \gamma_X) \tag{12.40}$$

where Γ_{XX} refers also to γ_{XX} as appropriate. This explains why spectral filtering was ineffective for the spin-blockaded biexciton cascade.

12.5.3 Radiative Cascades in Singly Charged Quantum Dots

We recently demonstrated [71] that radiative cascades occur also when the QD is charged. In this case, the intermediate, charged exciton level maybe metastable, spin blockaded from thermal relaxation. In the following, we present and discuss measurements performed on a positively charged QD.

The energy levels of a positively charged QD [37,38] containing up to three heavy holes and two electrons are schematically described in Figure 12.22a. The figure presents also the relevant radiative and non-radiative total-spin-conserving transitions between these levels. The two-photon radiative cascades start from the ground level of the three heavy holes and two electrons' state. The unpaired hole's spin projection along the growth axis determines the total spin of the two Kramers degenerate

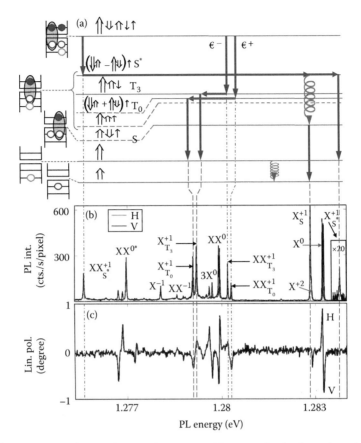

FIGURE 12.22 (See color insert.) (a) Schematic description of the energy levels of a singly positively charged QD. Vertical (curly) arrows indicate radiative (non-radiative) transitions between these levels. State occupation and spin wavefunctions are described to the left of each level where $\uparrow$ ($\Downarrow$) represents an electron (hole) with spin up (down). A short blue (long red) arrow represents a carrier in its first (second) level. S (T) stands for two holes' singlet (triplet) state and 0 (3) for $S_z=0$ ($S_z=\pm3$) total holes' pseudo-spin projection on the QD growth direction. The excited state singlet is indicated by S*. Only one out of two (Kramers) degenerate states is described. (b) Measured PL spectrum on which the actual transitions are identified. Transitions that are not discussed here are marked by gray letters. (c) Measured degree of linear polarization spectrum, along the in-plane symmetry axes of the QD. Positive (negative) value represents polarization along the QD's major (minor) axis. (Reprinted with permission from Poem, E., Kodriano, Y., Tradonsky, C., Gerardot, B.D., Petroff, P.M., and Gershoni, D., Radiative cascades from charged semiconductor quantum dots, *Phys. Rev. B*, 81, 085306, 2010. Copyright 2010 by the American Physical Society.)

states (for simplicity only one state is drawn in Figure 12.22a). Radiative recombination of first level e–h pair leaves three unpaired charge carriers within the QD. There are eight possible different spin configurations for the remaining carriers. These configurations form four energy levels of Kramers' pairs [37,38]. The three lowest levels are those in which the two unpaired holes are in spin-triplet states. Those states are separated from the highest energy level in which the holes are in a singlet spin state by the hole–hole isotropic exchange interaction, which is significantly stronger than the e–h exchange interaction. The later removes the degeneracy between the triplet states as shown in Figure 12.22a. The lowest triplet level cannot be reached optically. The optical transitions into the other levels are optically allowed. The circular polarizations of the emitted photons are indicated in the figure. They depend on the spins of the annihilated electron hole pair. The measured emission contains also linear components (see Figure 12.22c), due to the anisotropic e–h exchange interaction [37,38].

The relaxation proceeds by radiative recombination of the remaining first level e–h pair, leaving thus only one hole in its second level. The hole can then quickly relax nonradiatively to its ground level. There is a fundamental difference between the singlet and triplet intermediate states. While in the later, due to Pauli's exclusion principle, radiative recombination must occur before the excited hole can relax to its ground state (resulting in two "direct" cascades), in the former non-radiative relaxation of the excited hole state may occur prior to the radiative recombination (resulting in one "direct" and one "indirect" cascade).

In Figure 12.22b, we present the spectrum measured under nonresonant cw excitation with $1\,\mu W$ of HeNe laser light (1.96 eV). The corresponding degree of linear polarization is presented in Figure 12.22c. The spectral lines participating in the radiative cascades described in Figure 12.22a are clearly identified spectrally in the single QD PL and linear polarization spectra.

In Figure 12.23, we present the measured and calculated intensity correlation functions for photon pairs emitted in the four spin-conserving radiative cascades outlined in Figure 12.22a. The measured data clearly reveal the sequence of the radiative events, reassuring the interpretations of Figure 12.22.

In Figure 12.24, we present measured and calculated intensity correlation functions between different radiative cascades. Since spin blockade prevents the relaxation of the second level hole to its first level, they provide an estimate for the rate by which the holes' spin scatters [73]. In Figure 12.24a and c, we probe possible transitions from the singlet intermediate state S^* to the triplet T_0 and $T_{\pm 3}$ intermediate states, respectively.

In Figure 12.24b and d, we probe possible transitions from the triplet T_0 and $T_{\pm 3}$ intermediate states, respectively, to the singlet ground state S. Assuming that relaxation from the intermediate triplet states to the ground singlet states must be preceded by transition to the intermediate singlet states, these measurements provide quantitative estimation for the reverse of the processes described in Figure 12.24a and c. From the measured data in Figure 12.24, one clearly notes that transition between the two holes'

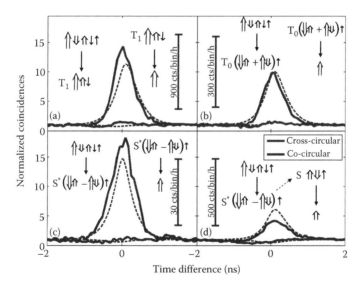

FIGURE 12.23 (See color insert.) Measured and calculated time-resolved, polarization-sensitive intensity correlation functions for the four radiative cascades described in Figure 12.22. The states involved in the first (second) photon emission are illustrated to the left (right) side of each panel. All symbols and labels are as in Figure 12.22. Solid blue (red) line stands for measured cross-(co-)circularly polarized photons. Dashed lines represent the corresponding calculated functions. The bar presents the acquisition rate in coincidences per time bin (80 ps) per hour. (Reprinted with permission from Poem, E., Kodriano, Y., Tradonsky, C., Gerardot, B.D., Petroff, P.M., and Gershoni, D., Radiative cascades from charged semiconductor quantum dots, *Phys. Rev. B*, 81, 085306, 2010. Copyright 2010 by the American Physical Society.)

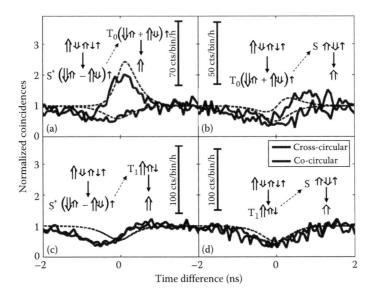

FIGURE 12.24 (See color insert.) Measured and calculated time-resolved, polarization-sensitive intensity correlation functions, across the radiative cascades. (a and c) Correlations between the singlet biexciton transition and the exciton transition from the T_0, (T_3) state. (b and d) Correlations between the T_0, (T_3) biexciton transition and the ground X^{+1} exciton transition. All symbols and labels are as in Figure 12.22. The meanings of all line types and colors are as in Figure 12.23. (Reprinted with permission from Poem, E., Kodriano, Y., Tradonsky, C., Gerardot, B.D., Petroff, P.M., and Gershoni, D., Radiative cascades from charged semiconductor quantum dots, *Phys. Rev. B*, 81, 085306, 2010. Copyright 2010 by the American Physical Society.)

singlet state to the $T_{\pm3}$ triplet state (Figure 12.24c) and vice versa (Figure 12.24d) are forbidden (within the radiative lifetime), while transitions between the singlet and the T_0 triplet states (Figure 12.24a) and vice versa (Figure 12.24b) are partially allowed.

This means that the holes spin projection on the QD's growth axis is conserved during the relaxation while their in-plane spin projection scatters [73]. The difference between the scattering rates from the singlet to triplet state and that from the triplet to singlet is due to the energy difference between these two states ($\sim$4 meV), which is much larger than the ambient thermal energy ($\sim$0.5 meV).

In order to calculate the expected correlation functions and to compare them to the measured ones, we used a classical rate equation model. In this model, we include all the states as described in Figure 12.22a, together with their Kramers conjugates. In addition, we include four more states representing charged multiexcitons up to six e–h pairs [53]. There are clear spectral evidences for processes in which the QD changes its charge state and becomes neutral due to optical depletion [74,75] (see Figure 12.22b). These observations are considered in our model by introducing one additional state that represents a neutral QD. The transition rates between the states include radiative rate ($\gamma_r = 1.25\,\text{ns}^{-1}$ deduced directly from the PL decay of the exciton lines) and non-radiative spin-conserving rate ($\Gamma_{S^*\to S} = 35\gamma_r$, deduced from the intensity ratios of the S^* and the S PL lines). We also include the rate for optical generation of e–h pairs ($G_e = 1\gamma_r$ forced by equating between the emission intensities of the biexciton and exciton spectral lines), the optical depletion, and recharging rates ($G_D = 4\gamma_r$ and $G_C = 0.1\gamma_r$ as deduced from the relevant line intensity ratios and correlation measurements between the neutral and charged exciton). The data clearly show that hole spin scattering rates, ($\Gamma_{S\rightleftarrows T_3}$), which do not conserve the spin projection on the QD's growth axis, are vanishingly small. Therefore, we set them to 0. In order to account for the observed correlations between singlet (S) and T_0 states, Figure 12.3, we fitted in-plane scattering rates [73] $\Gamma_{T_0\to S} = 0.6\gamma_r$ and $\Gamma_{S^*\to T_0} = 10\gamma_r$ (such processes still conserve the projection of the total spin along the growth axis). The ratio between these rates simply gives the temperature of the optically excited QD ($\sim$19 K).

The anisotropic EHEI mixes the T_0 and $T_{\pm3}$ states [37,38]. This makes the natural polarizations of the relevant transitions elliptical rather than circular. The mixing degree is obtained from the measured degree of linear polarization of the biexciton transitions [38]. Our model considers this mixing as well. It explains the nonzero correlations in co-circular polarizations.

The two in-plane hole's spin scattering rates that we fitted describe very well the 16 measured intensity correlation functions. The calculated functions (convoluted with the system response) are presented in Figures 12.23 and 12.24 by dashed lines.

Over all, we identified three direct and one indirect radiative cascades in singly charged QDs and demonstrated unambiguous correlations between the polarizations of the emitted photons and the spin of the remaining charge carrier. Our correlation measurements show that while holes' spin-projection conserving scattering rates are a few times faster than the radiative rates, spin-projection nonconserving rates are vanishingly small.

12.6 Summary

We discussed and reviewed PL spectroscopy from single semiconductor QDs, in general and radiative cascades from these nanostructures, in particular. For understanding in details the available rich experimental data we developed a theoretical many interacting carrier model. Our model, though relatively simple, describes very well the measured polarization-sensitive PL spectra. In particular, we were able to explain linear and circular polarization memory in quasi-resonant optical excitation of the QDs. We concluded this chapter with quantitative analysis of polarization-sensitive intensity correlation measurements of various biexciton–exciton radiative cascades in neutral and charged QDs. Emission of polarization entangled pairs of photons in these cascades was reviewed with strong emphasis on spectral filtering as a tool for distilling entanglement.

We believe that the insights and findings that we reviewed here be useful for future developments of semiconductor QDs as an important tool for quantum information processing technologies.

References

1. M. A. Kastner, Artificial atoms. *Phys. Today* **46**, 24 (1993).
2. R. C. Ashoori, Electrons in artificial atoms. *Nature* **379**, 413 (1996).
3. D. Gammon, Semiconductor physics: Electrons in artificial atoms. *Nature* **405**, 899 (2000).
4. P. J. Carson, G. F. Strouse, P. Michler, A. Imamoglu, M. D. Mason, and S. K. Buratto, Quantum correlation among photons from a single quantum dot at room temperature. *Nature* **406**, 968 (2000).
5. E. Hu, P. Michler, A. Kiraz, C. Becher, W. V. Schoenfeld, P. M. Petroff, L. Zhang, and A. Imamoglu, Quantum correlation among photons from a single quantum dot at room temperature. *Science* **290**, 2282 (2000).
6. C. Santori, M. Pelton, G. Solomon, Y. Dalem, and Y. Yamamoto, Triggered single photons from a quantum dot. *Phys. Rev. Lett.* **86**, 1502 (2001).
7. N. Akopian, N. H. Lindner, E. Poem, Y. Berlatzky, J. Avron, D. Gershoni, B. D. Gerardot, and P. M. Petroff, Entangled photon pairs from semiconductor quantum dots. *Phys. Rev. Lett.* **96**, 130501 (2006).
8. R. Hafenbrak, S. M. Ulrich, P. Michler, L. Wang, A. Rastelli, and O. G. Schmidt, Triggered polarization-entangled photon pairs from a single quantum dot up to 30 K. *New J. Phys.* **9**, 315 (2007).
9. D. Loss and D. P. DiVincenzo, Quantum computation with quantum dots. *Phys. Rev. A* **57**, 120 (1998).
10. C. H. Bennett and G. Brassard, Quantum cryptography: Public key distribution and coin tossing. In *Proc. IEEE Int. Conf. Comput. Syst. Signal Process*, Bangalore, India, December 1984.
11. B. Lounis and W. E. Moerner, Single photons on demand from a single molecule at room temperature. *Nature* **407**, 491 (2000).

12. H. Drexler, D. Leonard, W. Henson, J. P. Kotthaus, and P. M. Petroff, Spectroscopy of quantum levels in charge-tunable InGaAs quantum dots. *Phys. Rev. Lett.* **73**, 2252 (1994).

13. A. Szabo and N. S. Ostlund, *Modern Quantum Chemistry: Introduction to Advanced Electronic Structure Theory*. Dover, Mineola, New York (1996).

14. L. Fetter and J. D. Walecka, *Quantum Theory of Many Particle Systems*. McGraw-Hill, New York, 1971.

15. A. Barenco and M. A. Dupertuis, Quantum many-body states of excitons in a small quantum dot. *Phys. Rev. B* **52**, 2766 (1995).

16. E. L. Ivchenko and G. E. Pikus. *Super Lattices and Other Hetero Structures*, Volume 110 of Springer series in Solid State Sciences. Springer Verlag, Berlin, Germany (1997).

17. T. Takagahara, Theory of exciton doublet structures and polarization relaxation in single quantum dots. *Phys. Rev. B* **62**, 16840 (2000).

18. M. Bayer, G. Ortner, O. Stern, A. Kuther, A. A. Gorbunov, A. Forchel, P. Hawrylak, S. Fafard, K. Hinzer, T. L. Reinecke, S. N. Walck, J. P. Reithmaier, F. Klopf, and F. Schafer, Fine structure of neutral and charged excitons in self assembled In(Ga)As/(Al)GaAs quantum dots. *Phys. Rev. B* **65**, 195315 (2002).

19. L. W. Wang, A. J. Williamson, and A. Zunger, Theoretical interpretation of the experimental electronic structure of lens shaped self assembled InAs/GaAs quantum dots. *Phys. Rev. B* **62**, 12963 (2000).

20. S. Lee, L. Jönsson, J. W. Wilkins, G. W. Bryant, and G. Klimeck, Electron–hole correlations in semiconductor quantum dots with tight-binding wave functions. *Phys. Rev. B* **63**, 195318 (2001).

21. M. Zieliński, M. Korkusiński, and P. Hawrylak, Atomistic tight-binding theory of multiexciton complexes in a self-assembled InAs quantum dot. *Phys. Rev. B* **81**, 085301 (2010).

22. D. Gershoni, Eigen function expansion method for solving the quantum wire problem: Formulation. *Phys. Rev. B* **43**, 4011 (1991).

23. M. Grundmann, O. Stier, and D. Bimberg, InAs/GaAs pyramidal quantum dots: Strain distribution, optical phonons and electronic structure. *Phys. Rev. B* **52**, 11969 (1995).

24. H. Jiang and J. Singh. Strain distribution and electronic spectra of InAs/GaAs self assembled quantum dots: An eight band study. *Phys. Rev. B* **56**, 4696 (1997).

25. C. Pryor, Eight-band calculations of strained InAs/GaAs quantum dots compared with one-, four-, and six-band approximations. *Phys. Rev. B* **57**, 7190 (1998).

26. A. Kumar, S. E. Laux, and F. Stern, Electron states in a GaAs quantum dot in a magnetic field. *Phys. Rev. B* **42**, 5166 (1990).

27. U. Merkt, J. Huser, and M. Wagner, Energy spectra of two electrons in a harmonic quantum dot. *Phys. Rev. B* **43**, 7320 (1991).

28. E. Poem, J. Shemesh, I. Marderfeld, D. Galushko, N. Akopian, D. Gershoni, B. D. Gerardot, A. Badolato, and P. M. Petroff, Polarization sensitive spectroscopy of charged quantum dots. *Phys. Rev. B* **76**, 235304 (2007).

29. H. A. Kramers, General theory of paramagnetic rotation in crystals. *Proc. Acad. Sci. Amsterdam* **33**, 959 (1930).

30. V. Mlinar and A. Zunger, Effect of atomic-scale randomness on the optical polarization of semiconductor quantum dots. *Phys. Rev. B* **79**, 115416 (2009).

31. D. M. Bruls, J. W. A. M. Vugs, P. M. Koenraad, H. W. M. Salemink, J. H. Wolter, M. Hopkinson, M. S. Skolnick, F. Long, and S. P. A. Gill, Determination of the shape and indium distribution of low-growth-rate InAs quantum dots by cross-sectional scanning tunneling microscopy. *Appl. Phys. Lett.* **81**, 1708 (2002).

32. K. V. Kavokin, Symmetry of anisotropic exchange interactions in semiconductor nanostructures. *Phys. Rev. B,* **69**, 075302 (2004).

33. C. Cohen-Tannoudji, B. Diu, and F. Laloe, *Quantum Mechanics*. Wiley-Interscience, New York (1977).

34. E. Poem, Y. Kodriano, N. H. Lindner, C. Tradonsky, B. D. Gerardot, P. M. Petroff, and D. Gershoni, Accessing the dark exciton with light. *Nat. Phys.* **6**, 993 (2010).

35. H. Henry and K. Nassau, Lifetimes of bound excitons in CdS. *Phys. Rev. B* **1**, 1628 (1970).

36. S. Rodt, A. Schliwa, K. Pötschke, F. Guffarth, and D. Bimberg, Correlation of structural and few-particle properties of self-organized InAs/GaAs quantum dots. *Phys. Rev. B* **71**, 155325 (2005).

37. K. V. Kavokin, Fine structure of the quantum-dot trion. *Phys. Status Solidi (a)* **195**, 592 (2003).

38. I. A. Akimov, K. V. Kavokin, A. Hundt, and F. Henneberger, Electron–hole exchange interaction in a negatively charged quantum dot. *Phys. Rev. B* **71**, 075326 (2005).

39. B. Urbaszek, R. J. Warburton, K. Karrai, B. D. Gerardot, P. M. Petroff, and J. M. Garcia, Fine structure of highly charged excitons in semiconductor quantum dots. *Phys. Rev. Lett.* **90**, 247403 (2003).

40. E. Dekel, D. V. Regelman, D. Gershoni, E. Ehrenfreund, W. V. Schoenfeld, and P. M. Petroff, Cascade evolution and radiative recombination of quantum dot multiexcitons studied by time-resolved spectroscopy. *Phys. Rev. B* **62**, 11038 (2000).

41. G. Ramon, U. Mizrahi, N. Akopian, S. Braitbart, D. Gershoni, T. L. Reinecke, B. D. Gerardot, and P. M. Petroff, Emission characteristics of quantum dots in planar microcavities. *Phys. Rev. B* **73**, 205330 (2006).

42. R. I. Dzhioev, K. V. Kavokin, V. L. Korenev, M. V. Lazarev, B. Ya. Meltser, M. N. Stepanova, B. P. Zakharchenya, D. Gammon, and D. S. Katzer, Low-temperature spin relaxation in n-type GaAs. *Phys. Rev. B* **66**, 245204 (2002).

43. J. Hilton and C. L. Tang, Optical orientation and femtosecond relaxation of spin-polarized holes in GaAs. *Phys. Rev. Lett.* **89**, 146601 (2002).

44. M. E. Ware, E. A. Stinaff, D. Gammon, M. F. Doty, A. S. Bracker, D. Gershoni, V. L. Korenev, Ş. C. Bădescu, Y. Lyanda-Geller, and T. L. Reinecke, Polarized fine structure in the photoluminescence excitation spectrum of a negatively charged quantum dot. *Phys. Rev. Lett.* **95**, 177403 (2005).

45. D. Heiss, M. Kroutvar, J. J. Finley, and G. Abstreiter, Progress towards single spin optoelectronics using quantum dot nanostructures. *Solid State Commun.* **135**, 591 (2005).

46. E. Dekel, D. V. Regelman, D. Gershoni, E. Ehrenfreund, W. V. Schoenfeld, and P. M. Petroff, Radiative lifetimes of single excitons in semiconductor quantum dots—Manifestation of the spatial coherence effect. *Solid State Commun.* **117**, 395 (2001).

47. V. K. Kalevich, I. A. Merkulov, A. Yu. Shiryaev, K. V. Kavokin, M. Ikezawa, T. Okuno, P. N. Brunkov, A. E. Zhukov, V. M. Ustinov, and Y. Masumoto, Optical spin polarization and exchange interaction in doubly charged InAs self-assembled quantum dots. *Phys. Rev. B* **72**, 045325 (2005).

48. D. Gammon, E. S. Snow, B. V. Shanabrook, D. S. Katzer, and D. Park, Fine structure splitting in the optical spectra of single GaAs quantum dots. *Phys. Rev. Lett.* **76**, 3005 (1996).

49. M. Paillard, X. Marie, P. Renucci, T. Amand, A. Jbeli, and J.-M. Gérard, Spin relaxation quenching in semiconductor quantum dots. *Phys. Rev. Lett.* **86**, 1634 (2001).

50. I. Favero, G. Cassabois, C. Voisin, C. Delalande, Ph. Roussignol, R. Ferreira, C. Couteau, J. P. Poizat, and J.-M. Gérard, Giant optical anisotropy in a single InAs quantum dot in a very dilute quantum-dot ensemble. *Phys. Rev. B* **71**, 233304 (2005).

51. R. I. Dzhioev, B. P. Zakharchenya, E. L. Ivchenko, V. L. Korenev, Yu. G. Kusraev, N. N. Ledentsov, V. M. Ustinov, A. E. Zhukov, and A. F. Tsatsulnikov, Fine structure of excitonic levels in quantum dots. *J. Exp. Theor. Phys. Lett.* **65**, 804 (1997).

52. R. Hanbury Brown and R. Q. Twiss, A test of a new type of stellar interferometer on Sirius. *Nature* **178**, 1046 (1956).

53. D. V. Regelman, U. Mizrahi, D. Gershoni, E. Ehrenfreund, W. V. Schoenfeld, and P. M. Petroff, Semiconductor quantum dot: A quantum light source of multicolor photons with tunable statistics. *Phys. Rev. Lett.* **87**, 257401 (2001).

54. H.-P. Breuer and F. Petruccione, *The Theory of Open Quantum Systems*. Oxford University Press, New York (2002).

55. O. Benson, C. Santori, M. Pelton, and Y. Yamamoto, Regulated and entangled photons from a single quantum dot. *Phys. Rev. Lett.* **84**, 2513 (2000).

56. E. Moreau, I. Robert, L. Manin, V. Thierry-Mieg, J.-M. Gérard, and I. Abram, Quantum cascade of photons in semiconductor quantum dots. *Phys. Rev. Lett.* **87**, 183601 (2001).

57. C. Santori, D. Fattal, M. Pelton, G. S. Solomon, and Y. Yamamoto, Polarization-correlated photon pairs from a single quantum dot. *Phys. Rev. B* **66**, 045308 (2002).

58. D. Gerardot, S. Seidl, P. A. Dalgarno, R. J. Warburton, D. Granados, J. M. Garcia, K. Kowalik, O. Krebs, K. Karrai, A. Badolato, and P. M. Petroff, Manipulating exciton fine structure in quantum dots with a lateral electric field. *Appl. Phys. Lett.* **90**, 041101 (2007).

59. A. Muller, W. Fang, J. Lawall, and G. S. Solomon, Creating polarization-entangled photon pairs from a semiconductor quantum dot using the optical Stark effect. *Phys. Rev. Lett.* **103**, 217402 (2009).

60. R. M. Stevenson, R. J. Young, P. See, D. G. Gevaux, K. Cooper, P. Atkinson, I. Farrer, D. A. Ritchie, and A. J. Shields, Magnetic-field-induced reduction of the exciton polarization splitting in InAs quantum dots. *Phys. Rev. B* **73**, 033306 (2006).

61. S. Seidl, M. Kroner, A. Högele, K. Karrai, R. J. Warburton, A. Badolato, and P. M. Petroff, Effect of uniaxial stress on excitons in a self-assembled quantum dot. *Appl. Phys. Lett.* **88**, 203113 (2006).

62. R. J. Young, R. M. Stevenson, A. J. Shields, P. Atkinson, K. Cooper, D. A. Ritchie, K. M. Groom, A. I. Tartakovskii, and M. S. Skolnick, Inversion of exciton level splitting in quantum dots. *Phys. Rev. B* **72**, 113305 (2005).

63. A. Mohan, M. Felici, P. Gallo, B. Dwir, A. Rudra, J. Faist, and E. Kapon, Polarization-entangled photons produced with high-symmetry site-controlled quantum dots. *Nat. Photon.* **4**, 302 (2010).

64. J. S. Bell, On the Einstein-Podolsky-Rosen paradox. *Physics* **1**, 195 (1964).

65. J. F. Clauser, M. A. Horne, A. Shimoni, and R. A. Holt, Proposed experiment to test local hidden-variable theories. *Phys. Rev. Lett.* **23**, 880 (1969).

66. S. J. Freedman and J. F. Clauser, Experimental test of local hidden-variable theories. *Phys. Rev. Lett.* **28**, 938 (1972).

67. J. F. Clauser and M. A. Horne, Experimental consequences of objective local theories. *Phys. Rev. D* **10**, 526 (1974).

68. F. V. James, P. G. Kwiat, W. J. Munro, and A. G. White, Measurement of qubits. *Phys. Rev. A* **64**, 052312 (2001).

69. A. Peres, Separability criterion for density matrices. *Phys. Rev. Lett.* **77**, 1413 (1996).

70. Y. Kodriano, E. Poem, N. H. Lindner, C. Tradonsky, B. D. Gerardot, P. M. Petroff, J. E. Avron, and D. Gershoni, Radiative cascade from quantum dot metastable spin-blockaded biexciton. *Phys. Rev. B.* **82**, 155329 (2010).

71. E. Poem, Y. Kodriano, C. Tradonsky, B. D. Gerardot, P. M. Petroff, and D. Gershoni, Radiative cascades from charged semiconductor quantum dots. *Phys. Rev. B* **81**, 085306 (2010).

72. T. Warming, E. Siebert, A. Schliwa, E. Stock, R. Zimmermann, and D. Bimberg, Hole–hole and electron–hole exchange interactions in single InAs/GaAs quantum dots. *Phys. Rev. B* **79**, 125316 (2009).

73. E. Poem, S. Khatsevich, Y. Benny, I. Marderfeld, A. Badolato, P. M. Petroff, and D. Gershoni, Polarization memory in single quantum dots. *Solid State Commun.* **149**, 1493 (2009).

74. A. Hartmann, Y. Ducommun, E. Kapon, U. Hohenester, and E. Molinari, Few-particle effects in semiconductor quantum dots: Observation of multicharged excitons. *Phys. Rev. Lett.* **84**, 5648 (2000).

75. M. H. Baier, A. Malko, E. Pelucchi, D. Y. Oberli, and E. Kapon, Quantum-dot exciton dynamics probed by photon-correlation spectroscopy. *Phys. Rev. B* **73**, 205321 (2006).

13

Photoluminescence and Carrier Transport in Nanocrystalline TiO$_2$

Jeanne L. McHale
Washington State University

Fritz J. Knorr
Washington State University

13.1 Introduction

TiO$_2$ is a wide bandgap semiconductor that is earth-abundant, inexpensive, and of great interest for applications involving the capture and utilization of solar energy. The ability to easily acquire and exploit nanocrystalline TiO$_2$ preparations has created an explosion of interest in high surface area applications such as dye-sensitized solar energy conversion,[1] sensing,[2] hazardous waste remediation, and photocatalysis,[3] particularly photochemical fuel production[4] and photolysis of water.[5] However, the same high surface-area-to-volume ratio that makes TiO$_2$ nanoparticles desirable for applications results in enormous impact of surface properties that are challenging to control. The chemical properties of nanocrystalline surfaces and the presence of even small quantities of intrinsic defects and dopants can dictate the optical and electronic properties of powders and films of TiO$_2$ nanoparticles. Hence, details of sample preparation and handling can have strong effects on device performance.

As will be shown in this chapter, the localized states associated with surface defects are revealed by emission of light at sub-bandgap energies (visible and near-infrared [IR] wavelengths) upon excitation with photons at and above the energy of the bandgap, which is in the ultraviolet (UV) region. The intensity and wavelength range of this emission varies with phase, crystallinity, morphology, environment, and electrochemical potential. The shape of the emission spectrum can vary with incident power and may evolve in time, owing to the dynamics of electron trapping and the potential to saturate traps. When the emission results from recombination of oppositely charged free and trapped carriers, the energetic distribution of electron and hole traps can be determined from the wavelength range of the photoluminescence (PL) associated with these traps. The challenge then is to determine the molecular nature and the spatial distribution of these intra-bandgap states. Progress toward that goal will be presented later in the chapter. The trap-state distribution of nanocrystalline TiO$_2$ plays an important role in

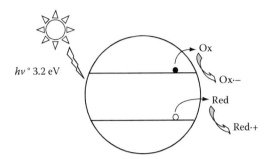

FIGURE 13.1 Photoredox reactions occurring at a TiO_2 (anatase) nanoparticle under bandgap illumination. Oxidizing agents (Ox) with redox potentials more positive than the conduction band edge, such as O_2, are reduced by conduction band electrons ("electron scavenging"), while reducing agents (Red) such as ethanol having redox potentials more negative than the valence band edge are oxidized by valence band holes ("hole scavenging").

carrier transport in solar photovoltaic cells based on this material. PL experiments are intimately tied to this transport for two reasons. Since radiative recombination rates depend on the spatial overlap of the wavefunctions of oppositely charged carriers, improved transport tends to diminish the PL intensity. In addition, carrier transport in nanocrystalline TiO_2 films is mediated by traps rather than macroscopic electric fields, and PL provides an experimental handle on the nature of these traps. In photovoltaic applications of nanocrystalline TiO_2, there is great interest in understanding how to optimize the shallow traps that favor transport while minimizing the density of deep traps, which impede transport and permit recombination into the surroundings. In photocatalysis applications, on the other hand, surface traps are important in order to permit charge transfer to the substrate.

Attempts to understand the photoluminescent properties of nanocrystalline TiO_2 from similar studies on the bulk crystal are hindered somewhat by the difference in preferred phase of the nanoparticulate (anatase) and bulk (rutile) materials. The predominant (100) and (110) facets of single crystal rutile have been the subject of numerous optical and scanning probe experiments. In a few studies, the optical properties of synthetically grown anatase single crystals have been revealed and these provide much insight into the nature of the nanocrystals. Yet commercial preparations of nanocrystalline TiO_2 are often mixtures of the two phases, and indeed these mixed-phase samples may be superior to either pure phase for photocatalysis, owing to spatial separation of electrons and holes. As will be shown, the distinct difference in the intensity and spectral distributions of PL in anatase and rutile makes PL a sensitive measure of interphasial carrier transport.

The cartoon in Figure 13.1 illustrates the key photocatalytic processes that make nanocrystalline TiO_2 so useful in applications. As shown in the figure, conduction band electrons and valence band holes produced by UV light are capable of respectively reducing or oxidizing molecules adsorbed at the surface. These interfacial electron transfer processes compete with radiative and nonradiative recombination of electron–hole pairs, hence PL and photocatalytic efficiency are linked. In addition, surface properties including defects may influence adsorbate binding and carrier trapping, which are important aspects of both catalytic activity and PL spectra. This chapter begins with a brief review of optical properties and surface defects of bulk TiO_2, then proceeds to the consideration of TiO_2 powders and supported catalysts before narrowing the focus to the rich and complex PL of TiO_2 nanoparticles, both in colloidal suspensions and as thin films. We will emphasize the connection between optical properties and applications.

13.2 Bulk TiO_2: Crystal Structure and Surface Properties

The thermodynamically favored phase of bulk TiO_2 is rutile, but the anatase and brookite crystalline modifications are also found in nature. We will focus our attention on the first two of these. A wealth of optical spectroscopy[6,7] and scanning probe microscopy[8,9] data exists for single crystal rutile and provides a basis for understanding the band structure and surface properties, respectively. There are relatively fewer studies of this nature for bulk anatase. Both are tetragonal crystal structures that can be built up from TiO_6 octahedra, which are packed more tightly in rutile, where each octahedron has 10 neighbors, than in anatase, where each octahedron has 8 neighbors. The bandgap of rutile is 3.0 eV compared to 3.2 eV for anatase, and the difference can be associated with a 0.2 eV lower conduction band edge for rutile.[10] These bandgaps correlate to near-UV wavelengths for the onset of absorption, 390 nm in the case of anatase and 410 nm for rutile. In both phases, bandgap emission is weak at room temperature as a result of the forbidden nature of the transition.[11,12] The large Stokes shift of the PL observed by illumination in the UV reveals that this emission results from intra-bandgap states.

13.2.1 Rutile

The near-IR PL of rutile peaks at about 810–840 nm (Figure 13.2) and was observed as early as 1968 by Addiss et al.,[6,7] who tentatively assigned it to a Ti^{3+} center. The transition has a low oscillator strength ($\sim10^{-4}$) and was observed at liquid nitrogen temperature. Addiss et al. also used thermoluminescence and thermally stimulated current measurements to locate a range of traps within 1 eV of the conduction band edge. More recently, emission of single crystal rutile has been widely studied by the Nakato group in the context of the photooxidation of water.[13,14] They assign the transition to the radiative recombination of conduction band electrons with holes trapped on the (100) and (110) crystal planes. Though hole traps of TiO_2 have frequently been assigned to surface hydroxyl groups,[15,16] Nakato et al. attribute the trapping sites to normally coordinated oxygen atoms, that is, oxygen atoms with three Ti neighbors. Since the emission wavelength translates to about 1.5 eV, the hole trap is located in the middle of the bandgap, but at an energy that is slightly different for the (100) and (110) surfaces, leading to PL peaking at 840 and 810 nm, respectively. The Nakato group exploited the PL of single crystal rutile to elucidate the mechanism for photooxidation of water, a process that competes with radiative recombination. The first step in this mechanism is the nucleophilic attack of a surface-trapped hole by a water molecule, followed by diffusion of the hole (h^+) to a bridging O atom at a step, kink, or terrace. As a result of the influence of these bridging O atoms, there is a strong decrease of the PL intensity on surface roughening in rutile when it is in contact with water.

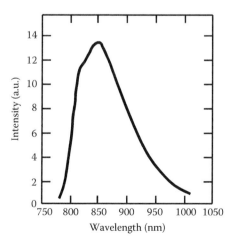

FIGURE 13.2 PL spectrum of rutile excited at 365 nm at liquid nitrogen temperature. (Reprinted with permission from Addiss, R.R., Jr., Ghosh, A.K., and Wakim, F.G., Thermally stimulated currents and luminescence in rutile, *Appl. Phys. Lett.*, 12, 397, 1968. Copyright 1968, American Institute of Physics.)

Surface oxygen vacancies of rutile are well known from scanning tunneling microscopy and are important in photocatalysis and sensing applications.[17,18] They are believed to be responsible for the intrinsic n-type nature of TiO_2.[19] As discussed for example by Kuznetsov et al.,[20] the removal of an oxygen atom and its associated −2 charge leads to defects such as Ti^{3+} centers and F^+ and F centers where one or two electrons, respectively, are trapped at vacancies. Other defects revealed by scanning probe microscopy include Ti^{4+} interstitials and structural defects such as step edges and grain boundaries.[21]

13.2.2 Anatase

Beside its slightly higher bandgap, the optical properties of anatase differ from those of rutile in that the latter has a much steeper band edge.[22,23] The presence of an Urbach tail in anatase crystals grown by chemical vapor transport was reported by Tang et al., who assigned it to self-trapped excitons rather than to defects or surface states on the basis of expected strong electron–phonon coupling in a polar semiconductor.[24] Similarly, broad PL emission peaking at 540 nm was observed at 10 K and also associated with self-trapped excitons. Tang et al. also observed anatase crystals to be dichroic based on the difference in the excitation spectrum of the PL for exciting light polarized parallel and perpendicular to the c-axis, where the former extends to higher energy than the latter.

Several recent papers have examined the time dependence of the PL decay of synthetically grown anatase crystals at cryogenic temperatures, where the nonradiative quenching mechanisms are suppressed and much brighter PL is observed. Goto et al. used 3.75 eV excitation to observe anatase PL at 10 K, which they determined to consist of fast (~180 ns) and slow (~10 µs) components.[25] Vacuum annealing resulted in an overall increase in the PL intensity and an increase in the contribution from the slow component. The peak in this emission was observed at 2.32 eV or about 540 nm. Watanabe et al.,[26] working at liquid He temperatures, also observed PL peaking in the vicinity of this wavelength and with two components to the decay, one of which was exponential and the other fit to a $1/t^n$ power law with $n \approx 1$. These decay components were attributed to direct and indirect pathways for the formation of the self-trapped exciton. Interestingly, the power-law component of the PL could be saturated at high power densities. Wakabayashi et al.[27] also looked at the dependence of the PL decay on the power density of the exciting radiation and found that higher powers resulted in faster decay.

We have recently observed PL emission from the (101) and (001) facets of natural anatase crystals, which are light yellow in color as a result of impurities.[28] We observe a very bright green emission from both surfaces, even at room temperature and in the presence of air. As will be shown in the following, the green PL arises from intra-bandgap states associated with oxygen vacancies. A variety of dopants have been shown to result in oxygen-vacancy-related color centers;[29] hence, the emission spectrum of doped anatase crystals has a similar shape to, but is much brighter than, that from the intrinsic defects present in more pure anatase crystals.

As compared to rutile, fewer single crystal scanning probe studies of anatase have been reported. The notable exception is the work of Diebold et al., who used STM to conclude that oxygen vacancies are less prevalent on the surface of anatase than on rutile.[30] Recent theoretical[31] and experimental[32] work both suggest that oxygen vacancies reside at subsurface layers of anatase (101), which nevertheless play an important role in the surface chemistry.

13.3 TiO_2 Powders and Supported TiO_2 for Photocatalysis

The importance of TiO_2 as a photocatalyst has been recognized since the discovery by Fujishima and Honda[5] that it could be used to split water. Photocatalytic applications of TiO_2 that predate the emergence of nanotechnology made use of TiO_2 powders for their high surface-area-to-volume ratio. Anpo et al. correlated the intensity of TiO_2 PL to photocatalytic activity of micron-sized TiO_2 particles nominally in the rutile phase.[33] They observed that the ability of contacting hydrocarbons to increase PL intensity was larger for hydrocarbons with lower ionization potential. This trend was explained using the "dead-layer model" of Ellis et al., in which more efficient separation of nascent electron–hole pairs within the space-charge layer of the semiconductor results in diminished PL.[34] Intrinsic n-type defects of TiO$_2$, associated with oxygen vacancies, result in upward band bending, which is enhanced by electron acceptors and diminished by electron donors. Hence, the ability of more easily ionized contacting molecules to flatten the bands and increase the PL. The importance of this dead-layer model in interpreting environmental effects on the PL intensity is greatly diminished in nanoparticles, which are not large enough to sustain potential drops of more than a few millielectronvolts.

In related work, Anpo et al. used the reaction of TiCl$_4$ with surface hydroxyl groups of porous glass to create highly dispersed TiO$_2$ for photocatalysis applications.[35] They observed blue PL at 77 K peaking at about 440 nm, which they assigned to radiative decay of a charge-transfer excited state O$^-$–Ti^{3+} of the (amorphous) anchored catalysis. In support of this assignment, vibrational structure with a spacing typical of the Ti–O stretch (720–820 cm^{-1}) was observed in the PL emission spectrum. On deposition of additional layers of TiO$_2$, weak x-ray diffraction features showed formation of anatase, and the observed PL intensity decreased while shifting to about 460 nm. The appearance of blue PL in samples of low crystallinity will be echoed in the next section where colloidal TiO$_2$ samples are considered. Anpo et al. also observed PL quenching on exposure to O$_2$ and N$_2$O through formation of O$_2^-$ and N$_2$O$^-$ adducts, respectively.

Fujihara et al.[36] measured the steady-state and time-resolved PL of aqueous suspensions of commercial photocatalyst powders in the rutile and anatase phases. In the case of rutile, the emission lifetime increases with increasing particle size, from about 0.03 ns for 64 nm particles to 1.6 ns for 3 μm particles. From this they concluded that nonradiative surface recombination competes with radiative electron–hole recombination. However, they observe rutile emission peaking at about 450 nm, rather than the near-IR PL typical of bulk and nanocrystalline rutile. This blue PL was assigned to an exciton localized on a TiO$_6$ octahedron and can also be observed in samples that are nominally in the anatase phase. The absence of near-IR PL in the data of Ref. [36] might be a consequence of low crystallinity of the commercial photocatalysts employed. The authors also report green PL from the commercial anatase powders, with emission maximum at about 500 nm. Though the sizes of the anatase particles were not reported, it was stated that no clear relation between particle size and emission lifetime could be found.

13.4 Nanoparticulate TiO$_2$

13.4.1 Suspensions and Films of Anatase Nanoparticles

The anatase phase of TiO$_2$ is thermodynamically more favorable than rutile for nano-sized particles,[37] though rutile nanoparticles can also be prepared in aqueous solution at very low pH.[38] The anatase phase is preferred for photovoltaic applications despite its larger bandgap as a result of better carrier transport, which in turn may result from better interparticle connections.[39] Both anatase and rutile nanoparticles display PL that is generally similar to that of their bulk phases. However, variations in the shape and intensity of the PL of nanocrystalline anatase for different samples are striking, and only recently has the basis for these variations begun to emerge. Even more interesting is the apparent communication between the PL from anatase and rutile in mixed-phase samples, and the complete absence of rutile PL in a common commercial preparation (Degussa P25), which contains 20%–25% rutile! As will be seen, the optical properties of rutile and anatase nanoparticles are similar but not identical to those of the bulk phases. The small Bohr radius of the electron in TiO$_2$ (on the order of 2 nm[40]) is less than typical nanoparticle dimensions, resulting in little importance of quantum confinement effects.

There are several reasons why there is much conflicting literature on the PL of nanocrystalline anatase. As an indirect semiconductor, the PL from anatase is inherently weak at room temperature. Bieber et al., for example, observed the temperature-dependent PL of P25 from 10 to 230 K, at which temperature the PL was barely observable.[41] Dispersions and films of TiO$_2$ nanoparticles are highly scattering and care must be taken to reduce artifacts from substrate fluorescence and stray excitation light that can masquerade as luminescence. As will be seen in the following sections, some of the luminescent traps can be saturated at high excitation power density, causing a change in the shape of the PL spectrum. At low power density and for low concentrations of defects, electron scavenging by oxygen effectively quenches the PL. As a surface phenomenon, the intensity and shape of the emission spectrum can be a strong function of the nanoparticle environment and surface ligands. Careful sample preparation and attention to experimental details for obtaining PL spectra are required in order to obtain reproducible data.

Though commercially available, anatase nanoparticles are readily prepared by a sol–gel route starting from precursors such as $TiCl_4$[42] or titanium alkoxides.[43] Titanium tetra-isopropoxide, $Ti(i\text{-}OPr)_4$, is a frequently used starting material for nanoparticle synthesis. The generic overall reaction is $Ti(OR)_4 + 2H_2O \rightarrow TiO_2 + 4ROH$. The sol–gel route can be used to obtain monolithic gels or colloidal suspensions, which are convenient for spectroscopy, but as prepared the particles are often amorphous. Colloidal TiO_2 nanoparticles sequestered in the water pools of a reverse micelle reveal a weak ($f \sim 0.002$) emission on UV excitation,[44] peaking at 445 nm with a shoulder at 540 nm. Similarly, TiO_2 gel monoliths were reported by Castellano et al.[45] to emit at about 400 nm with a lifetime shorter than 1 ns. It is worth noting for future reference that though the amorphous nanoparticles are no doubt rich in defects, they give little hint of the broad visible PL displayed by the nanocrystalline anatase, to be discussed in the following.

In a few studies, capping agents have been used to synthesize crystalline, rather than amorphous, colloidal particles of TiO_2, which show similar PL to that of amorphous nanoparticles. Pan et al.[46] used a two-phase thermal method to prepare colloidal TiO_2 particles capped with oleic acid and stearic acid and used Raman spectroscopy and x-ray diffraction to verify that the particles were crystalline anatase. Suspended in various solvents, these particles displayed band-edge PL at 370 nm, weaker emission at 465 nm assigned to a surface trapped state, and emission at 730 nm assigned to the recombination of mobile electrons with trapped holes (see Figure 13.3). Interestingly, the peak of the absorption spectrum of the suspended particles, but not that of the emission spectrum, is quite solvent dependent. A similar emission peaking in the blue was reported for crystalline colloidal anatase particles prepared from the reaction of benzyl alcohol and $TiCl_4$.[47]

Applications of nanocrystalline TiO_2 for solar photovoltaics require porous films of electrically connected particles. The porosity, typically on the order of 50%, allows for the hole transport medium of the

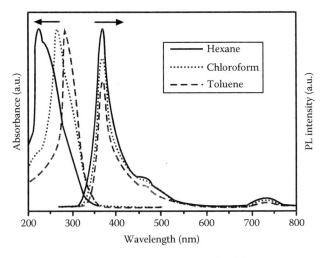

FIGURE 13.3 Absorption and emission spectra of crystalline colloidal anatase in various solvents. (Pan, D., Zhao, N., Wang, Q., Jiang, S., Ji, X., and An, L.: Facile synthesis of luminescent TiO_2 nanocrystals. *Adv. Mater.* 2005, 17, 1991–1995. Copyright Wiley-VCH Verlag GmbH & Co. KGaA. Reproduced with permission.)

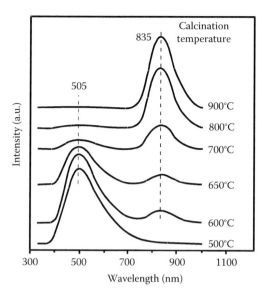

FIGURE 13.4 PL of nanocrystalline TiO$_2$ prepared by sol–gel route. The anatase PL at 505 nm decreases and the rutile PL at 835 nm increases with increasing calcination temperature. (Reprinted with permission from Shi, J., Chen, J., Feng, Z., Chen, T., Lian, Y., and Li, C., *J. Phys. Chem. C*, 111, 693–699, 2007. Copyright 2007 American Chemical Society.)

dye-sensitized solar cell (DSSC) to penetrate the film. In order to prepare porous films, surfactants are added to aqueous or ethanolic dispersions of TiO$_2$ nanoparticles prior to casting the film on conductive transparent glass. The films are then sintered at 450°C to decompose the insulating surfactants and to induce electrical contact between particles. The sintering process also improves the crystallinity and results in the desired anatase phase, while higher temperature (about 900°C) leads to conversion to the rutile phase.[48] The very different PL spectra of nanocrystalline anatase and rutile were exploited by Li et al.[49] to follow this phase transition, as shown in Figure 13.4, in order to investigate the phase dependence of photocatalytic properties of TiO$_2$ films. Note that the emission that peaks in the green and that seen in near-IR are from anatase and rutile phases, respectively, and are similar to the corresponding bulk phases.

Though huge variations are seen, prompting one author to state that "the emission of titania varies violently,"[50] many reports of the PL from nanocrystalline anatase show broad spectra that peak at a green wavelength. Zhang et al. synthesized TiO$_2$ nanoparticles in the ~7–32 nm size range and observed PL at 2.4–2.7 eV or about 520–460 nm.[51] In that work, the emission was observed even with visible light excitation, and the peak in the PL spectrum was found to shift to the blue when higher energy photons were used to excite the spectrum. Qian et al.[52] also observed changes in the peak wavelength and intensity of TiO$_2$ nanotubes (containing the anatase and brookite phases) as the excitation wavelength was varied. As λ_{exc} was increased from 360 to 480 nm, the PL intensity increased but maintained a peak at 546 nm. Further increase in excitation wavelength resulted in a monotonic increase in the wavelength of the maximum. The authors attribute this dependence to emission from two sets of traps located 2.3 and 1.8 eV above the valence band. The power dependence of the PL spectrum was also studied by Zhang et al.,[53] and sub-linear dependence of the PL intensity at excitation power above 0.3 W cm^{-2} was attributed to saturation of surface states. Similarly, Scepanovic et al.,[54] using anatase nanoparticles in the 30–70 nm size range, found the shape of the PL spectrum to vary with excitation wavelength. They fit the total emission to two Gaussians peaking at 2.16 eV (574 nm) and 2.43 eV (510 nm). Related to this work is that by Cavigli et al.,[55] which observed the time-resolved PL of 20–130 nm particles and found the early-time spectra to peak at shorter wavelengths.

Perhaps as disparate as the shape and intensity of the PL from nanocrystalline anatase is the variety of assignments for it in the literature. The tentative assignment of the green emission of bulk anatase to a self-trapped exciton has been echoed many times in the literature for nanocrystalline anatase. Others invoke oxygen vacancies, under-coordinated Ti^{3+}, and Ti^{3+} interstitials. In the next section, we describe our own studies of the effect of morphology, environment, and surface treatment on the PL of nanocrystalline anatase films and progress toward elucidating the molecular nature of photoluminescent traps.

13.4.2 Transport and Photoluminescence of Nanocrystalline TiO_2 Films

Intense interest in nanoporous TiO_2 films derives from their use for DSSCs, first reported by O'Regan and Grätzel in 1991.[56] Figure 13.5 shows the energy level scheme for this device, which is based on visible light harvesting by a dye (D) attached to the nanoparticle surface. Absorption of light results in a dye excited state with a redox potential $E°$ more negative than that of TiO_2 and consequently electron transfer (injection) into the conduction band is thermodynamically favored. The rate of injection, k_{inj}, occurs on a sub-picosecond timescale, fast enough to compete with radiative and nonradiative decay, $k_{rad} + k_{non}$, of the dye. For efficient collection of electrons at the conductive substrate, recombination of electrons with oxidized dye D^+ and redox mediator (k_{recomb}) should be slow and regeneration of the reduced form of the dye by reaction with iodide ion (k_{regen}) should be fast. V_{max} is the maximum achievable voltage and $\Delta G°$ is the driving force for injection. While V_{max} can be as large as the difference between the electrochemical potential of the conduction band edge and that of the redox mediator, recombination of injected electrons into the electrolyte and with oxidized dye reduces the value of the open-circuit voltage.

DSSCs show great promise as economical alternatives to silicon-based solar cells, but there continue to be hurdles to their full-scale manufacture. One of these is the reliance on volatile liquid phase electrolytes. Though solid-phase[57] and other less volatile[58] hole-transport media have also been reported, these

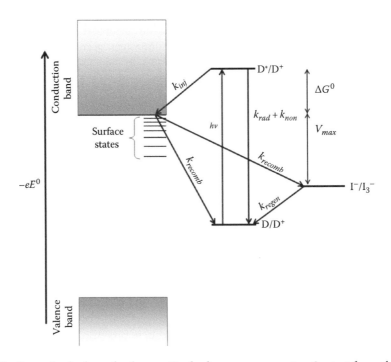

FIGURE 13.5 Energy level scheme for dye-sensitized solar energy conversion. See text for explanation of the symbols.

exhibit more recombination into the electrolyte and consequently lower energy conversion efficiency. Another problem is the inherently slow transport of electrons in the semiconductor film. Unlike bulk semiconductors, nanoparticle films of TiO$_2$ have no macroscopic field to drive the injected electrons toward the substrate, and inefficient trap-mediated diffusion results in a tortuous path from the front to the back of the film.[59-62] It is well known that deep traps, as dead ends for transport, are detrimental to performance, and oxygen vacancies are frequently suspected to be the culprits.[63] However, shallow traps lying within thermal energy of the conduction band edge participate in carrier transport, leading to several reports of procedures for enhancing these "good traps."[64,65] Similarly, empirical treatments such as exposure of TiO$_2$ films to aqueous TiCl$_4$, followed by sintering, are used to improve transport by a mechanism that is not yet completely understood.[66] Although particles in the anatase phase are preferred over rutile in the DSSC, commercial mixed-phase particles known as P25 are frequently employed. P25 is a mixture of about 75%–80% anatase and 25%–20% rutile prepared by flame pyrolysis of TiCl$_4$. As discussed later in the chapter, the microscopic structure of P25 has eluded characterization.

The shapes of anatase nanocrystals, like natural bulk crystals, reflect the relative thermodynamic stability of a truncated octahedral bipyramid consisting of eight {101} facets and two {001} facets.[30,67] Though the lower surface-energy {101} facets predominate in conventional preparations, nanocrystals with a high yield of {001} planes have recently been reported.[68,69] The ability to tailor the shape and nanocrystal morphology is significant in studies to understand surface traps and their role in transport and photocatalysis as the density of under-coordinated Ti ions varies with crystal facet. Indeed, all the Ti ions on the surface of the more reactive {001} facets are fivefold coordinated and thus likely sites for trapping electrons. The abundant {101} facets of conventional anatase nanocrystals, on the other hand, provide relatively efficient sites for photooxidative decomposition of organic compounds but show low activity for hydrogen evolution.[67] Recent theoretical and experimental work using TiO$_2$ "nanobelts" show that the (101) facets are more reactive to O$_2$, leading to reduced oxygen species.[70] Thus, the morphology of TiO$_2$ nanocrystals will be of interest in understanding how trap states influence the efficiency of photoredox reactions and solar energy conversion.

13.4.2.1 Influence of Surface Treatment by TiCl$_4$

Our initial forays into TiO$_2$ PL were motivated by a desire to understand the basis for improved performance of DSSCs containing TiCl$_4$-treated films. The treatment has been reported to improve DSSC characteristics by enhancing dye adsorption and electron injection,[66,71] by improving light harvesting through enhanced scattering,[72] and by increasing the electron lifetime and diffusion length.[73] We compared the PL of nanocrystalline films of 6 nm anatase or 25 nm P25 before and after TiCl$_4$ treatment, excited at 350 nm.[74] Surprisingly, we found the PL spectrum of untreated anatase to differ from that of P25, as shown in Figure 13.6, in that the broad PL of the former extends more to the red. In all films, we find a sharp emission in the blue at about 420 nm, which is unaffected by surface treatment. Since this emission wavelength is similar to that found in colloidal nanoparticles, which lack well-defined crystal facets, we speculate that this peak is associated with an excited state localized to an individual TiO$_6$ octahedron, perhaps a bulk self-trapped exciton. We focus henceforth on the longer wavelength emission, which will be shown to vary with surface treatment and nanoparticle environment. Surprisingly, the expected rutile PL at about 840 nm is absent from P25 despite its significant rutile content. However, on treatment with TiCl$_4$, the visible emission of P25 is quenched and emission characteristic of rutile appears. The same treatment applied to the anatase film only reduces the PL, particularly in the green, leaving a broad luminescent plateau extending to red wavelengths, and no PL characteristic of rutile appears.

It is supposed that TiCl$_4$ treatment improves carrier transport in nanoporous TiO$_2$ films by depositing TiO$_2$ in the regions between nanoparticles ("necking"), permitting better electrical connections. As previously mentioned, radiative recombination of electron–hole pairs requires them to overlap in space and time, such that better carrier transport would lead to diminished PL intensity. Indeed, we find that unsintered films of anatase nanoparticles exhibit much higher PL than sintered ones and

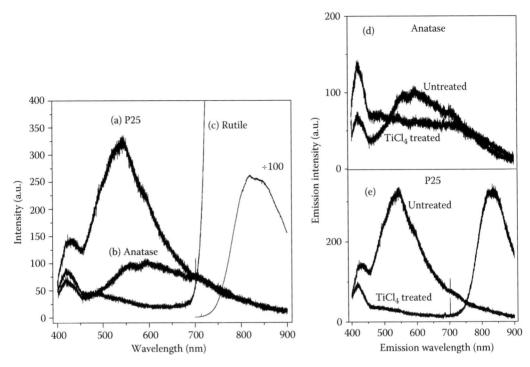

FIGURE 13.6 PL from nanocrystalline films of (a) P25, (b) anatase, and (c) rutile (left), and effect of $TiCl_4$ treatment on PL from (d) anatase and (e) P25 (right). (Reprinted with permission from Knorr, F.J., Zhang, D., and McHale, J.L., *Langmuir*, 23, 8686–8690, 2007. Copyright 2007 American Chemical Society.)

perform poorly in the DSSC owing to poor carrier transport.[75,76] Thus, better particle necking and improved interparticle carrier transport after $TiCl_4$ treatment would result in diminished PL. Such an explanation, however, cannot account for the different effects of $TiCl_4$ treatment on anatase and P25. Another explanation for improved transport in $TiCl_4$-treated TiO_2 is that it quenches surface traps, which impede transport and promote recombination. We therefore considered the existence of two distributions of luminescent traps in untreated anatase, only one of which is present in P25. This model is consistent with the results of electron paramagnetic resonance (EPR) studies of Hurum et al.,[77] which led to the same conclusion. The same group has also used EPR to conclude that there are unique traps states of P25 at the rutile–anatase interface.[78] In our model, pure anatase possesses a set of traps that gives rise to broad PL extending into the red, which is unaffected by $TiCl_4$, and a distribution that emits with a peak wavelength in the green, which is quenched or passivated by the treatment. The first set (which we call "red traps" based on the peak wavelength) is apparently passivated in P25 as is the near-IR emission of rutile, leaving only the "green traps." The ability of the treatment to form rutile when applied to P25 appears to be a natural consequence of seed particles of this phase. To interpret the different effects of $TiCl_4$ treatment on the two types of films, one must consider the microscopic structure of the mixed-phase P25 nanoparticles. This question also relates to the well-known superior photocatalytic ability of mixed-phase TiO_2, which probably results from separation of electrons and holes in the different phases.[77,79] There is also evidence that interphasial carrier transport in mixed-phase TiO_2 films is beneficial to DSSC performance.[80]

The nanometer scale structure of P25 has been addressed in the literature by electron microscopy and is not without controversy.[81–83] More recent experiments of Ohno et al.[82] as well as theoretical treatments[84] suggest that nanoparticles of the individual phases are intimately connected. Our own results also tend to argue against the existence of *separate* nanoparticles of anatase and rutile, since this would have led to PL spectra characteristic of both pure phases. Theory[84] suggests that the formation of rutile–anatase interfaces is driven by a tendency for Ti to complete its coordination sphere. All of the Ti ions on the (001) planes of anatase nanocrystals are five-coordinate, while on the more stable (101) facets that make up the majority of the surface, half are five-coordinate and half are six-coordinate. Could the red-emitting defects reside on these less stable (001) planes in anatase? The formation of interfaces between these surfaces and rutile (100) and (110) facets, where the luminescent rutile hole traps reside, might be the reason for the absence of both the red anatase PL and the near-IR rutile PL in P25. For further insight, we turn to the discussion of the influence of nanoparticle environment on the shape and intensity of the PL spectrum in the next section.

13.4.2.2 Influence of Contacting Media on Photoluminescence of Anatase and P25 Nanoparticle Films

There is great interest in optimizing the practical advantages of the DSSC by replacing the commonly used liquid solvent electrolytes by less volatile (gel, ionic liquid, or solid) hole-transport media.[85–87] Liquid electrolytes, however, permit faster diffusion of the redox mediator. Thus, the desirable features of low volatility and low viscosity are at cross purposes to one another. In addition, though a variety of solvents are capable of dissolving the common I^-/I_3^- redox mediator used in the DSSC, the nitrile solvents acetonitrile (ACN or CH_3CN) and methoxypropionitrile (MPN or $CH_3OCH_2CH_2CN$) are superior to other electrolyte solvents that have been tried.[88,89] In addition, it is desirable to be able to skip the costly sintering step that is required for efficient carrier transport in the TiO$_2$ film.[90] We therefore sought to understand the influence of contacting solvent and TiO$_2$ film heat treatment on the performance of DSSCs and the PL of (unsensitized) films.

Figure 13.7 shows the influence of electrolyte solvent on the wavelength-dependent incident photon-to-current conversion efficiencies, IPCE(λ), of a DSSC sensitized with the ruthenium dye N3, Ru(4,4′-dicarboxylic acid-2,2′-bipyridine)$_2$(NCS)$_2$. IPCE(λ) represents the number of electrons collected in the external circuit per incident photon of wavelength λ and is the product of three quantum efficiencies: (1) the light-harvesting efficiency, determined by the extinction of the sensitized film; (2) the efficiency of electron injection by excited-state dye, which is a function of the competition between the injection step and radiative and nonradiative decay of the excited state; and (3) the current collection efficiency, which is larger for faster transport and slower recombination. Note the much lower IPCE for the DSSC containing an unsintered anatase film, for which the collection efficiency is small owing to poor inter-particle carrier transport. We have found that for N3 on TiO$_2$ the light harvesting and electron injection efficiencies are not very dependent on contacting solvent, thus the data of Figure 13.7 suggest that the solvent influences transport and recombination. The diminished relative values of IPCE at more red wavelengths, though partly the result of solvent shifts in the driving force for electron injection, strongly suggest shorter electron diffusion lengths in the presence of the non-nitrile solvents. (See, e.g., the data of Jennings and Wang,[91] who used added Li^+ to lower the conduction band edge and compared their experimental IPCE data to simulated data calculated as a function of diffusion length.) To understand this conclusion, it should be pointed out that the solar cells used to collect the data of Figure 13.7 were illuminated through the TiO$_2$-coated conductive substrate. Since red light is absorbed less than shorter wavelength light (a consequence of the absorption spectrum of the N3 dye), electrons resulting from excitation at red wavelengths have farther to travel to reach the conductive substrate and are more susceptible to recombination into the electrolyte. The results suggest that contacting solvents influence carrier diffusion and recombination, perhaps through contribution of a solvent reorganizational energy barrier, which impedes interparticle electron transfer. Alternatively, solvents may interact directly with transport limiting traps on TiO$_2$, as several studies have suggested.[92,93] Another possibility is that the polarity of the nitrile group of "good" solvents presents a surface dipole that opposes electron recombination.[94]

As part of our ongoing investigation into the basis for this solvent dependence, we sought to investigate the environment dependence of the PL of nanocrystalline anatase and P25 films. Figure 13.8

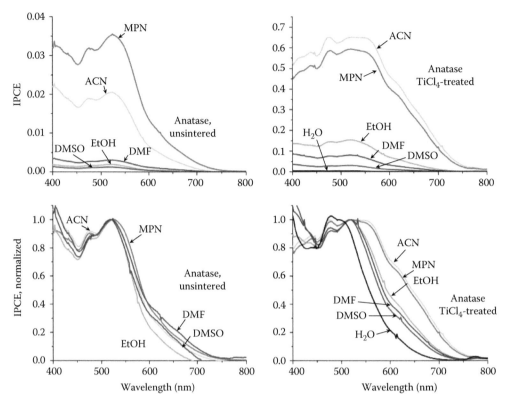

FIGURE 13.7 IPCE(λ) (top) and normalized IPCE(λ) (bottom) for DSSCs containing an N3-sensitized unsintered anatase film (left) and a TiCl$_4$-treated sintered anatase film (right) in the presence of various electrolyte solvents.

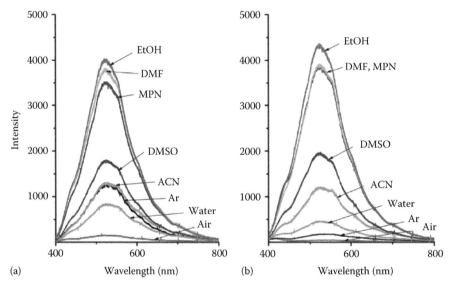

FIGURE 13.8 PL spectra, excited at 350 nm, of (a) an unsintered and (b) a sintered film of P25 in contact with ethanol (EtOH), dimethylformamide (DMF), methoxypropionitrile (MPN), dimethylsulfoxide (DMSO), acetonitrile (ACN), argon, water, and air. (Reprinted with permission from Knorr, F.J., Mercado, C.C., and McHale, J.L., *J. Phys. Chem. C*, 112, 12786–12794, 2008. Copyright 2008 American Chemical Society.)

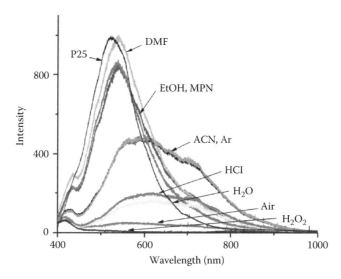

FIGURE 13.9 PL emission excited at 350 nm of a sintered anatase film in contact with different solvents, compared to a scaled PL spectrum of P25 in contact with argon. (Reprinted with permission from Knorr, F.J., Mercado, C.C., and McHale, J.L., *J. Phys. Chem. C*, 112, 12786–12794, 2008. Copyright 2008 American Chemical Society.)

shows the results of measuring the PL emission spectra of sintered and unsintered P25 films in contact with various media.[95] In the presence of air or water, the PL is greatly diminished compared to that in an argon atmosphere, owing to recombination, and overall greater PL is observed from the unsintered film in accord with inefficient interparticle carrier transport. In the presence of organic solvents ethanol (EtOH), dimethylformamide (DMF), and MPN, the PL intensity is greater and not strongly dependent on sintering. Among the organic solvents, ACN results in lower PL, an observation that might correlate to improved transport were it not for the more intense PL observed from the film in contact with the equally good DSSC solvent MPN. Also puzzling is the solvent dependence of the difference in PL from sintered versus unsintered films. Sintering in air at 450°C was found to decrease the intensity of the PL from a film in contact with argon, air or water, while little difference in intensity is observed for sintered and unsintered films in contact with organic solvents. It is reasonable to suspect that sintering reduces the number of oxygen vacancies, which are believed to give rise to the visible PL (indeed, vacuum annealing causes the PL intensity to increase[28,75]), but the above data suggest additional factors at work.

Further insight was provided by examining the PL spectra of a nanocrystalline anatase film in contact with a series of solvents, as shown in Figure 13.9. Surprisingly, in the presence of the same three solvents for which the P25 PL is strongest, EtOH, DMF, and MPN, the shape of the anatase emission spectrum is similar to that of P25. EtOH and DMF are more easily oxidized solvents, hence we considered the possibility that their influence on the anatase PL spectrum, and by inference that of MPN as well, could be the result of their ability to scavenge holes from the valence band. This same behavior would explain the ability of these solvents to increase the PL from P25, owing to a decrease in the competing nonradiative electron–hole recombination. In addition, oxidized alcohols have the capacity to undergo further oxidation by efficient injection of electrons into the conduction band (leading to "current doubling" on TiO$_2$ photoelectrodes[79,96]), which would also lead to increased PL intensity. On the other hand, water has been shown to deplete conduction band electrons in TiO$_2$,[97,98] resulting in diminished PL in the presence of water for both P25 and anatase films.

The above results lead us to propose the model shown in Figure 13.10. We conclude that the PL from nanocrystalline anatase is a superposition of two types of radiative recombination: that of mobile

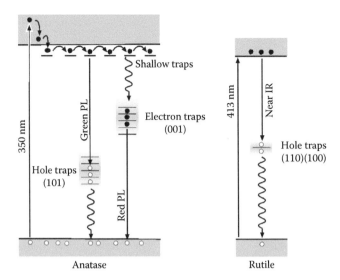

FIGURE 13.10 Model for the trap state distributions in anatase and rutile as determined from the breadth of the PL emission spectra. Wavy lines indicate nonradiative decay. (Adapted from Knorr, F.J. et al., *J. Phys. Chem. C*, 112, 12786, 2008.)

electrons, from the conduction band and shallow traps, with trapped holes (Type 1 PL) and that of trapped electrons with valence band holes (Type 2 PL):

$$\text{Type } 1 : e_{CB}^{-} + h_{tr}^{+} \rightarrow h\nu_{green}$$

$$\text{Type } 2 : h_{VB}^{+} + e_{tr}^{-} \rightarrow h\nu_{red}$$

With this definition the near-IR PL of rutile is Type 1. The shaded areas of Figure 13.10 represent a range of trap state energies estimated from the shape of the corresponding emission spectra. The ability of oxygen to act as an electron scavenger is well known,[99] explaining its ability to quench both types of anatase PL. Hole scavengers such as EtOH, on the other hand, shut off the Type 2 PL and result in P25-like PL from anatase.

The model depicted in Figure 13.10 may account for some aspects of the time- and power-dependent PL of nanocrystalline anatase discussed in Section 13.4.1. Following excitation at energies above the bandgap, hot electrons in the conduction band rapidly thermalize to the bottom of the band and are subsequently trapped on a sub-picosecond[100,101] timescale. Hence, the early time PL emission is blue-shifted, as reported in Ref. [55], owing to a larger relative contribution of Type 1 over Type 2 PL. We have also observed that increasing the laser power results in a blue shift in the PL spectrum of an anatase film, a result that could be accounted for by increasing relative contribution from the Type 1 PL as the traps responsible for Type 2 PL become saturated. The questions remain: What is the molecular nature of the two types of traps and why do only the green-emitting traps respond to TiCl$_4$ treatment? Further, why are the red traps absent in P25, and is their absence a consequence of the nanoparticle structure or carrier transport? We have speculated that the red and green traps are sequestered on different exposed facets of anatase nanocrystals, a hypothesis that derives some support from reports of experiments in which the products of photoreduction and photooxidation are imaged on the (101) and (001) facets, respectively.[102] However, it is not clear whether the photoredox reactions of Ref. [102] involve trapped carriers, and if so, one would have to conclude that electron traps located on the stable (101) anatase planes are somehow absent in P25. Another possibility is that hole traps, which respond to

TiCl₄ treatment are located on the surface, and the red-emitting electron traps are found in the bulk. The diminished red emission from bulk anatase, discussed in Section 13.2.2, argues against this hypothesis. It is also of interest to understand the basis for the breadth of the PL spectrum, which could result from phonons and from a heterogeneous distribution of trap energies. For example, density functional theory calculations suggest a variety of localized Ti^{3+} centers (five-coordinate, six-coordinate, interstitial, etc.) in bulk anatase, as well as delocalized states where the additional electron is shared by several titanium ions.[103] Finally, the role of grain boundaries as likely electron trap sites is important to consider.

13.4.2.3 Spectroelectrochemical Luminescence

We have recently used spectroelectrochemistry to observe the PL from an anatase film as a function of applied bias.[104] In aqueous solution, the electrochemical potential of the conduction band edge of TiO₂ is a well-known function of pH: $E_{cb}^{\circ} = -0.16 - 0.06\,\text{pH}$ versus NHE (normal hydrogen electrode).[105] Figure 13.11 shows the changes in the PL as a function of applied bias for an anatase film in contact with a pH 6 aqueous solution, along with a graph of the intensity at 600 nm as a function of applied voltage for the same film at three different pH values. In each case, the red PL shows an initial increase in intensity as the electrode potential is made more negative, followed by conversion to a "P25-like" emission spectrum with further increase in bias. As shown in Figure 13.11, the abrupt increase in PL intensity at 600 nm takes place in the vicinity of the TiO₂ conduction band edge. It is interesting that the entire PL spectrum at first increases in intensity with no change in shape. Direct electron injection into surface traps might have led to PL spectra, which are shifted to increasingly higher energies with increasing negative bias. Instead, the data suggest that the traps associated with Type 2 PL are populated via the conduction band, as depicted in Figure 13.10, such that the entire breadth of the spectrum is observed at all potentials short of those that cause the green Type 2 PL to begin to dominate. The onset of the P25-like emission spectrum occurs at a potential that is a few tenths of a volt more negative than the conduction band and is accompanied by a decrease in the PL at more red wavelengths.

In order to explain these results, we introduce some related work directed at uncovering the physical basis for the two trap distributions. Reduced Ti species (Ti^{3+} and Ti^{2+}) can be associated with localized defects such as oxygen vacancies and under-coordinated Ti. We have recently observed very broad red PL from TiO₂ nanosheets, which are rich in (001) texture.[106] The shape of the nanosheet PL is the same as that of a conventional anatase nanoparticle film in a non-hole scavenging environment. The (001) surface of anatase contains an abundance of fivefold coordinated Ti, which is a likely site for electron trapping. It is also less stable then the (101) plane and more likely to be absent in the mixed-phase P25

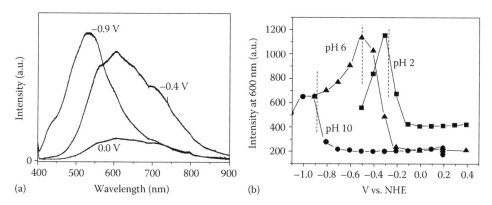

FIGURE 13.11 PL of nanocrystalline anatase film in aqueous solution as a function of applied potential, excited at 350 nm. (a) PL spectrum at pH 6 at 0.0, −0.4, and −0.9 V vs. NHE. (b) Emission intensity at 600 nm as a function of external applied potential at pHs 2, 6, and 10, in 0.1 M NaClO₄ supporting electrolyte. The vertical dashed lines represent the redox potential of the anatase conduction band edge.

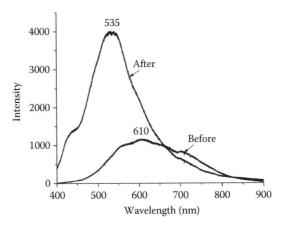

FIGURE 13.12 PL spectrum of a nanocrystalline anatase film before and after vacuum annealing, excited at 350 nm. (Adapted from McHale, J.L. et al., *Mater. Res. Soc. Proc.*, 1268-EE03-08, 2010.)

particles where it may be covered by the rutile phase. The resistance of the red traps to $TiCl_4$ treatment is easily understood if the (occupied) red-emitting traps are five-coordinate Ti_{5c}^{3+} from the (001) planes that recombine with valence band holes to form Ti_{5c}^{4+}. While $TiCl_4$ treatment may passivate defects such as oxygen vacancies by depositing fresh layers of TiO_2 on the nanocrystals, all of the Ti's on unreconstructed (001) planes, whether freshly deposited or not, are five-coordinate, and the surface treatment thus has no effect.

In addition, we have strong evidence that the green Type 1 PL arises from deep electron traps (in other words, "hole traps") associated with oxygen vacancies, for example, conversion of anatase PL to "P25-like" after vacuum annealing,[28] as shown in Figure 13.12. These oxygen vacancies are known to adsorb molecular oxygen,[107] which then scavenges electrons and quenches both types of PL. Thus, the ability of $TiCl_4$ to heal these defects by deposition of fresh TiO_2, at least initially, is also understood. However, it is necessary to resolve the following apparent paradox. Oxygen vacancies are widely held to result in the n-type behavior of TiO_2 and thus should lead to donor states within thermal energy of the conduction band edge, which are beneficial to transport. On the other hand, oxygen vacancies are detrimental to the DSSC performance and would appear to introduce deep trap levels.[108] However, it should be considered that each oxygen atom removed from the lattice leaves two electrons that must be accommodated in new states. Depending on the level of theory, Di Valentin et al.[103] were able to show that one of these states may be a relatively deep trap level corresponding to a localized Ti^{3+} ion, while the other is a delocalized state that accounts for the enhanced conductivity of n-type TiO_2. Thus, oxygen vacancies may be associated with both shallow and deep traps.

We have also recently observed strong PL from compact nanocrystalline anatase films, which is only weakly quenched by O_2, a result attributed to the much lower exposed surface area compared to nanoporous films.[109] These compact films, used as gas sensors that respond through changes in resistance, have a much greater number of grain boundaries than the nanoporous films used in the DSSC. These grain boundaries serve as traps, which inhibit transport and result in brighter PL. We also note that the appearance of green P25-like PL, in all cases we have examined, is always associated with a greater concentration of conduction band electrons, whether through applied bias, increasing intensity of incident light, the presence of n-type donors, or the presence of hole scavengers. This is in accord with our assignment of green PL to radiative recombination of mobile electrons with trapped holes. Summarizing the evidence to date, it appears that the electron traps responsible for the broad PL of anatase that peaks in the red are associated with five-coordinate Ti on (001) planes, while the green PL arises from hole traps on (101) planes and is associated with oxygen vacancies. Further support for associating the green traps with the (101) planes is presented in the next section.

13.4.2.4 Photoluminescence of TiO$_2$ Nanotubes

TiO$_2$ nanotubes prepared from anodization of Ti metal foil have proven to have superior transport properties when used in the DSSC.[110–113] Consisting of well-aligned nanotubes perpendicular to the conductive substrate, these films permit better charge collection as a result of reduced recombination rates, longer diffusion lengths and longer lifetimes than those in nanoparticle films.[113] While improved charge collection in nanotube films probably results from fewer grain boundaries and a direct path to the current collecting substrate, reduced density of surface traps has also been invoked.[114] Significantly, the walls of TiO$_2$ nanotubes are mostly (101) planes.[115]

We have prepared TiO$_2$ nanotubes from anodization of Ti foil in ethylene glycol solution containing 0.5% NH$_4$F and 2% water, resulting in nanotubes with inner and outer diameters of about 100 and 190 nm, as shown in the inset of Figure 13.13.[116] Nanotube samples prepared by anodization are initially amorphous but are converted to anatase by sintering at 450°C. The PL spectrum, as shown in Figure 13.13, peaks in the green at about 520–550 nm and is thus dominated by the Type 1 PL, that we have associated with hole traps. However, the observed PL is orders of magnitude less intense than that of typical nanocrystalline TiO$_2$ films. When initially prepared nanotubes were subjected to vacuum annealing, the PL intensity increased by a factor of approximately 50, with a slight change in shape arising from diminished PL in the red. The reasons for weak nanotube PL may be somewhat interrelated in that fewer defects in the nanotubes may contribute to improved transport, in addition to the advantages provided by the morphology of the nanotubes. Efficient transport inhibits the spatial overlap of electron and hole wavefunctions that is required for radiative recombination to take place. Since the weak nanotube PL is not strongly quenched by O$_2$, which we propose must adsorb at oxygen vacancies in order to scavenge electrons, the evidence is that the nanotubes have fewer transport-limiting defects than nanoparticles. The trap density of states of TiO$_2$ nanotubes has been reported to be shifted to lower energies compared to that of anatase nanoparticles,[113] consistent with the greater contribution of hole traps over electron traps in the nanotubes.

As reported in Ref. [116], the solvent dependence of nanotube PL is much less pronounced than that of nanoparticles. Though part of the reason for this could derive from poor penetration of the solvent into the film, careful experiments suggest that this is not the deciding factor. Rather, the influence of solvent on nanotube PL may be diminished as a result of faster transport. In nanoparticle films used for IPCE and PL measurements, it appears that interfacial solvents have varying abilities

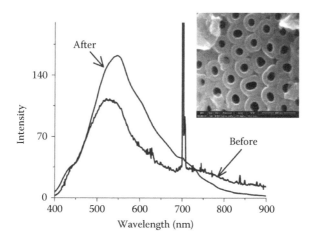

FIGURE 13.13 PL of TiO$_2$ nanotubes before and after vacuum annealing, excited at 350 nm, with SEM image as an inset, adapted from Ref. [116]. The data for the sample after annealing was divided by 50 to bring it on scale with the unannealed sample. The sharp peak at 700 nm is an artifact from the excitation light. The peak in the emission shifts from 530 to 546 nm after vacuum annealing.

to inhibit interparticle transport, resulting in strong dependence of both current collection and TiO_2 PL intensity on the identity of the solvent. Interparticle electron transport is much less important in the nanotube films, a conclusion that may bode well for the use of non-nitrile electrolyte solvents in nanotube-based DSSCs.

13.4.2.5 Interphasial Carrier Transport in Mixed-Phase TiO_2 Films

As shown in Ref. [95], the PL of a film of nanocrystalline rutile is only weakly dependent on contacting solvent and is only slightly quenched by air despite the fact that electron scavenging by O_2 is thermodynamically favored. Organic solvents in contact with the nanocrystalline rutile film tend to lead to slightly stronger PL compared to an argon environment, more so for Lewis bases (DMF, ACN, and MPN) than Lewis acids (EtOH, H_2O, and aqueous HCl). This is in contrast to results with anatase and P25 films where the PL is a strong function of nanoparticle environment. We speculate that this is a consequence of weaker solvent adsorption in the case of rutile. However, in mixed-phase films prepared from thermally annealing anatase at temperatures and for times sufficient to promote the incomplete phase transition to rutile, the PL from both phases becomes strongly dependent on the film environment.[95] As shown in Figure 13.14, films containing a small amount of rutile (about 5%, based on x-ray diffraction) in anatase are found to give PL from both phases, in strong contrast to the behavior of P25 where the rutile phase is not evident in the PL spectrum. Interestingly, however, the anatase phase PL in Figure 13.14 is more similar to that of P25, showing a peak at about 530 nm. Excitation at 350 nm (~3.5 eV) is sufficient to create electron–hole pairs in both phases, but the data show increasing near-IR PL from rutile in environments that tend to result in weaker anatase PL. Figure 13.14 presents strong evidence for the role of transport in determining the intensity of PL from TiO_2 nanoparticles. Since the conduction band edge of rutile is 0.2 eV lower than that of anatase, as depicted in Figure 13.10, rutile particles in a matrix of anatase can act as sinks for conduction band electrons. Note however that the total PL intensity in Figure 13.14 is not conserved, owing to variations in carrier scavenging superimposed on the trends in carrier transport. For example, the hole scavengers EtOH, DMF, and MPN

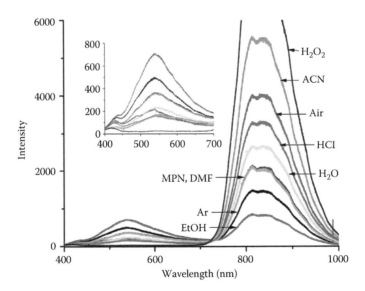

FIGURE 13.14 PL of a thermally annealed anatase film containing about 5% rutile content in contact with different environments. The inset shows the anatase emission on an expanded scale, in the presence of (in the order of decreasing intensity) EtOH, Ar, MPN ~ DMF, H_2O, air, HCl ~ ACN, and H_2O_2. (Reprinted with permission from Knorr, F.J., Mercado, C.C., and McHale, J.L., *J. Phys. Chem. C*, 112, 12786–12794, 2008. Copyright 2008 American Chemical Society.)

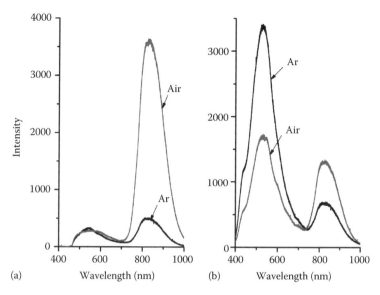

FIGURE 13.15 PL spectra of a thermally annealed film consisting of nanocrystalline anatase with approximately 5% rutile, excited at (a) 413 nm and (b) 350 nm.

(which have little effect on the PL intensity of a rutile film) result in weaker total PL than ACN, which is relatively inert to photoredox reactions on TiO_2. While EtOH, DMF, and MPN result in PL of similar intensity from an anatase film, the mixed-phase film shows higher visible PL in the presence of EtOH with the trend EtOH > MPN ≈ DMF. While the molecular and nanoscale principles that underlie the trends in Figure 13.14 remain to be uncovered, the results clearly show that the environment of a mixed-phase film can strongly influence carrier transport and recombination.

Another example of interphasial carrier transport is provided by the data in Figure 13.15, which shows the PL from the same mixed-phase film as Figure 13.14, but excited at both 413 nm (3.0 eV) and 350 nm (3.5 eV). While the latter can produce electron–hole pairs in both phases, excitation at 413 nm favors the excitation of the rutile phase. Nevertheless, 413 nm excitation clearly results in unexpected emission from the anatase phase. More puzzling is the effect of oxygen on the PL. With 413 nm excitation, the rutile component of the PL is much *larger* in air than in argon, while the same PL intensity from anatase is found in both argon and air environments. This is in contrast to the PL excited at 350 nm, where the anatase component shows the usual decreased intensity in air, which translates into greater rutile PL.

The results of Figures 13.14 and 13.15 are inconsistent with a model in which the rutile phase forms as a shell around the preexisting anatase nanoparticles, since this would have resulted in diminished dependence of the PL on the environment in accord with our results for pure rutile nanoparticles.[95] However, a model in which the rutile phase forms between or inside the anatase nanoparticles may explain the results provided that the thermal annealing results in anatase particle sizes large enough to permit a depletion layer and associated upward band bending. Consider the data in Figure 13.15 obtained at near-UV and UV excitation. 413 nm excitation is capable of exciting trap states in the Urbach tail of the anatase phase but creates conduction band electrons only in the rutile phase. The weak anatase PL at 413 nm, independent of oxygen or argon, could result from radiative recombination of holes in the anatase phase with mobile electrons from undepleted donors adjacent to the rutile phase. Oxygen as an electron acceptor would enhance the band bending in the outer anatase phase, making radiative recombination of mobile electrons with trapped holes in the rutile phase more favorable, hence an increase in near-IR PL in the presence of oxygen at 413 nm excitation. At 350 nm excitation, sufficient to excite mobile electron–hole pairs in both phases, oxygen scavenges conduction band electrons from anatase, resulting in transfer of holes to the inner rutile phase where they are trapped and then recombine radiatively with the conduction band electrons created there. Thus, oxygen results in enhanced PL from rutile in this mixed-phase film even though it has little effect when pure rutile nanoparticles are examined. We observed a similar effect in the PL of TiO_2 nanotubes having a small amount of rutile phase at the bottom of the nanotube layer.[116] Clearly PL experiments can reveal valuable information about the internal structure and interphasial carrier transport of mixed-phase TiO_2.

13.5 Conclusions

The weak luminescence from nanocrystalline TiO_2 is strongly linked to the transport and interfacial carrier transfer that makes these materials of interest in solar energy applications. The high surface area per unit volume and defects associated with dangling bonds emphasize the importance of surface defects in these materials, and PL spectroscopy provides a convenient and sensitive approach to studying these defects. This review of the complex PL behavior of TiO_2 nanoparticles is intended to reveal the rich dependence of PL on nanoparticle phase, environment, and morphology as well as experimental variables such as excitation power, temperature, and wavelength. PL spectroscopy will continue to play an important role in understanding and optimizing the performance of nanocrystalline TiO_2 for a myriad of applications.

Acknowledgments

The contributions of graduate students Jonathan Downing, Christopher Rich, Candy Mercado, Kritsa Chindanon, and postdoctoral researcher Dongshe Zhang, have been essential to the work from the authors' lab and are gratefully acknowledged. The support of the Department of Chemistry and the College of Science at Washington State University is greatly appreciated.

References

1. Hagfeldt, A.; Grätzel, M. Molecular photovoltaics, *Chem. Rev.* 2000, *33*, 269–277.
2. Fergus, J. W. Doping and defect association in oxides for use as oxygen sensors, *J. Mater. Sci.* 2003, *38*, 4259–4270.
3. Linsebigler, A. L.; Lu, G.; Yates, J. T., Jr. Photocatalysis on TiO_2 surfaces: Principles, mechanisms and selected results, *Chem. Rev.* 1995, *95*, 735–758.
4. Roy, S. C.; Varghese, O. K.; Paulose, M.; Grimes, C. A. Toward solar fuels: Photocatalytic conversion of carbon dioxide to hydrocarbons, *ACS Nano* 2010, *4*, 1259–1273.
5. Fujishima, A.; Honda, K. Electrochemical photolysis of water at a semiconducting electrode, *Nature* 1972, *238*, 37–38.
6. Addiss, R. R., Jr.; Ghosh, A. K.; Wakim, F. G. Thermally stimulated currents and luminescence in rutile, *Appl. Phys. Lett.* 1968, *12*, 397–400.
7. Ghosh, A. K.; Wakim, F. G.; Addiss, R. R., Jr. Photoelectronic processes in rutile, *Phys. Rev.* 1969, *184*, 979–988.
8. Li, M.; Hebenstreit, W.; Gross, L.; Diebold, U.; Henderson, M. A.; Jennison, D. R.; Schultz, P. A.; Sears, M. P. Oxygen-induced restructuring of the TiO_2(110) surface: A comprehensive study, *Surf. Sci.* 1999, *437*, 173–190.
9. Gong, X.-Q.; Selloni, A.; Batzill, M.; Diebold, U. Steps on anatase TiO_2(101), *Nat. Mater.* 2005, *5*, 665–670.
10. Barbe, C. J.; Arendse, F.; Camte, P.; Jirousek, M.; Lenzmann, F.; Shklover, V.; Grätzel, M. Nanocrystalline titanium oxide electrodes for photovoltaic applications, *J. Am. Ceram. Soc.* 1997, *80*, 3157–3171.
11. Daude, N.; Gout, C.; Jouanin, C. Electronic band structure of titanium dioxide, *Phys. Rev. B* 1978, *15*, 3229–3235.

12. Pascual, J.; Camassel, J.; Mathieu, H. Fine structure in the intrinsic absorption of titanium dioxide, *Phys Rev. B* 1978, *18*, 5606–5614.

13. Nakamura, R.; Nakato, Y. Primary intermediates of oxygen photoevolution reaction on TiO$_2$ (rutile) particles, revealed by in situ FTIR absorption and photoluminescence measurements, *J. Am. Chem. Soc.* 2004, *126*, 1290–1298.

14. Imanishi, A.; Okamura, T.; Ohashi. N.; Nakamura, R.; Nakato, T. Mechanism of water photooxidation in reaction on anatomically flat TiO$_2$ (rutile) (110) and (100) surfaces: Dependence of solution pH, *J. Am. Chem. Soc.* 2007, *129*, 11569–11578.

15. Bahnemann, D. W.; Henglein, A.; Spanhel, L. Detection of intermediates of colloidal titania-catalyzed photoreactions, *Faraday Discuss. Chem. Soc.* 1984, *78*, 151.

16. Thompson, T. L.; Yates, J. T., Jr. Monitoring hole trapping in photoexcited TiO$_2$(110) using a surface photoreaction, *J. Phys. Chem. B* 2005, *109*, 18230–18236.

17. Murray, P. W.; Leibsle, F. M.; Fisher, H. J.; Flipse, C. F. J.; Muryn, C. A.; Thornton, G. Observation of ordered oxygen vacancies on titania (100) (1X3) using scanning tunneling microscopy and spectroscopy, *Phys. Rev. B* 1992, *46*, 12877–12879.

18. Henderson, M. A. Structural sensitivity in the dissociation of water on TiO$_2$ single-crystal surface, *Langmuir* 1996, *12*, 5093–5098.

19. Kofstad, P. *Nonstoichiometry, Diffusion, and Electrical Conductivity in Binary Metal Oxides*, Wiley-Interscience, New York, 1972, pp. 139–145.

20. Kuznetsov, V. N.; Serpone, N. Visible light absorption by various titanium dioxide specimens, *J. Phys. Chem. B* 2006, *110*, 25203–25209.

21. Diebold, U. The surface science of titanium dioxide, *Surf. Sci. Rep.* 2003, *48*, 53–229.

22. Tang, H.; Berger, H.; Schmid, P. E.; Lévy, F.; Burri, G. Photoluminescence in TiO$_2$ anatase single crystals, *Solid State Commun.* 1993, *87*, 847–850.

23. Tang, H.; Berger, H.; Schmid, P.; Lévy, F. Optical properties of anatase (TiO$_2$), *Sol. State Commun.* 1994, *92*, 267–271.

24. Tang, H.; Lévy, F.; Berger, H.; Schmid, P. E. Urbach tail of anatase TiO$_2$, *Phys. Rev. B* 1995, *52*, 7771–7774.

25. Goto, M.; Harada, N.; Ijima, K.; Kunugita, H.; Ema, K.; Tsukamoto, M.; Ichikawa, N.; Sakama, H. Physics of semiconductors, *28th International Conference*, eds. W. Janysch, F. Schaffler, 2007, American Institute of Physics, New York.

26. Watanabe, M.; Sasaki, S.; Hayashi, T. Time-resolved study of photoluminescence in anatase TiO$_2$, *J. Lumin.* 2000, *87–89*, 1234–1236.

27. Wakabayashi, K.; Yamaguchi, Y.; Sekiya, T.; Kurita, S. Time-resolved luminescence spectra in colorless anatase TiO$_2$ single crystals, *J. Lumin.* 2005, *112*, 50–53.

28. McHale, J. L.; Rich, C. C.; Knorr, F. J. Trap-state photoluminescence of nanocrystalline and bulk TiO$_2$: Implications for carrier transport, *Mater. Res. Soc. Proc.* 2010, 1268-EE03-08.

29. Serpone, N. Is the band gap of pristine TiO$_2$ narrowed by anion- and cation-doping of titanium dioxide in second-generation photocatalysts? *J. Phys. Chem. B* 2006, *110*, 24287–24293.

30. Diebold, U.; Ruzycki, N.; Herman, G. S.; Selloni, A. One step toward bridging the materials gap: Surface studies of TiO$_2$ anatase, *Catal. Today* 2003, *85*, 93–100.

31. Aschauer, U.; He, Y.; Cheng, H.; Li, S.-C.; Diebold, U.; Selloni, A. Influence of subsurface defects on the surface reactivity of TiO$_2$: Water on anatase (101), *J. Phys. Chem. C* 2010, *114*, 1278–1284.

32. He, Y.; Dulub, O.; Cheng, H.; Selloni, A.; Diebold, U. Evidence for the predominance of subsurface defects on reduced anatase TiO$_2$(101), *Phys. Rev. Lett.* 2009, *102*, 106105.

33. Anpo, M.; Tomonari, M.; Fox, M. A. In-situ photoluminescence of TiO$_2$ as a probe of photocatalytic reaction, *J. Phys. Chem.* 1989, *93*, 7300–7302.

34. Meyer, G. J.; Lisenky, G. C.; Ellis, A. B. Evidence for adduct formation at the semiconductor gas interface. Photoluminescent properties of cadmium selenide in the presence of amines, *J. Am. Chem. Soc.* 1988, *110*, 4914–4918.

35. Anpo, M.; Aikawa, N.; Kubokawa, Y.; Che, M.; Louis, C.; Giamello, E. Photoluminescence and photo-catalytic activity of highly dispersed titanium oxide anchored onto porous Vycor glass, *J. Phys. Chem.* 1985, *89*, 5017–5021.

36. Fujihara, K.; Izumi, S.; Ohno, T.; Matsumura, M. Time-resolved photoluminescence of particulate TiO₂ photocatalysts suspended in aqueous solution, *J. Photochem. Photobiol. A* 2000, *132*, 99–104.

37. Zhang, H.; Banfield, J. F.; Thermodynamic analysis of phase stability of nanocrystalline titania, *J. Mater. Chem.* 1998, *8*, 2073–2076.

38. Finnegan, M. F.; Zhang, H.; Banfield, J. F. Phase stability and transformation in titania nanoparticles in aqueous solution dominated by surface energy, *J. Phys. Chem. C* 2007, *111*, 1962–1968.

39. Park, N.-G.; van de Lagemaat, J.; Frank, A. J. Comparison of dye-sensitized rutile- and anatase-based TiO₂ solar cells, *J. Phys. Chem. B* 2000, *104*, 8989–8994.

40. Serpone, N.; Lawless, D.; Khairutdinov, R. Size effects on the photophysical properties of colloidal anatase TiO₂ particles: Size quantization versus direct transitions in this indirect semiconductor? *J. Phys. Chem.* 1995, *99*, 16646–16654.

41. Bieber, H.; Gilliot, P.; Gallart, M.; Keller, N.; Keller, V.; Bégin-Colin, S.; Pighini, C.; Millot, N. Temperature dependent photoluminescence of photocatalytically active titania nanopowders, *Catal. Today* 2007, *122*, 101–108.

42. Zhang, G.; Roy, B. K.; Allard, L. F.; Cho, J. Titanium dioxide nanoparticles precipitated from low-temperature aqueous solution: 1. Nucleation, growth, and aggregation, *J. Am. Ceram. Soc.* 2008, *91*, 3875–3882.

43. Su, C.; Hong, B.-Y.; Tseng, C.-M. Sol-gel preparation and photocatalysis of titanium dioxide, *Catal. Today*, 2004, *96*, 119–126.

44. Ghosh, H. N.; Adhikari, S. Trap state emission from TiO₂ nanoparticles in microemulsion solutions, *Langmuir* 2001, *17*, 4129–4130.

45. Castellano, F. N.; Stipkala, J. M.; Friedman, L. A.; Meyer, G. J. Spectroscopic and excited state properties of titanium dioxide gels, *Chem. Mater.* 1994, *6*, 2123–2129.

46. Pan, D.; Zhao, N.; Wang, Q.; Jiang, S.; Ji, X.; An, L. Facile synthesis of luminescent TiO₂ nanocrystals, *Adv. Mater.* 2005, *17*, 1991–1995.

47. Niederberger, M.; Bartl, M. H.; Stucky, G. D. Benzyl alcohol and titanium tetrachloride—A versatile reaction system for the nonaqueous and low-temperature preparation of crystalline and luminescent titania nanoparticles, *Chem. Mater.* 2002, *14*, 4363–4370.

48. Zhang, J.; Li, M.; Feng, Z.; Chen, J.; Li, C. UV Raman spectroscopic study on TiO₂. I. Phase transformation at the surface and in the bulk, *J. Phys. Chem. B* 2006, *110*, 927–935.

49. Shi, J.; Chen, J.; Feng, Z.; Chen, T.; Lian, Y.; Li, C. Photoluminescence characteristics of TiO₂ and their relationship to the photoassisted reaction of water/methanol mixture, *J. Phys. Chem. C* 2007, *111*, 693–699.

50. Zhu, Y. C.; Ding, C. X. Investigation on the surface state of TiO₂ ultrafine particles by luminescence, *J. Solid State Chem.* 1999, *145*, 711–715.

51. Zhang, W. F.; Zhang, M. S.; Yin, Z.; Chen, Q. Photoluminescence in anatase titanium dioxide nanocrystals, *Appl. Phys. B* 2000, *70*, 261–265.

52. Qian, L.; Jin, Z. S.; Zhang, J. W.; Huang, Y. B.; Zhang, Z. J.; Du, Z. L. Studies of the visible-excitation luminescence of NTA-TiO₂(AB) with single-electron-trapped oxygen vacancies, *Appl. Phys. A* 2005, *80*, 1801–1805.

53. Zhang, W. F.; Zhang, M. S.; Yin, Z. Microstructures and visible photoluminescence of TiO₂ nanocrystals, *Phys. Stat. Sol. A* 2000, *179*, 319–327.

54. Scepanovic, M.; Dohcevic-Mitrovic, Z. D.; Hinic, I. Grujic-Brojcin, M.; Stanisic, G.; Popovic, Z. V. Photoluminescence of laser-synthesized titanium dioxide nanopowders, *Mater. Sci. Forum* 2005, *494*, 265–270.

55. Cavigli, L.; Bogani, F.; Vinattieri, A.; Faso, V.; Baldo, G. Volume versus surface-mediated recombination in anatase TiO₂ nanoparticles, *J. Appl. Phys.* 2009, *106*, 053516/1–8.

56. O'Regan, B.; Grätzel, M. A low-cost, high-efficiency solar cell based on dye-sensitized colloidal titanium dioxide films, *Nature* 2001, *414*, 338–344.

57. O'Regan, B.; Schwartz, D. T. Large enhancement in photocurrent efficiency caused by UV illumination of the dye-sensitized heterojunction TiO$_2$/RuLĽNCS/CuSCN: Initiation and potential mechanisms, *Chem. Mater.* 1998, *10*, 1501–1509.

58. Paulsson, H.; Kloo, L.; Hagfeldt, A.; Boschloo, G. Electron transport and recombination in dye-sensitized solar cells with ionic liquid electrolytes, *J. Electroanal. Chem. B* 2006, *586*, 56–61.

59. De Jongh, P. E.; Vanmaekelbergh, D. Trap-limited electronic transport in assemblies of nanometer-size TiO$_2$ particles, *Phys. Rev. Lett.* 1996, *77*, 3427–3430.

60. Zaban, A.; Meier, A.; Gregg, B. A. Electric potential distribution and short-range screening in nanoporous TiO$_2$ electrodes, *J. Phys. Chem. B* 1997, *101*, 7985–7990.

61. Kopidakis, N.; Schiff, E. A.; Park, N.-G.; van de Lagemaat, J.; Frank, A. J. Ambipolar diffusion of photocarriers in electrolyte-filled nanoporous TiO$_2$, *J. Phys. Chem. B* 2000, *104*, 3930–3936.

62. Bailes, M.; Cameron, P. J.; Lobato, K.; Peter, L. M. Determination of the density and energetic distribution of electron traps in dye-sensitized nanocrystalline solar cells, *J. Phys. Chem. B* 2005, *109*, 15429–15435.

63. Weidmann, W.; Dittrich, Th.; Konstantinova, E.; Lauermann, I.; Uhlendorf, I.; Koch, F. Influence of oxygen and water related surface defects on the dye sensitized TiO$_2$ solar cell, *Sol. Energy Mat. Sol. Cells* 1999, *56*, 153–165.

64. Gregg, B. A.; Chen, S.-G.; Ferrere, S. Enhanced dye-sensitized photoconversion efficiency via reversible production of UV-induced surface states in nanoporous TiO$_2$, *J. Phys. Chem. B* 2003, *107*, 3019–3029.

65. Wang, Q.; Zhang, Z.; Zakeeruddin, S. M.; Grätzel, M. Enhancement of the performance of dye-sensitized solar cell by formation of shallow transport levels under visible light illumination, *J. Phys. Chem. C* 2008, *112*, 7084–7092.

66. O'Regan, B. C.; Durrant, J. R.; Sommeling, P. M.; Bakker, N. J. Influence of TiCl$_4$ treatment on nanocrystalline TiO$_2$ films in dye-sensitized solar cells. 2. Charge density, band edge shifts, and quantification of recombination losses at short circuit, *J. Phys. Chem. C* 2007, *111*, 14001–14010.

67. Amano, F.; Yasumoto, T.; Prieto-Mahaney, O.-O.; Uchida, S.; Shibayama, T.; Ohtani, B. Photocatalytic activity of octahedral single-crystalline mesoparticles of anatase Ti(IV) oxide, *Chem. Commun.* 2009, 2311–2313.

68. Yang, H. G.; Sun, C. H.; Qiao, S. Z.; Zou, J.; Liu, G.; Smith, S. C.; Cheng, H. M.; Lu, G. Q. Anatase TiO$_2$ single crystals with a large percentage of reactive facets, *Nature* 2008, *453*, 638–642.

69. Wu, B.; Guo, C.; Zheng, N.; Xie, Z.; Stucky, G. D. Non-aqueous production of nanostructured anatase with high-energy facets, *J. Am. Chem. Soc.* 2008, *130*, 17563–17567.

70. Wu, N.; Wang, J.; Tafen, D. N.; Wang, H.; Zheng, J.-G.; Lewis, J. P.; Liu, X.; Leonard, S. S.; Manivannan, A. Shape-enhanced photocatalytic activity of single-crystalline anatase TiO$_2$ (101) nanobelts, *J. Am. Chem. Soc.* 2010, *132*, 6679–6685.

71. Sommeling, P. M.; O'Regan, B. C.; Haswell, R. R.; Smit, H. J. P.; Bakker, N. J.; Smits, J. J. T.; Kroon, J. M.; van Oosmalen, J. A. M. Influence of a TiCl$_4$ post-treatment on nanocrystalline TiO$_2$ films in dye-sensitized solar cells, *J. Phys. Chem. B* 2006, *110*, 19191–19197.

72. Park, N.-G.; Schlichtörl, G.; Van de Lagemaat, J.; Cheong, H. M.; Mascarenhas, A. Dye-sensitized TiO$_2$ solar cells: Structural and photoelectrochemical characterization of nanocrystalline electrodes formed through hydrolysis of TiCl$_4$, *J. Phys. Chem. B* 1999, *103*, 3308–3314.

73. Barnes, P. R. F.; Anderson, A. Y.; Koops, S. E.; Durrant, J. R.; O'Regan, B. C. Electron injection efficiency and diffusion length in dye-sensitized solar cells derived from incident photon conversion efficiency measurements, *J. Phys. Chem. C* 2009, *113*, 1126–1136.

74. Knorr, F. J.; Zhang, D.; McHale, J. L. Influence of TiCl$_4$ treatment on surface defect photoluminescence in pure and mixed-phase nanocrystalline TiO$_2$, *Langmuir* 2007, *23*, 8686–8690.

75. Zhang, D.; Downing, J. A.; Knorr, F. J.; McHale, J. L. Room-temperature preparation of nanocrystalline TiO$_2$ films and the influence of surface properties on dye-sensitized solar energy conversion, *J. Phys. Chem. B* 2006, *110*, 21890–21898.

76. Mori, S.; Sunahara, K.; Fukai, Y.; Kanzaki, T.; Wada, Y.; Yanagida, S. Electron transport and recombination in dye-sensitized TiO_2 solar cells fabricated without sintering process, *J. Phys. Chem. C* 2008, *112*, 20505–20509.

77. Hurum, D. C.; Agrios, A. G.; Gray, K. A.; Rajh, T.; Thurnauer, M. C. Explaining the enhanced photocatalytic activity of Degussa P25 mixed-phase TiO_2, *J. Phys. Chem. B* 2003, *107*, 4545–4549.

78. Hurum, D. C.; Gray, K. A.; Rajh, T.; Thurnauer, M. C. Recombination pathways in the Degussa P25 formulation of TiO_2: Surface versus lattice mechanisms, *J. Phys. Chem. B* 2005, *109*, 977–980.

79. Kho, Y. K.; Iwase, A.; Teoh, W. Y.; Madler, L.; Kudo, A.; Amal, R. Photocatalytic H_2 evolution over TiO_2 nanoparticles: The synergistic effect of anatase and rutile, *J. Phys. Chem. C* 2010, *114*, 2821–2829.

80. Li, G.; Richter, C.; P.; Milot, R. L.; Cai, L.; Schmuttenmaer, C. A.; Crabtree, R. H.; Brudvig, G. W.; Batista, V. S. Synergistic effect between anatase and rutile TiO_2 nanoparticles in dye-sensitized solar cells, *Dalton Trans.* 2009, *45*, 10078–10085.

81. Bickley, R. I.; Gonzalez-Carreno, T.; Lees, J. S.; Palmisano, L.; Tilley, R. J. D. A structural investigation of titanium dioxide photocatalysts, *J. Solid State Chem.* 1991, *92*, 178–190.

82. Ohno, T.; Sarukawa, K.; Tokieda, K.; Matsumura, M. Morphology of a TiO_2 photocatalyst (Degussa P25) consisting of anatase and rutile crystalline phases, *J. Catal.* 2001, *203*, 82–86.

83. Datye, A. K.; Riegel, G.; Bolton, J. R.; Huang, M.; Prairie, M. R. *J. Solid State Chem.* 1995, *115*, 236–239.

84. Deskins, N. A.; Kerisit, S.; Rosso, K. M.; Dupuis, M. Molecular dynamics characterization of rutile-anatase interfaces, *J. Phys. Chem. C* 2001, 111, 9290–9298.

85. Mohmeyer, N.; Wang, P.; Schmidt, H. W.; Zakeeruddin, S. M.; Grätzel, M. Quasi-solid-state dye sensitized solar cells with 1,3:2,4-di-O-benzylidene-d-sorbitol derivatives as low-molecular weight organic gelators, *J. Mater. Chem.* 2004, *14*, 1950–1959.

86. Gorlov, M.; Kloo, L. Ionic liquid electrolytes for dye-sensitized solar cells, *Dalton Trans.* 2008, 2655–2666.

87. Li, B.; Wang, L. D.; Kang, B. N.; Wang, P.; Qiu, Y. Review of recent progress in solid-state dye-sensitized solar cells, *Sol. Energy Mater. Sol. Cells* 2006, *90*, 549–573.

88. Pollard, J. A.; Zhang, D.; Downing, J. A.; Knorr, F. J.; McHale, J. L. Solvent effects on interfacial electron transfer from Ru(4,4′-dicarboxylic acid-2,2′-bipyridine)$_2$(NCS)$_2$ to nanoparticulate TiO_2: Spectroscopy and solar photoconversion, *J. Phys. Chem. B* 2005, *109*, 11443–11452.

89. Fukui, A.; Komiya, R.; Yamanaka, R.; Islam, A.; Han, L. Effect of redox electrolyte in mixed solvents on the photovoltaic performance of a dye-sensitized solar cell, *Sol. Energy Mater. Sol. Cells* 2006, *90*, 649–638.

90. Halme, J.; Boschloo, G.; Hagfeldt, A.; Lund, P. Spectral characteristics of light harvesting, electron injection, and steady-state charge collection in pressed TiO_2 dye solar cells, *J. Phys. Chem. C* 2008, *112*, 5623–5637.

91. Jennings, J. R.; Wang, Q. Influence of lithium ion concentration on electron injection, transport, and recombination in dye-sensitized solar cells, *J. Phys. Chem. C* 2010, *114*, 1715–1724.

92. Schwanitz, K.; Weiler, U.; Hunger, R.; Mayer, T.; Jaegermann, W. Synchrotron-induced photoelectron spectroscopy of the dye-sensitized nanocrystalline TiO_2/electrolyte interface: Band gap states and their interaction with dye and solvent molecules, *J. Phys. Chem. C* 2007, *111*, 849–854.

93. Haque, S. A.; Tachibana, Y.; Willis, R. L.; Moser, J. E.; Grätzel, M.; Klug, D. Parameters influencing charge recombination kinetics in dye-sensitized nanocrystalline titanium dioxide films, *J. Phys. Chem. B* 2000, *104*, 538–547.

94. Rühle, S.; Greenshtein, M.; Chen, S.-G.; Merson, A.; Pizem, H.; Sukenik, C. S.; Cahen, D.; Zaban, A. Molecular adjustment of the electronic properties of nanoporous electrodes in dye-sensitized solar cells, *J. Phys. Chem. B* 2005, *109*, 18907–18913.

95. Knorr, F. J.; Mercado, C. C.; McHale, J. L. Trap state distributions and carrier transport in pure and mixed-phase TiO_2: Influence of contacting solvent and interphasial electron transfer, *J. Phys. Chem. C* 2008, *112*, 12786–12794.

96. Miyake, M.; Yoneyama, H.; Tamura, H. Two-step oxidation reactions of alcohols on an illuminated rutile electrode, *Chem. Lett.* 1976, *6*, 635–640.

97. Panyatov, D. A.; Yates, J. T., Jr. Depletion of conduction band electrons in TiO₂ by water chemisorption—IR spectroscopic studies of the independence of Ti-OH frequencies on electron concentration, *Chem. Phys. Lett.* 2005, *410*, 11–17.

98. Rühle, S.; Dittrich, T. Recombination controlled signal transfer through mesoporous TiO₂ films, *J. Phys. Chem. B* 2006, *110*, 3883–3888.

99. Semenikhin, O. A.; Kazarinov, V. E.; Jiang, L.; Hashimoto, K.; Fujishima, A. Suppression of surface recombination on TiO₂ anatase in aqueous solutions containing alcohol, *Langmuir* 1999, *15*, 3731–3737.

100. Colombo, D. P., Jr.; Bowman, R. M. Does interfacial charge transfer compete with charge carrier recombination? A femtosecond diffuse reflectance investigation of TiO₂ nanoparticles, *J. Phys. Chem.* 1996, *100*, 18445–18449.

101. Skinner, D. E.; Colombo, D. P., Jr.; Cavaleri, J. J.; Bowman, R. M. Femtosecond investigation of electron trapping in semiconductor nanoclusters, *J. Phys. Chem.* 1995, *99*, 7853–7856.

102. Murakami, N.; Kurihara, Y.; Tsubota, T.; Ohno, T. Shape-controlled anatase titanium(IV) oxide particles prepared by hydrothermal treatment in the presence of polyvinyl alcohol, *J. Phys. Chem. C* 2009, *113*, 3062–3069.

103. Di Valentin, C.; Pacchioni, G.; Selloni, A. Reduced and n-type doped TiO₂: Nature of Ti³⁺ species, *J. Phys. Chem. C* 2009, *113*, 20543–20552.

104. Knorr, F. J.; McHale, J. L. Spectroelectrochemical photoluminescence investigation of trap states in nanocrystalline TiO₂, submitted.

105. Enright, B.; Redmond, F.; Fitzmaurice, D. Spectroscopic determination of flatband potentials for polycrystalline TiO₂ electrodes in mixed solvent systems, *J. Phys. Chem.* 1994, *98*, 6195–6200.

106. Chindanon, K.; Knorr, F. J.; Usmani, S. M.; Ichimura, A.; McHale, J. L., Location of hole and electron traps on nanocrystalline anatase TiO₂, submitted.

107. Bilmes, S. A.; Mandelbaum, P.; Alvarez, F.; Victoria, N. M. Surface and electronic structure of titanium dioxide photocatalysis, *J. Phys. Chem. B* 2000, *104*, 9851–9858.

108. Wang, K.-P.; Teng, H. Structure-intact TiO₂ nanoparticles for efficient electron transport in dye-sensitized solar cells, *Appl. Phys. Lett.* 2007, *91*, 173102/1–3.

109. Mercado, C. C.; Seeley, Z.; Bandyopadhyay, A.; Bose, S.; McHale, J. L. Photoluminescence of dense nanocrystalline titanium dioxide thin films, submitted.

110. Varghese, O. K.; Paulose, M.; Grimes, C. A. Long vertically-aligned titania nanotubes on transparent conducting oxide for highly efficient solar cells, *Nat. Nanotechnol.* 2009, *4*, 592–597.

111. Zhu, K.; Neale, N. R.; Miedaner, A.; Frank, A. J. Enhanced charge collection efficiencies and light-scattering using vertically oriented TiO₂ nanotubes arrays, *Nano Lett.* 2007, *7*, 69–74.

112. Grimes, C. A. Synthesis and application of highly oriented arrays of TiO₂ nanotubes, *J. Mater. Chem.* 2007, *17*, 1451–1457.

113. Mohammadpour, R.; Iraji Zad, A.; Hagfeldt, A.; Boschloo, G. Comparison of trap-state distribution and carrier transport in nanotubular and nanoparticulate TiO₂ electrodes for dye-sensitized solar cells, *ChemPhysChem* 2010, *11*, 2140–2145.

114. Shankar, K.; Bandara, J.; Paulose, M.; Wietsasch, H.; Varghese, O. K.; Mor, G. K.; LaTempa, T. J.; Thelakkat, M.; Grimes, C. A. Highly efficient solar cells using TiO₂ nanotube arrays sensitized with a donor-antenna dye, *Nano Lett.* 2009, *8*, 1654–1659.

115. Mor, G. K.; Varghese, O. K.; Paulose, M.; Shankar, K.; Grimes, C. A. A review on highly ordered, vertically oriented TiO₂ nanotube arrays: Fabrication, materials properties, and solar energy applications, *Sol. Energy Mater. Sol. Cells* 2006, *90*, 2011–2075.

116. McHale, J. L.; Mercado, C. C. Defect photoluminescence of TiO₂ nanotubes, *Mater. Res. Soc. Proc.* 2010, 1268-EE03-10.

14

Photoluminescence Spectroscopy of Single Semiconductor Nanoparticles

Takashi Tachikawa
Osaka University

Tetsuro Majima
Osaka University

14.1 Introduction

Semiconductor nanoparticles including dots, rods, wires, and tubes have received much attention because of their applications, such as sensors, detectors, light-emitting diodes, photovoltaic cells, and photocatalysis.[1] Photoluminescence (PL) is a particularly useful tool for the identification of crystal structures, particle size, and defect states, because the positions of PL peaks are strongly dependent on these structural characteristics.[2] In addition, temporal changes of PL intensity and spectrum give us information about the underlying reaction chemistry. Recently, single-molecule (single-particle) fluorescence spectroscopy is emerging as an important technique for studying the photophysical and photochemical processes of all types of systems from simple dye molecules to luminescent quantum dots (QDs).[3-5] In this chapter, we would like to outline the spectral and kinetic characteristics of PL from single semiconductor nanoparticles, mainly TiO$_2$ nanoparticles. Nanostructured TiO$_2$ photocatalysts have been extensively studied and used for the water-splitting reaction that produces hydrogen, the degradation of organic pollutants, the surface wettability conversion, etc.[6,7] The PL bands originating from defects in the bulk and/or on the surface of TiO_2 were observed in the visible region with numerous "photon bursts" under 405 nm laser irradiation. From the single-molecule kinetic analysis of the bursts, it was found that the quenching reaction of trapped electrons by molecular oxygen follows a Langmuir–Hinshelwood mechanism. In addition, a novel spectroscopic method, i.e., single-molecule spectroelectrochemistry, was utilized to explore the nature of the defect states inherent in TiO_2. The spatially resolved PL imaging techniques thus enable us to ascertain the location of the luminescent active sites that are related to the heterogeneously distributed defects. The single-particle spectroscopy was also applied to investigate defect-mediated PL dynamics of Eu^{3+}-doped TiO_2 nanoparticles. It was revealed that free excitons in the photoirradiated TiO_2 host can excite both interior and surface Eu^{3+} ions, while trapped excitons at the surface only excite the latter. Furthermore, it is of great interest to explore a novel composite system allowing an efficient electron flow at the heterogeneous interface between semiconductor nanoparticles, such as CdS, CdSe, and CdTe QDs, and organic (inorganic) materials. Mechanistic studies of the electron transfer (ET) processes on semiconductor nanoparticles using single-molecule (single-particle) spectroscopies will be introduced.

14.2 Single-Molecule Fluorescence Spectroscopy

14.2.1 Experimental Methods

Single-molecule fluorescence spectroscopy and imaging are emerging as powerful techniques for exploring the photodynamics of all types of molecular systems from simple dye molecules to fluorescent proteins in various environments.[3–5] Until now, the necessity for observing very weak emissions from a single fluorophore has required the development of advanced detection techniques, such as total internal reflection fluorescence microscopy (TIRFM) and confocal fluorescence microscopy.[8] The former technique is advantageous in visualizing the fluorescence from single molecules or particles immobilized or located at the interface (e.g., a glass surface) with a very low background noise using an evanescent field, and thus has been applied to the investigation of the temporal dynamics of biomolecules (e.g., DNA, proteins, and enzymes) labeled with dyes, which provides information that is useful for revealing various biological functions at the molecular level.[9] The latter, confocal microscopy, is widely used for the detection of a molecule freely diffusing in solution, often referred to as fluorescence correlation spectroscopy (FCS), which enables us to investigate and clarify the binding interactions, such as protein–ligand binding and DNA hybridization, by measuring the correlation time of diffusing molecules into the focal volume. These techniques for single-molecule fluorescence detection have advantages superior to the conventional ones that rely on the bulk sample, providing us with opportunities such as the ultimate high sensitivity, the possible observations of the properties hidden in ensemble measurements, and eliminating the need for synchronization. Therefore, the single-molecule (single-particle) fluorescence spectroscopy has been applied to elucidate the inherent features of heterogeneous catalyses, such as the dynamics of reagent molecules in mesoporous silica,[10] the mechanism of the hydrolysis or redox reaction occurring on the surface of the layered double hydroxide crystal,[11] and gold nanoparticles.[12]

14.2.2 Total Internal Reflection Fluorescence Microscopy

TIRFM can be used to observe fluorophores attached onto glass surfaces, biomolecules, and living cells.[9] The illumination method utilized for the excitation of fluorophores in TIRFM is conceptually simple.[13] When the excitation light for fluorophores is incident above some critical angle upon the glass/liquid interface, the light is totally internally reflected and generates a thin electromagnetic field, so-called *evanescent field*, in a medium with the same wavelength as the incident light. When the electromagnetic wave is permitted to penetrate into only a limited depth as explained by Maxwell's equations, the intensity of the transmitted wave, I_T, is given by

$$I_T = |A_T|^2 \exp\left[-z\frac{4\pi}{\lambda_2}\sqrt{\frac{\sin^2\theta}{n^2}-1}\right] \tag{14.1}$$

where A_T, z, λ_2, and θ are the amplitude of the electric field, the perpendicular distance from the interface, the wavelength in medium 2, and the incidence angle, respectively, and $n = n_2/n_1(\sin\theta_c)$, where n_1, n_2, and θ_c are refractive indices for media 1, 2, and the critical angle, respectively. The intensity of the evanescent field decays exponentially with the distance from the glass surface. The penetration depth d, i.e., $1/e$ value in Equation 14.1, is also given by

$$d = \frac{\lambda}{4\pi\sqrt{n_1^2\sin^2\theta-n_2^2}} \tag{14.2}$$

where λ $(=n_2\lambda_2)$ is the wavelength of the incident light in vacuum.

Two different technical solutions for TIR illumination, i.e., prism-type and objective-type TIRM, have been established. The optical configuration of objective-type TIRFM based on an inverted microscope is illustrated in Figure 14.1. A high numerical aperture (NA) oil-immersion objective lens is mounted on an inverted microscope. A linear or circular-polarized laser beam is focused by a lens on the back focal plane of the objective. By shifting the laser position, the path of the incident laser light is shifted from the center to the edge of the objective. At the center position, the microscope can be used

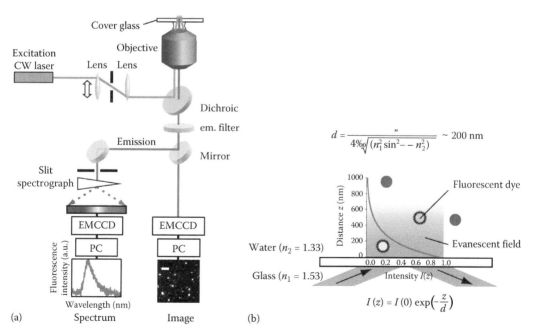

FIGURE 14.1 (a) Illustration of the experimental setup for the TIRFM. The movement of the incident laser light can control the angle of incidence at the interface, enabling the switching from the epi-fluorescence to TIRF excitation. (b) The principle of evanescent wave excitation for single-molecule fluorescence experiments ($\lambda_{ex} = 532$ nm). Only molecules near the interface (within the penetration depth, $d \approx 200$ nm) can be efficiently excited, and then emit the detectable fluorescence.

as a standard *epi*-fluorescence microscope. In this experimental setup, light emitted from a continuous wave (CW) solid or ion laser (typical output power is >10 mW) passing through an objective lens was totally reflected at the cover glass–air interface to obtain an evanescent field, which can excite single dye molecules on the cover glass.

The fluorescence emission from single dye molecules or particles is collected using the same objective, filtered by a dichroic mirror and a suitable long-pass or band-pass filter, and imaged by a sensitive charge-coupled device (CCD) camera, such as an electron-multiplying CCD (EM-CCD) or intensified CCD camera. The recorded images can be processed using freely available software, such as ImageJ (http://rsb.info.nih.gov/ij/).

To measure the single-molecule spectrum, only the emission that passed through a slit is brought into a spectrograph and imaged by an EM-CCD or intensified CCD camera. The spectrum is typically integrated for a few seconds to obtain enough signal-to-noise ratio and analyzed by a personal computer.

14.3 Photoluminescence of Single Semiconductor Nanoparticles

14.3.1 Luminescent Quantum Dots

Nano-sized (spherical) particles, also referred to as QDs, represent a class of quantum-confined objects in which carrier motion is restricted in all three dimensions.[1] This three-dimensional quantum confinement results in discrete size-dependent absorption and emission spectra and redox properties. In particular, luminescent QDs, such as CdS, CdSe, and CdTe QDs, can be easily modified with various functional groups during or after preparation, which make them multifunctional, and have been expected for many attractive device applications, such as light-emitting diodes, solar cells, and bioimaging and sensing applications.[14–18] Recent reports of multiple exciton generation, or carrier multiplication, by one absorbed photon in some QDs offer an exciting possibility to dramatically improve the efficiency of QD-based solar cells.[19,20] To facilitate the charge transport across the heterogeneous interface between QDs and anodes, typically a nanocrystalline TiO_2 or ZnO film, the possible rate-limiting factors must be addressed and suitably optimized. Therefore, understanding the dynamics of charge carriers generated in QDs is an important matter for using them to design those devices. Some examples of photochemical reactions on the surface of semiconductor nanoparticles will be discussed in Section 14.5.

A typical PL image measured during 532 nm laser excitation of single QDs (Invitrogen, Qdot 605) is shown in Figure 14.2a. The integration time per one frame was 33 ms. Evidence for the detection of the individual QDs comes both from the observed strong fluorescence intermittency as well as the scaling of the number of luminescent spots. A typical single-particle PL spectrum is also shown in Figure 14.2b. The spectral shape significantly varies from particle to particle, but there was no significant peak shift in the PL spectrum during the course of the experiment. As demonstrated in Figure 14.2c, individual QDs exhibit the fluorescence intermittency, also called as photoblinking phenomenon, in which the fluorescence intensity drops to background in a single step but returns to the original intensity after a brief time. Numerous studies have developed model mechanisms for the blinking wherein a QD switches between the on and off states via charging events.[21] A recently developed mechanism is introduced in Section 14.5.

14.3.2 Defect Emission from Semiconductor Nanoparticles

TiO_2 nanoparticles show a very weak emission in the UV and visible regions in ambient air at room temperature. Therefore, despite instrumental developments, there have been only a few reports in the literature on the direct observation of the emission from individual TiO_2 nanoparticles by fluorescence microscopy.[22–24] Recently, Tachikawa and coworkers discovered a dramatic increase in the emission intensity of TiO_2 nanoparticles by changing the atmosphere from air to an inert gas during

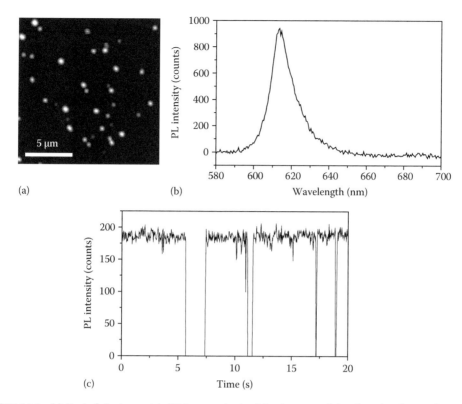

FIGURE 14.2 (a) Typical single-particle PL images obtained for QDs immobilized on the glass surface. (b) The typical PL spectrum observed for a single QD. (c) The PL intensity trajectory observed for a single QD.

the photoirradiation with a 405 nm laser (Figure 14.3a).[22] They could clarify the mechanism of the defect-mediated PL dynamics of TiO_2 nanoparticles at the single-particle level. The quantitative examination of the spectral and kinetic characteristics clearly revealed that the PL bands originating from defects in the bulk and/or on the surface appeared in the visible region with numerous photon bursts by the photoirradiation in Ar gas atmosphere. The spectral measurements of individual nanoparticles also show the growth of a broad PL band in the visible region (500–750 nm), as depicted in Figure 14.3b.

Historically, the visible emission of TiO_2 is considered to be ascribed to the self-trapped excitons localized on the TiO_6 octahedra[25] or oxygen vacancies.[26] For instance, Sekiya et al. reported that the visible PL spectrum consists of three components centered at about 517, 577, and 636 nm, which are assigned to excitons bound to partially reduced Ti ions, self-trapped excitons through an exciton–lattice interaction, and oxygen vacancies, respectively.[27,28] Recently, McHale et al. observed the PL spectra of nanocrystalline TiO_2 in the anatase and rutile phases, and in mixed-phase samples at room temperature, and found that the total PL of anatase is shown to be a superposition of transitions involving spatially separated trapped electrons and holes.[29] In short, they suggested that the green PL with a peak around 525 nm is due to the transition between the mobile electrons (those in the conduction band or in shallow bulk traps) and trapped holes, and the red PL with a peak around 600 nm is due to that between the deeply trapped electrons and valence band holes. According to their assignment, at least two different light-emitting species due to the trapped charges are involved in the visible PL bands, and their distribution gradually changes from green emission to red emission during the photoirradiation.[22]

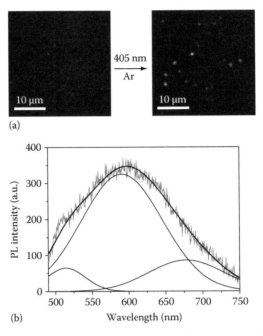

(a)

(b)

FIGURE 14.3 (a) Typical PL images observed during the 405-nm laser excitation for single TiO_2 nanoparticles under air (left) and Ar (right) atmospheres. (b) Typical PL spectrum of a single TiO_2 nanoparticle in Ar atmosphere.

14.3.3 Photoactivation Dynamics of Defect Emission from TiO_2 Nanoparticles

Photoactivation of the TiO_2 emission, i.e., an order of magnitude or more increase in intensity, seems to be similar to that frequently observed for luminescent nanocolloids, such as CdSe and CdTe QDs, but their mechanisms are completely different from each other.[30-32] In the present system, light absorption in the spectral region of the intrinsic absorption ($h\nu > 3.2\,eV$) and in the bands corresponding to the color centers (extrinsic absorption, $h\nu < 3.2\,eV$) generates free carriers that induce the formation of trapped carriers and surface oxygen species, such as the superoxide radical anion ($O_2^{\bullet-}$), which oxidizes the color centers.

According to the literature,[33] the main feature of the kinetics during visible-light irradiation is the dependence of the absorbance of the sample on the number of color centers (N) as follows:

$$TiO_2 + h\nu\,(405\,nm) \rightarrow e_{free}^- + h_{free}^+ \text{ (intrinsic absorption)} \tag{14.3}$$

$$e_{free}^- \rightarrow e_{tr}^- \text{ (color centers)} \tag{14.4}$$

$$h_{free}^+ \rightarrow h_{tr}^+ \tag{14.5}$$

$$e_{tr}^- \text{ (color centers)} + h\nu\,(405\,nm) \rightarrow e_{free}^- + h_{free}^+ \text{ (extrinsic absorption)} \tag{14.6}$$

$$e_{free}^- + h_{free/tr}^+ \rightarrow PL \tag{14.7}$$

$$e_{tr}^- \text{ (color centers)} + h_{free/tr}^+ \rightarrow PL \text{ (defect emission)} \tag{14.8}$$

$$e_{free}^- + O_2 \rightarrow O_2^{\bullet-} \tag{14.9}$$

Visible color centers, most probably, trapped electrons (e_{tr}^-) in the vacancy defect sites, are generated by the intrinsic and/or extrinsic excitations of TiO_2 under the 405 nm laser irradiation. Both e_{free}^- and e_{tr}^- can then recombine with the photogenerated holes $(h_{free/tr}^+)$ to produce the PL in the UV and visible regions, respectively. The quenching of e_{free}^- by O_2 molecules consequently results in the decreased PL. Based on the reaction equations, the main feature of the kinetics during visible-light irradiation is described as the dependence of the absorbance of the sample on the number of color centers (N). The differential equation for the formation (k^+) and deactivation (k^-) rates of the color centers is given by

$$\frac{dN}{dt} = k^+ N - k^- N^2 \tag{14.10}$$

The k^+ value significantly depends on the light intensity, the absorbance of the color centers, and the quantum yield of the photoreaction. On the other hand, k^- depends on these same factors in addition to the oxygen content available in the gas-phase environment. Equation 14.10 is of the form of the Bernoulli equation and is converted into a first-order linear differential equation as given by

$$N(t) = \frac{1}{k^-/k^+ - (k^-/k^+ - 1/N_0)\exp(-k^+t)} \tag{14.11}$$

where N_0 is the number of color centers that exist prior to irradiation. As shown in Figure 14.4, the observed PL intensity trajectories for single nanoparticles were well reproduced by Equation 14.11, verifying the validity of the model. Assuming that the absorption cross section and PL quantum yield of color centers are constant, the N values increased by at least 10 times after the photoactivation and the k^+ value was determined to be $0.05 \pm 0.01\,s^{-1}$ for both processes, while the k^- values were 0.5 ± 0.2 and $15 \pm 5\,s^{-1}$ for the photoactivation (trace (a)) and deactivation processes (trace (b)), respectively. The remarkable difference in k^- should be due to the different oxygen concentration in the gas phase. Similar PL activation and deactivation behaviors were observed for other nano-sized TiO_2 materials, such as TiO_2 nanowires.[23]

14.3.4 Electric Potential–Induced PL from Individual TiO_2 Nanowires

A novel spectroscopic method, i.e., single-molecule spectroelectrochemistry, is utilized to explore the nature of the luminescent defects present in single TiO_2 nanowires.[34] This method is a powerful

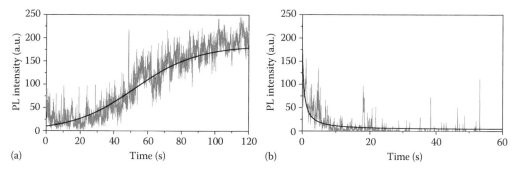

FIGURE 14.4 Trajectories of PL intensities under Ar (a) and air (b) atmospheres (bin time is 33 ms). Solid lines indicate the kinetic traces calculated using Equation 14.11.

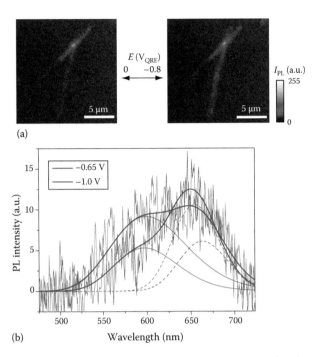

(a)

(b)   Wavelength (nm)

FIGURE 14.5 (See color insert.) (a) PL images of single TiO$_2$ nanowires acquired without (left) and with (right) applied potential (E) of −0.8 V vs. Ag wire (= −0.37 V vs. NHE). Upon the applied potential of −0.8 V, a significant increase in the PL intensity was observed. (b) PL spectra observed at different times. The solid lines indicate the Gaussian distributions fitted to the spectra.

new technique for studying electrochemical kinetics in highly heterogeneous systems and allows us to measure electrochemical kinetics of many individual materials at a time. In a typical experiment, the potential of the working electrode, i.e., an indium tin oxide (ITO) thin film fabricated on a cover glass, was linearly scanned at a scan rate of 0.1 V s^{-1} with simultaneous measuring of the PL images of many single wires.[23]

Figure 14.5a shows the PL images of single TiO$_2$ nanowires acquired without (left) and with (right) the applied potential (E) of −0.8 V vs. silver wire quasi-reference electrode (QRE) (= −0.37 V vs. NHE). Upon the applied potential of −0.8 V, a noticeable increase in the PL intensity was observed, while no PL was detected at all without the 405 nm laser irradiation. Interestingly, the wire lying over another one on the ITO surface has a much lower response to the potential, indicating the importance of physical contact between the wire and ITO on the potential-induced PL, and moreover, the possibility that the electron transport between physically contacted wires is inefficient. Additionally, one can distinguish the wires that have less contact with the ITO, because no significant PL quenching due to the ET from the photoexcited TiO$_2$ to the ITO was observed for these wires in the absence of the applied potential. This interpretation is supported by the fact that a remarkable suppression of the photoactivation was observed for the TiO$_2$ nanoparticles on the ITO surface.

In addition, as shown in Figure 14.5b, the spectra are roughly divided into two components, i.e., those with a peak at around 600 and 660 nm. These results infer that several sites for the electron trapping are included in the potential-induced PL process. The flat band potentials have been estimated to be −1.2 ∼ −1.1 V vs. QRE (= −0.76 ∼ −0.66 V vs. NHE) in acetonitrile containing 0.1 M LiClO$_4$.[35,36] As is well known, numerous electron trap sites lie just below the conduction band of TiO$_2$. For a nanocrystalline TiO$_2$ electrode consisting of sintered anatase particles, the dependence of the charge accumulation rate constant on the initially applied potential has been used to infer the existence of an intraband state about 0.7 eV below the conduction band edge.[37] The subsequent spectroelectrochemical investigations of the nanocrystalline TiO_2 electrodes prepared from Degussa P25 (anatase [80%] and rutile [20%]) indicated the presence of traps at ca. 0.5 eV below the conduction band edge.[38] Similar studies of electrode films composed of small anatase particles showed that traps are present at about 0.5 eV below the conduction band edge.[39] For the TiO_2 nanowire, therefore, the energy level of trapped electrons is considered to be −0.5 to −0.7 V vs. QRE (= −0.1 to −0.3 V vs. NHE), which is quite consistent with the potential that begins to activate the PL obtained by single-particle electrochemical spectral measurements. Thus, the ET from free or shallowly trapped electrons to the ITO electrode (about 0 V vs. NHE)[40] is energetically possible, supporting the significant PL quenching of TiO_2 nanowires in contact with the ITO surface in the absence of the applied potential. On the other hand, the application of a negative potential results in the accumulation of (trapped) electrons in TiO_2. In addition, the slow PL decay was observed after the removal of the potential. These results infer that deeply trapped electrons are stable until exposure to the laser irradiation during which the photostimulated detrapping occurred to generate mobile electrons, followed by ET to the ITO. To fully understand the detrapping phenomenon of trapped electrons, the temperature and laser power effects on the luminescence characteristics must be examined.

14.4 Photoluminescence of Single Lanthanide-Doped Semiconductor Nanoparticles

14.4.1 Lanthanide-Doped Metal Oxide Nanoparticles

Lanthanide-doped materials are finding use in a wide variety of applications in optics as gain media for amplifiers and lasers and as biolabels, white-light emitters, and full-color phosphors for displays.[41] Since direct excitation of the parity-forbidden intra-f-shell lanthanide ion crystal-field transitions is inefficient, it is anticipated that the luminescence of RE ions incorporated in the wide band-gap semiconductor lattice (e.g., ZnO and TiO_2) could be sensitized efficiently by exciton recombination in the host. The various types of defect states have been considered to play an important role in energy transfer between TiO_2 and the activator Eu^{3+} ions. For example, with the increase of annealing temperature, PL intensity of visible emissions due to Eu^{3+} ions first increases, and then decreases and reaches a maximum when annealing temperature is 700°C.[42] In this sense, the luminescence of Eu^{3+} depends critically on their doping locations in the host. However, the mechanism of the energy transfer process from the defect energy levels of the host to dopants has not yet been clarified owing to several difficulties, such as the inhomogeneous distribution of ions in the material. Recently, the PL dynamics of undoped TiO_2 and Eu^{3+}-doped TiO_2 (TiO_2:Eu^{3+}) nanoparticles was investigated using single-particle PL spectroscopy.[22] The photostimulated formation of emissive defects at the TiO_2 surface and the defect-mediated PL of the doped Eu^{3+} were examined at the single-particle level.

14.4.2 Local Environment around Lanthanide Ions Doped in Metal Oxides

Figure 14.6a shows the typical PL spectra (bold line) of individual luminescent spots below the diffraction limit of ~150 nm for TiO_2:Eu^{3+} nanoparticles ([Eu^{3+}] = 0.5 atom%) in ambient air. The single-particle spectral measurements revealed that the PL bands at around 590 and 615 nm are attributable to the transitions from the $^5D_0$ level to the $^7F_1$ and $^7F_2$ levels of Eu^{3+}, respectively.[43] The 5% of all particles showed a fairly strong blue-shifted emission, which might be assigned to $Eu_2Ti_2O_7$ particles,[43] because the fraction of such particles increased with increasing concentration to 5 atom% of doped Eu^{3+}. Since the $^5D_0 \rightarrow {}^7F_2$ transition is an electrically allowed transition, it is very sensitive to the surroundings of the Eu^{3+} ion, whereas the magnetically allowed $^5D_0 \rightarrow {}^7F_1$ transition is almost not influenced. Consequently, the relative intensity (area) ratio of $^5D_0 \rightarrow {}^7F_2$ to $^5D_0 \rightarrow {}^7F_1$, the so-called R value, provides information about the breaking of centrosymmetry and the degree of disorder around the Eu^{3+} ions.[41]

As shown in Figure 14.6b, a very wide R distribution was obtained. This result clearly indicates that the local environment around the doped Eu^{3+} ions in TiO_2 is quite different between individual

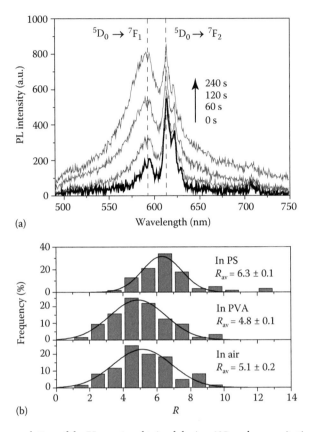

(a)

(b)

FIGURE 14.6 (a) Time evolution of the PL spectra obtained during 405 nm laser excitation for a single TiO_2:Eu^{3+} nanoparticle under an Ar atmosphere. (b) Histograms of the R values obtained for single TiO_2:Eu^{3+} nanoparticles spin-coated on the cover glass (dry sample; RH = 40%), embedded in a poly(vinyl alcohol) (PVA) film, and octadecyltrimethoxysilane-modified nanoparticles embedded in a polystyrene (PS) film.

nanoparticles. According to the average particle size (~10 nm), the number of Eu^{3+} ions per single particle and interion distance were roughly estimated to be 230 ions and 0.8 nm on an average, respectively. The luminescence properties are significantly influenced by dopant pair formation.[44,45] For example, the emission from higher ^5D_j levels of Eu^{3+} or Tb^{3+} is quenched by cross-relaxation processes in pairs, whereas this is not observed for single ions.[41] Based on the mathematical probabilistic theory (Stein–Chen Poisson approximation),[46] the probability for pair-state formation was calculated to be 26%. These estimates imply that several emitting sites exist in the interior region and at the surface of the nanoparticles.

The TiO_2 nanoparticles are usually covered with hydroxyl groups as well as physisorbed water molecules. Dossot and coworkers recently reported that the R value of ~4 obtained for the Eu^{3+}-doped glass sample can be assigned to the partially hydroxylated Eu^{3+} ions with Eu-OH bonds.[47] In fact, almost the same distribution of R values was obtained for the TiO_2:Eu^{3+} nanoparticles in a poly(vinyl alcohol) (PVA) film, confirming that the TiO_2 surface is covered with hydroxyl groups in ambient air. By modifying the TiO_2 surface with octadecyltrimethoxysilane, the distribution of R values was obviously shifted to a higher value and became narrower, when compared to that for the PVA-coated sample. From these findings, it was concluded that the Eu^{3+} ions located at the surface have a lower R value than those in the interior region of the TiO_2 host in ambient air.

Figure 14.6a also shows the time evolution of the PL spectra observed for a single TiO_2:Eu^{3+} nanoparticle under an Ar atmosphere. Only the PL bands due to Eu^{3+} were observed immediately after the laser

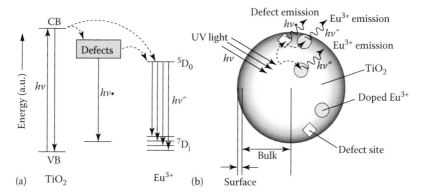

FIGURE 14.7 (a) Energy diagram for the charge and energy transfer reactions induced by the photoexcitation of TiO$_2$:Eu^{3+} nanoparticles. VB and CB denote valence and conduction bands, respectively. The charge trapping and energy transfer processes are indicated by the dotted and broken arrows, respectively. (b) Schematic illustration of the energy transfer from the TiO$_2$ host to the doped Eu^{3+} ions. The Eu^{3+} ions located on the surface of the TiO$_2$ host are sensitized efficiently by charge recombination of the trapped carriers.

irradiation, and then, the PL band from the trapped exciton appeared in the visible region (500–750 nm) and increased over time (see Sections 14.3.2 and 14.3.3 for details). It should be noted that the PL bands attributed to the $^5D_0 \rightarrow {^7}F_1$ and $^5D_0 \rightarrow {^7}F_2$ transitions increased and decreased with the irradiation time, respectively (Figure 14.6a). Furthermore, the time evolution of the *R* value was synchronized with the photoactivation event. This contains information difficult or impossible to obtain from ensemble experiments since each particle behaves differently.

14.4.3 Defect-Mediated Energy Transfer

Based on the above results and discussion, the PL mechanism was proposed as summarized in Figure 14.7. First, visible color centers, most probably, trapped electrons in the vacancy defect sites, are generated by the intrinsic and/or extrinsic excitations of TiO$_2$ nanoparticles under the 405 nm laser irradiation. Both free and trapped electrons can then recombine with the photogenerated holes to produce the PL in the UV and visible regions, respectively. The quenching of free electrons by O$_2$ molecules consequently results in the decreased PL from the trapped excitons. Free excitons should excite both the interior and surface-located Eu^{3+} ions, while trapped excitons at the surface would only excite the surface-located Eu^{3+} ions. This interpretation is supported by the fact that the photoactivation of color centers in the TiO$_2$ host, i.e., the formation of trapped excitons at the defect sites, is accompanied by a significant decrease in *R* (Figure 14.6a). These findings and methodologies would provide further insight into the mechanisms of energy and charge transfer reactions in these hybrid nanomaterial systems.

14.5 Interfacial Electron Transfer on Single Semiconductor Nanoparticles

14.5.1 Electron Transfer from TiO$_2$ to Surface-Adsorbed Oxygen Molecules

As mentioned in Section 14.3.3, it appears that a large number of color centers formed in the bulk and/or on the surface are deactivated with a characteristic lifetime. In fact, the photoactivation of TiO$_2$ emission was accompanied by numerous photon bursts and their frequency of appearance gradually increased until saturation of the PL intensity occurred. The most direct way to evaluate the burst behavior is to record the emission intensity as a function of time, to distinguish between the on times (τ_{on}) and

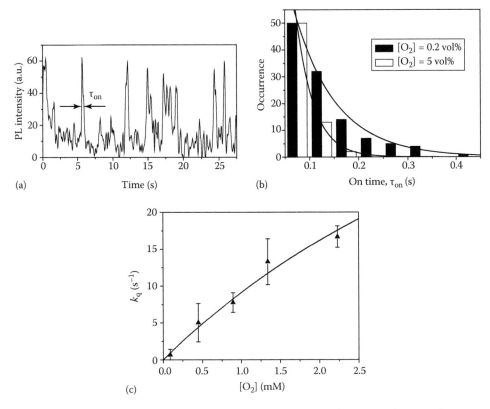

FIGURE 14.8 (a) PL bursts observed for a single TiO_2 nanowire under 405 nm laser irradiation in Ar. (b) Histogram of on time (τ_{on}) determined for the bursts under the oxygen concentrations of 0.2 and 5 vol%. The integration time per one frame was 50 ms. The solid lines indicate the single-exponential fits. Inset shows the oxygen concentration dependence of average quenching rate constant (k_q), which is calculated from $1/\tau_{on} - 1/\tau_{on}^0$, where τ_{on} and τ_{on}^0 are the on times of the bursts in the presence and absence of oxygen, respectively. The solid line is obtained from Equation 14.12, and the fitting parameters are as follows: $K_{ad} = 150\,M^{-1}$ and $[O_2]_{bs} = 68\,M$.

background signals by means of a threshold chosen to be 3 σ greater than the noise levels, and to measure the distribution of τ_{on} as a histogram. As shown in Figure 14.8a, most of the PL bursts disappeared within a few hundred milliseconds under an Ar atmosphere ($[O_2] = 0.2$ vol%). The possible explanations for the limited lifetime of the bursts are as follows: (1) the charge recombination of the color centers with the photogenerated holes, (2) the scavenging of the color centers by residual oxygen molecules or adventitious impurities, and (3) photostimulated detrapping of the color centers to give mobile electrons.[48] In order to elucidate the mechanism of the burst-like PL phenomenon, the oxygen concentration dependence was examined. The experimental and analytical results in Figure 14.8b indeed demonstrate a significant decrease in τ_{on} from 85 ± 6 to 36 ± 2 ms, which are derived by a single-exponential fitting, by increasing the oxygen concentration up to 5 vol%. The frequency of the burst generation rapidly decreased with the increasing oxygen concentration ($[O_2] > 5$ vol%), and eventually only one or no counts during the course of the experiment (15 s) was observed in an air atmosphere ($[O_2] = 21$ vol%). These observations are indicative of a second-order reaction between electrons in TiO_2 and oxygen molecules and a saturation of available oxygen binding sites on the TiO_2 surface.

Let us consider in more detail the quenching dynamics of the photon bursts in accordance with the Langmuir–Hinshelwood equation, which was successfully used to explain the photocatalytic reduction of oxygen molecules adsorbed on the nanocrystalline TiO_2 films.[49]

According to the model, the average quenching rate of the bursts by oxygen is described by

$$k_q = k_q^0 [O_2]_{ad} = \frac{1}{\tau_{on}} - \frac{1}{\tau_{on}^0} \tag{14.12}$$

where

k_q^0 is the second-order reaction rate constant for ET

$[O_2]_{ad}$ is the concentration of adsorbed oxygen molecules on the TiO_2 surface

τ_{on}^0 is the on time of the bursts in the absence of oxygen

The τ_{on}^0 value was estimated to be 91 ± 6 ms from the intercept of the plots of τ_{on}^{-1} versus the concentration of oxygen in the gas phase. As shown in Figure 14.8c, the plots of k_q as a function of the oxygen concentration in the gas phase can be fitted by a simple Langmuir adsorption isotherm:

$$[O_2]_{ad} = \frac{K_{ad} [O_2][O_2]_{bs}}{1 + K_{ad}[O_2]} \tag{14.13}$$

where

K_{ad} is the equilibrium adsorption constant for the oxygen binding

$[O_2]_{bs}$ is the concentration of the oxygen-binding sites

Based on this equation, we obtained the K_{ad} value of $150 \pm 50\,M^{-1}$, which is on the same order as those reported elsewhere.[49,50] Moreover, the quenching yield of the bursts by oxygen, which is defined as $\phi_q = 1 - (\tau_{on}/\tau_{on}^0)$, increased to ca. 60% as the oxygen concentration increased to 5 vol%. It should be noted here that the electron half-times in the nanocrystalline TiO_2 film are reported to be about 60 and 550 ms in the presence (5 vol%) and absence of oxygen, respectively.[49] A comparison with the single-molecule analysis data enables us to deduce that the τ_{on} is not only explained by the scavenging of the mobile electrons by oxygen molecules, but also by other factors. At the present stage, it is considered that the bursts originated from the radiative recombination of holes and electrons at the color centers, such as the oxygen vacancies (VO) with one or two trapped electrons, i.e., the F^+ or F center.[51,52]

14.5.2 Electron Transfer from QDs to Surface-Adsorbed Aromatic Compounds

The photoinduced ET has not received much attention at the single-molecule level,[53–57] since the fluorescence of organic dye molecules is significantly quenched leaving no signal to be detected. To probe the ET dynamics at the single-particle level, QDs are good candidates because their optical properties are superior to conventional organic fluorophores in many aspects, such as prominence PL, high photostability, and multiplexing capability. Numerous investigations have focused on the dynamics of charge carriers in single QDs (bare QDs, core/shell QDs such as CdSe/ZnS, and ligand-capped QDs) in order to answer how they become charged.[21] However, a few studies have been performed on the interfacial ET dynamics in single luminescent QD–chromophore (molecular acceptor) conjugates until now.[58–60] Using a single-molecule fluorescence microscope, one can get more information about the electronic interaction and photoinduced ET reactions between individual QDs and adsorbates at the heterogeneous interface.

When a pyromellitimide compound with a carboxylic acid (PI-CA) was modified as an electron acceptor on the surface of CdTe QDs, the number of luminescent spots remarkably decreased, although the intensity histogram did not change. Typical PL intensity trajectories of single CdTe QDs with and without PI-CA are shown in Figure 14.9a. The intensity distributions are almost undistinguishable from

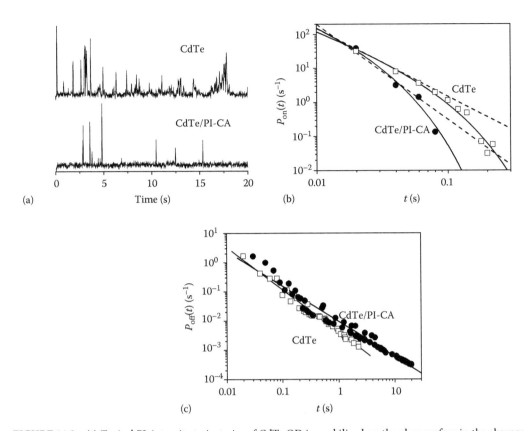

FIGURE 14.9 (a) Typical PL intensity trajectories of CdTe QD immobilized on the glass surface in the absence and presence of PI-CA in chloroform. The integration time per one frame was 20 ms. (b) On-time probability density, $P_{on}(t)$, obtained for single QDs. Solid lines indicate the best fits based on the equation: $P(t) = At^{-1.5}\exp(-\Gamma t)$, where A is the scaling coefficient. Dotted lines indicate the simple power law fits. (c) Off-time probability density, $P_{off}(t)$, obtained for a single QD. Solid lines indicate the simple power law fits.

each other. However, the luminescence of single CdTe QDs with PI-CA appeared less frequently than that of unmodified QDs as well as QDs with linker groups, i.e., propionic acid.

Individual CdTe QDs excited with a 405 nm laser show a fast blinking with an on time of several tens of milliseconds in chloroform. The histograms of the on and off times for single CdTe QDs modified on a glass surface with and without quenchers were analyzed and compared. The characteristic decay times ($\tau_{on/off}$) for the on and off events dramatically decreased and increased by modification of the PI-CA molecules on the surface of the single QDs, respectively.

At sufficiently long times, the histograms have only one or no counts per bin time due to finite counting statistics. Therefore, the probability density, $P(t)$, is analyzed by weighting each point in the on/off histograms by the average time (Δt_{av}) between nearest neighbor event bins using the following equation[61]:

$$P_i(t) = \frac{N_i(t)}{N_{total}} \times \frac{1}{\Delta t_{av}} (i = \text{on or off}) \qquad (14.14)$$

Recently, Marcus et al. reported a nonadiabatic ET theory with a diffusion-controlled ET (DCET) model for the PL blinking of QDs based on the ET processes between a QD and its localized surface states (charge trapping sites).[62–65] To better match the shape of the on-time distributions, we fit them to a truncated power law predicted by the DCET theory:

$$P_i(t) \approx \frac{\sqrt{t_{c,i}}}{2\sqrt{\pi}} t^{-1.5} \exp(-\Gamma_i t) (i = \text{on or off}) \tag{14.15}$$

where

t_c is the critical time that is a function of the electronic coupling strength and other quantities
Γ is the saturation rate

With the assumption that two parabolas along a reaction coordinate, which represent the free energy curves of the light-emitting and dark states, have the same curvature, Γ_{on}, is given by

$$\Gamma_{on} = \frac{(\lambda + \Delta G_{ET})^2}{8 t_{diff} \lambda k_B T} \tag{14.16}$$

where

λ is the reorganization energy
ΔG_{ET} is the free energy change for the ET
t_{diff} is the diffusion correlation time constant for motion on a parabolic energy surface

As shown by the solid lines in Figure 14.9b, this function well matches the shape of the on-time distributions rather than the simple power law fits. It is noteworthy that the $1/\Gamma_{on}$ values decrease by about one-third due to the conjugation with PI-CA (from 0.055 to 0.020). For the off-time histograms, it was found that m_{off} (1.2) for the CdTe/PI-CA system is lower than those (1.5) for the CdTe and CdTe/propionic acid systems, indicating that the apparent decrease in the number of luminescent spots is partially due to the lengthening of the off time (Figure 14.9c).

According to the DCET model, where a more negative value for ΔG_{ET} should lead to an increase in Γ_{on}, i.e., a larger bending, our experimental results are qualitatively explained by the fact that ΔG_{ET} estimated for the CdTe/PI-CA (−0.91 eV) is significantly higher than that (−0.079 eV) for the ET with surface states just below the edge of the conduction band.[66] However, Scholes et al. recently reported a very small λ value (0.020 eV) for the photoinduced ET from CdTe to CdSe, which is probably due to the weak exciton–phonon coupling in QDs.[67] The fact implies that the ET reaction between the QD and its surface states might occur in the Marcus inverted region since $\lambda < -\Delta G_{ET}$.[68] Meanwhile, the internal and solvent λ values are calculated to be 0.24[69] and 0.50 eV for the formation of PI$^{\cdot-}$ in chloroform, respectively. The total λ of 0.74 eV is also slightly lower than $-\Delta G_{ET}$ (0.91 eV) for the CdTe/PI-CA system. If these calculations are correct, the opposite dependence of ΔG_{ET} upon Γ_{on} should be observed. To more quantitatively discuss the connection between the experimental and theoretical results, one would need additional data with shorter time bins and further information about the energetic and spatial distributions of the surface trap states, hopping of the charge carriers into and out of these states, the λ values (both internal and solvent) for trapped electrons (holes) at the surface, conformational changes in the adsorbed molecules, etc.

14.5.3 Intermittent Electron Transfer Reaction

The power-law decay in a single QD is presumably associated with temporal fluctuations in the energy differences between the ground and light-emitting states and between the light-emitting and dark states. The former corresponds to the spectral diffusion of a QD emission as reported by Empedocles and Bawendi,[70] although a notable spectral change was not observed during the course of the above experiment. On the other hand, the latter can be attributed to several properties, such as the local dielectric environments, the conformation of the adsorbed molecules, the vibronic coupling between the QDs and adsorbed molecules, etc. These make it rather difficult for ensemble-averaged measurements to analyze the ET dynamics at the heterogeneous interfaces.

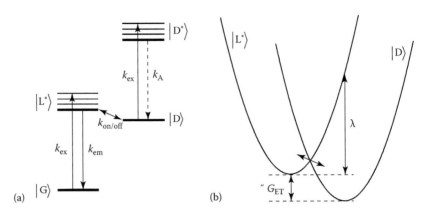

FIGURE 14.10 (a) Energy levels in the DCET model. (b) Parabolic potential surfaces based on the DCET model. Transitions from the ground state, $|G\rangle$, to the light-emitting state, $|L^*\rangle$, or from the dark state, $|D\rangle$, to the excited dark state, $|D^*\rangle$, are driven by incident light. Transitions from $|L^*\rangle$ to $|G\rangle$ are primarily radiative, whereas transitions from $|D^*\rangle$ to $|D\rangle$ are primarily nonradiative. k_A denotes the rate constant for the radiationless Auger processes. The transition between $|L^*\rangle$ and $|D\rangle$ represents the bottleneck charge separation and recombination processes with rate constants $k_{on/off}$ when the system is at reaction coordinate.

The obtained shorter on times and longer off times are possibly explained by the fact that the electrons generated in the QDs were trapped by the surface-modified PI-CA molecules as mentioned above. A fast forward ET reaction, followed by the back ET in the subnanosecond time region, is expected when the electronic interaction between the CdTe QD and PI chromophore is strong enough. During these reaction cycles, the PL of the CdTe QD should be completely quenched and show the off state because the charged QD is dark as illustrated in Figure 14.10. If their electronic interplay is interrupted, the PL should recover and show the on state. A significant low total PL intensity before single CdTe/PI-CA conjugates bleach infer the importance of the dark state on the photodecomposition processes during the photochemical cycles.

Another explanation is that the dark state is attributed to the long-lived charge-separated state, i.e., $PI^{\bullet-} + CdTe(h^+)$, since the trapping of electrons leaves behind holes in the core or on the surface of the QDs. However, both the intensity and blinking characteristics of the single QDs did not show an obvious change when compared to those in the absence of 1 mM DABCO as a hole scavenger, suggesting that this mechanism would be excluded.

Lian et al. reported that ET activity between single CdSe/ZnS QDs and molecular acceptors, Fluorescein 27, undergoes large fluctuation.[58] They speculated that this fluctuation is caused by the fluctuation of adsorbate adsorption conformation on QD and/or the change of charge states in QDs, which may affect the driving force and electronic coupling strength. A similar mechanism was previously proposed for the photosensitized interfacial ET processes between single dye molecules, such as Coumarin 343 and Cresyl Violet, and TiO$_2$ nanoparticles.[55] Thus, the blinking characteristics observed for single CdTe QDs modified with PI-CA are due to the energy diffusions away from and back to a resonance condition fulfilled by the energy of the acceptor states resulting in the intermittent changes in the interfacial ET redox turnover rates.

14.6 Conclusions

The quantitative examination of the spectral and kinetic characteristics revealed that the PL bands originating from defects in the bulk and/or on the surface of TiO$_2$ nanoparticles appeared in the visible region with numerous photon bursts by the photoirradiation with a 405 nm laser in an inert gas atmosphere. The spatially resolved PL imaging techniques will enable us to ascertain the location of the active sites and to elucidate the reaction dynamics of charge carriers in the TiO_2 nanostructures with various morphologies, e.g., nanotubes, nanowires, nanosheets, etc. Such information is useful not only for exploration, but also for the development of TiO_2-based solar cells and photocatalysts, and provides us insights not available from other methods.

The defect-mediated PL dynamics of pure and Eu^{3+}-doped TiO_2 nanoparticles and the interfacial ET dynamics between semiconductor nanoparticles and organic (inorganic) compounds have been investigated at the single-particle level. The interfacial ET processes are closely associated with the spatial heterogeneities of the nanoscale local environments, the temporal fluctuations in the conformation of adsorbates, and the electronic coupling between the adsorbed molecules and the rough surfaces of the semiconductors. Forthcoming studies will focus on the mechanistic details to resolve the above-mentioned issues.

In conclusion, it is stated that single-molecule, single-particle experiments can provide novel information for elucidating the mechanism of heterogeneous chemical reactions on the surface of semiconductor nanoparticles and for designing or identifying new materials for applications in a variety of areas.

References

1. Alivisatos, A. P. *J. Phys. Chem.* 1996, *100*, 13226–13239.
2. Brus, L. E. *J. Chem. Phys.* 1983, *79*, 5566–5571.
3. Moerner, W. E.; Orrit, M. *Science* 1999, *283*, 1670–1676.
4. Kulzer, F.; Orrit, M. *Annu. Rev. Phys. Chem.* 2004, *55*, 585–611.
5. Tinnefeld, P.; Sauer, M. *Angew. Chem. Int. Ed.* 2005, *44*, 2642–2671.
6. Thompson, T. L.; Yates, J. T., Jr. *Chem. Rev.* 2006, *106*, 4428–4453.
7. Chen, X.; Mao, S. S. *Chem. Rev.* 2007, *107*, 2891–2959.
8. Moerner, W. E.; Fromm, D. P. *Rev. Sci. Instrum.* 2003, *74*, 3597–3619.
9. Cornish, P. V.; Ha, T. *ACS Chem. Biol.* 2007, *2*, 53–61.
10. Zuerner, A.; Kirstein, J.; Doeblinger, M.; Braeuchle, C.; Bein, T. *Nature* 2007, *450*, 705–708.
11. Roeffaers, M. B. J.; Sels, B. F.; Uji-i, H.; De Schryver, F. C.; Jacobs, P. A.; De Vos, D. E.; Hofkens, J. *Nat.* 2006, *439*, 572–575.
12. Xu, W.; Kong, J. S.; Yeh, Y.-T. E.; Chen, P. *Nat. Mater.* 2008, *7*, 992–996.
13. Axelrod, D. *Traffic* 2001, *2*, 764–774.
14. Colvin, V. L.; Schlamp, M. C.; Allvisatos, A. P. *Nature* 1994, *370*, 354–357.
15. Steckel, J. S.; Snee, P.; Coe-Sullivan, S.; Zimmer, J. P.; Halpert, J. E.; Anikeeva, P.; Kim, L.-A.; Bulovic, V.; Bawendi, M. G. *Angew. Chem. Int. Ed.* 2006, *45*, 5796–5799.
16. Kamat, P. V. *J. Phys. Chem. C* 2008, *112*, 18737–18753.
17. Medintz, I. L.; Uyeda, H. T.; Goldman, E. R.; Mattoussi, H. *Nat. Mater.* 2005, *4*, 435–446.
18. Gill, R.; Zayats, M.; Willner, I. *Angew. Chem. Int. Ed.* 2008, *47*, 7602–7625.
19. Schaller, R. D.; Sykora, M.; Jeong, S.; Klimov, V. I. *J. Phys. Chem. B* 2006, *110*, 25332–25338.
20. Ellingson, R. J.; Beard, M. C.; Johnson, J. C.; Yu, P.; Micic, O. I.; Nozik, A. J.; Shabaev, A.; Efros, A. L. *Nano Lett.* 2005, *5*, 865–871.
21. Gomez, D. E.; Califano, M.; Mulvaney, P. *Phys. Chem. Chem. Phys.* 2006, *8*, 4989–5011.
22. Tachikawa, T.; Ishigaki, T.; Li, J.-G.; Fujitsuka, M.; Majima, T. *Angew. Chem. Int. Ed.* 2008, *47*, 5348–5352.
23. Tachikawa, T.; Majima, T. *J. Am. Chem. Soc.* 2009, *131*, 8485–8495.
24. Jeon, K.-S.; Oh, S.-D.; Suh, Y. D.; Yoshikawa, H.; Masuhara, H.; Yoon, M. *Phys. Chem. Chem. Phys.* 2009, *11*, 534–542.
25. Tang, H.; Berger, H.; Schmid, P. E.; Levy, F.; Burri, G. *Solid State Commun.* 1993, *87*, 847–850.
26. Serpone, N.; Lawless, D.; Khairutdinov, R.; Pelizzetti, E. *J. Phys. Chem.* 1995, *99*, 16655–16661.
27. Sekiya, T.; Kamei, S.; Kurita, S. *J. Lumin.* 2000, *87–89*, 1140–1142.

28. Sekiya, T.; Tasaki, M.; Wakabayashi, K.; Kurita, S. *J. Lumin.* 2004, *108*, 69–73.

29. Knorr, F. J.; Mercado, C. C.; McHale, J. L. *J. Phys. Chem. C* 2008, *112*, 12786–12794.

30. Cordero, S. R.; Carson, P. J.; Estabrook, R. A.; Strouse, G. F.; Buratto, S. K. *J. Phys. Chem. B* 2000, *104*, 12137–12142.

31. Wang, Y.; Tang, Z.; Correa-Duarte, M. A.; Liz-Marzan, L. M.; Kotov, N. A. *J. Am. Chem. Soc.* 2003, *125*, 2830–2831.

32. Javier, A.; Strouse, G. F. *Chem. Phys. Lett.* 2004, *391*, 60–63.

33. Kuznetsov, V. N.; Serpone, N. *J. Phys. Chem. C* 2007, *111*, 15277–15288.

34. Palacios, R. E.; Fan, F.-R. F.; Bard, A. J.; Barbara, P. F. *J. Am. Chem. Soc.* 2006, *128*, 9028–9029.

35. Redmond, G.; Fitzmaurice, D. *J. Phys. Chem.* 1993, *97*, 1426–1430.

36. Staniszewski, A.; Morris, A. J.; Ito, T.; Meyer, G. J. *J. Phys. Chem. B* 2007, *111*, 6822–6828.

37. Redmond, G.; Fitzmaurice, D.; Graetzel, M. *J. Phys. Chem.* 1993, *97*, 6951–6954.

38. Boschloo, G. K.; Goossens, A. *J. Phys. Chem.* 1996, *100*, 19489–19494.

39. Boschloo, G.; Fitzmaurice, D. *J. Phys. Chem. B* 1999, *103*, 2228–2231.

40. Parker, I. D. *J. Appl. Phys.* 1994, *75*, 1656–1666.

41. Blasse, G.; Grabmaier, B. C. *Luminescent Materials*, Springer, Berlin, Germany, 1994.

42. Jia, C. W.; Xie, E. Q.; Zhao, J. G.; Sun, Z. W.; Peng, A. H. *J. Appl. Phys.* 2006, *100*, 023529/023521–023529/023525.

43. Li, J.-G.; Wang, X.; Watanabe, K.; Ishigaki, T. *J. Phys. Chem. B* 2006, *110*, 1121–1127.

44. Riwotzki, K.; Haase, M. *J. Phys. Chem. B* 1998, *102*, 10129–10135.

45. Blasse, G. *Prog. Solid State Chem.* 1988, *18*, 79–171.

46. Suyver, J. F.; Meester, R.; Kelly, J. J.; Meijerink, A. *Phys. Rev. B Condens. Matter Mater. Phys.* 2001, *64*, 235408/235401–235408/235406.

47. Grausem, J.; Dossot, M.; Cremel, S.; Humbert, B.; Viala, F.; Mauchien, P. *J. Phys. Chem. B* 2006, *110*, 11259–11266.

48. Shkrob, I. A.; Sauer, M. C., Jr. *J. Phys. Chem. B* 2004, *108*, 12497–12511.

49. Peiro, A. M.; Colombo, C.; Doyle, G.; Nelson, J.; Mills, A.; Durrant, J. R. *J. Phys. Chem. B* 2006, *110*, 23255–23263.

50. Addamo, M.; Augugliaro, V.; Coluccia, S.; Faga, M. G.; Garcia-Lopez, E.; Loddo, V.; Marci, G.; Martra, G.; Palmisano, L. *J. Catal.* 2005, *235*, 209–220.

51. Serpone, N. *J. Phys. Chem. B* 2006, *110*, 24287–24293.

52. Lei, Y.; Zhang, L. D.; Meng, G. W.; Li, G. H.; Zhang, X. Y.; Liang, C. H.; Chen, W.; Wang, S. X. *Appl. Phys. Lett.* 2001, *78*, 1125–1127.

53. Lu, H. P.; Xie, X. S. *J. Phys. Chem. B* 1997, *101*, 2753–2757.

54. Holman, M. W.; Liu, R.; Adams, D. M. *J. Am. Chem. Soc.* 2003, *125*, 12649–12654.

55. Biju, V.; Micic, M.; Hu, D.; Lu, H. P. *J. Am. Chem. Soc.* 2004, *126*, 9374–9381.

56. Tachikawa, T.; Cui, S.-C.; Tojo, S.; Fujitsuka, M.; Majima, T. *Chem. Phys. Lett.* 2007, *443*, 313–318.

57. Bell, T. D. M.; Stefan, A.; Masuo, S.; Vosch, T.; Lor, M.; Cotlet, M.; Hofkens, J.; Bernhardt, S.; Muellen, K.; van der Auweraer, M.; Verhoeven, J. W.; De Schryver, F. C. *ChemPhysChem* 2005, *6*, 942–948.

58. Issac, A.; Jin, S.; Lian, T. *J. Am. Chem. Soc.* 2008, *130*, 11280–11281.

59. Tachikawa, T.; Majima, T. *J. Fluoresc.* 2007, *17*, 727–738.

60. Cui, S.-C.; Tachikawa, T.; Fujitsuka, M.; Majima, T. *J. Phys. Chem. C* 2010, *114*, 1217–1225.

61. Kuno, M.; Fromm, D. P.; Hamann, H. F.; Gallagher, A.; Nesbitt, D. J. *J. Chem. Phys.* 2001, *115*, 1028–1040.

62. Tang, J.; Marcus, R. A. *Phys. Rev. Lett.* 2005, *95*, 107401/107401–107401/107404.

63. Tang, J.; Marcus, R. A. *J. Chem. Phys.* 2005, *123*, 054704/054701–054704/054712.

64. Tang, J.; Marcus, R. A. *J. Chem. Phys.* 2005, *123*, 204511/204511–204511/204516.

65. Pelton, M.; Smith, G.; Scherer, N. F.; Marcus, R. A. *Proc. Natl. Acad. Sci. U.S.A.* 2007, *104*, 14249–14254.

66. Wang, X.; Yu, W. W.; Zhang, J.; Aldana, J.; Peng, X.; Xiao, M. *Phys. Rev. B Condens. Matter Mater. Phys.* 2003, *68*, 125318/125311–125318/125316.
67. Scholes, G. D.; Jones, M.; Kumar, S. *J. Phys. Chem. C* 2007, *111*, 13777–13785.
68. Marcus, R. A.; Sutin, N. *Biochim. Biophys. Acta Rev. Bioenerg.* 1985, *811*, 265–322.
69. Mi, Q.; Chernick, E. T.; McCamant, D. W.; Weiss, E. A.; Ratner, M. A.; Wasielewski, M. R. *J. Phys. Chem. A* 2006, *110*, 7323–7333.
70. Empedocles, S. A.; Bawendi, M. G. *Science* 1997, *278*, 2114–2117.

15

Biological Applications of Photoluminescent Semiconductor Quantum Dots

Oleg Kovtun
Vanderbilt University

Sandra J. Rosenthal
Vanderbilt University

15.1 Introduction

Recent advances in photoluminescence-based techniques for molecular biology have allowed investigators to study fundamental biological processes in living systems with unprecedented spatiotemporal resolution in real time. The dynamic nature of biological processes in conjunction with instrumentation progress demanded that new classes of improved probes be developed to allow access to spatial and temporal scales that have largely remained undisturbed by scientific inquiry with conventional biochemical means [1–5]. Currently, there are several classes of photoluminescent probes available to biologists including, but not limited to, small organic dyes [5,6], genetically encoded fluorescent proteins [7,8], metal–ligand complexes [9,10] silver [11,12], carbon [13,14], and silicon [15] nanoparticles, and semiconductor nanocrystals [16–21], termed as quantum dots (QDs). Among these, QDs have already proven to be valuable tools in biological inquiry since their introduction to the field in 1998 [18,19].

QDs are nanometer-sized semiconductor nanocrystals with the exciton Bohr radius on the order of or smaller than that of the bulk semiconductor material [22]. The quantum confinement effects lead to the increase of band gap and the appearance of discrete energy states, and together with the robust inorganic nature of the QDs give rise to their unique photophysical properties [22]. QDs are characterized by excellent brightness that is a product of large molar absorption coefficients, typically ranging between $100,000–1,000,000\,M^{-1}\,cm^{-1}$ at the first excitonic absorption band, and high photoluminescence (PL) quantum yields (QYs), with values reported close to unity for CdSe/CdZnS nanocrystals emitting in the visible range (400–700 nm) [16,23–25]. They have broad absorption spectra that gradually decrease toward longer wavelengths and narrow, Gaussian, emission spectra independent of excitation wavelength. The full width at half maximum (FWHM) of the emission spectra are ~25–40 nm in the case of

CdS, CdSe, and CdTe QDs and slightly higher values of ~70–90 nm for PbS and PbSe QDs [26–29]. The position of the emission peak is size dependent and shifts to longer wavelengths as the QD size increases. Size tunability of the emission spectra combined with broad absorption spectra enables and considerably simplifies multiplexing, simultaneous imaging of several fluorophores using a single excitation source [30–33]. The robust inorganic nature of QDs and effective surface passivation strategies render them resistant to photodegradation under continuous illumination and physiological conditions [29,31]. Long-term dynamic imaging at high frame rates fully takes advantage of their excellent photostability. PL lifetime of QDs (>10 ns) is significantly longer than conventional organic dye lifetimes (1–4 ns) [6,34]. Time-gated imaging and fluorescence lifetime imaging microscopy (FLIM) techniques exploit such lifetime differences to achieve higher contrast and signal-to-noise ratio (SNR) [35–37]. However, the complicated multi-exponential fluorescence decay behavior of QDs and significant variations of QD spectral properties within a single batch are intrinsic limitations for fluorescence lifetime-based imaging techniques [34,38]. Also, with the two-photon absorption cross sections larger than those of any other photoluminescent probe, QDs are an attractive choice for multiphoton in vivo imaging applications [39]. Overall, it is the ability to tune QD photophysical properties through synthetic control and surface modification that makes them an exciting class of photoluminescent probes for biological applications.

15.1.1 Materials

The peak emission wavelength of QDs depends on the material composition in addition to nanocrystal size and the degree of dispersity. Therefore, it is possible to obtain desirable QD optical properties for a given biological application by varying the material constituents of the semiconductor nanocrystal. Cadmium selenide (CdSe) is the most commonly used material for QD preparation. CdSe-based QDs are characterized by high QYs and effectively span the entire visible light range (420–660 nm) [24,26,29,40–42]. Other QDs include CdS (II/VI type) that feature blueshifted peak emission wavelengths extending into the near-ultraviolet region (350–400 nm) and CdTe (II/IV type) that allow biological investigation in the far-red and near-infrared (NIR) spectral regions (660–750 nm) [29,43–46]. III/V and I-IV-VI$_2$ QDs such as InP, ternary In$_x$Ga$_{1-x}$P, and CuInS$_2$/ZnS have been prepared in attempt to develop more environmentally friendly alternatives to Cd-based semiconductor nanocrystals [47–52]. Also, syntheses of NIR-emitting PbS and PbSe QDs (≥700 nm) have been reported [26,27]. NIR QD probes offer a promising potential in deep tissue and in vivo imaging by dramatically improving tissue penetration depth and minimizing cellular damage from the excitation source.

All of the QD materials mentioned above suffer from low QY and poor thermal and photochemical stability under physiologically relevant conditions if not properly surface passivated. Encapsulation of a core particle in a shell of wider-bandgap semiconductor material significantly improves QD PL QY and stability by eliminating the presence of reactive surface trap states that induce unfavorable nonradiative relaxation processes and physical degradation [26,40–42]. A one- or two-layer inorganic shell was shown to increase PL QY by up to 300% [40]. As the thickness of the inorganic shell increases, the core particle exhibits higher resistance to degradation. In the case of CdSe cores, CdS/ZnS is the surface-passivating material of choice to give CdSe/CdZnS core/shell QDs a reported PL QY in excess of 85% [23].

15.1.2 Solubilization

QDs must satisfy several requirements to be compatible with biological systems. They must be rendered water soluble and stable in physiologically relevant pH and ionic strength ranges, possess functional surface elements that confer biological specificity, display minimal nonspecific interactions, and retain their optical properties after post-preparative surface modifications. Since core/shell semiconductor nanocrystal synthesis typically takes place in high-temperature, organic, nonpolar solvents, the as-prepared nanocrystals have no intrinsic water solubility due to the hydrophobic nature of the organic capping ligands (trioctylphosphine oxide, TOPO; hexadecylamine, HDA) [29,42]. Currently, the main approaches to rendering as-prepared QDs water soluble are (i) ligand cap exchange (LCE), (ii) encapsulation in a heterofunctional polymer shell, and (iii) encapsulation in a silica shell. In the LCE approach, bifunctional ligands, which contain a point of attachment to the core/shell nanocrystal surface and a hydrophilic moiety to aid in aqueous dispersion, displace native organic capping ligands. Surface attachment of the ligand occurs via a thiol (–SH), amine (–NH$_2$), or phosphine (–PH$_2$) functionality, and such groups as hydroxyl (–OH), carboxyl (–COOH), methoxy (–OCH$_3$), and polyethylene glycol (PEG) [(–OCH$_2$CH$_2$–)$_n$] provide the hydrophilic interface that is necessary to render QD water soluble. Representative examples of the ligands used in LCE approach are demonstrated in Figure 15.1. Thiolated carboxylic acids are commonly chosen as the capping ligands due to the strong binding affinity of thiols to the QD surface metal atoms [20]. In the case of the polymer encapsulation approach, native

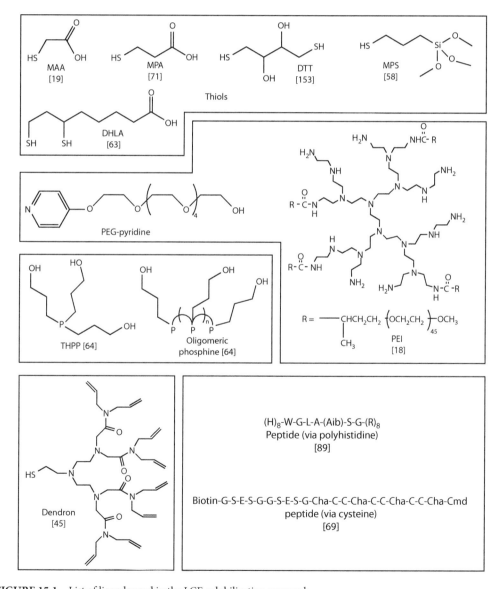

FIGURE 15.1 List of ligands used in the LCE solubilization approach.

TOPO/HDA are retained and associate with added amphiphilic polymer via hydrophobic interactions [53–55]. The aliphatic chains of TOPO/HDA protrude into the surrounding environment and tightly interdigitate with the aliphatic chains of the added amphiphilic polymer, phospholipid, or amphiphilic polysaccharide through hydrophobic interactions (Figure 15.2). The outer hydrophilic backbone of the resulting polymer shell aids in the aqueous dispersity and colloidal stability of the encapsulated water-soluble QD and presents functionalities necessary for further conjugation (e.g., COO⁻). In the silica encapsulation approach, as-prepared core/shell nanocrystals are coated with an inert silica shell via techniques based on the adapted Stöber method or the water-in-oil reverse microemulsion method [15,56–62]. The Stöber method is based on the alkaline hydrolysis and condensation of tetraethyl orthosilicate (TEOS) in ethanol:water mixtures and involves surface exchange of the native QD capping

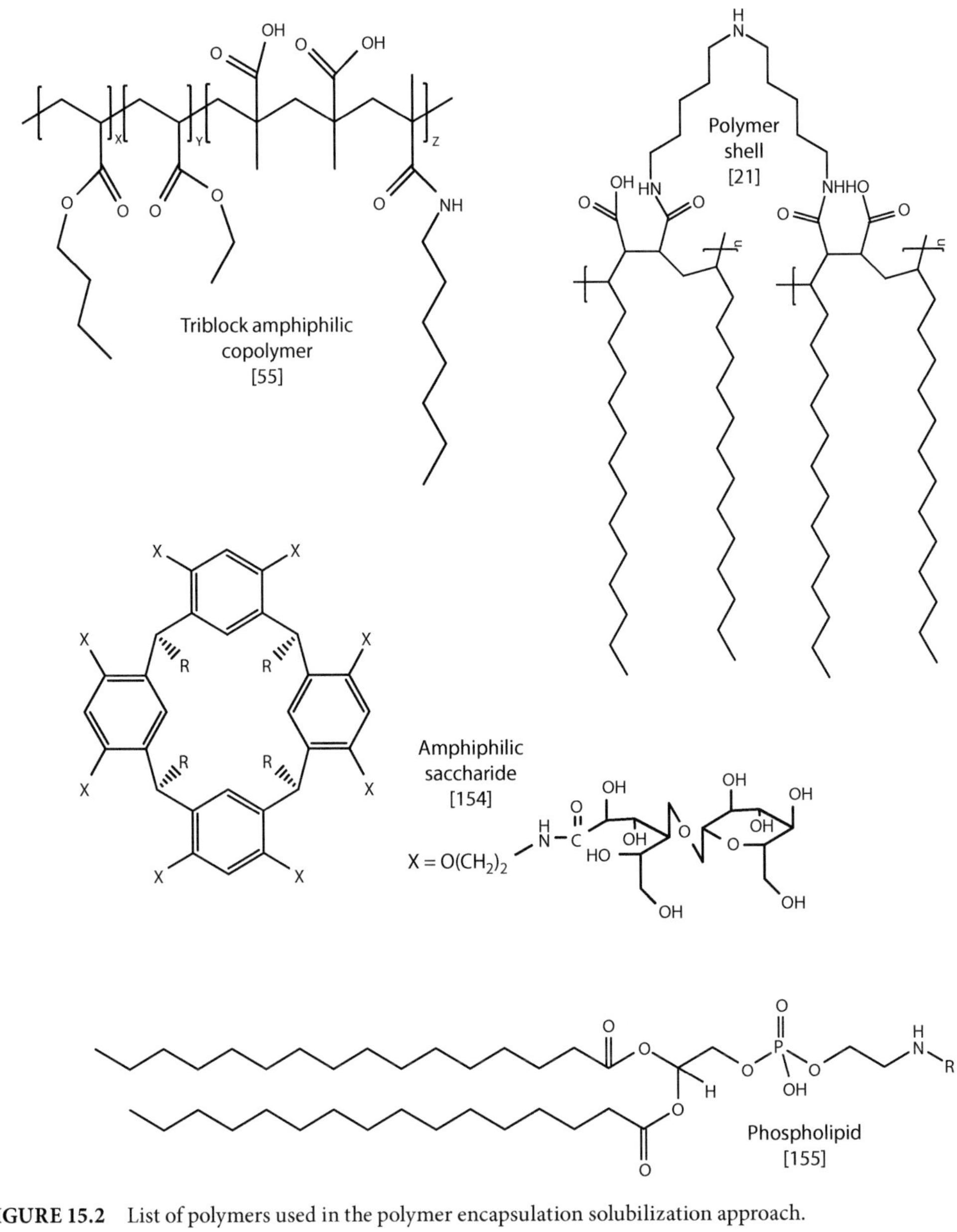

FIGURE 15.2 List of polymers used in the polymer encapsulation solubilization approach.

ligands with a silane coupling agent such as 3-mercapto-1-propane sulfonic acid (MPS). In the micro-emulsion approach, QDs do not act as seeds for silica growth but are instead incorporated into silica spheres formed by the hydrolysis and condensation of TEOS at the water/oil interface.

Each of the QD solubilization approaches mentioned above is associated with distinct advantages and drawbacks and may be chosen depending on the requirements a particular biological system imposes on the photoluminescent probe. Ligand-exchanged QDs are known to suffer from significant QY loss and poor long-term colloidal stability in aqueous buffers due to the surface-altering nature of the ligand-exchange process and the dynamic character of QD–ligand bonds [20]. QD stability may be improved through the use of polydentate ligands with multiple attachment sites to the nanocrystal surface. For example, dihydrolipoic acid (DHLA) and oligomeric phosphines were reported to dramatically prolong the shelf life of QDs in comparison to monodentate thiol ligands [63,64]. In addition, the lack of tools to determine the efficiency of the LCE process and the final surface coverage remains a significant obstacle to broader use of ligand-exchanged QDs in biological applications [20]. Amphiphilic polymer encapsulation has thus far been the preferred solubilization strategy since polymer-coated QDs retain the native passivating ligands and do not exhibit significant loss of PL QY. In addition, the use of an amphiphilic polymer shell results in excellent QD stability over a wide range of pH and aqueous buffer concentrations [65]. Similar to the polymer shell, silica shell provides QDs with improved colloidal stability and preserves QD optical properties. However, in contrast to LCE approach, silica shell and polymer encapsulation significantly increase hydrodynamic diameter (HD) of QDs, with typical HD values of 30–40 nm [53,65]. Increased post-encapsulation QD size places limits on the applicability of QDs to size-sensitive biological applications such as single-QD tracking (SQT), intracellular delivery, and fluorescence resonance energy transfer (FRET).

An alternative approach to prepare water-soluble QDs is synthesis directly in aqueous medium [43,66,67]. This allows to avoid any post-preparative solubilization modifications and possibly to eliminate large decreases in QY associated with such processing. There have been a few reports of aqueous preparations of QDs capped with 3-mercaptoethanol, thioglycerol, thioglycolic acid (TGA), and gluta-thione [67,68]. Although a water-based approach is simpler and cost effective in comparison to the traditional TOPO synthetic methodology, QDs prepared in water tend to suffer from lower QYs and broad emission spectra, which severely limits their utility in biology.

15.1.3 Functionalization

QDs must be rendered water soluble in such a way that they contain or allow subsequent conjugation of functional surface elements, which enable specific recognition of biological targets. In addition, it is desirable to obtain QD conjugates that display minimal nonspecific interaction with the surrounding biological environment. Several methodologies have been developed to confer biological specificity to QDs and can be grouped into covalent conjugation and noncovalent interaction strategies (Figure 15.3).

In covalent conjugation strategy, a reactive group (–COOH, –NH$_2$, –SH) at the QD surface is coupled to a biomolecule containing a compatible reactive group with the use of a cross-linker molecule. Coupling agents such as 1-ethyl-3-(3-dimethylaminopropyl) carbodiimide (EDC) and succinimidyl-4-(N-maleimidomethyl) cyclohexane-1-carboxylate (SMCC) are commonly used to conjugate –COOH and –NH$_2$ groups and –NH$_2$ and –SH groups, respectively (Figure 15.3a and b). Such covalent conjugation methodology has been employed to couple a variety of biomolecules to QDs including proteins, peptides, nucleic acids, and small molecules (Figure 15.4). Also, direct attachment of biologically active molecules to the QD surface during ligand exchange may also be used to impart QDs with biological specificity. For example, Pinaud et al. coated CdSe/ZnS core/shell QDs with synthetic phytochelatin-related α peptides that contained a cysteine-rich adhesive domain for direct conjugation to QD surface Zn atoms [69]. Gomez et al. used a combination of mercaptopropionic acid and thiolated RGD peptide to modify the surface of CdS QDs and subsequently target $\alpha_v\beta_3$ integrin receptors in rat neuroblastoma

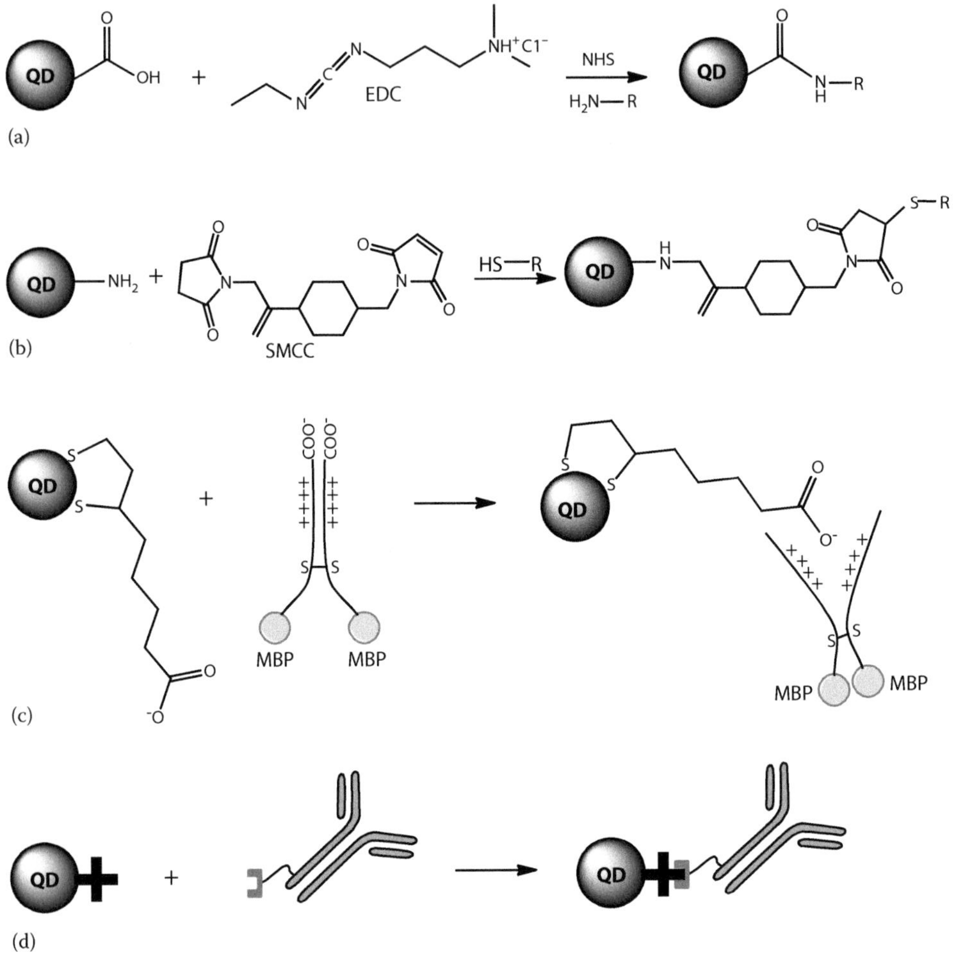

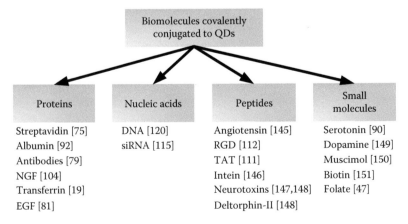

FIGURE 15.3 QD functionalization strategies. EDC/SMCC covalent conjugation strategies are shown in (a) and (b). Electrostatic assembly of MBP protein on DHLA-capped QDs is shown in (c) [64]. Conjugation of streptavidin-conjugated QDs to biotinylated antibody to give an immunofluorescent probe is shown in (d).

Biomolecules covalently
conjugated to QDs

Proteins	Nucleic acids	Peptides	Small molecules
Streptavidin [75]	DNA [120]	Angiotensin [145]	Serotonin [90]
Albumin [92]	siRNA [115]	RGD [112]	Dopamine [149]
Antibodies [79]		TAT [111]	Muscimol [150]
NGF [104]		Intein [146]	Biotin [151]
Transferrin [19]		Neurotoxins [147,148]	Folate [47]
EGF [81]		Deltorphin-II [148]	

FIGURE 15.4 Biomolecules covalently conjugated to the QD surface via EDC/SMCC methodology.

PC12 and rat neonatal cortical cells [70]. In another example, Mitchell and colleagues used alkylthiol-capped oligonucleotides to functionalize CdSe/ZnS core/shell nanocrystals and study their interactions with complementary Au–DNA hybrids [71].

Among all the classes of biomolecules successfully conjugated to QDs, PEG and tetrameric protein streptavidin (Sav) deserve a special mention. PEG is a water-soluble, flexible, organic polymer and is generally considered to be chemically inert and have low toxicity [72]. Due to their excellent biocompatibility, PEG groups were utilized to address the issue of nonspecific binding to the surrounding environment reported for QDs, especially those functionalized with carboxylic acid groups. Bentzen et al. conjugated PEG chains with average molecular weight of 2000 Da to amphiphilic polyacrylic acid-coated QDs functionalized with carboxylic acid groups (AMP-QDs) [73]. PEG surface modification significantly reduced nonspecific interactions of AMP-QDs with different cell lines used in the study. In another study, Chattopadhyay et al. observed that PEGylated QDs effectively address the issue of poor separation of dimly and brightly staining cell populations in flow cytometry [33]. Also, passivation with PEG was reported to minimize nonspecific complexation of bovine serum albumin (BSA) with QDs [74]. Currently, surface passivation with PEG has been employed in a large variety of QD architectures including QDs commercially available from Invitrogen® (Carlsbad, CA).

In 2002, Goldman et al. prepared avidin-conjugated QDs and used them in fluoroimmunoassays in conjunction with antibodies chemically tagged with biotin (biotinylation), a small organic molecule with a high affinity for avidin [75]. The authors emphasized the general utility and large potential of such mixed covalent conjugation approach whereby avidin is used as a bridge between inorganic semiconductor fluorophores and biotinylated probes. Streptavidin is a homotetrameric protein with a molecular mass of ~52.8 kDa [76]. Each streptavidin tetramer is capable of binding four biotins, and streptavidin–biotin interaction is one of the strongest and most stable noncovalent interactions known in nature (dissociation constant, $K_D \sim 10^{-15}$ M; half-life, greater than several days) [77]. Glycosylated, positively charged avidin and deglycosylated neutravidin are similar, evolutionary unrelated biotin-binding proteins that may be used instead of streptavidin [78]. Due to excellent stability, tight interaction, and minimal nonspecific binding, streptavidin–biotin assembly has become a very popular basis for QD biological targeting approach. In 2002, Wu et al. targeted the breast cancer marker Her2 with the QD–streptavidin (QD–strep) conjugates together with humanized anti-Her2 antibody and biotinylated goat antihuman IgG (Figure 15.5a) [79]. They also used the multicolored QD–streptavidin conjugates to visualize actin filaments, microtubules, and nuclear antigens in 3T3 mouse fibroblast cells. In both cases, little to no nonspecific binding was observed when the streptavidin-conjugated QDs were incubated with the cells only. Dahan et al. used the QD–strep conjugates to investigate the diffusion dynamics of individual glycine receptors (GlyR) in the neuronal plasma membrane in an SQT experiment [80]. Again, specific labeling was achieved through the use of a QD–strep conjugate, a biotinylated secondary antibody (Fab fragment), and a GlyR-specific primary antibody. In another example, Lidke et al. relied on biotinylated epidermal growth factor (EGF) and the QD–strep conjugates to elucidate a previously unreported mechanism of retrograde transport to the cell body of the erbB1 receptor tyrosine kinase [81].

Although biotinylated antibodies are frequently used in conjunction with QD–strep conjugates for immunofluorescent staining, there are several drawbacks associated with the antibody-based labeling strategy (Figure 15.3d) [82,83]. In particular, the final size of the QD–antibody conjugate is large (~50 nm) compared to the labeling target, and the antibody–target interaction often suffers from poor stability. It should also be noted that the generation of an efficient antibody against the extracellular epitope of the biological target of interest is typically a costly and complicated procedure that does not ensure a high-affinity final product. In an attempt to address the weaknesses of the antibody-based labeling, Howarth et al. developed a direct, enzymatic, site-specific cell surface protein targeting strategy [84]. In this strategy, a fifteen amino acid acceptor peptide sequence (AP) is genetically fused to either C- or N-terminus of the protein, and biotin ligase (BirA) is used to biotinylate a lysine side chain within the AP sequence. The attached biotin then serves as a handle for streptavidin-conjugated QDs. Such

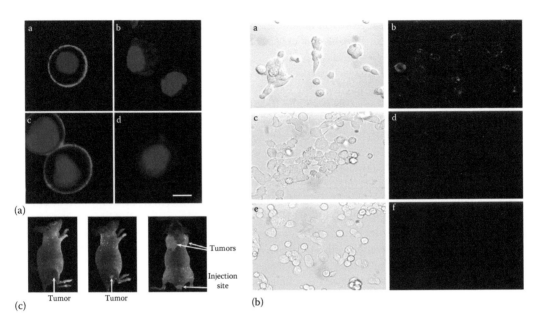

FIGURE 15.5 (See color insert.)   QDs in cellular and whole animal in vivo imaging. Immunofluorescent labeling of extracellular Her2 cancer marker with antibody-conjugated QDs is shown in (a). (Reprinted by permission from MacMillan Publishers Ltd. *Nat. Biotechnol.*, Wu, X., Liu, H. et al., Immunofluorescent labeling of cancer marker Her2 and other cellular targets with semiconductor quantum dots, 21(1), 41–46, Copyright 2003.) Visualization of hSERT in HEK-293T cells with serotonin-conjugated CdSe/ZnS core/shell nanocrystals is shown in (b). (Reprinted with permission from Rosenthal, S.J., Tomlinson, I. et al., Targeting cell surface receptors with ligand-conjugated nanocrystals, *J. Am. Chem. Soc.*, 124(17), 4586–4594. Copyright 2002 American Chemical Society.) The use of QDs to target cancer tumors in living mice is shown in (c). (Reprinted by permission from MacMillan Publishers Ltd. *Nat. Biotechnol.*, Gao, X., Cui, Y. et al., In vivo cancer targeting and imaging with semiconductor quantum dots, *Nat. Biotechnol.*, 22(8), 969–976, 2004. Copyright 2004.)

strategy was employed by Howarth et al. to label α-amino-3-hydroxy-5-methyl-4-isoxazolepropionate (AMPA) receptors in neurons and subsequently observe plasma membrane trafficking of the AMPA–QD complex in the crowded synaptic cleft. In a more recent example, Sun et al. utilized BirA to biotinylate calmodulin subunits of myosin X molecular motor and subsequently visualized the movement of myosin X along actin filaments and bundles with QD–strep conjugates (Figure 15.6b) [85].

Alternative specific targeting approaches have recently emerged and are based on the expression of high-affinity fusion tags (CrAsH, Halotag, polyhistidine) by the cellular target [86–89]. QDs are directed to the target by conjugation with the complementary fusion tag-recognition elements. Our group relies on conjugation of small-molecule ligands to confer biological specificity to QDs [83,90]. In 2002, we conjugated serotonin (5-HT), a monoamine neurotransmitter, to CdSe/ZnS core/shell nanocrystals and measured the electrophysiological response of the interactions between QD–5-HT conjugates and 5-HT$_3$ receptor. Also, we utilized the QD conjugates to visualize human serotonin transporter protein (hSERT) transiently expressed in human embryonic kidney (HEK) cells (Figure 15.5b). The small-molecule strategy is a cost-effective alternative to antibody-based immunofluorescent labeling that results in significantly smaller final QD size. However, one must ensure that the site of attachment on the small molecule does not alter its biological function.

Biological specificity can also be introduced to QDs via adsorption and electrostatic self-assembly in addition to covalent coupling. Mattoussi et al. constructed two-domain maltose binding protein-basic leucine zipper (MBP-zp) fusion protein and conjugated it to DHLA-capped CdSe/ZnS core/shells (Figure 15.3c) [61]. The MBP-zp fusion protein self-assembled on the surface of

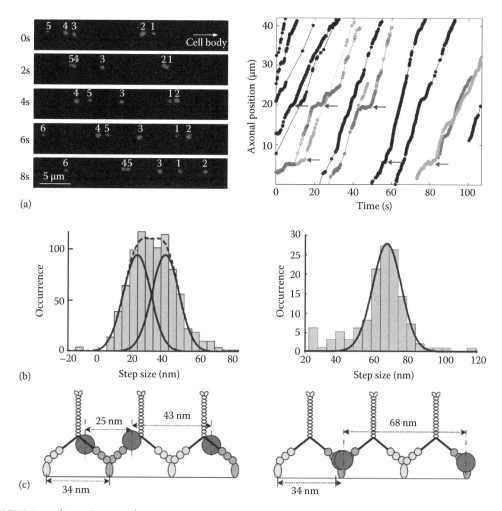

FIGURE 15.6 (See color insert.) SQT examples. (a) Cui et al. observed unidirectional retrograde transport of endosomes, containing QDs conjugated to NGF dimer, along the neuronal axon. (Reprinted by permission from Cui, B., Wu, C. et al., One at a time, live tracking of NGF axonal transport using quantum dots, *Proc. Natl. Acad. Sci. U.S.A.*, 104(34), 13666–13671, 2007. Copyright 2007 National Academy of Sciences, U.S.A.) (b) Sun et al. visualized the movement of biotinylated myosin X along actin filaments and bundles with QD–streptavidin conjugates. (Reprinted by permission from MacMillan Publishers Ltd. *Nat. Struct. Mol. Biol.*, Sun, Y., Sato, O. et al., Single-molecule stepping and structural dynamics of myosin X, *Nat. Struct. Mol. Biol.*, 17(4), 485–491, 2010. Copyright 2010.)

DHLA-capped QDs through electrostatic interactions between negatively charged DHLA carboxyl groups and positively charged leucine zippers of the MBP-zp fusion protein. Goldman et al. used the highly basic leucine zipper domain as the basis for the engineered molecular adaptor PG-zb protein, which served as a bridge for subsequent conjugation of IgG antibody to DHLA-capped CdSe/ZnS core/shell nanocrystals [91]. In the report by Hanaki et al., a series of 10 serum albumins were nonspecifically adsorbed on the surface of CdSe/ZnS core/shells coated with 11-mercaptoundecanoic acid (MUA) [92]. QDs complexed with sheep serum albumin (SSA) were found to be the most stable in aqueous solution.

In conclusion, there exist several surface modification strategies that render QDs water soluble and biocompatible. Each strategy is associated with distinct advantages and disadvantages. Therefore, the surface modification approach must be carefully chosen as it determines photophysical properties of solubilized and functionalized QDs and ultimately the outcome of the experiment. In addition, rigorous control experiments must be performed to determine whether the biological function of the molecule conjugated to QDs is retained.

15.2 Biological Applications

15.2.1 Cellular Labeling

Fluorescent labeling of extra- and intracellular components is the area in which QDs excel. Their high brightness ensures cellular component visualization with high SNR and eliminates the need for a large number of fluorophores to produce a pronounced signal [93]. As a consequence, QDs have been successfully used to label a tremendous number of cellular components both external and internal to the plasma membrane in various types of fixed and live cells (Figure 15.5a and b; Table 15.1). Their excellent resistance to degradation and photobleaching in physiologically relevant conditions enables microscopy image acquisition for extended periods of time under constant illumination without significant image quality loss. Such superior photostability has been often referred to as the most impressive feature of QDs. Indeed, it opens the door to individual protein, endosome, virus, or whole-cell long-term dynamic tracking experiments with unprecedented three-dimensional (3D) spatial and temporal resolution. The ability to visualize and analyze the dynamic interactions between viruses and host cells was recently demonstrated by Joo et al. who tagged human immunodeficiency virus (HIV) with QD–strep conjugates via site-specific BirA biotinylation and examined the kinetics of QD–HIV complex internalization into mammalian cells [94]. Single virus tracking experiments offer an opportunity to visualize and elucidate the molecular details of the viral infection process. In the case of whole-cell tracking, QDs may be used to monitor transplanted and stem cell location, survival, and differentiation. For example, Schormann et al. attempted to track and examine metastatic behavior of QD-tagged MCF-7

TABLE 15.1 Internal and External Cellular Components Labeled with QDs

Internal cellular components	
Nucleus [55]	Prominent membrane-bounded organelle that contains DNA organized into chromosomes
Mitochondria [79]	Membrane-bounded organelle that carries out oxidative phosphorylation and produces most of the ATP in eukaryotic cells
Synaptic vesicles [103]	50 nm spherical membrane-bounded organelles, storing neurotransmitter molecules and mediating neuronal signaling at chemical synapses
Actin filaments [79]	Helical protein filament formed by polymerization of globular actin molecules that determines cell shape and is necessary for whole-cell locomotion
Microtubules [79]	Long hollow cylindrical structures composed of the protein tubulin, controlling intracellular transport
Kinesin, myosin [85,151]	Molecular motor proteins that use the energy of ATP hydrolysis to move along cytoskeletal filaments
External cellular components	
SERT [90]	Plasma membrane transporter protein that is responsible for reuptake of serotonin from the synaptic space into the presynaptic neuron and is a major target for selective serotonin reuptake inhibitors (SSRIs)
erbB/HER [81]	Receptor protein kinases that are central to cellular signaling and are upregulated in certain cancer types
AMPAR [84]	Plasma membrane ionotropic glutamate receptor that mediates fast excitatory synaptic transmission in the central nervous system (CNS)
GlyR [80]	Transmembrane inhibitory GlyR in CNS
K^+ channel [152]	Transmembrane protein that controls a wide variety of cell functions
PSMA [55]	Transmembrane glycoprotein that is a marker of prostate cancer

human breast cancer cells transplanted into healthy mice. Tracking studies at the single QD level will be described in detail in Section 15.2.2 [95].

In contrast to conventional photoluminescent probes, QDs are ideal candidates for multiplexing labeling experiments owing to their broad absorption spectra and size-tunable, narrow, Gaussian emission spectra. They effectively eliminate the need for multiple excitation sources and convoluted spectral compensation algorithms. In 2007, Chattopadhyay et al. were able to resolve 17 fluorescence emissions including eight QD emissions in a polychromatic flow cytometry experiment, the most impressive demonstration of QD multiplexing abilities to date [33]. In this study, 9 antibody-conjugated organic dyes and 8 pMHCI antigen- and antibody-conjugated QDs were featured in a 17-color staining panel to immunophenotype antigen-specific T cells. The resolution of the eight QD emissions was achieved with minimal spectral compensation requirements.

15.2.2 Single-QD Tracking

SQT has emerged as a powerful technique to investigate individual dynamics rather than an ensemble average behavior of QD-tagged single molecules. Careful analysis of SQT data allows elucidation of the molecular details of various biological processes, in particular, membrane trafficking of cell surface proteins [93]. The visualization and interpretation of the dynamic interaction with the biological surroundings have been reported for QD-labeled growth factors, cell surface receptors, membrane lipids, molecular motors, and synaptic vesicles. In a typical SQT experiment, one must label the biological target of interest with a QD, introduce a biological stimulus, observe the effect via image time-series acquisition, apply an algorithm to identify and locate single QD positions in each frame, link QD positions in successive frames to generate trajectories, and analyze the obtained trajectories (Figure 15.7). The x–y position of the diffraction-limited QD spot can be located by 2D Gaussian fitting of its point spread function (PSF) via fluorescence imaging with one nanometer accuracy (the FIONA technique) [96]. The localization accuracy is dependent upon the standard deviation of the PSF and the number of photons detected from a single QD. Very bright particles like QDs significantly improve SNR image profile and can be located with high accuracy (as low as ~10 nm for a single QD) [93]. The next step is to link the centers of single QD spots across adjacent frames of the entire image time series to generate QD trajectories. The overwhelming majority of studies so far focused on particle dynamics in an xy coordinate system. However, it will not be long before 3D SQT will become the norm, with the emergence of 3D tracking techniques [97–99]. There are several tracking software packages available online that can be used as a starting point to develop a QD tracking algorithm and obtain individual trajectories.

When the tracking portion of the SQT experiment is completed, generated trajectories are analyzed to extract information about the QD dynamic behavior including displacement, velocity, and diffusion coefficient. Generally, QD mean-square displacement (MSD) is calculated as a function of time to reduce the noise of the experimental trajectory [100]. The resulting MSD curve is used to determine whether the QD-tagged biological target undergoes Brownian, directed, confined, or anomalous diffusive behavior [101]. The extracted mode of motion parameters are subjected to rigorous statistical analysis and subsequent biological interpretation. Several examples that clearly demonstrate the utility of SQT for elucidating the molecular mechanisms underlying the dynamic behavior of QD-tagged biological structures are described below.

In 2007, Cui et al. observed unidirectional retrograde transport of endosomes, containing QDs conjugated to nerve growth factor (NGF) dimer, along the neuronal axon (Figure 15.6a) [102]. The QD–NGF complexes were internalized by endosomes at the distal axon in a 1:1 stoichiometry and subsequently transported to the cell body in a characteristic stop-and-go movement pattern. In 2009, Zhang et al. published a controversial report on the mechanism of presynaptic neuronal transmission in which they subjected single QD-loaded individual synaptic vesicles to external stimulus and then observed vesicular response [103]. As a result, it was established that the transient vesicular fusion and reuse

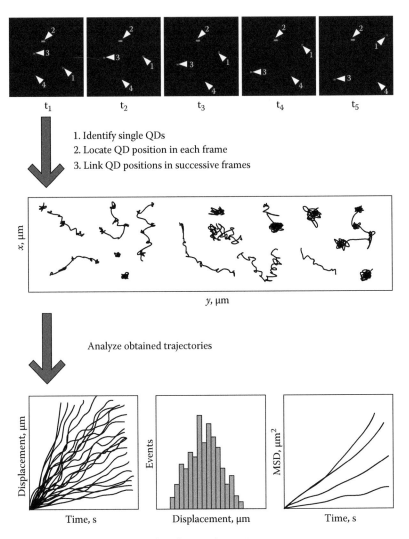

FIGURE 15.7 Scheme of data processing and analysis in the SQT experiment.

(kiss-and-run) is the preferred mechanism for presynaptic transmission as opposed to the full-collapse vesicular fusion. SQT is a particularly useful technique for elucidating dynamics of cell-surface-associated proteins such as potassium and CFTR channels, GABA, NMDA, Gly, and AMPA neurotransmitter receptors, and integrins [93]. In a specific example, Heine et al. showed that postsynaptic AMPA glutamate receptors (AMPAR) undergo fast lateral diffusion (>0.25 μm²/s), which permits efficient replacement of desensitized receptors with the new, functional ones [104]. Such AMPAR exchange facilitates rapid recovery of the depressed synaptic transmission that can be slowed down by constraining the movement of AMPAR within the synapse. Clearly, SQT allows detailed investigation of the aspects of neurotransmission with unprecedented spatial and temporal resolution.

In another remarkable display of their utility in the field of single-particle tracking, QDs were used to reveal the details of intracellular movement of kinesin and myosin motor proteins along cytoskeletal microtubules and actin filaments, respectively. In particular, Pierobon et al. visualized and determined the size of individual steps myosin V takes in the hand-over-hand walking motion along the cytoplasmic actin filaments in living HeLa cells [105]. To observe this phenomenon, myosin–QD constructs were loaded into the cytoplasm of live HeLa cells via nonspecific endocytosis technique and imaged with a wide-field epi-fluorescence microscope (Figure 15.6b). Molecular motor tracking experiments clearly demonstrated that QDs offer a promising potential in reporting intracellular dynamic activity of endogenous molecules. Existing strategies for intracellular delivery of QDs will be discussed in the following section.

15.2.3 Intracellular Delivery and Therapeutics

Successful use of QDs to probe membrane dynamics of extracellular proteins and intracellular motion of individual molecular motors in the cytoplasm of live cells leaves no doubt that QDs have a large potential to become intracellular activity photoluminescent reporters of choice. In addition to intracellular SQT and imaging various disease markers, QDs may be utilized as drug delivery vehicles or tags of conventional drug carriers. For example, Manabe et al. studied the effects of QD-conjugated captopril, an antihypertensive drug, on the blood pressure of hypertensive rats. QD-captopril conjugates and captopril alone were reported to decrease rat blood pressure to similar degrees 30 min after administration [106]. However, the hypotensive effect of the conjugates disappeared after the initial 60 min window. Additional experiments are required to shed light on the mechanism of therapeutic action of QD conjugates. Recently, Bagalkot et al. demonstrated parallel combination disease marker- and drug release-sensing elements in QD architecture. QDs were covalently conjugated with RNA aptamers (A10), which contained an intercalated chemotherapeutic agent doxorubicin [107]. The A10 RNA aptamer enabled specific targeting of prostate-specific membrane antigen (PSMA), while RNA-associated doxorubicin quenched fluorescence. Restored QD fluorescence due to the slow release of doxorubicin in the intracellular environment provided the means to monitor the therapeutic process in real time. Treatment of PSMA-positive cells with the QD–RNA–doxorubicin conjugates induced apoptosis and resulted in significant decrease of cell viability. In another instance, small interfering RNA (siRNA) molecules were covalently linked to the QD surface in an attempt to silence eGFP protein expression [71]. QDs may also be used to visualize drug delivery via other drug carriers as demonstrated by Jia et al. They used QDs to visualize intracellular delivery of polyethyleneimine (PEI)-coated carbon nanotubes that served as a vector for an antisense oligodeoxynucleotide sequence, a hydrophilic therapeutic agent [108].

The future of intracellular applications of QDs depends on the efficient cytoplasmic delivery. Delehanty et al. group QD intracellular delivery strategies into three broad categories: (1) passive delivery, (2) facilitated delivery, and (3) active delivery [109]. In passive delivery, hydrophilic QDs undergo nonspecific endosomal sequestration due to electrostatic QD–membrane interactions. Jaiswal et al. visualized nonspecific endocytosis that occurred when HeLa cells were incubated with ~500 nM DHLA-coated QDs for several hours [110]. Although passive delivery is attractive in terms of simplicity, the cell-type-independent internalization and the inability to escape endosomal sequestration remain significant challenges.

Facilitated delivery is based upon QD functionalization with peptides, proteins, small molecules, lipids, and polymers. Typically, intracellular delivery is achieved through the initial interaction of the QD conjugate with a specific receptor and the subsequent QD–receptor complex endosomal sequestration. Alternatively, QD conjugates undergo endocytosis due to electrostatic interactions between charged surface groups and plasma membrane. The positively charged TAT peptide and the arginine–glycine–aspartate (RGD) tripeptide are two prominent examples of peptides that mediate intracellular delivery of QDs. High positive charge of the TAT peptide due to the presence of a linear polyarginine (lysine) chain permits electrostatic interaction with negatively charged receptors and subsequent receptor-mediated endocytosis [111]. On the other hand, the RGD peptide was reported to mediate QD delivery via binding to the membrane receptors known as integrins [112,113]. Similar to peptide–QD conjugates, protein–QD and small molecule–QD conjugates take advantage of receptor endocytosis, with QD specificity for a given cell membrane marker introduced by protein (small molecule). Antibodies, cholera toxin B, transferrin, and folate conjugated to QDs have all been reported to undergo endocytosis after binding to the corresponding cell membrane proteins [19,47,109,114,].

Lipids and polymers have also been used to deliver QDs into cells. Derfus et al. reported the internalization of the liposomes containing commercial transfection agent Lipofectamine 2000 self-assembled on the surface of negatively charged QDs [115]. Several reports describe the use of phospholipid-based micelles engulfing QDs in the central pocket [109]. Bruchez et al. anchored the endosomal-disrupting PEG–PEI copolymers on the QD surface and observed subsequent cytoplasmic release of the QD conjugates [18]. While lipid- and polymer-mediated delivery usually results in a higher degree of uptake, the relatively large size of the final QD conjugate and the inability to control QD concentration and loading efficiency are likely to prevent this strategy from wide utilization.

In contrast to passive and facilitated delivery, active delivery techniques such as electroporation and microinjection rely on the application of mechanical stress to the cell membrane. In the electroporation technique, a brief electric pulse applied to the membrane temporarily permeabilizes the lipid bilayer and results in the influx of QDs into the cytoplasm. The advantages of electroporation are that a large number of cells can be processed simultaneously in a relatively inexpensive procedure. However, the QDs tend to aggregate near the membrane entry, which is undesirable for intracellular reporting applications. Also, a strong electrical shock applied to the cell suspension may result in reduced cell viability. In the microinjection technique, cells are first visualized with a fluorescent microscope and then a thin-glass tube is used to inject a QD solution either into the nucleus or the cytoplasm of the cell [116]. While QDs are typically dispersed at the injection site and diminished cell mortality is reported, microinjection is associated with high cost and low potential for high throughput. However, both active delivery techniques enable one to escape endosomal sequestration that typically occurs in other strategies.

Overall, it is evident that the intracellular fate of QDs depends on the QD size, surface coating, ability to escape endosomal capture, and specific intracellular localization. Much progress remains to be made in the field of intracellular delivery of QDs as the existing strategies are largely inadequate to meet the requirements for efficient drug delivery and intracellular dynamics sensing.

15.2.4 FRET-Based Biosensing

Förster or fluorescence resonance energy transfer (FRET) has emerged as a powerful technique for monitoring biological events at the nanometer scale (1–10 nm) including ligand-receptor binding, protein–protein interactions, and biomolecule conformational changes [117,118]. FRET-based applications require a short distance between donor (D) and acceptor (A) molecules and finite spectral overlap between the emission spectrum of D and the absorption spectrum of A. In their extensive review, Medintz and Mattoussi discuss the utility of QD photophysical properties in FRET-based biosensing applications [119]. In particular, the size dependency of the QD emission profiles allows to tune the D–A spectral overlap and maximize energy transfer efficiency. Another useful property of QDs is the ability to form multivalent constructs. This becomes useful when several A molecules are attached to one QD, thereby improving FRET signal. While broad absorption profiles of QDs allow multiplexing experiments and significantly improve SNR in the FRET experiment by reducing direct excitation contribution to the FRET signal, they render QDs ineffective acceptor fluorophores [119]. Therefore, a typical QD-FRET configuration includes an organic dye molecule serving as the acceptor fluorophore. Another serious limitation associated with QD-based FRET is the QD size. The energy transfer efficiency has a sixth order dependence on the D–A separation distance and is defined as follows:

$$E = \frac{k_{D-A}}{k_{D-A} + \tau_D^{-1}} = \frac{R_0^6}{R_0^6 + r^6}, \tag{15.1}$$

where
    k_{D-A} is the rate of energy transfer between D and A
    τ_D is the exciton radiative lifetime of D
    R_0 is the Förster radius
    r is the center-to-center separation distance between D and A

Surface modification and functionalization of QDs significantly increases the final conjugate size and results in large D–A separation distances leading to poor FRET rates and low energy transfer efficiency. Nevertheless, QD-dye FRET pairs have been successfully utilized by many groups to sense nucleic acid hybridization, detect the presence of ions in solution, monitor enzyme activity, and assess competitive receptor binding kinetics. Several biosensing applications of QD-FRET are discussed below.

In the first example, Zhang and Johnson used QD-FRET to study an aspect of the HIV-1 virus replication [120]. Specifically, they attempted to monitor sequence-driven interaction between the arginine-rich fragment of Rev, a HIV-1 regulatory protein, and the RNA-based Rev responsive element, RRE IIB RNA. The experimental setup included streptavidin-coated QDs conjugated with biotinylated RRE and the Rev fragment labeled with the Cy5 dye. To monitor RRE–Rev interactions, QD PL was recorded as a function of Rev peptide concentration, and significant QD quenching was reported with increasing peptide concentration. The disruptive effect of neomycin B presence on the RRE–Rev association translated into significantly lower FRET rates. Such a QD-FRET configuration may be used as a template for FRET-based drug discovery platform targeting viral replication. In the second example of DNA sensing, Gill et al. attempted to detect a Texas Red-labeled DNA sequence with a thiolated complementary DNA sequence conjugated to CdSe/ZnS core/shell nanocrystals [121]. Measurements of QD PL quenching and Texas Red PL were used to detect the hybridization process between the labeled complementary DNA. When DNase I enzyme was added to the solution, QD PL partial recovery was observed as the result of DNA hybrid degradation and simultaneous Texas Red separation.

By coupling a dye with the pH-sensitive absorption profile to QDs, it is possible to accurately monitor pH changes in real time. In the report by Snee et al., squaraine dye was covalently conjugated to polymer-encapsulated QDs, and the FRET interactions were measured as a function of the solution pH [122]. As a result, the ratiometric dependence of the QD and dye emission peaks is revealed and it demonstrated the utility of such QD-dye assembly. In another example, Medintz et al. were able to detect the presence of the maltose sugar in solution with positively charged maltose-binding proteins (MBP) self-assembled on the DHLA-capped QD surface [123]. Cy3-labeled MBP was specifically prebound to maltose analog β-cyclodextrin (BCD) tagged with Cy3.5 dark quencher prior to self-assembly. Excitation of the QD resulted in Cy3 FRET excitation that was then fully quenched by the BCD–Cy3.5 complex. Addition of maltose to the solution resulted in the displacement of BCD–Cy3.5 complex, with Cy3 PL recovery measured in a dose-dependent manner. Thus, it is possible to develop QD-protein biosensors for FRET-based sensitive nutrient detection.

One of the most exciting QD-FRET applications to date is the use of QDs in photodynamic therapy (PDT). PDT has emerged as a cancer therapeutic tool alternative to surgical treatment and is widely used as treatment for several types of cancer, in particular basal cell carcinoma. In a general PDT process, a photosensitizing agent (PS) is photoexcited, and its excitation energy is transferred to a proximal triplet oxygen molecule to generate singlet oxygen. Singlet oxygen belongs to the family of reactive oxygen species (ROS) that are able to induce apoptotic damage of the cells in the immediate vicinity of ROS generation. By coupling the PS molecules to QDs in the QD-FRET-enhanced PDT, singlet oxygen generation QY can be significantly increased as a result of the excellent UV or IR energy-harvesting capability of QDs and subsequent FRET excitation of the PSs [124–126]. In addition, it is possible to anchor several membrane cancer marker-specific antibodies and PSs to the QD surface to achieve specific targeting and imaging of cancer cells along with highly localized therapeutic action of PSs (Figure 15.8) [124]. In the most recent proof-of-concept experiment, Tsay et al. conjugated two PSs, Rose Bengal and chlorine e6, to CdSe/ZnS QDs, with lysine-terminated phytochelatin-related peptides serving as a rigid linker between the QD surface and the PS molecule [126]. Singlet oxygen generation was achieved either by indirect FRET-based excitation (355 nm) or direct laser excitation of PSs on the QD surface (532 nm) and detected spectroscopically at 1270 nm by measuring oxygen phosphorescence at 1270 nm. Singlet oxygen QYs as high as 0.17 and 0.31 were reported for QD–Rose Bengal conjugates and QD–chlorin e6 conjugates, respectively. Interestingly, QYs from indirect excitation were much higher for QD–Rose Bengal complexes suggesting FRET is the primary mechanism of singlet oxygen generation.

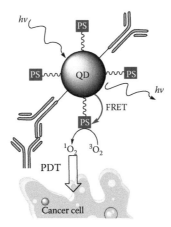

FIGURE 15.8 Achieving specific targeting in PDT. A cancer cell is first exposed to a primary antibody that specifically binds to an extracellular cancer marker. Then the cancer cell is incubated with multivalent QDs conjugated to secondary antibody and photosensitizer molecule. (From Bakalova, R. et al., *Nano Lett.*, 4(9), 1567–1573.)

The opposite was true for QD–chlorin e6 complexes, indicating that FRET-based excitation of chlorine e6 was considerably lower. Further experiments are needed to pinpoint the basis for such differences between the QD–PS conjugates.

Performing FRET at the single molecule level offers the ability to monitor individual molecular interactions with increased sensitivity and temporal resolution when compared to ensemble experiments. In a general solution-phase, single QD-FRET experiment, emission intensity bursts of the donor and acceptor molecules are measured after the defined focal volume of the solution is excited through a high numerical aperture objective [127]. Zhang et al. utilized a single QD-FRET biosensor for accurate and highly sensitive detection of DNA sequences in solution [128]. The Cy5-labeled target DNA sequence was captured by hybridization with the biotinylated complementary DNA sequence introduced into the solution. The Cy5- and biotin-terminated DNA hybrid was then mixed with streptavidin-conjugated QDs, and the resulting solution with QD–DNA complexes was delivered into a glass microcapillary to minimize the sample volume for FRET analysis. The sensing responsivity of the QD-FRET DNA biosensor was determined to be several orders of magnitude higher than that of the dye-based molecular beacon.

Once again, the unique photophysical properties make QDs an attractive choice for the FRET-based biosensing and therapeutic applications. Clearly, there are significant challenges associated with QD-FRET, specifically, QD–acceptor separation distances and QD–acceptor association multivalency. Addressing these issues will ensure that QD-FRET continues to be actively explored in a biological setting.

15.2.5 In Vivo Deep Tissue Imaging

Until recently, in vivo deep tissue fluorescence imaging relied on conventional dyes and suffered from poor tissue penetration depth, low SNR intensity profiles, and short-lived fluorescence signal. QDs offer several distinct advantages to the field of deep tissue imaging [21,55]. First, their high brightness and the ability to shift their peak emission wavelengths to the far red and NIR regions of the EM spectrum result in dramatic reduction of cellular autofluorescence, significant improvement of the SNR intensity profiles and tissue penetration depth. Second, their excellent photostability enables longer circulation times in the endothelial system and allows long-term imaging without the significant image quality loss. Third, their large two-photon absorption cross sections permit the use of low-energy NIR excitation sources minimizing tissue damage and improving SNR profiles. Fourth, QD-based in vivo deep tissue imaging is a safer, cost-effective alternative to radiation-based imaging modalities. So far, intravitally injected QDs have been successfully used to image blood vessels, detect cancer tumors, and map lymph nodes in live animals.

In the work by Smith et al., a series of QDs of varying sizes and surface coatings were intravitally injected into the chick chorioallantoic membrane (CAM) blood vessels, a model system developed for studying angiogenesis and blood vessel formation in vivo [129]. Visualization of the chick blood vessels with NIR-emitting QDs produced a stronger, more uniform signal and eliminated virtually all of the background fluorescence compared to FITC-dextran, a dye-based vasculature visualization agent. It should also be noted that multiple QD injections produced no deleterious effects on the development of the chick embryo. In a more recent example, Smith B. R. et al. covalently attached ~30–50 RGD peptides to NIR-emitting QDs and injected QD–RGD complexes into the tail vein of live mice [130]. The authors exploited intravital microscopy to monitor QD–RGD binding to integrins $\alpha_v\beta_3$ within the tumor neo-vasculature in real time. Interestingly, only QD–RGD aggregates and not single QD–RGD complexes were observed to bind to the tumor neovasculature. Also, it was reported that QDs were cleared from the vasculature via reticuloendothelial system within 1.5 h post administration. In another study describing the utility of QDs in the field of angiography, Larson et al. administered water-soluble QDs into living mice via tail vein injection and were able to clearly visualize QD-containing vasculature at a skin depth of ~100 μm via two-photon (900 nm) excitation microscopy [39].

In 2004, Gao et al. described the use of amphiphilic triblock copolymer-encapsulated QDs to visualize human prostate cancer tumors in living nude mice (Figure 15.5c) [55]. PEG chains and PSMA-specific antibody were conjugated to polymer-coated QDs to achieve passive and active targeting of cancer cells, respectively. QD complexes were administered into living animals via tail vein injection and were allowed a 24 h circulation period for QD accumulation in the tumor. The acquired spectral fluorescence images of whole animals clearly demonstrated QD accumulation in the prostate cancer tumors. Nonspecific activity of QD–PSMA antibody conjugates was determined by histological examination of six different host organs, showing nonspecific QD uptake in the liver and the spleen. The rate of uptake and retention of QD–PEG conjugates was significantly lower and was attributed to small HD of the conjugate.

Kim et al. utilized type II NIR-emitting (840–860 nm) QDs as surgical aids during the sentinel lymph node (SLN) mapping procedure in the mouse and pig (Figure 15.9) [132]. After subcutaneous administration of 400 pmol of QDs on the thigh of the pig, it was possible to quickly localize the position of the SLN at the depth of ~1 cm in real time. The localization was achieved using a low fluence rate of 5 mW/cm² of NIR excitation light. According to Kim et al., NIR QDs considerably simplify the surgeon's task of identifying the SNL and performing complete resection during the cancer surgery.

Despite multiple successful demonstrations of their use in vivo, QDs face significant challenges before they can be approved for clinical use. The large size of QD–conjugates promotes QD retention in the liver and the spleen and prevents renal clearance, which imposes a 5–6 nm maximum size requirement [132]. In addition, toxicity remains a widely discussed issue and will be addressed in a separate section.

15.2.6 Multimodal Imaging

Although QD-based fluorescence microscopy has demonstrated potential for in vivo molecular imaging, the tissue penetration depth is limited by the working distance of the microscope objective and excitation light scattering [133]. However, it is possible to enhance the visualization of QDs in vivo by introducing an additional property, such as paramagnetism, which allows molecular imaging in multiple modes. Paramagnetic QDs can serve as both photoluminescent probes and magnetic resonance imaging (MRI) contrast agents in a highly complementary bimodal molecular imaging mode. Several paramagnetic QD architectures have been developed. Yi et al. reported the synthesis of silica-coated nanocomposites of magnetic nanoparticles and QDs [62]. γ-Fe$_2$O$_3$ nanoparticles and

Color video   NIR fluorescence   Color-NIR merge

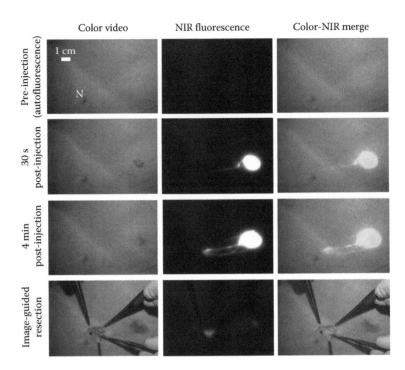

FIGURE 15.9 (See color insert.)   SLN mapping surgical procedure aided by NIR-emitting type II QDs. (Reprinted by permission from MacMillan Publishers Ltd. *Nat. Biotechnol.*, Kim, S., Lim, Y.T. et al., Near-infrared fluorescent type II quantum dots for sentinel lymph node mapping, 22(1), 93–97, 2004. Copyright 2004.)

CdSe QDs were prepared separately and then encapsulated in a silica shell via reverse microemulsion. The resulting nanocomposites retained the paramagnetic properties of γ-Fe_2O_3 and the optical properties of CdSe. Mulder and colleagues prepared paramagnetic QDs by coating CdSe/ZnS core/shells with a phospholipid micelle composed of a PEGylated phospholipid, PEG-DSPE, and a Gd-containing paramagnetic lipid, Gd-DTPA-BSA [134]. Chelated complexes of Gd are widely used as MRI contrast agents. The paramagnetic QDs were then functionalized with integrin-specific RGD peptide and were used to image human umbilical vein endothelial cells. In another example, Yang et al. functionalized silica-coated Mn-doped CdS/ZnS core/shells with Gd via capture of Gd^{III} ions by TSPETE, a metal-chelating silane [135]. Inductively coupled plasma (ICP) analysis was utilized to determine the average number of Gd^{III} ions on the QD surface to be ~107. QD–Gd complexes possessed large proton relaxivities and retained yellow PL emission of the original QDs. In conclusion, the bimodal character of paramagnetic QDs makes them useful MRI contrast agents with the capability of parallel optical detection.

15.2.7 Toxicity

QD core semiconductor constituents, such as Cd, Se, and Te, are highly toxic in their bulk form, and their adverse effects have been well documented. In particular, Cd exposure is associated with the increased rates of cancer, birth defects, and endocrine disruption [136]. In addition, Cd is known to facilitate the formation of ROS, causing subsequent oxidative damage. Recently, several reports presented evidence that Cd expresses genotoxic activities in mammalian cells and animals, specifically chromosomal damage and gene expression modulation [137]. There is sufficient evidence establishing the toxic effects of II–IV bulk semiconductors to raise concerns about possible deleterious impact of their nanoscale counterparts on the biological systems.

Numerous reports on the cytotoxicity associated with the utilization of QDs in biological setting have been published to date. In his comprehensive review in 2006, Hardman et al. summarized the existing state of knowledge on in vivo toxicity and biological fate of QDs [133]. Hardman concluded that QD toxicity ultimately depends on the effectiveness of the nanocrystal surface passivation, which directly influences QD size, charge, and stability. Indeed, several recently published toxicity studies have shown that QD toxicity can be minimized through complete surface passivation with an appropriate choice of inorganic shell or surface coating. In 2009, Pelley et al. published another extensive review, in which they updated, expanded, and put Hardman's 2006 review in a regulatory context [138]. Particular attention was paid to biological and environmental fate of QDs and a striking lack of studies showing QD long-term effects. Currently, QD toxicity still remains a widely discussed issue with many questions remaining to be answered.

Inherent toxicity of traditional QD core material constituents is a serious obstacle to the clinical utilization of QDs. To address this issue, III–V and I–III–VI$_2$ materials have been employed to replace toxic II–IV semiconductor elements in the QD architecture. Unfortunately, new materials suffer from sub-par performance compared to their traditional counterparts, and there have not been any reports establishing their reduced toxicity until recently.

15.2.8 Limitations

QDs have undoubtedly demonstrated their utility to the field of biological investigation. However, there are several distinct limitations associated with QDs that must be overcome before QDs can become the routine biological photoluminescent probes of first choice. First, the final size of biocompatible, water-soluble, functionalized QD conjugates can easily exceed 20 nm in diameter and several hundred kDa in molecular mass [93,139]. This is especially true in the case of polymer-encapsulated streptavidin-coated QDs conjugated to large biotinylated antibodies. The large size of QD probes may impair labeled protein trafficking in SQT, cause reduced FRET efficiency between donor and acceptor molecules, and restrict access to crowded cellular locations. In addition, large HD precludes QDs from being cleared from the body via renal filtration or urinary excretion [132,138]. Current strategies to reduce QD size are aimed at the development of new compact surface encapsulants, shorter linkers, and conjugation strategies as well as the reduction of QD multivalency.

Although multivalency has been successfully used in several instances to maximize the utility of QDs, it can cause QD aggregation due to cross-linking during conjugation and impair receptor mobility by cross-linking cell surface proteins. The ability to synthetically control QD valency will effectively eliminate issues due to cross-linking and significantly reduce the HD of QD conjugates. In an attempt to generate monovalent QDs, Howarth et al. conjugated single monovalent streptavidin (mSa) with a femtomolar biotin binding site to core/shells coated with DHLA-PEG$_8$-COOH and used the QD–mSa conjugates to visualize membrane mobility of individual LDL receptors [140].

Single QD PL emission intensity alternates between dark and bright states, and such fluorescent intermittency has been referred to as blinking (Figure 15.10a). Several models have been proposed to explain the mechanistic basis underlying the blinking phenomenon of single QDs [141,142]. Although blinking may be used to distinguish single fluorophores from aggregates, it significantly complicates trajectory reconstruction in SQT and may not be a reliable indicator of single molecule fluorophores due to the dependence of the intermittency rate on experimental parameters and biological environment [93]. To reduce "blinking," a thick inorganic shell (CdS, CdZnS) with reduced lattice mismatch is grown around the QD core (Figure 15.10b) [143,144]. Although these thick shells dramatically reduce fluorescent intermittency of QDs, they result in the HD increase, which is undesirable for biological applications.

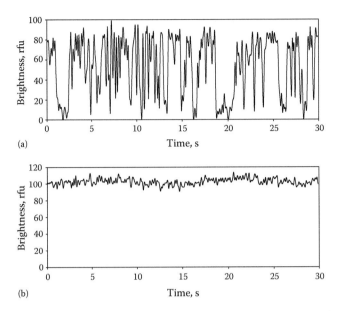

FIGURE 15.10 Fluorescent intermittency of a single semiconductor nanocrystal. Typical fluorescence spectrum of a single QD is shown in (a). Random alterations between dark and bright states constitute blinking. A fluorescence spectrum for a single, non-blinking CdSe core nanocrystal coated with a 14-layer thick CdS shell is shown in (b). Here, the QD spends the entire time interval in the bright state.

15.3 Conclusion

It is clear that QDs have emerged as an attractive class of photoluminescent probes since their introduction to the field in 1998. Recent advances in nanocrystal surface chemistry resulted in more compact and more stable QDs. Their relatively small size, excellent stability at physiologically relevant conditions, and unique optical properties permit biological investigation of cellular processes with unprecedented spatiotemporal resolution in real time. While QDs have already been established as powerful biological imaging agents, their use in the fields of diagnostics and targeted drug delivery is being actively explored. Utilizing their spectral characteristics, QDs can potentially form the basis of multiplexed fluorescence assays that examine individual protein–protein interactions and ultimately interrogate cell signaling pathways. Also, their robust inorganic nature renders QDs resistant to degradation and photobleaching for long periods of time and thereby guarantees a bright future for QDs as reporters of dynamic activity of biological molecules at the single-molecule level.

Acknowledgments

This work was supported by the National Institute of Health under Grant EB003728. O. K. and S. J. R. would like to thank Dr. James McBride for providing QD samples to examine the dependency of blinking frequency on the shell thickness.

References

1. Fernandez-Suarez, M. and A. Y. Ting (2008). Fluorescent probes for super-resolution imaging in living cells. *Nat. Rev. Mol. Cell Biol.* **9**(12): 929–943.
2. Haraguchi, T. (2002). Live cell imaging: Approaches for studying protein dynamics in living cells. *Cell Struct. Funct.* **27**(5): 333–334.

3. Lippincott-Schwartz, J. and G. H. Patterson (2003). Development and use of fluorescent protein markers in living cells. *Science* **300**(5616): 87–91.

4. Sako, Y. and T. Yanagida (2003). Single-molecule visualization in cell biology. *Nat. Rev. Mol. Cell Biol.* **Suppl**: SS1–5.

5. Zhang, J., R. E. Campbell et al. (2002). Creating new fluorescent probes for cell biology. *Nat. Rev. Mol. Cell. Biol.* **3**(12): 906–918.

6. Resch-Genger, U., M. Grabolle et al. (2008). Quantum dots versus organic dyes as fluorescent labels. *Nat. Methods* **5**(9): 763–775.

7. Heim, R. and R. Y. Tsien (1996). Engineering green fluorescent protein for improved brightness, longer wavelengths and fluorescence resonance energy transfer. *Curr. Biol.* **6**(2): 178–182.

8. Shaner, N. C., P. A. Steinbach et al. (2005). A guide to choosing fluorescent proteins. *Nat. Methods* **2**(12): 905–909.

9. Piszczek, G. (2006). Luminescent metal-ligand complexes as probes of macromolecular interactions and biopolymer dynamics. *Arch. Biochem. Biophys.* **453**(1): 54–62.

10. Terpetschnig, E., H. Szmacinski et al. (1995). Metal-ligand complexes as a new class of long-lived fluorophores for protein hydrodynamics. *Biophys. J.* **68**(1): 342–350.

11. Maretti, L., P. S. Billone et al. (2009). Facile photochemical synthesis and characterization of highly fluorescent silver nanoparticles. *J. Am. Chem. Soc.* **131**(39): 13972–13980.

12. Patel, S. A., C. I. Richards et al. (2008). Water-soluble Ag nanoclusters exhibit strong two-photon-induced fluorescence. *J. Am. Chem. Soc.* **130**(35): 11602–11603.

13. Barnard, A. S. (2009). Diamond standard in diagnostics: Nanodiamond biolabels make their mark. *Analyst.* **134**(9): 1751–1764.

14. Yang, S. T., X. Wang et al. (2009). Carbon dots as nontoxic and high-performance fluorescence imaging agents. *J. Phys. Chem. C. Nanomater. Interfaces.* **113**(42): 18110–18114.

15. Ow, H., D. R. Larson et al. (2005). Bright and stable core-shell fluorescent silica nanoparticles. *Nano Lett.* **5**(1): 113–117.

16. Alivisatos, A. P. (1996). Semiconductor clusters, nanocrystals, and quantum dots. *Science* **271**(5251): 933–937.

17. Alivisatos, A. P., W. Gu et al. (2005). Quantum dots as cellular probes. *Annu. Rev. Biomed. Eng.* **7**: 55–76.

18. Bruchez, M., Jr., M. Moronne et al. (1998). Semiconductor nanocrystals as fluorescent biological labels. *Science* **281**(5385): 2013–2016.

19. Chan, W. C. and S. Nie (1998). Quantum dot bioconjugates for ultrasensitive nonisotopic detection. *Science* **281**(5385): 2016–2018.

20. Medintz, I. L., H. T. Uyeda et al. (2005). Quantum dot bioconjugates for imaging, labelling and sensing. *Nat. Mater.* **4**(6): 435–446.

21. Michalet, X., F. F. Pinaud et al. (2005). Quantum dots for live cells, in vivo imaging, and diagnostics. *Science* **307**(5709): 538–544.

22. Brus, L. E. (1984). Electron-electron and electron-hole interactions in small semiconductor crystallites: The size dependence of the lowest excited electronic state. *J. Chem. Phys.* **80**(9): 4403–4409.

23. McBride, J., J. Treadway et al. (2006). Structural basis for near unity quantum yield core/shell nanostructures. *Nano Lett.* **6**(7): 1496–1501.

24. Yu, W. W., L. Qu et al. (2003). Experimental determination of the extinction coefficient of CdTe, CdSe, and CdS nanocrystals. *Chem. Mater.* **15**(14): 2854–2860.

25. Kucur, E., F. M. Boldt et al. (2007). Quantitative analysis of cadmium selenide nanocrystal concentration by comparative techniques. *Anal. Chem.* **79**(23): 8987–8993.

26. Dabbousi, B. O., J. Rodriguez-Viejo et al. (1997). (CdSe)ZnS core-shell quantum dots: synthesis and characterization of a size series of highly luminescent nanocrystallites. *J. Phys. Chem. B.* **101**(46): 9463–9475.

27. Du, H., C. Chen et al. (2002). Optical properties of colloidal PbSe nanocrystals. *Nano Lett.* **2**(11): 1321–1324.

28. Brumer, M., A. Kigel et al. (2005). PbSe/PbS and PbSe/PbSe core/shell nanocrystals. *Adv. Funct. Mater.* **15**(7): 1111–1116.

29. Murray, C. B., D. J. Norris et al. (1993). Synthesis and characterization of nearly monodisperse CdE (E = sulfur, selenium, tellurium) semiconductor nanocrystallites. *J. Am. Chem. Soc.* **115**(19): 8706–8715.

30. Chan, W. C. W., D. J. Maxwell et al. (2002). Luminescent quantum dots for multiplexed biological detection and imaging. *Curr. Opin. Biotech.* **13**(1): 40–46.

31. Jaiswal, J. K., H. Mattoussi et al. (2003). Long-term multiple color imaging of live cells using quantum dot bioconjugates. *Nat. Biotech.* **21**(1): 47–51.

32. Stroh, M., J. P. Zimmer et al. (2005). Quantum dots spectrally distinguish multiple species within the tumor milieu in vivo. *Nat. Med.* **11**(6): 678–682.

33. Chattopadhyay, P. K., D. A. Price et al. (2006). Quantum dot semiconductor nanocrystals for immunophenotyping by polychromatic flow cytometry. *Nat. Med.* **12**(8): 972–977.

34. Grecco, H. E., K. A. Lidke et al. (2004). Ensemble and single particle photophysical properties (two-photon excitation, anisotropy, FRET, lifetime, spectral conversion) of commercial quantum dots in solution and in live cells. *Microsc. Res. Tech.* **65**(4–5): 169–179.

35. Bastiaens, P. I. H. and A. Squire (1999). Fluorescence lifetime imaging microscopy: Spatial resolution of biochemical processes in the cell. *Trends Cell Biol.* **9**(2): 48–52.

36. Dahan, M., T. Laurence et al. (2001). Time-gated biological imaging by use of colloidal quantum dots. *Opt. Lett.* **26**(11): 825–827.

37. Gadella, Jr., T. W. J., T. M. Jovin et al. (1993). Fluorescence lifetime imaging microscopy (FLIM): Spatial resolution of microstructures on the nanosecond time scale. *Biophys. Chem.* **48**(2): 221–239.

38. Schlegel, G., J. Bohnenberger et al. (2002). Fluorescence decay time of single semiconductor nanocrystals. *Phys. Rev. Lett.* **88**(13): 137401.

39. Larson, D. R., W. R. Zipfel et al. (2003). Water-soluble quantum dots for multiphoton fluorescence imaging in vivo. *Science* **300**(5624): 1434–1436.

40. Markus, G., Z. Jan et al. (2008). Stability and fluorescence quantum yield of CdSe/ZnS quantum dots. influence of the thickness of the ZnS shell. In *Fluorescence Methods and Applications: Spectroscopy, Imaging, and Probes*, O. S. Wolfbeis (ed.), Annals of the New York Academy of Sciences 1130, Wiley, New York, pp. 235–241.

41. Talapin, D. V., I. Mekis et al. (2004). CdSe/CdS/ZnS and CdSe/ZnSe/ZnS core/shell/shell nanocrystals. *J. Phys. Chem. B.* **108**(49): 18826–18831.

42. Talapin, D. V., A. L. Rogach et al. (2001). Highly luminescent monodisperse CdSe and CdSe/ZnS nanocrystals synthesized in hexadecylamine/trioctylphosphine oxide/trioctylphospine mixture. *Nano Lett.* **1**(4): 207–211.

43. Bao, H., Y. Gong et al. (2004). Enhancement effect of illumination on the photoluminescence of water-soluble CdTe nanocrystals: Toward highly fluorescent CdTe/CdS core-shell structure. *Chem. Mater.* **16**(20): 3853–3859.

44. Chen, Y. and Z. Rosenzweig (2002). Luminescent CdS quantum dots as selective ion probes. *Anal. Chem.* **74**(19): 5132–5138.

45. Lemon, B. I. and R. M. Crooks (2000). Preparation and characterization of dendrimer-encapsulated CdS semiconductor quantum dots. *J. Am. Chem. Soc.* **122**(51): 12886–12887.

46. Mews, A., A. V. Kadavanich et al. (1996). Structural and spectroscopic investigations of CdS/HgS/CdS quantum-dot quantum wells. *Phys. Rev. B.* **53**(20): R13242.

47. Bharali, D. J., D. W. Lucey et al. (2005). Folate-receptor-mediated delivery of InP quantum dots for bioimaging using confocal and two-photon microscopy. *J. Am. Chem. Soc.* **127**(32): 11364–11371.

48. Kim, S.-W., J. P. Zimmer et al. (2005). Engineering InAsxP1-x/InP/ZnSe III-V alloyed core/shell quantum dots for the near-infrared. *J. Am. Chem. Soc.* **127**(30): 10526–10532.

49. Micic, O. I., J. R. Sprague et al. (1995). Synthesis and characterization of InP, GaP, and GaInP2 quantum dots. *J. Phys. Chem.* **99**(19): 7754–7759.

50. Xie, R., D. Battaglia et al. (2007). Colloidal InP nanocrystals as efficient emitters covering blue to near-infrared. *J. Am. Chem. Soc.* **129**(50): 15432–15433.

51. Pons, T., E. Pic et al. (2010) Cadmium-free CuInS2/ZnS quantum dots for sentinel lymph node imaging with reduced toxicity. *ACS Nano* **4**(5): 2531–2538.

52. Torimoto, T., T. Adachi et al. (2007). Facile synthesis of ZnS-AgInS2 solid solution nanoparticles for a color-adjustable luminophore. *J. Am. Chem. Soc.* **129**(41): 12388–12389.

53. Hezinger, A. F., J. Tessmar et al. (2008). Polymer coating of quantum dots—A powerful tool toward diagnostics and sensorics. *Eur. J. Pharm. Biopharm.* **68**(1): 138–152.

54. Pellegrino, T., L. Manna et al. (2004). Hydrophobic nanocrystals coated with an amphiphilic polymer shell: A general route to water soluble nanocrystals. *Nano Lett.* **4**(4): 703–707.

55. Gao, X., Y. Cui et al. (2004). In vivo cancer targeting and imaging with semiconductor quantum dots. *Nat. Biotechnol.* **22**(8): 969–976.

56. Darbandi, M., R. Thomann et al. (2005). Single quantum dots in silica spheres by microemulsion synthesis. *Chem. Mater.* **17**(23): 5720–5725.

57. Gerion, D., F. Pinaud et al. (2001). Synthesis and properties of biocompatible water-soluble silica-coated CdSe/ZnS semiconductor quantum dots. *J. Phys. Chem. B.* **105**(37): 8861–8871.

58. Hu, X., P. Zrazhevskiy et al. (2009). Encapsulation of single quantum dots with mesoporous silica. *Ann. Biomed. Eng.* **37**(10): 1960–1966.

59. Rong, H. et al. (2007). Core/shell fluorescent magnetic silica-coated composite nanoparticles for bioconjugation. *Nanotechnology* **18**(31): 315601.

60. Selvan, S. T., T. T. Tan et al. (2005). Robust, non-cytotoxic, silica-coated CdSe quantum dots with efficient photoluminescence. *Adv. Mater.* **17**(13): 1620–1625.

61. Thomas, N. and M. Paul (2004). Single quantum dots in spherical silica particles13. *Angew. Chem. Int. Ed.* **43**(40): 5393–5396.

62. Yi, D. K., S. T. Selvan et al. (2005). Silica-coated nanocomposites of magnetic nanoparticles and quantum dots. *J. Am. Chem. Soc.* **127**(14): 4990–4991.

63. Mattoussi, H., J. M. Mauro et al. (2000). Self-assembly of CdSe/ZnS quantum dot bioconjugates using an engineered recombinant protein. *J. Am. Chem. Soc.* **122**(49): 12142–12150.

64. Kim, S. and M. G. Bawendi (2003). Oligomeric ligands for luminescent and stable nanocrystal quantum dots. *J. Am. Chem. Soc.* **125**(48): 14652–14653.

65. Mazumder, S., R. Dey, M. K. Mitra, S. Mukherjee, and G. C. Das (2009). Review: Biofunctionalized quantum dots in biology and medicine. *J. Nanomater.*, vol. 2009, Article ID 815734, 17pp, 2009.

66. Peng, H., L. Zhang et al. (2007). Preparation of water-soluble CdTe/CdS core/shell quantum dots with enhanced photostability. *J. Lumin.* **127**(2): 721–726.

67. Wuister, S. F., I. Swart et al. (2003). Highly luminescent water-soluble CdTe quantum dots. *Nano Lett.* **3**(4): 503–507.

68. Zheng, Y., Z. Yang et al. (2007). Aqueous synthesis of glutathione-capped ZnSe and ZnCdSe alloyed quantum dots. *Adv. Mater.* **19**(11): 1475–1479.

69. Pinaud, F., D. King et al. (2004). Bioactivation and cell targeting of semiconductor CdSe/ZnS nanocrystals with phytochelatin-related peptides. *J. Am. Chem. Soc.* **126**(19): 6115–6123.

70. Gomez, N., J. O. Winter et al. (2005). Challenges in quantum dot-neuron active interfacing. *Talanta* **67**(3): 462–471.

71. Mitchell, G. P., C. A. Mirkin et al. (1999). Programmed assembly of DNA functionalized quantum dots. *J. Am. Chem. Soc.* **121**(35): 8122–8123.

72. Brannon-Peppas, L. (2000). *Poly(ethylene glycol): Chemistry and Biological Applications*: J.M. Harris and S. Zalipsky, editors, American Chemical Society, Washington, DC, 1997, 489 pp. *J. Control. Release* **66**(2–3): 321–321.

73. Bentzen, E. L., I. D. Tomlinson et al. (2005). Surface modification to reduce nonspecific binding of quantum dots in live cell assays. *Bioconj. Chem.* **16**(6): 1488–1494.

74. Warnement, M. R., I. D. Tomlinson et al. (2008). Controlling the reactivity of amphiphilic quantum dots in biological assays through hydrophobic assembly of custom PEG derivatives. *Bioconj. Chem.* **19**(7): 1404–1413.

75. Goldman, E. R., E. D. Balighian et al. (2002). Avidin: A natural bridge for quantum dot-antibody conjugates. *J. Am. Chem. Soc.* **124**(22): 6378–6382.

76. Chaiet, L. and F. J. Wolf (1964). The properties of streptavidin, a biotin-binding protein produced by streptomycetes. *Arch. Biochem. Biophys.* **106**: 1–5.

77. Weber, P. C., D. H. Ohlendorf et al. (1989). Structural origins of high-affinity biotin binding to streptavidin. *Science* **243**(4887): 85–88.

78. Wilchek, M. and E. A. Bayer (1990). Introduction to avidin–biotin technology. *Methods Enzymol.* **184**: 5–13.

79. Wu, X., H. Liu et al. (2003). Immunofluorescent labeling of cancer marker Her2 and other cellular targets with semiconductor quantum dots. *Nat. Biotechnol.* **21**(1): 41–46.

80. Dahan, M., S. Levi et al. (2003). Diffusion dynamics of glycine receptors revealed by single-quantum dot tracking. *Science* **302**(5644): 442–445.

81. Lidke, D. S., P. Nagy et al. (2004). Quantum dot ligands provide new insights into erbB/HER receptor-mediated signal transduction. *Nat. Biotechnol.* **22**(2): 198–203.

82. George, N., H. Pick et al. (2004). Specific labeling of cell surface proteins with chemically diverse compounds. *J. Am. Chem. Soc.* **126**(29): 8896–8897.

83. Warnement, M. R., I. D. Tomlinson et al. (2007). Fluorescent imaging applications of quantum dot probes. *Curr. Nanosci.* **3**: 273–284.

84. Howarth, M., K. Takao et al. (2005). Targeting quantum dots to surface proteins in living cells with biotin ligase. *Proc. Natl. Acad. Sci. U.S.A.* **102**(21): 7583–7588.

85. Sun, Y., O. Sato et al. Single-molecule stepping and structural dynamics of myosin X. *Nat. Struct. Mol. Biol.* **17**(4): 485–491.

86. Genin, E., O. Carion et al. (2008). CrAsH-quantum dot nanohybrids for smart targeting of proteins. *J. Am. Chem. Soc.* **130**(27): 8596–8597.

87. Yan, Z., S. Min-kyung et al. (2006). Halotag protein-mediated site-specific conjugation of bioluminescent proteins to quantum dots13. *Angew. Chem. Int. Ed.* **45**(30): 4936–4940.

88. Hauser, C. T. and R. Y. Tsien (2007). A hexahistidine-Zn2+-dye label reveals STIM1 surface exposure. *Proc. Natl. Acad. Sci. U.S.A.* **104**(10): 3693–3697.

89. Roullier, V., S. Clarke et al. (2009). High-affinity labeling and tracking of individual histidine-tagged proteins in live cells using Ni2+ tris-nitrilotriacetic acid quantum dot conjugates. *Nano Lett.* **9**(3): 1228–1234.

90. Rosenthal, S. J., I. Tomlinson et al. (2002). Targeting cell surface receptors with ligand-conjugated nanocrystals. *J. Am. Chem. Soc.* **124**(17): 4586–4594.

91. Goldman, E. R., G. P. Anderson et al. (2002). Conjugation of luminescent quantum dots with antibodies using an engineered adaptor protein to provide new reagents for fluoroimmunoassays. *Anal. Chem.* **74**(4): 841–847.

92. Hanaki, K.-i., A. Momo et al. (2003). Semiconductor quantum dot/albumin complex is a long-life and highly photostable endosome marker. *Biochem. Biophys. Res. Commun.* **302**(3): 496–501.

93. Pinaud, F., S. Clarke et al. Probing cellular events, one quantum dot at a time. *Nat. Methods* **7**(4): 275–285.

94. Joo, K.-I., Y. Lei et al. (2008). Site-specific labeling of enveloped viruses with quantum dots for single virus tracking. *ACS Nano* **2**(8): 1553–1562.

95. Schormann, W., F. Hammersen et al. (2008). Tracking of human cells in mice. *Histochem. Cell Biol.* **130**(2): 329–338.

96. Yildiz, A. and P. R. Selvin (2005). Fluorescence imaging with one nanometer accuracy: Application to molecular motors. *Acc. Chem. Res.* **38**(7): 574–582.

97. McHale, K., A. J. Berglund et al. (2007). Quantum dot photon statistics measured by three-dimensional particle tracking. *Nano Lett.* **7**(11): 3535–3539.

98. Pavani, S. R. P., M. A. Thompson et al. (2009). Three-dimensional, single-molecule fluorescence imaging beyond the diffraction limit by using a double-helix point spread function. *Proc. Natl. Acad. Sci. U.S.A.* **106**(9): 2995–2999.

99. Toprak, E., H. Balci et al. (2007). Three-dimensional particle tracking via bifocal imaging. *Nano Lett.* **7**(7): 2043–2045.

100. Saxton, M. J. and K. Jacobson (2003). Single-particle tracking: Applications to membrane dynamics. *Annu. Rev. Biophys. Biomol. Struct.* **26**(1): 373–399.

101. Saxton, M. J. (2007). A biological interpretation of transient anomalous subdiffusion. I. Qualitative model. **92**(4): 1178–1191.

102. Cui, B., C. Wu et al. (2007). One at a time, live tracking of NGF axonal transport using quantum dots. *Proc. Natl. Acad. Sci. U.S.A.* **104**(34): 13666–13671.

103. Zhang, Q., Y. Li et al. (2009). The dynamic control of kiss-and-run and vesicular reuse probed with single nanoparticles. *Science* **323**(5920): 1448–1453.

104. Heine, M., L. Groc et al. (2008). Surface mobility of postsynaptic AMPARs tunes synaptic transmission. *Science* **320**(5873): 201–205.

105. Pierobon, P., S. Achouri et al. (2009). Velocity, processivity, and individual steps of single myosin v molecules in live cells. **96**(10): 4268–4275.

106. Manabe, N., A. Hoshino et al. (2005). Toward the in vivo study of captopril-conjugated quantum dots. *Nanobiophotonics and Biomedical Applications II*, San Jose, CA, pp. 272–282, SPIE.

107. Bagalkot, V., L. Zhang et al. (2007). Quantum dot-aptamer conjugates for synchronous cancer imaging, therapy, and sensing of drug delivery based on bi-fluorescence resonance energy transfer. *Nano Lett.* **7**(10): 3065–3070.

108. Jia, N., Q. Lian et al. (2007). Intracellular delivery of quantum dots tagged antisense oligodeoxy-nucleotides by functionalized multiwalled carbon nanotubes. *Nano Lett.* **7**(10): 2976–2980.

109. Delehanty, J., H. Mattoussi et al. (2009). Delivering quantum dots into cells: Strategies, progress and remaining issues. *Anal. Bioanal. Chem.* **393**(4): 1091–1105.

110. Jaiswal, J. K., E. R. Goldman et al. (2004). Use of quantum dots for live cell imaging. *Nat. Methods* **1**(1): 73–78.

111. Ruan, G., A. Agrawal et al. (2007). Imaging and tracking of tat peptide-conjugated quantum dots in living cells: New insights into nanoparticle uptake, intracellular transport, and vesicle shedding. *J. Am. Chem. Soc.* **129**(47): 14759–14766.

112. Cheng, Z., Y. Wu et al. (2005). Near-infrared fluorescent RGD peptides for optical imaging of integrin alphavbeta3 expression in living mice. *Bioconj. Chem.* **16**(6): 1433–1441.

113. Cai, W., D.-W. Shin et al. (2006). Peptide-labeled near-infrared quantum dots for imaging tumor vasculature in living subjects. *Nano Lett.* **6**(4): 669–676.

114. Chakraborty, S. K., J. A. J. Fitzpatrick et al. (2007). Cholera toxin b conjugated quantum dots for live cell labeling. *Nano Lett.* **7**(9): 2618–2626.

115. Derfus, A. M., A. A. Chen et al. (2007). Targeted quantum dot conjugates for siRNA delivery. *Bioconj. Chem.* **18**(5): 1391–1396.

116. Derfus, A. M., W. C. W. Chan et al. (2004). Intracellular delivery of quantum dots for live cell labeling and organelle tracking. *Adv. Mater.* **16**(12): 961–966.

117. Sekar, R. B. and A. Periasamy (2003). Fluorescence resonance energy transfer (FRET) microscopy imaging of live cell protein localizations. *J. Cell Biol.* **160**(5): 629–633.

118. Selvin, P. R. (2000). The renaissance of fluorescence resonance energy transfer. *Nat. Struct. Biol.* **7**(9): 730–734.

119. Medintz, I. L. and H. Mattoussi (2009). Quantum dot-based resonance energy transfer and its growing application in biology. *Phys. Chem. Chem. Phys.* **11**(1): 17–45.

120. Zhang, C. Y. and L. W. Johnson (2006). Quantum-dot-based nanosensor for RRE IIB RNA-Rev peptide interaction assay. *J. Am. Chem. Soc.* **128**(16): 5324–5325.

121. Gill, R., I. Willner et al. (2005). Fluorescence resonance energy transfer in CdSe/ZnS DNA conjugates: Probing hybridization and DNA cleavage. *J. Phys. Chem. B.* **109**(49): 23715–23719.

122. Snee, P. T., R. C. Somers et al. (2006). A ratiometric CdSe/ZnS nanocrystal pH sensor. *J. Am. Chem. Soc.* **128**(41): 13320–13321.

123. Medintz, I. L., A. R. Clapp et al. (2003). Self-assembled nanoscale biosensors based on quantum dot FRET donors. *Nat. Mater.* **2**(9): 630–638.

124. Bakalova, R., H. Ohba et al. (2004). Quantum dots as photosensitizers? *Nat. Biotechnol.* **22**(11): 1360–1361.

125. Bakalova, R., H. Ohba et al. (2004). Quantum dot anti-CD conjugates: Are they potential photosensitizers or potentiators of classical photosensitizing agents in photodynamic therapy of cancer? *Nano Lett.* **4**(9): 1567–1573.

126. Tsay, J. M., M. Trzoss et al. (2007). Singlet oxygen production by peptide-coated quantum dot-photosensitizer conjugates. *J. Am. Chem. Soc.* **129**(21): 6865–6871.

127. Pons, T., I. L. Medintz et al. (2006). Solution-phase single quantum dot fluorescence resonance energy transfer. *J. Am. Chem. Soc.* **128**(47): 15324–15331.

128. Zhang, C.-Y., H.-C. Yeh et al. (2005). Single-quantum-dot-based DNA nanosensor. *Nat. Mater.* **4**(11): 826–831.

129. Smith, J. D., G. W. Fisher et al. (2007). The use of quantum dots for analysis of chick CAM vasculature. *Microvasc. Res.* **73**(2): 75–83.

130. Smith, B. R., Z. Cheng et al. (2008). Real-time intravital imaging of RGD-quantum dot binding to luminal endothelium in mouse tumor neovasculature. *Nano Lett.* **8**(9): 2599–2606.

131. Kim, S., Y. T. Lim et al. (2004). Near-infrared fluorescent type II quantum dots for sentinel lymph node mapping. *Nat. Biotechnol.* **22**(1): 93–97.

132. Hardman, R. (2006). A toxicologic review of quantum dots: toxicity depends on physicochemical and environmental factors. *Environ. Health Perspect.* **114**(2): 165–172.

133. Evans, C. L., E. O. Potma et al. (2005). Chemical imaging of tissue in vivo with video-rate coherent anti-stokes Raman scattering microscopy. *Proc. Natl. Acad. Sci. U.S.A.* **102**(46): 16807–16812.

134. Mulder, W. J., R. Koole et al. (2006). Quantum dots with a paramagnetic coating as a bimodal molecular imaging probe. *Nano Lett.* **6**(1): 1–6.

135. Yang, H., S. Santra et al. (2006). Gd[III]-functionalized fluorescent quantum dots as multimodal imaging probes. *Adv. Mater.* **18**(21): 2890–2894.

136. Flick, D. F., H. F. Kraybill et al. (1971). Toxic effects of cadmium: A review. *Environ. Res.* **4**(2): 71–85.

137. Bertin, G. and D. Averbeck (2006). Cadmium: Cellular effects, modifications of biomolecules, modulation of DNA repair and genotoxic consequences (a review). *Biochimie* **88**(11): 1549–1559.

138. Pelley, J. L., A. S. Daar et al. (2009). State of academic knowledge on toxicity and biological fate of quantum dots. *Toxicol. Sci.* **112**(2): 276–296.

139. Jaiswal, J. K. and S. M. Simon (2004). Potentials and pitfalls of fluorescent quantum dots for biological imaging. *Trends Cell Biol.* **14**(9): 497–504.

140. Howarth, M., D. J. F. Chinnapen et al. (2006). A monovalent streptavidin with a single femtomolar biotin binding site. *Nat. Methods* **3**(4): 267–273.

141. Nirmal, M., B. O. Dabbousi et al. (1996). Fluorescence intermittency in single cadmium selenide nanocrystals. *Nature* **383**(6603): 802–804.

142. Frantsuzov, P. A., S. Volkán-Kacsó et al. (2009). Model of fluorescence intermittency of single colloidal semiconductor quantum dots using multiple recombination centers. *Phys. Rev. Lett.* **103**(20): 207402.

143. Mahler, B., P. Spinicelli et al. (2008). Towards non-blinking colloidal quantum dots. *Nat. Mater.* **7**(8): 659–664.

144. Wang, X., X. Ren et al. (2009). Non-blinking semiconductor nanocrystals. *Nature* **459**(7247): 686–689.
145. Tomlinson, I. D., J. N. Mason et al. (2005). Peptide-conjugated quantum dots: Imaging the angiotensin type 1 receptor in living cells. *Method. Mol. Biol.* **303**: 51–60.
146. Charalambous, A., M. Andreou et al. (2009). Intein-mediated site-specific conjugation of quantum dots to proteins in vivo. *J. Nanobiotechnol.* **7**(1): 9.
147. Orndorff, R. L., M. R. Warnement et al. (2008). Quantum dot ex vivo labeling of neuromuscular synapses. *Nano Lett.* **8**(3): 780–785.
148. Zhou, M., E. Nakatani et al. (2007). Peptide-labeled quantum dots for imaging GPCRs in whole cells and as single molecules. *Bioconj. Chem.* **18**(2): 323–332.
149. Clarke, S. J., C. A. Hollmann et al. (2006). Photophysics of dopamine-modified quantum dots and effects on biological systems. *Nat. Mater.* **5**(5): 409–417.
150. Gussin, H. A., I. D. Tomlinson et al. (2006). Binding of muscimol-conjugated quantum dots to GABAC receptors. *J. Am. Chem. Soc.* **128**(49): 15701–15713.
151. Courty, S. B., C. Luccardini et al. (2006). Tracking individual kinesin motors in living cells using single quantum-dot imaging. *Nano Lett.* **6**(7): 1491–1495.
152. O'Connell, K. M. S., A. S. Rolig et al. (2006). Kv2.1 potassium channels are retained within dynamic cell surface microdomains that are defined by a perimeter fence. *J. Neurosci.* **26**(38): 9609–9618.
153. Pathak, S., S.-K. Choi et al. (2001). Hydroxylated quantum dots as luminescent probes for in situ hybridization. *J. Am. Chem. Soc.* **123**(17): 4103–4104.
154. Osaki, F., T. Kanamori et al. (2004). A quantum dot conjugated sugar ball and its cellular uptake. on the size effects of endocytosis in the subviral region. *J. Am. Chem. Soc.* **126**(21): 6520–6521.
155. Dubertret, B., P. Skourides et al. (2002). In vivo imaging of quantum dots encapsulated in phospholipid micelles. *Science* **298**(5599): 1759–1762.

Index

S